AAPG Treatise of Petroleum Geology
Reprint Series

The American Association of Petroleum Geologists
gratefully acknowledges and appreciates the leadership and support
of the AAPG Foundation in the development of the
Treatise of Petroleum Geology.

TRAPS AND SEALS II
STRATIGRAPHIC/CAPILLARY TRAPS

COMPILED BY
NORMAN H. FOSTER
AND
EDWARD A. BEAUMONT

TREATISE OF PETROLEUM GEOLOGY
REPRINT SERIES, NO. 7

PUBLISHED BY
THE AMERICAN ASSOCIATION OF PETROLEUM GEOLOGISTS
TULSA, OKLAHOMA 74101, U.S.A.

Library of Congress Cataloging-in-Publication Data

Traps and seals / compiled by Norman H. Foster and Edward A.
Beaumont.   p.   cm. — (Treatise of petroleum geology reprint
series ; no. 6-7)
   Includes bibliographies.
   Contents: 1. Structural/fault-seal and hydrodynamic traps. 2.
Stratigraphic/capillary traps.
   ISBN 0-89181-405-1 (v. 1): $34.00
   ISBN 0-89181-406-x (v. 2)
   1. Rock traps (Hydraulic engineering) 2. Petroleum—Geology. 3.
Gas, Natural—Geology. I. Foster, Norman H. II. Beaumont, E. A.
(Edward A.) III.—Series: Treatise of petroleum geology reprint
series: no. 6, 7.
QE611,T73 1988
553.2'8—dc19   88-10471
                                                        CIP

TRAPS AND SEALS II

STRATIGRAPHIC/CAPILLARY TRAPS

TABLE OF CONTENTS

TRAPS AND SEALS I

STRUCTURAL/FAULT-SEAL TRAPS

HYDRODYNAMIC TRAPS

INTRODUCTION

This reprint belongs to a series of reprint volumes which in turn are part of the *Treatise of Petroleum Geology*. The *Treatise of Petroleum Geology* was born during a discussion we had at the Annual AAPG Meeting in San Antonio in 1984. When our discussion ended, we had decided to write a state-of-the-art textbook in petroleum geology, directed not at the student, but at the practicing petroleum geologist. The project to put together one textbook gradually evolved into a series of three different publications: the Reprint Series, the Atlas of Oil and Gas Fields, and the Handbook of Petroleum Geology; collectively these publications are known as the *Treatise of Petroleum Geology*. Using input from the Advisory Board of the Treatise, we designed this entire effort so that the set of publications will represent the state of the art in petroleum exploration knowledge and application. The Reprint Series collects the most up-to-date, previously published literature; the Atlas is a collection of detailed field studies to illustrate all the various ways oil and gas are trapped; and the Handbook is a professional explorationist's guide to the latest knowledge in the various areas of petroleum geology and related fields.

Papers in the various volumes of the Reprint Series are meant to complement the chapters of the Handbook. Papers were selected mainly on the basis of their usefulness today in petroleum exploration and development. Many "classic papers" that led to our present state of knowledge are not included because of space limitations.

We divided the general topic of traps into two volumes: (I) Structural/Fault-Seal and Hydrodynamic Traps, and (II) Stratigraphic/Capillary Traps. Papers in these two volumes deal mainly with the mechanics of trapping hydrocarbons and with trap types. Methods of exploring for traps are covered in the Reprint volume: Methods of Exploration.

Traps are the end product of many processes: the maturation and migration of petroleum, and the formation and preservation of the porous rocks that form the reservoirs and the non-porous rocks that are the seals. There are time and space requirements. A potential trap must have been present during petroleum migration. Pathways must be available from the source to the trap, and reservoir rocks and seal rocks must be in the proper configuration to one another. Without any one of these elements, a trap cannot exist. Primarily, the geometry of a trap is determined by two factors: structure and stratigraphy. The two volumes of the *Traps and Seals* reprint divide papers on traps and seals along these two lines.

Volume I mainly contains papers dealing with structural/fault seal traps, but also includes papers that discuss the controversial subject of hydrodynamic trapping. Volume II contains papers that describe different stratigraphically controlled trap types, the preservation of porosity, and the importance of capillarity in trapping hydrocarbons.

Edward A. Beaumont
Tulsa, Oklahoma

Norman H. Foster
Denver, Colorado

Treatise of Petroleum Geology
Advisory Board

*Deceased

STRATIGRAPHIC/CAPILLARY TRAPS

American Association of Petroleum Geologists Bulletin
V. 53, No. 1 (January 1969) pp. 3-29.

Rationale for Deliberate Pursuit of Stratigraphic, Unconformity, and Paleogeomorphic Traps[1]

MICHEL T. HALBOUTY

Consulting Geologist, Petroleum Engineer, Independent
Producer and Operator, Houston, Texas 77027

Abstract Most basins contain facies changes, unconformities with resulting truncated beds, and buried erosional or constructive surfaces such as reefs, hills, channels, barrier sand bars, and other such phenomena—which form the basic requirements for the creation of subtle traps.

If folding, normal faulting, thrusting, and the formation of salt ridges and domes are added to the picture of an evolving but continuously filling basin, the resultant structural and stratigraphic patterns become much more complex. However, no matter how complex the history, those stratigraphic relations and lithologic changes which are conducive to the formation of stratigraphic, unconformity, and paleogeomorphic traps remain.

When hydrocarbon is expelled (primary migration) by pressure and heat from sediments which contain source material into adjacent reservoir rocks, it migrates through carrier beds (secondary migration) into sealed reservoirs, or traps. As long as the conditions necessary for secondary migration of a substantial amount of petroleum exist, migration will continue along strike and updip until all migrating hydrocarbons are either trapped in the subsurface or have escaped at the surface. As the petroleum moves, it will be captured by all traps—stratigraphic, unconformity, paleogeomorphic, structural, or a combination of these—which are in the path of migration.

Because paleogeomorphic, unconformity, and stratigraphic traps are related (1) to older geologic surfaces, (2) to the location of strata on and directly below an unconformity surface, and (3) to lithologic changes within and laterally adjacent to a stratum, it is suggested that, in general, the conditions which produce most subtle traps are present before development of structural traps. If migration of hydrocarbons through a particular region were to take place *before* structural movements, all petroleum trapped during this early migration would be in subtle traps.

Because subtle traps generally are formed as a result of constantly recurring depositional patterns which usually precede, or may be associated with, contemporaneous structural movement, petroleum basins probably contain more subtle traps than structural traps. Although much petroleum has migrated into structural traps, possibly more has accumulated in the earlier (and contemporaneously) formed subtle traps. Because subtle traps probably contain the large undiscovered domestic reserves needed for the future, explorationists must make the purposeful search for such traps an essential and substantial part of their exploration policy.

INTRODUCTION

The large onshore reserves of the future in this country will be found in stratigraphic, unconformity, and paleogeomorphic traps[2]—which are referred to herein as subtle traps.

The logic behind such a broad statement is based on two observations:

1. Large petroleum accumulations do occur in stratigraphic traps (*e.g.*, the Jackson trend of South Texas and the Hugoton field, Kansas); in unconformity traps (*e.g.*, the East Texas field); and in paleogeomorphic traps (such as those associated with the buried hills in the Kraft-Prusa district, Kansas, and in most of the Mississippian oil fields in southeastern Saskatchewan, Canada).

2. Most of the relatively easy-to-find structural traps in explored basins of the conterminous United States probably have been found by intensive past exploration; however, geologists agree that large reserves are yet to be found in our domestic basins.

If both statements are valid, then a major part of the total petroleum originally contained in structural traps in these basins already has been discovered, and, if large reserves remain undiscovered in these petroleum provinces, it is logical to assume that they are contained in subtle traps, which probably occur in large numbers in all basins.

OCCURRENCE OF SUBTLE TRAPS

Most basins contain facies changes, unconformities with resulting truncated beds, and buried erosional or constructive surfaces such as reefs, hills, barrier sand bars, channels, and other related geologic phenomena—which form the basic requirements for the creation of subtle traps. Basins differ in lithologic characteristics; some may contain principally sandstones and shales, others carbonate rocks, and others combinations of these. The ultimate controls on lithologic characteristics are the sources of the deposits and the environments of deposition.

[1] Manuscript received, January 14, 1971; revised, July 2, 1971. Reprinted from AAPG *Bulletin*, 1972, v. 56, no. 3. Introductory paper presented before the Special Stratigraphic-Trap Exploration Session at the Annual Meeting of The American Association of Petroleum Geologists, March 28-31, 1971, Houston, Texas.

I thank James J. Halbouty and A. A. Meyerhoff for their contribution to this paper.

[2] In previous lectures and papers I have referred to "subtle traps" as stratigraphic and/or paleogeomorphic. In this paper I have added the "unconformity" trap to the classification.

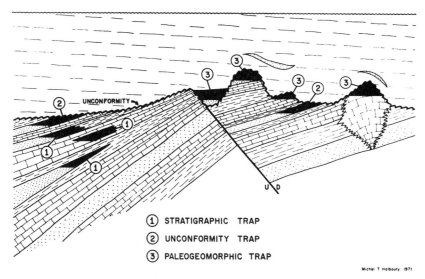

FIG. 1—Idealized portrayal of types of subtle traps. Note relations of stratigraphic traps to lateral changes in facies and relations of unconformity and paleogeomorphic traps to erosional surfaces and relief features buried beneath them.

In general, lateral lithologic changes within a given stratum are the result of crossing of depositional environmental boundaries, the nature of the provenance for the sediments, the type and configuration of the depositional basin, the depth of water at the site of deposition, *etc.* Moreover, there may be postdepositional effects on porosity and permeability within that stratum because of chemical and other reactions.

Stratigraphic traps are formed from lithologic changes which occur during and after deposition (Fig. 1). Included in this category are (1) depositional traps which are caused by lateral facies change from permeable to impermeable beds, or by pinchout of a reservoir rock which is not represented laterally by contemporaneously deposited strata, and (2) postdepositional traps caused by cementation or changes which take place in a reservoir bed after deposition. It should be intuitively clear—in view of the numerous possibilities for depositional and postdepositional changes within strata—that the conditions essential to the formation of stratigraphic traps must occur, and probably *abound,* in all basins.

Unconformity traps are formed where an impermeable bed on an erosional surface is in direct contact with a reservoir rock lying below the erosional surface and provides a seal for the trap. These traps usually occur in association with angular unconformities and *should be differentiated* from those paleogeomorphic traps which occur below erosional surfaces (*e.g.,* in a buried hill).

Since unconformity traps are a consequence of cycles of deposition, structural movement, and erosion, they should be prevalent in all petroleum basins. Their size can range from small local traps to regional ones such as the East Texas field.

Paleogeomorphic traps are formed where ancient subaqueous or land surfaces and the relief features on them (which are produced by geomorphic processes such as erosion, channeling, deposition, *etc.*) are buried by younger strata of a different lithology. Buried reefs, channels, barrier sand bars, and hills on erosional surfaces are examples of such features. Paleogeomorphic traps may occur above or below the erosion surface and may be directly or indirectly related to the buried feature. It is difficult to imagine an ancient basin that does not contain in at least one area, or within at least one stratigraphic interval, numerous and varied buried features which, directly or indirectly, may serve as traps for petroleum.

In the Gulf Coast embayment, the hundreds of salt domes may be considered to be mobile, paleogeomorphic features—impervious buried hills which, during geologic time, are capable not only of changing shape but also of moving laterally and vertically. At each position during their formation, they affect the depositional and structural placement of beds around and above them.

In contrast, a buried hill associated with ancient eroded topography is a stable feature. Its effect on depositional patterns produces a consistency in pattern for the duration of time that the feature is being buried. After burial and compaction of that feature, the depositional pattern changes because the buried hill is no longer an important factor in the basin of deposition. However, the mobile salt dome usually has a much more complex effect on the surrounding and overlying strata. The extent of its effect depends on the mobility of the dome. An unusually mobile salt mass, in contrast to the stable, stationary buried hill, can affect depositional patterns for long periods of geologic time.

If folding, normal faulting, thrusting, and the formation of salt ridges and domes are added to the picture of an evolving but continuously filling basin, the resultant structural and stratigraphic patterns become much more complex. However, no matter how complex the history, those stratigraphic relations and lithologic changes which are conducive to the formation of stratigraphic, unconformity, and paleogeomorphic traps remain.

Subtle traps which develop prior to the beginning of structural movement in a region are, of course, subject to the effects of such movement. For example, a stratigraphic trap which originally consisted of a sandstone pinchout along strike on a monocline may be uplifted and superimposed above or on the flank of an anticline or dome; it may become part of a deep synclinal area; it may be faulted; or it may be tilted laterally with respect to the previous strike.

Whatever the structural effect, any petroleum which had accumulated in the original pinchout trap could, as a result of subsequent tectonic processes, migrate from its original position and relocate in another place or acquire a different pattern of accumulation within the same trap, or the accumulation could escape as a result of erosion which might accompany substantial uplift.

PETROLEUM MIGRATION AND ACCUMULATION IN SUBTLE TRAPS

There is general agreement among geologists that petroleum, in some form, is expelled by pressure and heat from sediments containing source materials into adjacent reservoir rocks (primary migration), and that hydrocarbons migrate through carrier beds into sealed reservoirs, or traps (secondary migration). In some fields it appears that little if any secondary migration occurred because impermeable strata surround the reservoir (closed reservoir).

The location and differential gravitational separation of oil, gas, and water within closed

anticlines are sufficient proof that secondary migration does occur along carrier beds into reservoir rocks. Otherwise, petroleum would be required to originate from within the reservoir beds in the trap.

There are differences of opinion among geologists concerning the question of how far petroleum can migrate from the source area to the reservoir. The writer's opinion is that, as long as the requirements for secondary migration of a substantial amount of petroleum exist —pressure, heat, regional and local tilt, good-quality carrier beds with continuity to the reservoir rock—migration will continue along strike and updip until all migrating hydrocarbons have been trapped in the subsurface or have escaped at the surface.

Therefore, if stratigraphic, unconformity, paleogeomorphic, structural, or combination traps are in the path of migration, petroleum will be trapped in them. However, that part of the migrating oil or gas which is *not* trapped will continue updip to the surface through outcropping carrier beds.

Gulf Coast geologists are familiar with the similarity of electric-log resistance patterns or "kicks" for long distances along strike in wells drilled in the Anahuac (Miocene-Oligocene) marine shale wedge (also in other marine shale) of the Texas-Louisiana Gulf Coast. This subsurface shale wedge approximately parallels the present shoreline and increases in thickness from a few feet updip to several thousands of feet toward the Gulf of Mexico. The wedge separates the basal Miocene from the upper Frio (Oligocene); it has been considered to be the host rock for the source materials which generated the large petroleum accumulations found in sandstones of both the Miocene and Frio.

The similarity of electric-log patterns within the Anahuac shale wedge across long distances obviously means that each distinctive electric-log interval possesses electrical-response properties which are nearly constant through extended distances. Each such interval, therefore, must possess uniform lithologic characteristics through long distances along depositional strike, or in whatever direction the log pattern remains constant.

If a stratum's electrical-response patterns are related to the lithology, then the similarity of electric-log patterns across a long distance indicates that *this particular widespread stratum is a result of more or less simultaneous deposition of the same materials, including those from which oil and gas are generated.* Moreover, for the Anahuac, these materials—lithologic and organic—were derived from the same sources and were deposited in a widespread, uniform marine environment.

It is very probable, therefore, that the Anahuac shale was deposited in widespread layers, some of which contained petroleum-generating materials. As the nature of the deposited material and the environment changed, the characteristics of the strata changed accordingly; these changes are reflected by different electric-log patterns in different strata.

If primary hydrocarbon migration can begin from source materials which are as widespread as they apparently were within the Anahuac shale, then it is apparent that secondary petroleum migration from such beds, once begun, can indeed traverse a very wide area if carrier rocks are contiguous.

As long as the conditions which control generation and migration are favorable, hydrocarbons will continue through geologic time to move across a wide geographic area, except where structural features such as regional noses, synclines, and faults divert the direction of movement. As the petroleum moves along strike and updip, it will be captured by all traps —stratigraphic, unconformity, paleogeomorphic, structural, or a combination of these— which are in the path of the migrating hydrocarbons. These traps may be adjacent to, near, or distant from the area where secondary migration is initiated.

Thus:

1. Petroleum migration *can* be widespread along strike and petroleum *can* move long distances updip, depending on the availability of source materials and on contiguous carrier rocks.
2. Migrating petroleum does not move preferentially into structural traps.
3. Petroleum will be captured along its migratory path by whatever phenomenon provides proper trapping conditions—*as long as such conditions are present at the time of migration.*

Because paleogeomorphic, unconformity, and stratigraphic traps are related (1) to older geologic surfaces, (2) to the location of strata on and directly below an unconformity surface, and (3) to lithologic changes within and laterally adjacent to a stratum, it is suggested that, in general, the conditions which produce most subtle traps are present *before* development of structural traps. If migration of petroleum through a particular region were to take place before structural movements, *all* trapped petroleum would be in subtle traps. If both subtle and structural traps were formed *prior* to migration, both types would become reservoirs. In either case, structural movements, even if post-migration, could influence accumulations formed previously in subtle traps and thereby change the location of the petroleum originally contained in such traps.

Consider the East Texas field, more than 45 mi (75+ km) long and as wide as 12 mi (20+ km; Fig. 2). The migration apparently was widespread laterally. If the east flank of the East Texas basin had been an approximate mirror image of the west flank, where the Woodbine Sandstone (Lower Cretaceous) beds crop out, most of the great reserves in that field might have escaped to the surface. The eastward migration through the Woodbine sandstones would have allowed the escape of all the migrating petroleum, except for that which might have been caught by existing structural and subtle traps. However, because the truncated eastern edge of the Woodbine Sandstone beds is wedged between impermeable beds on the west flank of the Sabine uplift, the widespread eastward migration was halted and the East Texas accumulation was formed.

The pertinent question may well be asked: Are there as many subtle traps in petroleum basins as there are structural traps? The answer is that there probably are *more,* because subtle traps generally are formed as a result of constantly recurring depositional patterns which usually precede structural movement, or which may be associated with contemporaneous structural growth.

Huge amounts of petroleum have been generated in petroleum basins. Much of it has been lost because of absence of trapping conditions; yet much has been trapped in structural anomalies and probably more has accumulated in subtle traps. The problem is to *find* these subtle reservoirs.

IMPORTANCE OF SUBTLE-TRAP EXPLORATION

Exploration for subtle traps has been neglected by the petroleum industry simply because structures are easier to find with present exploratory methods. Only recently has exploratory thinking begun to be oriented toward the subtle reservoir.

Recent events in the Middle East and North Africa have proved, much to the chagrin of some U.S. politicians, that foreign oil is not always going to be cheap, and that the United

Stratigraphic, Unconformity, and Paleogeomorphic Traps

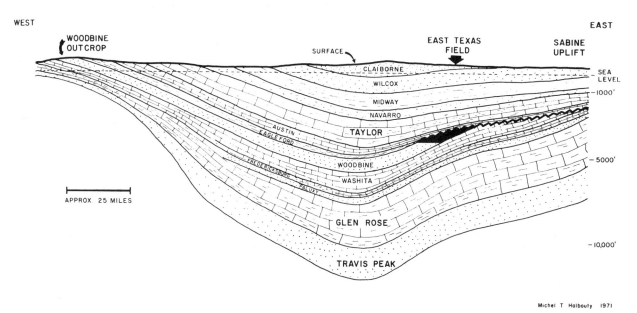

Fig. 2—Generalized cross section, East Texas basin, showing wedging of eroded Woodbine Sandstone between Austin and Washita limestones, forming trap for East Texas field on west flank of Sabine uplift. Note outcrop of Woodbine beds west of East Texas basin.

States cannot depend on such reserves in future emergencies because they can be cut off by circumstances beyond this country's political control.

The great need of the United States and of the domestic petroleum industry is to find a reliable source of oil and gas reserves. A *reliable* source necessarily is domestic. However, domestic reserves have been declining in the face of a continually growing demand for petroleum products.

Reserves are declining because exploration and drilling in this country have been decreasing for several years, and most of the exploratory effort continues to be directed toward the structural trap. Structures which remain undiscovered are becoming more difficult to find because they are smaller or less discernible than those structures found in the past.

Explorationists must make the *purposeful* search for subtle traps an essential and substantial part of their exploration policy. They will be required to make use of all basic earth studies which would shed light on ancient geologic environments that favored the formation of stratigraphic, unconformity, or paleogeomorphic traps.

If the thesis presented here is correct, it may be assumed that thousands of subtle traps, containing petroleum accumulations ranging in size from inconsequential to supergiant, are yet undiscovered in United States petroleum basins.

REFERENCES

Cram, I. H., ed., 1971, Future petroleum provinces of the United States—their geology and potential: Am. Assoc. Petroleum Geologists Mem. 15, 1,496 p.

Halbouty, M. T., 1967, Hidden trends and features: Gulf Coast Assoc. Geol. Socs. Trans., v. 17, p. 2–23.

——— 1968, Giant oil and gas fields in United States: Am. Assoc. Petroleum Geologists Bull., v. 52, no. 7, p. 1115–1151.

——— 1969, Hidden trends and subtle traps in Gulf Coast: Am. Assoc. Petroleum Geologists Bull., v. 53, no. 1, p. 3–29.

——— *et al.*, in press, Geology and environmental factors affecting giant fields: Am. Assoc. Petroleum Geologists Bull.

Martin, Rudolf, 1966, Paleogeomorphology and its application to exploration for oil and gas (with examples from Western Canada): Am. Assoc. Petroleum Geologists Bull., v. 50, no. 10, p. 2277–2311.

National Petroleum Council, 1970, Future petroleum provinces of the United States: 180 p.

BULLETIN OF THE AMERICAN ASSOCIATION OF PETROLEUM GEOLOGISTS
VOL. 50, NO. 10 (OCTOBER .1966), P. 2058-2067, 13 FIGS., 3 TABLES

THE OBSCURE AND SUBTLE TRAP[1]

A. I. LEVORSEN[2]
Tulsa, Oklahoma

ABSTRACT

The tremendous expanding demand for petroleum and its products that continues to develop means that we must take a hard look at our future sources of petroleum supply. In spite of the fact that most exploration has been and is directed toward the search for petroleum in local structural traps, many of the largest oil and gas pools in the Western Hemisphere are trapped by non-structural phenomena. Structural traps are so obvious that they are the first to be tested, but we are now facing a situation in which the supply of structural traps in the United States seems to be limited; untested anticlines are becoming more difficult to find. Does this indicate an impending shortage of petroleum? The answer would seem to be *no*—but this means that the search will have to be for more obscure and subtle trapping situations. The search will continue for the purely structural trap, out there will be added stratigraphic variations and fluid-flow phenomena, all operating either together or independently.

We have "stumbled" into many great non-structural oil and gas pools while looking for purely structural traps, but the time seems to have arrived when we must start looking directly for combination traps of all kinds involving different proportions of structure, stratigraphic change, and fluid-flow phenomena. Such traps may contain very large petroleum pools, as past experience has shown.

There are in the Rocky Mountain region many such untested potential combinations of large structures, stratigraphic changes, and favorable fluid-flow conditions to justify the belief that this region has a continued great future as a petroleum-producing region of importance to our national needs. The fact that the Rocky Mountain Section of the Association is dedicating a full meeting to the obscure and subtle trap is a sure indication of a change in our thinking. Once we start actively to look for traps that combine structure, stratigraphic change, and fluid phenomena instead of to look only for local structure, there is no reason why discoveries in the United States should not continue to meet the demand. The Rocky Mountain region has as bright a future for petroleum discovery as any other.

INTRODUCTION

A study of the abstracts for this meeting suggests that there are many oil pools in non-structural traps in the Rocky Mountain region. For decades the search has been concentrated in looking for structural traps—local anticlines, domes, and faulted folds. As structural traps become more difficult to find, some believe that this implies that the end of our supply of petroleum is coming. The program to follow, however, shows an awareness of the fact that there are other traps than the purely structural—in other words, the exploration road is beginning to turn. Local and regional structures still are important, but their effect is lessened and in some cases overshadowed by stratigraphic and fluid anomalies. We are entering the era of combination traps—where structure, stratigraphy, and fluid phenomena *combine* to make the trap.

[1] Read before the Rocky Mountain Section of the Association at Billings, Montana, September 27, 1965, by Orlo E. Childs. This keynote address was prepared by Dr. Levorsen. The illustrations were prepared from Dr. Levorsen's slides by Ezio Fanelli and Arnold Ewing. Manuscript received, September 27, 1965; accepted, March 30, 1966.

[2] Consulting petroleum geologist. Deceased, July 16, 1965.

COMBINATION TRAPS

GENERAL

There are several aspects of combination traps that should be mentioned. First, they generally are obscure, and a real detective ability is required to recognize the commonly subtle geologic changes that are significant. It is a lot of fun searching for them. However, we must be careful not to get into the situation of the lipstick manufacturer. He had the bright idea of making a lipstick that would glow in the dark—but it didn't sell. He found out that half the fun was in the search!

A second aspect of combination traps is their size. In the Western Hemisphere they hold some of the largest oil and gas pools yet discovered. A few examples will show the size of the stakes involved.

		Bbls. Recoverable	
1. Bolivar Coastal field, Lake Maracaibo, Venezuela	17 billion	(Fig. 1)
2. Poza Rica field, Mexico	2 billion	(Fig. 2)
3. East Texas field, U.S.A.	6 billion	(Fig. 3)
4. Pembina field, Alberta, Canada	2 billion	(Fig. 4)

In pools such as these, there is little or no local

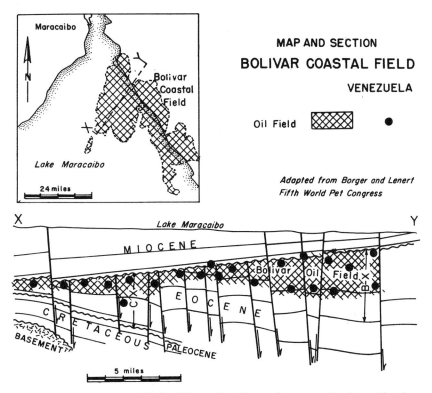

FIG. 1.—Bolivar Coastal field, Lake Maracaibo, Venezuela: example of combination trap.

structural anomaly; the position of the pool is determined chiefly by some stratigraphic and fluid anomaly—by a combination of several elements, no one of which is sufficient to trap a pool but which together are adequate.

For the third aspect of the combination trap we might consider an area that has been developed for a much longer period of time and more intensively than the Rocky Mountain region. This is eastern Kansas and eastern Oklahoma, where Eugene Weirich (1953) has discussed the oil pools trapped in the shelf deposits of the lower Pennsylvanian System. There are many hundreds of oil pools in this region, as shown on the map (Fig. 5), which are located within the Cherokee and Atoka units of the Pennsylvanian rock column.

This is the region in the Mid-Continent where the search for anticlines had its greatest boost in the 1910s and 1920s and where, today, in county after county, there is hardly a non-producing square mile without one or more dry holes. The Pennsylvanian sandstone bodies of this region are almost completely developed. Yet Weirich tells me that three quarters of all the pools shown on this map are in stratigraphic-type traps—sand-

stone patches, lenses, bars, channel sandstone, reefs, facies changes, and truncated reservoir rocks.

Closer to those of you from the United States Rockies is western Canada. I have been told that the number of pools in western Canadian stratigraphic traps is about the same as in the Cherokee and Atoka of Kansas and Oklahoma—three-fourths to four-fifths—but that the amount of oil produced is even greater; 90–95% is stratigraphic.

In developing the principles of oil accumulation, we should consider the possibility that many of the pools located on anticlines and other local structures are simply accumulations localized by younger folding of larger, pre-existing oil pools or were localized by pre-existing lenses and reefs that already had accumulated a pool of oil. If, for example, local folds had been superimposed on the East Texas pool or the Pembina pool, they would have been full of oil. However, anticlines in the vicinity of both are barren because they do not overlie a reservoir rock or a stratigraphic anomaly able to trap oil.

Early availability of a stratigraphic trap to accumulate a pool probably is a most significant

9

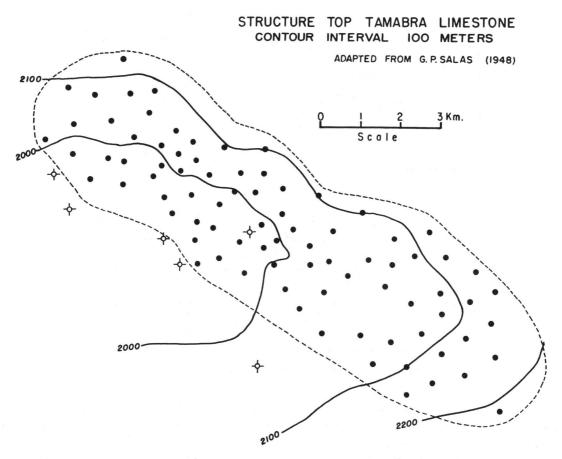

FIG. 2.—Poza Rica field, Mexico: example of combination trap.

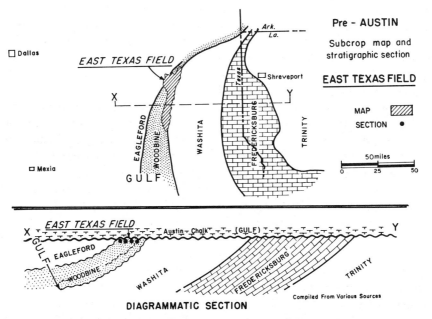

FIG. 3.—East Texas field: example of combination trap.

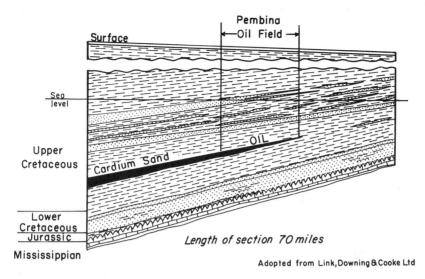

FIG. 4.—Pembina field, Alberta, Canada: example of combination trap.

and important feature. In short the trap was there and ready to receive the petroleum when first covered and sealed and when the fluids were squeezed out of the surrounding shales. Furthermore, the stratigraphic trap generally was available for a pool of petroleum to accumulate millions of years before local structures were formed.

My first experience with stratigraphic traps came early. In 1919 I was sent to Butler County, Kansas, to look for structures like those trapping the Augusta and Eldorado pools. The Fox-Bush field was being developed in a sandstone at the base of the Pennsylvanian, and I started to work near it. About 4 mi. away, I discovered an anticline several miles long, with 20 ft. of closure and about 50 ft. of structural relief. Were we excited! This was especially true because the Fox-Bush

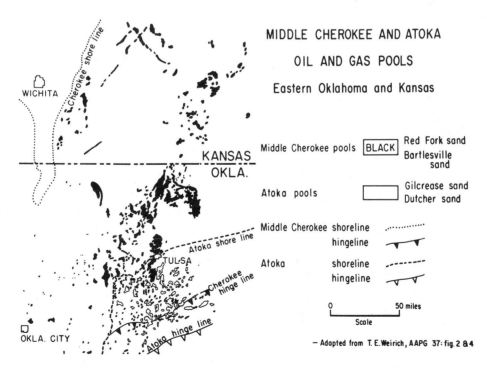

FIG. 5.—Oil pools trapped in shelf deposits of lower part of Pennsylvanian System.

11

field had no local structure associated with it. Management told us that it was all right to buy some royalty (my part came to around $50 for a 1/128th, as I remember it).

Drilling of the No. 1 Steinhoff well was begun on the closure, and I went to Minnesota to be married. On our return, my wife and I stopped in a hotel in Kansas City, the headquarters town for the company. A wire from one of the boys in the field came saying, "Steinhoff No. 1 looks like a 100-barrel-a-day well. Shut down the tankage." I can still see my wife and me dancing around the hotel room! I could not find any paper so used the back of our marriage license to write down the figures: 100 times 30 times 12 times 10 years times $2.75 a barrel—well, it came to a lot of money and the figures are still on the license!

We traveled to the well as fast as possible only to find that it had made all its oil during the first hour or so. When tanks were erected and the well drilled deeper, it was found that shale of the Pennsylvanian rested directly on the Mississippian limestone, and the small spurt of oil came from the unconformity zone. We had the structure all right, but Fox-Bush had the sandstone—and the oil!

Thus we see that the search for combination traps can be a lot of fun; they may be very large, and the pools in them are very common where exploration drilling has been active and aggressive.

As I see it, the three chief elements that combine in varying proportions to form a trap that will hold an oil pool are structure, stratigraphic variations, and fluid flow. Each occurs in a great variety of forms, gradients, and sizes. Some are shown in Tables I and II.

STRUCTURE

Some of the varieties of structure that form petroleum traps or combine with other elements to form traps are shown in Table I. The structural conditions that are associated with oil and gas

TABLE I. VARIETIES OF STRUCTURE WHICH FORM PETROLEUM TRAPS

Anticline, dome, fold, terrace, monocline, *etc.*
Fault, fault complex, intersecting faults, fractures
All combinations
All scales from regional to local
All gradients of structural change from low to high

pools include nearly every type and size that deform sedimentary rocks. Every oil-and-gas-pool trap contains some element of structure. Such elements may determine the trap or may be only incidental to the trap.

STRATIGRAPHIC VARIATIONS

Some of the stratigraphic variations of reservoir rocks that trap or aid in the trapping of petroleum are shown in Table II. Many variations in the stratigraphy and composition of the sedimentary rocks may become either a dominant or a partial cause of a trap. An example of the potential of one of these variations—the truncated. eroded, and overlapped secondary-type trap—is shown in Figure 6, a pre-Pennsylvanian subcrop map of the area around Oklahoma City. Half a dozen oil pools are situated along the updip truncated edges of the pre-Pennsylvanian formations in the area, and none has a local structure. This subcrop map may be expanded to cover the entire area of the pre-Pennsylvanian of the United States, and Figure 7 gives an idea of the potential of the many thousands of miles of truncated pre-Pennsylvanian sediments which are available to explore.

TABLE II. STRATIGRAPHIC VARIATIONS OF RESERVOIR ROCKS WHICH TRAP OR AID IN TRAPPING PETROLEUM

Primary: uniform lithologic character to facies change
Secondary: truncated, eroded, overlapped
            Solution channels
            Cementation
            Fracture systems
All scales from regional to local
All gradients of stratigraphic change from low to high

FLUID FLOW

The third element is the trapping effect of the fluid in the sediments, especially the rate and direction of flow. Some of the trapping effects are shown in Table III. They come under the general subject of hydrodynamics.

The term hydrodynamics comes from the Greek *hydor* (water) and the Greek *dynamikos* (powerful). From my observation of the impact of the term on the explorationist, I would judge that a realistic translation might be "your guess is as good as mine."

The reason I say this is not because of the principles involved in hydrodynamics, for they

TABLE III. FLUID FACTORS WHICH TRAP OR
AID IN TRAPPING PETROLEUM

Wide composition variation, chemical and physical
Gradients, change direction and rate with:
    Folding, faulting, mountain building
    Regional tilting, erosion, overlap
    Volcanism, heating, cooling
    Solution and recementation
All scales from regional to local, and all rates and
    directions. Always changing with normal geologi-
    cal processes

are sound, but rather because of their application to petroleum exploration. Much of the reason for the confusion and misunderstanding is that the industry is not yet willing to obtain the necessary fluid-pressure information. Just try to get a company to take pressure measurements on the water-bearing sandstone bodies found in a dry hole! Yet these measurements are essential if there is to be a quantitative approach to hydrodynamics. At present, with the exception of a few areas, the approach is qualitative and subjective.

A crude illustration of the importance of hy-drodynamics as a trapping agency is shown in Figure 8. In *A*, water is flowing down the tube and buoyant corks or balloons are rising against the flow. It might be said to be in dynamic equilibrium. In *B*, however, a slight constriction is placed in the tube, and in the constricted zone the water flows faster. It flows enough faster to cause the rising balloons to congregate below the constriction. We might liken the tube to the permeable sedimentary layer and the balloons to buoyant oil and gas. The oil and gas congregate below the constriction until enough buoyancy develops to force them through the faster-flowing water within the constricted zone. The faster flow of the water is a trapping mechanism.

The trapping effect of fluid flow is determined by several variables—such as the density and the relative amounts of water, oil, and gas; amount of the constriction; and pressure gradient, which controls the rate of flow of the water. Thus even a small constriction or change in slope may be sufficient to increase the rate of flow down the dip or to decrease the up-the-dip buoyancy effect

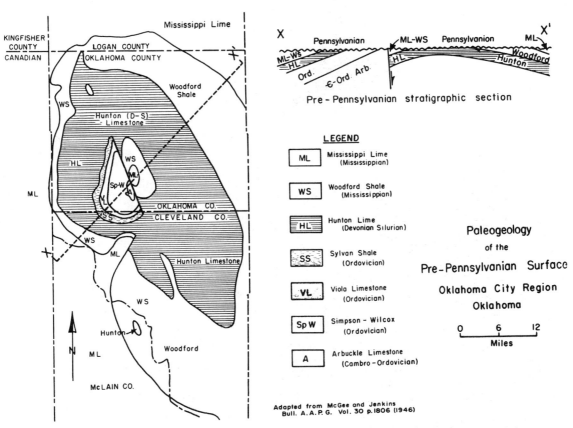

LEGEND

ML	Mississippi Lime (Mississippian)
WS	Woodford Shale (Mississippian)
HL	Hunton Lime (Devonian Silurian)
SS	Sylvan Shale (Ordovician)
VL	Viola Limestone (Ordovician)
SpW	Simpson – Wilcox (Ordovician)
A	Arbuckle Limestone (Cambro-Ordovician)

Paleogeology
of the
Pre-Pennsylvanian Surface
Oklahoma City Region
Oklahoma

0   6   12
Miles

Adapted from McGee and Jenkins
Bull. A.A.P.G. Vol. 30 p.1806 (1946)

FIG. 6.—Pre-Pennsylvanian subcrop map, Oklahoma City region: example of truncated, eroded, and overlapped secondary trap.

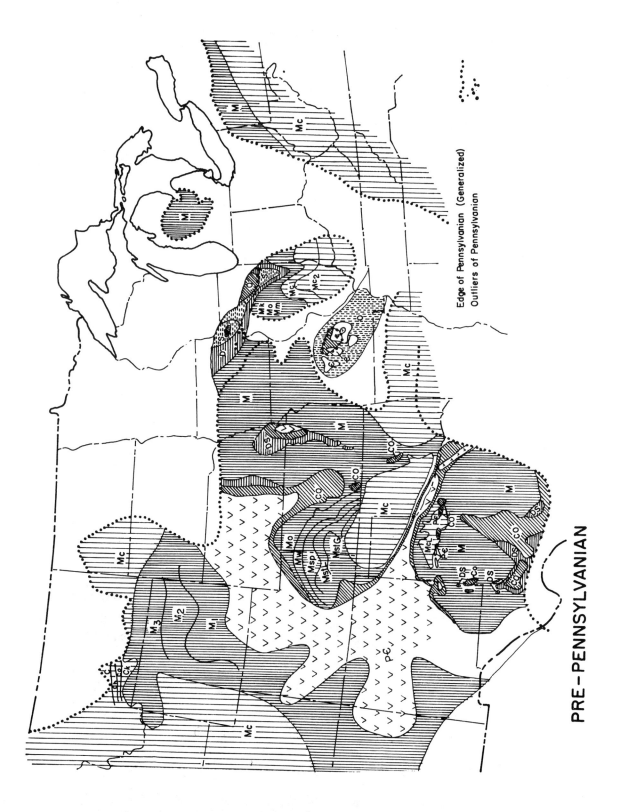

PRE–PENNSYLVANIAN

Edge of Pennsylvanian (Generalized)
Outliers of Pennsylvanian

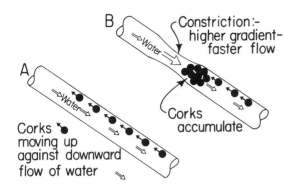

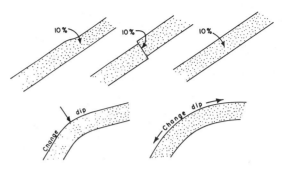

FIG. 8.—Hydrodynamics as a trapping agency.

FIG. 9.—Structural and stratigraphic changes that increase fluid flow downdip or decrease up-the-dip buoyancy. These changes affect fluid equilibrium and, as a result, trap oil.

until the equilibrium is upset and the oil is trapped. Some examples are shown in Figure 9.

ROCKY MOUNTAIN REGION

There are several pools in the Rocky Mountains that occur in a structural environment such as that shown in Figure 10. They may be trapped by some stratigraphic variation or by a down-the-dip fluid gradient. There is more to the situation than is shown in Figure 10, for when we extend the area mapped it is seen to be down the flank of a local anticline (Fig. 11). One question which might be asked is this: are pools such as these in structural traps or are they in hydrodynamic traps? The curving contours suggest that the slight arching on the flank of the anticline may be enough to provide favorable gradient conditions.

In each of the trapping elements, we are deal-

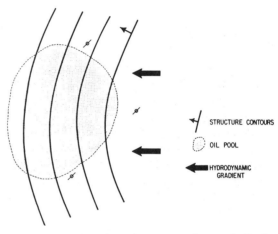

FIG. 10.—Structural environment of several Rocky Mountain pools.

FIG. 7.—Pre-Pennsylvanian map of United States.

M	Lower and middle Mississippian
Mk, Mm	Kinderhook, Meramec
M_1, M_2, M_3	Lower, middle, upper Madison
Mo, Msp, Mw	Osage, Spergen, Warsaw
MstG, MstL	Ste. Genevieve, St. Louis
Mc	Upper Mississippian (Chester)
Mc_1, Mc_2	Lower, upper Chester
Mct, Mch, McO, McK	Chester—Tyler, Heath, Otter, Kibbey
D	Devonian
S	Silurian
O	Ordovician
CO	Cambro-Ordovician
C	Cambrian

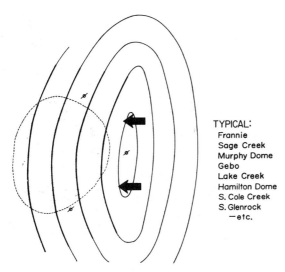

TYPICAL:
Frannie
Sage Creek
Murphy Dome
Gebo
Lake Creek
Hamilton Dome
S. Cole Creek
S. Glenrock
—etc.

FIG. 11.—View of structural environment of several Rocky Mountain pools, covering larger area than shown in Figure 10.

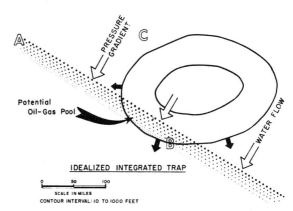

IDEALIZED INTEGRATED TRAP

SCALE IN MILES
CONTOUR INTERVAL: 10 TO 1000 FEET

FIG. 12.—Idealized example of combination or integrated oil traps.

ing with gradients of different kinds—structural gradients, stratigraphic gradients, and fluid gradients. They may combine in all proportions on all scales, and in all directions, and some of these combinations become traps that hold pools of oil and gas. Such traps are called combination or integrated traps; an idealized example is shown in Figure 12. The relative trapping effect may range from proportions of one-third each for structure,

stratigraphy, and fluids to 10–10–80 or 40–40–20, or any other combination.

Figure 13 is an attempt to classify some of the pools in the Rocky Mountain area, which have been described in geologic literature, by using a triangle with the corners labeled "structural," "stratigraphic," and "fluid pressure." The dots at the corners and along the sides of the triangle show the pools, and the blank area shows the remaining potential to be explored for combination traps.

The crests of the many anticlines of the Rocky Mountain region have been explored, and some

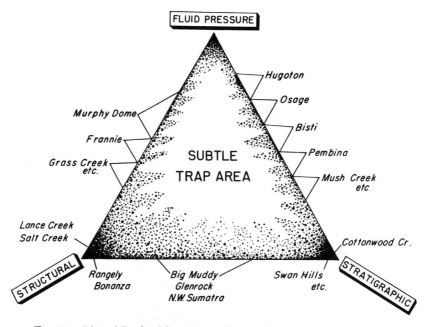

FIG. 13.—Plot of Rocky Mountain pools according to trapping controls.

16

anticlines have been found to be productive. These pools would correspond to the pools at the "structural" corner of the triangle. There is sufficient arching, minor faulting, and change of rate of slope on the flanks of the anticlines with dry holes at the crest, however, to combine with favorable stratigraphic and hydrodynamic elements to trap many undiscovered combination pools. Every dry anticline should be re-examined on all flanks to the deepest parts of the bounding synclines.

Similarly, there are thousands of miles of stratigraphic variations that occur within the stratigraphic section of the Rocky Mountain region that might combine with favorable structural and fluid phenomena to make traps—some of which might well be of large size.

Finally, with the many wide differences in elevation of each of the aquifers—or potential petroleum-reservoir rocks—there must be many sloping potentiometric surfaces. These indicate the existence of a fluid flow, which, combined with even minor changes of stratigraphic permeability and structural attitude, can provide a trap that will hold an oil or gas pool.

It is going to take some really fine detective work to find these undiscovered pools, and it is going to be fun. There may be many of them, and some may be very large. All I can say is, let's get our detective badges out, shine them up, and get going!

SELECTED BIBLIOGRAPHY

Borger, H. D., and E. F. Lenert, 1959, The geology and development of the Bolivar Coastal field at Maracaibo, Venezuela: Proc. 5th World Petroleum Cong., New York, Sec. 1, Paper 26, p. 481–498.

Levorsen, A. I., 1960, Paleogeologic maps: San Francisco and London, W. H. Freeman and Co., 174 p.

McGee, D. A., and H. D. Jenkins, 1946, West Edmond oil field, central Oklahoma: Am. Assoc. Petroleum Geologists Bull., v. 30, p. 1797–1829.

Salas, G. P., 1948, Geology and development of Poza Rica oil field, Veracruz, Mexico: Oil and Gas Jour., v. 47, no. 25 (Oct. 21), p. 129.

——— 1949, Geology and development of Poza Rica oil field, Veracruz, Mexico: Am. Assoc. Petroleum Geologists Bull., v. 33, p. 1385–1409.

Weirich, T. E., 1953, Shelf principle of oil origin, migration, and accumulation: Am. Assoc. Petroleum Geologists Bull., v. 37, p. 2027–2045.

American Association of Petroleum Geologists Memoir 16,
Stratigraphic Oil and Gas Fields, copyright 1972, pp. 14-28.

Stratigraphic-Trap Classification[1]

GORDON RITTENHOUSE

Shell Oil Company
Houston, Texas 77001

Abstract A trap for hydrocarbons requires the simultaneous existence of (a) a reservoir, (b) an isolated region of low potential in the reservoir, and (c) a barrier (or seal) with high enough entry pressure to retain a commercially producible volume of hydrocarbons. Three kinds of traps exist—structural, stratigraphic, and hydrodynamic. All three kinds have a reservoir bounded by a barrier but differ in what causes the isolated area of low potential. In classification of hydrocarbon accumulations, the conditions that determined the present location of the accumulation should be used where they can be ascertained.

In the stratigraphic-trap classification suggested here, primary emphasis has been placed on usability—*i.e.,* will the groupings help in the search for new hydrocarbon accumulations, and is the suggested terminology simple and descriptive enough to be accepted? A classification using the time relations between barrier and reservoir was considered and rejected.

The suggested classification starts with the simple concept that stratigraphic traps are adjacent to unconformities or they are not. For traps that are not adjacent to unconformities, the reservoir and barrier may be (I) primary (depositional, usually facies-related) or (II) wholly or in part secondary (diagenetic). Those traps in contact with unconformities may be (III) below the unconformity surface or (IV) above it, or (V) both below and above it. This approach uses some of Levorsen's ideas and eliminates some inconsistencies in his classification. Subdivision of these four major classes (facies-change traps, diagenetic traps, traps below unconformities, and traps above unconformities) allows more precise description of the different types of traps.

INTRODUCTION

What is a stratigraphic trap? Before one can classify such traps, one must decide what they are.

Although the existence of nonstructural traps was recognized as early as 1880 by Carll, and "Reservoirs closed because of varying porosity of rock" were distinguished by Wilson (1934) in his classification of oil and gas reservoirs, the term "stratigraphic trap" was proposed first by Levorsen (1936), who stated (p. 534):

[1] Manuscript received, March 25, 1971.

Published with permission of Shell Oil Company. The writer wishes to thank R. E. Farmer, A. W. Bally, D. B. MacKenzie, R. J. Dunham, and Howard Gould for reviewing the manuscript and Howard Gould for suggestions on classification made when preparation of this paper originally was discussed. The helpful suggestions and comments of many others with whom the ideas on classification were discussed are gratefully acknowledged also.

A stratigraphic trap may be defined as one in which a variation in the stratigraphy is the chief confining element in the reservoir which traps the oil.

In differentiating stratigraphic from structural traps, he explained (p. 524) that, in stratigraphic traps,

... the dominant trap-forming element is a wedging or pinching-out of the sand or porous reservoir rock, a lateral gradation from sand to shale or limestone, an uplift, truncation and overlap, or similar variation in the stratigraphic sequence.

Under this definition, there would be general agreement that a pod of porous and permeable sandstone completely surrounded by shale of essentially the same age and completely filled with oil or gas is a stratigraphic trap. However, if the pod were not completely filled, depositional or regional dip might determine where in the pod the hydrocarbons occur. Depositional tilt certainly would be considered as stratigraphic control; regional tilt, however, would add a structural element. Such regional tilt generally has been accepted as a component—usually a necessary component—of stratigraphic trapping.

Local, in contrast to regional, structural movements provide complications. The pod might coincide with the culmination of an anticline, might be restricted to one flank, or might be in a syncline—or the pod could have been separated into two or more parts by faulting. In his foreword to *Stratigraphic Type Oil Fields,* Levorsen (1941, p. x) clarified this by stating:

A stratigraphic pool is bounded on one or more than one side by non-porosity, whatever the cause, unless the non-porosity is altogether coincident with local structural deformation. Thus, a field would not be included if due to an interruption of stratigraphic continuity because of faulting, nor would one in which the porosity was a result of fracturing and brecciation be considered as a stratigraphic type pool. Neither would pools in which the area of accumulation was determined by a local uplift or deformation be classified as stratigraphic even though the reservoir rocks were pinched or wedged out.[2]

[2] Concerning porosity due to fracturing, Levorsen does not distinguish between fracturing caused by, and coincident with, local structural deformation and that which is not. A distinction between the two is made in this paper (see succeeding sections).

Under this definition, as commonly applied, the accumulation in the pod on the crest of the anticline would be considered as structural and the accumulation on the flank as either structural or, under Levorsen's (1954) later classification, combined stratigraphic-structural. However, what if the hydrocarbons had accumulated before the local structural deformation occurred and their positions in the pod had not been shifted materially because of it? The accumulations thus would be stratigraphically controlled and the relation to structure would be entirely coincidental. The words *determined by a local uplift or deformation* (italics added) in Levorsen's 1941 definition clearly cover this possibility, but in practice the geographic and not the time relations have been used. It may be important, not only in classification but also in exploration, to determine the time relations.

Until 1966 there also would have been general agreement that a carbonate reef, such as the Redwater reef in Alberta, was a stratigraphic trap, although the hydrocarbon distribution in it is controlled in part by regional tilt. In that year, however, Martin (1966, p. 2278) pointed out that reefs, erosion surfaces, and other types of reservoir rocks that are bounded laterally by air or water at the time of their formation become traps only as a result of subsequent deposition of younger strata adjacent to (and above) them. He proposed the term "paleogeomorphic" for such traps and believed the term "stratigraphic trap" should be limited to those traps caused by lateral change in reservoir properties within a given (single) stratum. If Martin's proposal were accepted, most traps previously considered as stratigraphic would be in his paleogeomorphic category. Should such traps, or some part of them, be considered a separate category of traps, or are they a kind of stratigraphic trap?

Hydrodynamics is another factor that should be reconsidered in deciding, "What *is* a stratigraphic trap?" Levorsen (1954, p. 142) included one hydrodynamic aspect when he expanded his earlier definition of a stratigraphic trap by stating: "The pool may rest on an underlying water table, which may be either level or tilted. . . ." However, hydrodynamics may have effects other than just tilting the hydrocarbon-water contact. As pointed out by Hill *et al.* (1961) and McNeal (1965, p. 325), it is possible that an updip pinchout in a stratum might hold hydrocarbons if the water flow is downdip, but might not hold them if the water flow

is updip or if the water is static. Should such a trap be considered as stratigraphic?

Another factor that needs reconsideration is fracturing. In his 1941 discussion, Levorsen specifically excluded pools in which the porosity was the result of fracturing and brecciation. In 1954, however, he included the Santa Maria field in California as a stratigraphic trap even though "The porosity is the result of the fracturing of the brittle Monterey shales (Miocene) and siltstones . . ." (p. 243). Also, in the Spraberry trend in West Texas, where fractures and production are not related to local structures, the fractures are considered to act as "feeders" from rocks that are too impermeable to have extensive commercial production if unfractured (Wilkinson, 1953). In defining a stratigraphic trap, should we be concerned with how the porosity or permeability of the reservoir developed, or should we be concerned with what controls the boundaries of the trap itself? It is my contention that the boundary controls should be the determining factor. Thus, fracture-porosity traps unrelated to local structure would, indeed, be stratigraphic traps.

TYPES OF TRAPS

Before considering structural, stratigraphic, paleogeomorphic, and hydrodynamic aspects of traps, it is appropriate to determine the basic requirements for a trap. Hubbert (1953, p. 1954) summarized trapping of hydrocarbons in terms of energy potentials as follows:

Oil and gas possess energy with respect to their positions and environment which, when referred to unit mass, may be termed the potential at any given point of the fluid considered. When the potential of a specified fluid in a region of underground space is not constant, an unbalanced force will act upon the fluid, driving it in the direction in which its potential decreases. Hence, oil and gas in a dispersed state underground migrate from regions of higher to those of lower energy levels, and come ultimately to rest in positions which constitute traps, where their potentials assume locally minimum or least values. In nearly all cases traps for petroleum are regions of low potential which are enclosed jointly by regions of higher potential and impermeable barriers.

As to permeability barriers, Hubbert (1953, p. 1979) stated:

Reference to Table I shows that the capillary pressure of oil in a shale is the order of tens of atmospheres, while in a sand it drops to the order of tenths. Hence a slug of oil extending across such a boundary would be expelled from the shale into the sand by an unbalanced pressure of the order of tens of atmospheres.

This formidable energy barrier, therefore, makes a shale-sand interface appear as a surface of unidirectional conductivity to oil (or gas). Across such a boundary the oil can flow in the direction from the shale to the sand without hindrance other than viscous drag; in the opposite direction it can not flow at all unless a pressure is applied to the oil in the sand greater than the opposing capillary pressure against the oil in the shale.

There can, of course, be all gradations in pressures that permit entry of oil or gas—from pressures in shales through those in siltstones to those in shaly sandstones and sandstones of increasing grain size. Sandstones of the same median grain size may differ in entry pressure because of sorting, partial compaction, or partial cementation. As hydrocarbons accumulate, the pressure upward in the accumulation will increase, until in some situations the entry pressure of some part of the formerly impermeable barrier is exceeded and hydrocarbons will pass through it. Equilibrium is established and, if migration continues, hydrocarbons leave the trap at the same rate at which they enter it. Thus, the extent of difference in entry pressure between reservoir and barrier may control the height of the hydrocarbon column and, in consequence, the lateral extent of the accumulation. The same effect may result if fractures in the barrier rock or faults bounding the accumulation have lower entry pressures than the unfractured barrier rock. Hill *et al.* (1961) and Smith (1966) and others have discussed this aspect of hydrocarbon trapping.

Thus, except in a relatively few cases, a hydrocarbon trap requires the simultaneous existence of (a) a reservoir, (b) an isolated region of low potential in the reservoir, and (c) a barrier (or seal) with high enough entry pressure to retain a commercially producible volume of hydrocarbons. An isolated pod of sandstone *completely* filled with hydrocarbons may be an exception to (b); there appear to be no exceptions to (a) and (c).

It should be emphasized that high entry pressure (c, above) is not synonymous with low porosity or nonporosity. Some shales that form trap barriers, for example, are more porous than the adjacent reservoirs that contain the hydrocarbons. The size of the pores or the size and shape of the connections between them, not the amount of porosity, are the important factors.

In order to distinguish various types of traps, we need to decide what the term "stratigraphic" means. Should we limit ourselves to a "given stratum" as suggested by Martin (1966), or is the term broader? Stratigraphy has been defined as:

"1. That branch of geology which treats of the formation, composition, sequence, and correlation of the stratified rocks as parts of the earth's crust. 2. That part of the descriptive geology of an area or district that pertains to the discrimination, character, thickness, sequence, age, and correlation of the rocks of the district. (La Forge)" (*Glossary of Geology and Related Sciences,* Am. Geol. Inst., J. V. Howell, chm., 1957, p. 281)

"a The arrangement of strata, esp. as to position and order of sequence." (Webster's New International Dictionary, Unabridged, 2nd ed., 1956, p. 2491)

It seems clear from these definitions that *stratigraphy,* and consequently *stratigraphic,* is a broader term applying to strata and not just to a bed or group of beds that constitute a stratum. Nor are there any restrictions in time relations between rocks that may be laterally or vertically adjacent. Under this broader interpretation, paleogeomorphic traps are kinds of stratigraphic traps. In retrospect, one might wish that another term—perhaps "permeability trap"—had been proposed instead of stratigraphic trap. However, stratigraphic trap, or "strat trap," is so widely accepted and used that an attempt to change to a more descriptive term now probably would lead only to confusion.

From the foregoing discussion it appears that three basic kinds of traps exist—namely, structural, stratigraphic, and hydrodynamic—and that there may be combinations of any two or of all three kinds. The three basic kinds have a reservoir bounded by a barrier with high enough entry pressure to retain a commercially producible volume of hydrocarbons. Each kind also is in an isolated area of low potential, but they differ as to what causes the isolation. In a structural trap, isolation results from local structural deformation; in a stratigraphic trap it results from a nonstructural lateral change in entry pressure that creates the barrier; and in a hydrodynamic trap it results from the rate of water flow. Regional dip may be a component of stratigraphic traps; change in regional dip (terracing) may be a component of hydrodynamic traps.

Traps of all three categories may be filled to capacity or be partially filled or may contain no hydrocarbons. As a colleague of mine has put it, "A trap is a trap, whether or not it has a mouse in it" (W. C. Finch, personal commun.). Those traps containing hydrocarbons might well be designated as structural, stratigraphic, or hydrodynamic pools or accumulations; those containing no hydrocarbons, as potential hydrocarbon traps.

In classification of hydrocarbon accumulations, the conditions that determined the present location of the accumulation should be used if they can be determined. Thus, a stratigraphically trapped pool (such as an accumulation in an isolated pod of sandstone) which happens to be located on a post-accumulation local uplift would be classified as a stratigraphic pool. If, in contrast, accumulation occurred after the uplift and hydrocarbons migrated into the trap because of its locally high structural position, the controlling factor would be structure. In other places, where accumulations were trapped due to stratigraphic factors and later local uplift completely or materially shifted the position of the accumulations within the reservoir, the accumulations would be structural or combined stratigraphic-structural.

In some cases, the relative times of accumulation and structural growth will not be known. If no attempt has been made to determine these relative times, there is no basis for classification and none should be attempted. If an attempt has been made but the results are inconclusive, I suggest either "structural(?)" or "stratigraphic(?)," depending on which appears more likely. This method allows a judgment based on the weight of evidence available, but the "(?)" alerts others to the uncertainties involved.

The present position of many pools is the result of some combination of structure, stratigraphy, and hydrodynamics, and it seems appropriate, as previously suggested by others, that these traps be designated as combinations. The distinctions suggested by Sanders (1943), appropriately expanded to include hydrodynamics, would appear to provide a good basis for designating such combination traps. Certainly, the relative importance of structural, stratigraphic, and hydrodynamic factors needs to be recognized and clearly indicated.

BASES FOR STRATIGRAPHIC-TRAP CLASSIFICATION

Bases that have been suggested or used in describing or classifying stratigraphic traps include:

1. Time of trap formation, *i.e.*, primary—a direct product of the depositional environment—vs. secondary—developed after deposition and diagenesis of the reservoir; mainly unconformity traps.
2. Kind of reservoir rock, *i.e.*, clastic and igneous vs. chemical.
3. Kind of porosity, *i.e.*, interparticle vs. leached vs. fracture.

4. Genesis of the reservoir rock, *i.e.*, alluvial vs. bar vs. dune, *etc.*
5. Relation to regional dip, *i.e.*, open—not dependent on regional dip—vs. closed—where one boundary results from regional dip.
6. Geometry of the reservoir rock, *i.e.*, shoestring sands, *etc.*
7. The way the impermeable barrier formed, *i.e.*, low original permeability (deposition) vs. diagenetic plugging of the reservoir pores by tar, clay, or mineral cement.

The real problem with stratigraphic-trap classification is that it, like Topsy, has just "growed." Originally, differentiation of traps was between structural and nonstructural. As knowledge grew, more kinds of nonstructural traps were recognized and attempts were made to fit them into preexisting broad subdivisions.

Probably the best-known classification is that of Levorsen (1954), which is summarized briefly below.

I. Primary stratigraphic traps—formed during the deposition and/or diagenesis of the rock. These include
  A. Lenses and facies of clastic rocks
  B. Lenses of volcanic rock
  C. Stratigraphic traps in chemical rocks
    1. Porous facies
    2. Porous mound- or lens-shaped carbonate masses
II. Secondary (unconformity) stratigraphic traps—resulting from some stratigraphic anomaly or variation that developed after deposition and diagenesis of the reservoir rock; almost everywhere associated with unconformities. Traps above and below the unconformities are included as secondary.

There are several inconsistencies in this classification and its application. One is that Levorsen would include as primary many traps that are wholly or in part of secondary diagenetic origin. Examples are traps resulting from secondary dolomitization or those due to cementation.

A second inconsistency is inclusion of stratigraphic traps above and below unconformities in the same category. Levorsen (1954, p. 239) stated:

Traps bounded by an unconformity are broadly classed as stratigraphic, and they are also classed as secondary stratigraphic because they are formed after the lithification and diagenesis of the reservoir rock.

However, many traps above unconformities have had no significant lithification or diagenesis of the *reservoir* rock or the barrier that overlies it. For many traps above unconformities, the unconformity merely provides part of the impermeable barrier. If the rock below the unconformity is sufficiently permeable and extends far enough updip, there is no trap. It is true that in some places the barrier may be a former reservoir rock in which permeability has been reduced by diagenetic processes, but more commonly the rock below the unconformity was originally of low permeability and has remained so. In many other traps above unconformities, relief on the erosional surface has controlled where reservoir beds were deposited.

I believe the suggested classification of stratigraphic traps will eliminate these inconsistencies. In the succeeding discussion and in the suggested classification, primary emphasis has been placed on usability—*i.e.,* will the groupings help in searching for new hydrocarbon accumulations, and is the suggested terminology simple and descriptive enough to be accepted?

Extensive consideration was given first to use of the time relations between barrier and reservoir as a first-order subdivision. Because the existence of a barrier above an oil or gas accumulation is common to all traps, the relation of the reservoir to the *lateral* barrier—which restricts both updip and sideways movement of the hydrocarbon out of the trap—would usually be the critical one.

Three simple relations are possible:

1. Barrier and reservoir formed at the same time;
2. Barrier formed before the reservoir;
3. Barrier formed after the reservoir.

For such simple relations, a reservoir-barrier basis for classification would have practical advantages in exploring for stratigraphic traps. If barrier and reservoir were formed at the same time, and thus were genetically related deposits, they could be studied together as a genetic couplet and a facies-"model" concept of alluvial, deltaic, shallow-marine, or turbidite deposition could be applied if the expected reservoirs were sandstones. In contrast, different exploratory concepts would be used (relation 2) where the barrier formed before the reservoir, as where a

youthful valley cut in shale was filled with sand, or (relation 3) where the barrier formed after the reservoir, as where a hill or an organic reef was buried by mud or other sediment which forms relatively impermeable rocks.

Many stratigraphic traps, however, have more complex reservoir-barrier relations because different parts of the barrier formed at different times. In addition to the three simple relations, four combinations are required to satisfy all possible relations. They are:

4. Barrier formed partly before and partly at the same time as the reservoir;
5. Barrier formed partly after and partly at the same time as the reservoir;
6. Barrier formed partly before and partly after the reservoir;
7. Barrier formed partly before, partly at the same time, and partly after the reservoir.

When these four combinations were explored, two serious complicating factors were found. The first is illustrated in Figure 1. If the hydrocarbon column is short (Fig. 1A), the barrier to lateral migration is entirely later, or younger, than the reservoir (relation 3). In contrast, if the hydrocarbon column is long (Fig. 1B), the barrier is formed partly by the post-unconformity shales, partly by the stratigraphically younger limestone, and partly by the stratigraphically older shale (relation 7). In this and other cases, the length of the hydrocarbon column would determine the classification. As a further complication, one might reasonably ask how a *potential* trap, one not containing any hydrocarbons, would be classified.

The second complicating factor is illustrated by Figure 2. The hydrocarbons in the bar (Fig. 2A) deposited on an essentially horizontal unconformity surface are restrained from lateral migration by genetically related open-marine and lagoonal shales (relation 1). After tilting (Fig. 2B), however, the rock below the uncon-

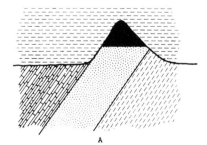 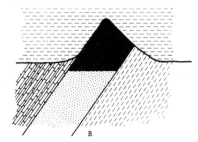

Fig. 1—Relation of age of barrier to hydrocarbon column height. **A.** For short columns, barrier is formed entirely by post-unconformity shales. **B.** For long columns, barrier is partly post-unconformity shales, partly pre-unconformity, post-reservoir limestone, and partly pre-unconformity, pre-reservoir shales.

formity has become part of the barrier (relation 4). Had the pre-unconformity rock been sufficiently permeable, the trap capacity would have been reduced. In this and other cases, regional tilting would determine the classification.

These two complications were among the factors that led to rejection of time relations between barrier and reservoir as a basis for stratigraphic-trap classification.

SUGGESTED CLASSIFICATION

What I now believe to be the best classification system uses some, but not all, of Levorsen's ideas. I suggest starting with the simple concept that stratigraphic traps are either adjacent to unconformities, or they are not. Those traps in contact with unconformities[3] can be below the unconformity surface, above it, or both. For traps not adjacent to unconformities, either (I) the reservoir and barrier both may be primary (depositional), or (II) the reservoir or the barrier may be wholly or in part secondary (diagenetic). Most of the primary traps not related to unconformities consist of genetic juxtapositions of coarse and fine (very high- and low-permeability) sediments; consequently, I suggest that they be designated "facies-change traps." This designation will help to distinguish them from traps adjacent to unconformities (classes III and IV) that also have primary (depositional) reservoirs but commonly have one or more boundaries not genetically related. Four major classes are each split into two subclasses as shown in Table 1, and these subclasses are subdivided further as shown in Table 2.

[3] The term "unconformity" is defined (AGI *Glossary of Geology and Related Sciences,* 1957, p. 308) as "A surface of erosion or nondeposition—usually the former—that separates younger strata from older rocks." "Unconformity" as used herein would not refer to a depositional break or hiatus of assumed minor duration (a diastem) during which erosional modification of the surface was minor.

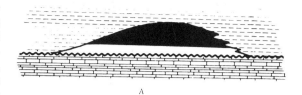

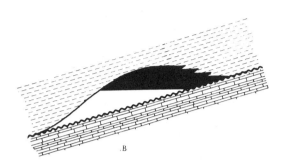

FIG. 2—Diagrammatic cross sections of barrier bar showing control of accumulation by facies-related shales before regional tilting (A) and by facies-related shales and pre-unconformity limestones after regional tilting (B).

I. Facies-Change Traps

The suggested first-order subdivision of facies-change traps is based on the depositional origin of the reservoir rock, *i.e.,* whether significant transport of particles by currents has occurred. "Current-transported" thus would imply mechanical transportation of particles or fragments of the reservoir rock to the site of deposition by water or wind currents. The mineral composition of the grains would not be critical; they could be quartz, feldspar, rock fragments, skeletal or nonskeletal carbonate particles, volcanic glass, or some combination of these. Some carbonate rocks in which traps occur thus would be in this current-transported category. "Not current-transported" would imply little or no transportation of particles, or movement due to gravity only.

Table 1. Major Subdivisions in Proposed Stratigraphic-Trap Classification

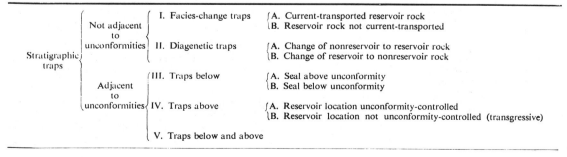

Stratigraphic traps	Not adjacent to unconformities	I. Facies-change traps	A. Current-transported reservoir rock B. Reservoir rock not current-transported
		II. Diagenetic traps	A. Change of nonreservoir to reservoir rock B. Change of reservoir to nonreservoir rock
	Adjacent to unconformities	III. Traps below	A. Seal above unconformity B. Seal below unconformity
		IV. Traps above	A. Reservoir location unconformity-controlled B. Reservoir location not unconformity-controlled (transgressive)
		V. Traps below and above	

23

Table 2. Suggested Stratigraphic-Trap Classification

Not Adjacent to Unconformities

I. Facies-change traps
  A. Current-transported reservoir rock
    1. Eolian
      a. Dune (coastal, inland)
      b. Eolian-sheet
    2. Alluvial-fan
    3. Alluvial-valley
      a. Braided-stream
      b. Channel-fill
      c. Point-bar
    4. Deltaic (lacustrine, bay)
      a. Distributary-mouth bar
      b. Deltaic-sheet
      c. Distributary channel-fill
      d. Finger-bar
    5. Nondeltaic coastal (lacustrine, bay)
      a. Beach
      b. Barrier-bar
      c. Spit, hook, *etc.*
      d. Tidal-delta
      e. Tidal-flat
    6. Shallow-marine
      a. Tidal-bar
      b. Tidal-bar belt
      c. Sand-belt
      d. Washover
      e. Shelf-edge
      f. Shallow-winnowed-crestal
      g. Shallow-winnowed-flank
      h. Shallow-turbidite
    7. Deep-marine
      a. Marine-fan
      b. Deep-turbidite
      c. Deep-winnowed-crestal
      d. Deep-winnowed-flank
  B. Reservoir rock not current-transported
    1. Gravity
      a. Slump
    2. Biogenic carbonate
      a. Stratigraphic reef
        1. Shelf-margin
        2. Mound (patch-reef, mud, algal, *etc.*)
      b. Blanket (crinoidal, tidal-flat, lagoonal, *etc.*)
II. Diagenetic traps
  A. Nonreservoir to reservoir rock
    1. Replacement (and leached)
      a. Dolomitized shelf-margin
      b. Dolomitized mound (patch-reef, mud, algal, *etc.*)
      c. Dolomitized blanket (crinoidal, tidal-flat, *etc.*)
      d. Dolomitized current-transported deposit (facies or lithologic type)
    2. Leached
      a. Leached shelf-margin
      b. Leached mound (patch-reef, mud, algal, *etc.*)
      c. Leached blanket (crinoidal, tidal-flat, *etc.*)
      d. Leached current-transported deposit (facies or lithologic type)
    3. Brecciated
    4. Fractured (lithologic type)
  B. Reservoir to nonreservoir rock
    1. Compaction
      a. Physical compaction
      b. Chemical compaction
    2. Cementation

Adjacent to Unconformities

III. Traps below unconformities
  A. Seal above unconformity
    1. Topography young
      a. Valley-flank
      b. Valley-shoulder
    2. Topography mature
      a. Crestal
      b. Dip-slope

Table 2. (*Continued*)

      c. Escarpment
      d. Valley
    3. Topography old
      a. Beveled
  B. Seal below unconformity
    1. Mineral cement (anhydrite, calcite, *etc.*)
    2. Tar-seal
    3. Weathering product (weathered-feldspar, weathered-tuff, *etc.*)
IV. Traps above unconformities
  A. Reservoir location unconformity-controlled
    1. Two sides
      a. Valley-fill
      b. Canyon-fill
      c. Blowout-fill
    2. One side (buttress)
      a. Lake-cliff
      b. Coastal-cliff (fault–coastal-cliff)
      c. Valley-side (fault–valley-side)
      d. Hill-flank (fringing-reef, mound, blanket, *etc.*)
      e. Structure-flank (fringing-reef, mound, blanket, *etc.*).
  B. Reservoir location not unconformity-controlled (transgressive)
    Facies terms followed by (*unconformity*) or (unconformity) where applicable (See text for explanation.)
V. Traps below and above unconformities

A second-order subdivision of current-transported reservoir rock can be made on the basis of depositional process or environment, a third-order subdivision on type or location of deposit, and a fourth-order subdivision, if needed, on lithology. Thus, under the "facies-change, current-transported" category (Table 1, class IA), one possibility would be a trap in which the reservoir rock is a shallow-marine (second order) tidal-bar belt (third order) of oolitic lithology (fourth order), which would be called an "oolitic tidal-bar–belt trap" (Table 2, class A6b). Facies change, current transport, and shallow-marine conditions would be implied.

Many of the suggested terms are in common usage and require no definitions or explanation. For others, additional comments appear desirable.

It is recognized that additional subdivisions will be required as our knowledge of sedimentary rocks and stratigraphic traps in them increases. We know much more about deltaic and interdeltaic deposits, for example, than about shallow- and deep-marine accumulations, largely because of the availability for detailed study of modern counterparts. Subdivision of such sedimentary complexes as alluvial fans, tidal deltas, and marine fans may prove desirable if many stratigraphic traps are found in them. It seems advisable to propose a framework in which such subdivisions can be made later as needed.

It is also recognized that inclusion of some terms, particularly dune and eolian sheet, in the facies-change category is questionable, because such bodies probably would be deposited most commonly on unconformity surfaces. Also, where dunes or eolian sheets are deposited during regressions and thus are not unconformity-related, the overlying units usually would not be genetically related.

Deltaic and nondeltaic coastal deposition occurs at or near a land-water interface. This interface may be between the land and a lake, bay, lagoon, estuary, or the ocean. Making a distinction between these types of interfaces may be important in searching for stratigraphic traps. The "(lacustrine, bay)" following both deltaic and nondeltaic coastal in Table 2 means that these terms should be included in the description where appropriate. Because most deltaic and nondeltaic coastal stratigraphic traps will probably be in deposits near the land-ocean interface, it is proposed that this be inferred and that no prefix be used.

Explanations of some of the terms used in the facies-change category in Table 2 are given below.

A6a. *Tidal bar*—Present-day examples described and illustrated by Off (1963).

A6b. *Tidal-bar belt*—Separated from other tidal bars by formation at a major slope break where tidal currents are concentrated by embayments (Ball, 1967).

A6c. *Sand belt*—Controlled by a major slope break, but without concentration of tidal currents (Ball, 1967). Differs in position from shelf-edge sands by being built up on the platform edge rather than accumulating at or below the break in slope. At Cat Cay, the example cited in the Bahamas by Ball, the carbonate sand is of local origin.

A6d. *Washover deposit*—Composed of debris washed over and accumulated behind barriers, reefs, or low islands. Composed commonly of carbonate sand or coarser debris.

A6e. *Shelf-edge deposit*—A sand-body type postulated by Rich (1951). Probably caused by relative lowering of sea level and transport of preexisting sand-sized shelf sediments seaward to and over the shelf edge (Fig. 3). Lehner (1969, p. 2469) used the terms "foreset beds" and "spillover beds" for sediments of Wisconsin age on the Texas shelf edge.

A6f, A7c. *Winnowed-crestal deposit*—Would result from winnowing of the fines from a coarse-fine particle admixture on the crest of a growing dome or anticline (Fig. 4).

A6g, A7d. *Winnowed-flank deposit*—Would be due to similar winnowing, but by stronger currents that would remove the sand or shell debris from the crest

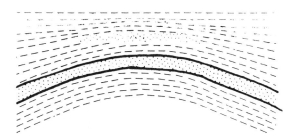

Fig. 4—Accumulation of sand over growing structure, as result of winnowing.

and deposit it on one or more flanks (Fig. 5). It is possible that winnowed deposits might occur also in other environments. If so, that environment may be used as a prefix.

A6h, A7b. *Shallow and deep turbidites*—Types separated at 600-ft (100 fm) water depth, shallow turbidites being on the shelf (or in lakes or bays) and deep turbidites at greater depths.

A7a, A7b. *Marine fan and deep turbidite*—Types separated on the basis of position, the marine fans being at a break in slope where velocities are reduced (thus being submarine equivalents of alluvial fans; Nelson et al., 1970), and the deep turbidites filling depressions on the slope (Lehner, 1969) or covering basin or ocean floors.

B1a. *Slump deposit*—Would result from mass movement of sand bodies, usually shelf-edge and associated sands, down submarine slopes (Lehner, 1969). Talus adjacent to a carbonate buildup also would be a slump accumulation, but, because of the close association with other facies in such buildups, such talus is included with the buildup rather than being designated a type of slump.

B2. *Biogenic carbonate deposit*—Non-current-transported carbonate rocks that originally had commercial porosity and permeability and have retained it. Some diagenetic enhancement may have occurred, but it is not critical to making the rock a reservoir. Actually, there may be few traps in this category, because most traps in carbonate rocks owe a critical part of their porosity and permeability to diagenetic processes.

Although some excellent ideas on carbonate-rock and porosity classification have been advanced in the past decade (Ham, 1962; Choquette and Pray, 1970), they do not appear to provide a suitable basis for classification of

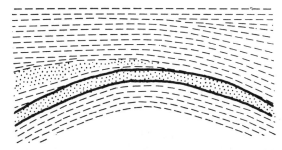

Fig. 5—Accumulation of sand over flank of growing structure as result of winnowing and lateral transport of sand.

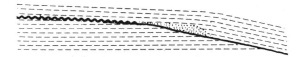

Fig. 3—Sand body formed at edge of shelf owing to relative lowering of sea level.

stratigraphic traps. I propose that non–current-transported carbonate rocks in which traps occur be considered as either "stratigraphic reefs" as defined by Dunham (1970) or "blankets" (buildups vs. sheets), and that stratigraphic reefs be subdivided into "shelf margins" and "mounds." The shelf margins would be elongate or arcuate and would separate facies of different types on the two sides; mounds would tend to be equidimensional, usually would be surrounded on all sides by the same facies (whether all lagoonal or all open marine), and might or might not have marginal and interior facies of the same type. This facies differentiation is a modification of that suggested by Heckel (1970), and I believe it can be useful in exploration.

The position of shelf margins may be inherited from preexisting topography or structure, or it may result from the buildup itself. Although such large carbonate buildups as the Central Basin platform or the Bahama Banks might be considered oversized mounds or be placed in a separate category, I suggest classifying their edges as shelf margins.

Stratigraphic reefs large enough to contain commercial accumulations of hydrocarbons will almost everywhere be combinations (complexes) of carbonate types, both laterally and vertically. The proportion of organisms having the potential to build wave-resistant structures may vary widely. In initial exploration for, and attempts to find extensions of, shelf-margin accumulations, the type of carbonate and its wave resistance normally will be of less importance than determining the position of the margin and the location of highs on it. Even where diagenetic processes have formed the porosity, the conditions that allow these processes to operate probably will be more important than the carbonate types. For example, whether exposure permitted vadose leaching may be more significant than whether the carbonate was a boundstone or an early-cemented packstone. In contrast, in exploring for mounds, and particularly in searching for other mounds after the first has been discovered, the type of sediment or organisms may be relatively more important. For these reasons, I suggest that some modifying terms be used for mounds but have no suggestions at this time for meaningful subdivisions of shelf margins.

The term "blanket" is suggested for sheetlike non–current-transported carbonate deposits.[4]

[4] Biostrome, a widely used term, is a type of blanket deposit with a large skeletal component.

Blankets may have no close association with a buildup or may be associated with one. For example, a landward traverse across an elongate buildup might show the shelf margin flanked seaward by a blanket and landward by a continuous or discontinuous band of washover or other current-transported carbonate sediment. These bodies might grade landward into blankets of lagoonal and tidal-flat deposits. To distinguish different kinds of blankets, subdivision is suggested on the basis of dominant environment (tidal flat, lagoonal, etc.) and/or components (crinoidal, pelletal, etc.).

II. Diagenetic Traps

Diagenetic traps not associated with unconformities may be formed during or soon after deposition or after considerable burial and perhaps after extensive lithification. They may occur either (A) where a nonreservoir rock has been changed to a reservoir rock and the unaltered or less extensively altered nonreservoir rock serves as an upper and/or a lateral barrier, or (B) where a reservoir rock has been changed partly to a nonreservoir rock and the altered part forms all or part of the barrier.

At least four processes—replacement, leaching, brecciation, and fracturing—can produce a reservoir rock from a nonreservoir rock. Replacement and leaching both require movement of water, but the effects differ. Both may operate concurrently. In replacement, the moving water brings with it dissolved matter which, under the prevailing surface or subsurface temperature and pressure, reacts with the preexisting rock. If new minerals of greater density and lesser volume are formed, new pore space may be created. Actually, rearrangement of existing pore space may be more important than creation of new pores because of volume decrease. For example, replacement in a porous but slightly permeable calcareous mud may result in larger pores and larger connections between them. Leaching that occurs concurrently with the replacement may further enhance porosity and permeability. Local, and in some places regional, dolomitization of limestones forms most such "replacement" traps.[5]

[5] The diagenetic changes during dolomitization may be complex. Dolomitization of a calcareous mud or wackestone first might selectively change a nonreservoir rock to a reservoir rock; continued dolomitization might reduce the porosity and change part of that reservoir rock to nonreservoir rock. Where this has happened, it may be difficult to determine whether the remaining reservoir rock should be classified under replacement or cementation, or whether a combination of both is involved.

In contrast, the major effect of the moving water may be as a solvent—to dissolve and carry away the rock or certain parts of it. Where the solution involves selective removal of some constituents of the nonreservoir rock, intergranular, oomoldic, or other types of fabric-selective porosity may result (Choquette and Pray, 1970). However, where the solution of the nonreservoir rock is nonselective, vugular, channel, or cavern porosity may form. Unaltered or less altered rocks above, lateral to, and in places stratigraphically below the reservoir prevent updip hydrocarbon migration. Where solution is at or near an unconformity and the seal is at least partially post-unconformity in age, the trap would be unconformity-related. In contrast, where the seal is not associated with the unconformity surface, the trap would be in the diagenetic category even though water that formed it may have moved downward or laterally from an erosion surface.

It is not meant to imply that the nonreservoir rock is completely devoid of porosity or permeability before replacement or leaching. In carbonate rocks particularly, fractures, differences in facies, and/or differential cementation may control the transmissibility of the rock to waters or the effectiveness of these waters in producing diagenetic changes.

Brecciated reservoirs also may result from solution, where such solution removes carbonate, anhydrite, or salt from large enough areas to permit collapse and brecciation of interbedded or overlying rocks. Mounds or other topographic features that penetrate upward into the solution zones may accentuate the brecciation.

Fractured reservoirs, in contrast, would be tectonic in origin but would not be the result of local structural deformation. The lateral termination of the fracturing may be due to change in subregional stresses or to change in rock ductility to a less easily fractured rock—i.e., from dolomite to limestone, or from cherty to less cherty rock. The overlying barrier also would be a less easily fractured rock.

Diagenetic changes of a reservoir-type rock to a barrier may result from compaction or cementation, or a combination of both. Compaction may be either physical—as where relatively ductile, usually lithic grains are plastically deformed and squeezed into the adjacent pore space by the weight of overburden—or chemical—as where part of the rock is removed by solution. Such solution may occur at points of contact between quartz, chert, or other hard grains, or along stylolitic seams.

Local redeposition of material so dissolved may reduce the amount of pore space still more. Some lateral variation in original grain size or composition usually would be required to allow formation of such barriers. They would be formed at depth, and, consequently, any relation to unconformities would be coincidental.

Cementation may reduce the pore space selectively in some parts of a reservoir rock but not in others. This cementation may be penecontemporaneous, or it may occur at depth where compaction waters moving upward or meteoric waters moving downward reach a critical temperature and/or pressure.

III. Traps Below Unconformities

Impermeable beds above unconformities form part or all of the barrier to vertical or lateral migration of hydrocarbons from many stratigraphic traps that occur below unconformities. However, for some traps below unconformities, part or all of the barrier is formed by diagenetic processes that are unconformity-dependent. Some barriers of this latter type, such as tar seals, may be just below the unconformity surface; others, such as those resulting from pore filling by mineral cements, may extend a considerable distance downdip. Because of its importance in exploration, differentiation of post-unconformity (depositional) seals from unconformity-related (diagenetic) seals seems desirable.

I suggest that the nondiagenetic traps below unconformities be differentiated on the basis of maturity of the unconformity surface, i.e., whether the erosional surface was in a young, mature, or old stage when buried. Exploration methods for the different types will differ.

In the young stage, narrow, steep-sided valleys are eroded into flat or gently dipping strata, some of which could be reservoirs. If these valleys are filled or partially filled later with relatively impermeable sediments, and some tilting occurs, the impermeable valley deposits could become a barrier to updip hydrocarbon migration. If the reservoir rock abuts only against the valley side, the impermeable bed above it will form part of the barrier, and only later deposition of relatively impermeable beds in the valley adjacent to the reservoir rock is needed to prevent updip migration. Such "valley-flank" traps (Fig. 6) may be differentiated from "valley-shoulder" traps, where the reservoir rock formed part of the surface of low relief into which the valley was trenched and where nonreservoir beds were deposited

FIG. 6—Valley-flank accumulation against impermeable sediments filling youthful valley.

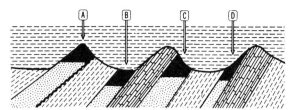

FIG. 8—Accumulations in crestal (A), valley (B), escarpment (C), and dip-slope (D) positions below mature erosion surface.

both laterally in the valley and on the low-relief surface above (Fig. 7).

If the unconformity surface is mature, the reservoir rocks may occupy topographically high, intermediate, or low positions, as shown in Figure 8. Those topographically high—including hills, cuestas, and mesas—may be termed "crestal." Those that are intermediate—*i.e.*, having less resistance to erosion than non-reservoir rocks that form the topographic highs—may be termed "dip-slope" or "escarpment" deposits. Those topographically low may be designated "valley" deposits. It seems probable that many stratigraphic traps in fractured and/or weathered igneous rocks would be in the "crestal" class.

For traps below old-age surfaces of slight topographic relief, the term "beveled" is suggested, because dipping beds have been truncated by an erosion surface of lesser slope. In exploration for the "young" and "mature" classes of traps, the application of geomorphic concepts as advocated by Martin (1966) and others may be very useful; however, for "beveled" traps below unconformities, other exploration methods are required.

In classification of stratigraphic traps, what role should lithology or the primary or secondary origin of the porosity or permeability play when the seal is above the unconformity—*i.e.*, "post-unconformity"? It ordinarily will be known from downdip penetrations of the section whether the porosity and permeability are primary. (Stratigraphic reefs may be an exception to this generalization.) The existence of widespread secondary porosity, as from dolomitization, generally will be known also. In these cases, the location of the subcrop and the

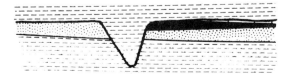

FIG. 7—Valley-shoulder accumulation against impermeable sediments both in valley and on low-relief surface to right.

topographic expression of the porous and permeable beds should be determined.

In contrast, diagenetic processes related to the unconformity which produce or enhance porosity and/or permeability at and near the unconformity may be significant in exploration. Also, fracture belts of subregional extent (not associated with local structure) which traverse the areas of interest may be of exploratory significance, particularly if rocks of different lithologies fracture to different extents. It seems desirable, therefore, to prefix a diagenetic or lithologic-diagenetic descriptive term where local secondary porosity and/or permeability development is important. Thus, for example, we might have "fractured igneous–crestal," "leached oolite–valley," or "leached subgraywacke–dip-slope" traps. In my opinion, such word descriptions are far better than a combined numerical-alphabetical or decimal system that would require frequent reference to a master code, though such a system might be more desirable for computer usage. Even so, this compound-word terminology may prove too cumbersome and, if so, may have to be abandoned in favor of written supplementary descriptions.

Other unconformity-dependent diagenetic processes may decrease rather than increase porosity and/or permeability; thus, the permeability of what was once a reservoir rock may be reduced sufficiently to make it a barrier to hydrocarbon migration. This change may occur in three ways: (1) by introduction and localized deposition of mineral cements such as anhydrite, carbonate, or silica; (2) by conversion of oil to tar; and (3) by weathering of feldspar or other materials to clay minerals. Recognition of such diagenetic barriers may be important in exploration because (1) accumulations may be downdip from the more obvious post-unconformity barrier, and (2) accumulations may be present where sandstones or other reservoir-type beds overlie the unconformity. Thus, traps with diagenetic seals are separated from traps

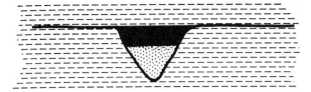

FIG. 9—Valley-fill (or canyon-fill) accumulation.

FIG. 11—Lake-cliff or coastal-cliff accumulation.

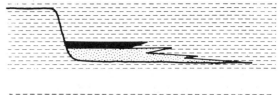

FIG. 12—Valley-side accumulation.

with seals above the unconformity. Those traps in which the seal is formed by deposition of mineral cements may be designated by the pore-filling minerals, of which the more common is probably anhydrite; those by degradation of oil, as tar seals; and those by weathering, by the original mineral or rock name—such as "weathered feldspar," "weathered lithic sandstone," or "weathered tuff."

IV. Traps Above Unconformities

Unconformity surfaces with considerable local relief may control the distribution of potential reservoir rocks. In contrast, unconformities with slight local relief may exert little, if any, control on the distribution of potential reservoir beds above them. Factors controlling sea level or regional subsidence may be much more important. This control or lack of it is suggested as a basis for classifying stratigraphic traps above unconformities. Exploration methods used to locate traps of the two kinds would differ.

The unconformity-controlled traps may be subdivided further on the basis of *extent* of control of the reservoir by the unconformity. In youthful valleys or submarine canyons, the potential reservoir rock may extend without interruption from one side of the valley or canyon to the other—*i.e.,* the reservoir is limited on two sides (and commonly its base) by relatively impermeable rocks below the unconformity surface. For such traps the terms "valley-fill" and "canyon-fill" (Fig. 9) are suggested. Although no examples are known, "blowout-fill" traps (Fig. 10), formed by sand deposition in wind-eroded depressions, should be included in this category.

Other unconformity-controlled reservoirs are limited on only one side by the unconformity,

and the other boundary is a facies-related rock of low permeability. This category of traps includes what Levorsen (1954) and others have called "buttress sands," as well as some low-relief organic buildups. The unconformity surface against which the reservoir terminates may be a lake or coastal cliff, the side of a valley, an isolated hill, or an eroded structural uplift. The suggested designations for such traps are "lake-cliff," "coastal-cliff" (Fig. 11), "valley-side" (Fig. 12), "hill-flank" (Fig. 13), and "structure-flank" (Fig. 14) traps. McCubbin (1969) showed good examples of coastal-cliff traps.

If the reservoir rock is not current-transported, an appropriate modifying term can be used in conjunction with the unconformity term. Such accumulations are probably restricted to "hill-flank" and "structure-flank" positions and might be fringing reefs, mounds, or blankets. The crinoidal mound on the flank of an eroded Ordovician fold in the Todd field (Levorsen, 1954, Fig. 6–26, p. 215) is a good illustration of a carbonate reservoir rock in which the unconformity is an important trapping factor. Most stratigraphic reefs, although initiated in hill-flank or structure-flank or other

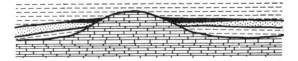

FIG. 13—Hill-flank accumulations.

FIG. 10—Blowout-fill accumulation.

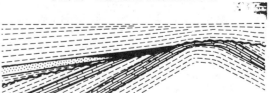

FIG. 14—Structure-flank accumulation.

positions on unconformity surfaces, normally would grow so high that the part abutting the unconformity would form only a minor part of the trap, even if tilting occurred later. Therefore, most stratigraphic-reef traps would not be included in this category.

It is possible that fault scarps may control *post-fault* distribution of reservoir rocks and also form one boundary of some traps. How should such traps be classified?

The scarps may be inland and, if so, may control, or partially control, the drainage system and be modified by river erosion; they may be at the coastline and may control the location of the land-water interface and be modified by wave erosion; or they may be submarine and not modified. The inland and coastal fault scarps are associated with unconformities; thus, they might be considered special types of "valley-side" or "coastal-cliff" unconformity traps to which the prefix "fault" would be added. Although the submarine fault scarps might not be associated with unconformities, I suggest that they be included here rather than be put in a separate class that would have only one representative. The suggested designation is "submarine fault-flank trap."

Actually, there may be so few traps in these fault categories that including them in a stratigraphic-trap classification may be academic. Usually, continued growth or later reactivation of the faults would control the location of the hydrocarbon accumulations or create new barrier-to-reservoir relations. Therefore, the trap would be structural rather than stratigraphic. Only if the hydrocarbons migrated into the trap before reactivation of the fault, and if the location of the accumulation in the trap were not modified substantially, could the trap be considered stratigraphic.

Nearly all traps above unconformity surfaces with essentially no relief result from transgression, and the distribution of reservoir rocks is controlled by factors other than the relief on the unconformity surface. It seems likely that such traps will be mainly of coastal or shallow-marine origin. The alluvial, deltaic, and deep-marine types commonly will not be adjacent to such low-relief unconformities. Except for some carbonate rocks, reservoir beds deposited during regression normally will be separated from the unconformity surface by nonreservoir rocks.

What we are really concerned with here is a kind of facies trap in which the reservoir body is deposited directly on a surface of uncon-

formity rather than on sediments of nearly the same age. Theoretically, if the reservoirs are bars or other deposits with considerable relief, it may make no difference whether the rock below the unconformity is permeable or impermeable to hydrocarbons, because the hydrocarbons would have no contact with it (Fig. 2A). The unconformity is unimportant. Exploration methods used for locating such traps would be the same as those used in searching for facies traps.

In contrast, for low-relief deposits such as beach sands, or for those higher relief deposits which have been tilted, the permeability of the rocks underlying the unconformity may be critical in determining whether a trap exists and/or how large it is. Location of such traps requires exploration methods used for facies traps combined with other methods that will provide the required information on the permeability and attitude of the rocks below the unconformity.

What terminology should be used for traps above unconformities of essentially no relief? Should a distinction be made, and a different terminology be used, for traps where the unconformity does or does not form a part of the barrier to updip or sideways migration? To answer the second question first, I suggest that a distinction be made where possible but, for practical reasons, no complicating separate terminology be used. It seems probable that there will be few traps where the unconformity is unimportant, and that those which do exist will have limited capacity. Regional tilting of even half a degree in the right direction would make the permeability of rocks below the unconformity significant for bars or other deposits a mile or more in length. Of equal or greater importance, however, is the fact that, in searching for traps above unconformities of essentially no relief, one would wish to know in advance whether the unconformity might be a limiting factor. Consequently, exploration methods used for facies traps would have to be supplemented with others that would provide the required information on the permeability and attitude of the rocks below the unconformity.

For traps above unconformity surfaces of essentially no relief, it is suggested that the appropriate facies-trap term be followed by "unconformity" in parentheses. If the permeability of the rocks below the unconformity is, or may be, critical to the existence or size of the trap, the word "unconformity" would be underlined or italicized; if demonstrably not critical, "unconformity" would not be underlined or italicized. Thus, the designation might be "barrier-bar (*unconformity*)" or "barrier-bar (unconformity)."

V. Traps Below and Above Unconformities

Where a connection exists between reservoir rocks that overlie and underlie an unconformity, the hydrocarbon column may bridge that unconformity. Although both the seal and the reservoir may be in part above the unconformity, a major part of the accumulation may be below it. For such pools, a dual terminology that describes both reservoirs seems desirable. Thus, in a valley-side–plus–escarpment trap, most of the hydrocarbon is below the unconformity. The order would be determined by the relative volume of hydrocarbons in each type of reservoir—that with the lesser volume being first.

COMBINATION STRATIGRAPHIC TRAPS

Most stratigraphic traps above unconformities involve a combination of facies and unconformities, and thus might be considered combination stratigraphic traps. Facies also may control the lateral extent of some diagenetic traps and traps below unconformities. In the diagenetic traps, however, this facies control is usually closely related to the operation of the diagenetic process that changes nonreservoir to reservoir rock or reservoir to nonreservoir rock. Replacement, solution, fracturing, or compaction that result in traps occur preferentially in some facies. Therefore, a combination terminology does not seem necessary for diagenetic traps. In contrast, where solution, fracturing, and possibly replacement have produced or enhanced porosity and/or permeability below unconformities, a diagenetic or diagenetic-lithologic terminology seems desirable, as proposed in the preceding section of this paper.

For traps below unconformities, a combined term may be desirable where facies control the lateral extent of the trap. The appropriate facies term might be added in parentheses. Thus, where a barrier bar has been truncated, the designation might be "beveled (barrier-bar) trap."

Should traps occurring at the updip intersection of two unconformities be considered as combinations? The dual-unconformity relations may be of two kinds—one in which the reservoir lies below both unconformities and one in which it is between them. In the first type, exemplified by the West Edmond pool, Oklahoma (Levorsen, 1954, p. 243, 623), a different direction of regional tilting between the earlier and later periods of erosion resulted in the reservoir extending farther updip at the intersection. No separate terminology for traps of this type seems necessary.

In the second type, the lower unconformity will be significant only if it controls the location of reservoir rocks deposited on it. Where the location of reservoir rocks is controlled by the older unconformity, truncation of these beds and later deposition of an impermeable bed above would form the trap and determine its location. The original unconformity-controlled distribution is modified by the later erosion. It is suggested that, for this relation, "truncated" be prefixed to the appropriate term for the trap above an unconformity; "truncated" rather than "beveled" is suggested, because the upper unconformity surface may have more relief than is implied by beveling.

DISCUSSION

The classification of stratigraphic traps suggested here appears to eliminate some of the inconsistencies in systems used previously. The number of factors involved in stratigraphic trapping is large, however, and their relative importance may vary in different traps. Consequently, no classification method can be completely definitive without having a very large number of subclasses—a number approaching, if not equal to, the total number of stratigraphic traps. Therefore, a compromise is necessary. As a result, some traps will fit neatly into classification pigeonholes and others will not.

The suggested classification represents such a compromise, proposing few enough subdivisions to be acceptable. Furthermore, because the purpose of a stratigraphic-trap classification is economic—to help in finding (and developing) hydrocarbon accumulations—those factors that I believe will help in searching for such accumulations have been emphasized. Others may disagree with my emphasis. If, however, this paper stimulates thought about trapping factors, their relative importance, and their implications regarding exploration methods, it will have served a useful purpose.

REFERENCES CITED

American Geological Institute (Howell, J. V., chm.), 1957, Glossary of geology and related sciences: Natl. Research Council Pub. 501, 325 p.

Ball, M. M., 1967, Carbonate sand bodies of Florida and the Bahamas: Jour. Sed. Petrology, v. 37, no. 2, p. 556–591.

Carll, J. F., 1880, The geology of the oil regions of

Warren, Venango, Clarion, and Butler Counties: 2d Pennsylvania Geol. Survey, v. 3, 482 p.

Choquette, P. W., and Pray, L. C., 1970, Geologic nomenclature and classification of porosity in sedimentary carbonates: Am. Assoc. Petroleum Geologists Bull., v. 54, no. 2, p. 207–250.

Dunham, R. J., 1970, Stratigraphic reefs versus ecologic reefs: Am. Assoc. Petroleum Geologists Bull., v. 54, no. 10, p. 1931–1932.

Ham, W. E., ed., 1962, Classification of carbonate rocks: Am. Assoc. Petroleum Geologists Mem. 1, 279 p.

Heckel, P. H., 1970, Organic carbonate buildups in epeiric seas: some theoretical aspects (abs.): Am. Assoc. Petroleum Geologists Bull., v. 54, no. 5, p. 851–852.

Hill, G. A., Colburn, W. A., and Knight, J. W., 1961, Reducing oil-finding costs by use of hydrodynamic evaluations, in Petroleum exploration, gambling game or business venture: Inst. Econ. Petroleum Explor., Devel., and Property Evaluation, Internat. Oil and Gas Educ. Center, Dallas, Texas, March 16–17: Englewood, New Jersey, Prentice-Hall, Inc., p. 38–69.

Hubbert, M. K., 1953, Entrapment of petroleum under hydrodynamic conditions: Am. Assoc. Petroleum Geologists Bull., v. 37, no. 8, p. 1954–2026.

Lehner, Peter, 1969, Salt tectonics and Pleistocene stratigraphy on continental slope of northern Gulf of Mexico: Am. Assoc. Petroleum Geologists Bull., v. 53, no. 12, p. 2431–2479.

Levorsen, A. I., 1936, Stratigraphic versus structural accumulation: Am. Assoc. Petroleum Geologists Bull., v. 20, no. 5, p. 521–530.

—— ed., 1941, Stratigraphic type oil fields: Am. Assoc. Petroleum Geologists, 902 p.

—— 1954, Geology of petroleum: San Francisco, W. H. Freeman and Co., 703 p.

Martin, R., 1966, Paleogeomorphology and its application to exploration for oil and gas (with examples from Western Canada): Am. Assoc. Petroleum Geologists Bull., v. 50, no. 10, p. 2277–2311.

McCubbin, D. G., 1969, Cretaceous strike-valley sandstone reservoirs, northern New Mexico: Am. Association Petroleum Geologists Bull., v. 53, no. 10, p. 2114–2140.

McNeal, R. P., 1965, Hydrodynamics of the Permian Basin, in Fluids in subsurface environments—a symposium: Am. Assoc. Petroleum Geologists Mem. 4, p. 308–326.

Nelson, C. H., et al., 1970, Development of the Astoria Canyon-fan physiography and comparison with similar systems: Marine Geology, v. 8, p. 259–291.

Off, T., 1963, Rhythmic linear sand bodies caused by tidal currents: Am. Assoc. Petroleum Geologists Bull., v. 47, no. 2, p. 324–341.

Rich, J. L., 1951, Three critical environments of deposition and criteria for recognition of rocks deposited in each of them: Geol. Soc. America Bull., v. 62, p. 1–20.

Sanders, C. W., 1943, Stratigraphic type oil fields and proposed traps: Am. Assoc. Petroleum Geologists Bull., v. 27, no. 4, p. 539–550.

Smith, D. A., 1966, Theoretical considerations of sealing and non-sealing faults: Am. Assoc. Petroleum Geologists Bull., v. 50, no. 2, p. 363–374.

Wilkinson, W. M., 1953, Fracturing in Spraberry reservoir, West Texas: Am. Assoc. Petroleum Geologists Bull., v. 37, no. 2, p. 250–265.

Wilson, W. B., 1934, Proposed classification of oil and gas reservoirs, in W. E. Wrather and F. H. Lahee, eds., Problems of petroleum geology (Sidney Powers memorial volume): Am. Assoc. Petroleum Geologists, p. 433–445.

The American Association of Petroleum Geologists Bulletin
V. 59, No. 6 (June 1975), P. 939-956, 17 Figs., 1 Table

Capillary Pressures in Stratigraphic Traps[1]

ROBERT R. BERG[2]

College Station, Texas 77843

Abstract Capillary pressures between oil and water in rock pores are responsible for trapping oil, and the height of oil column, z_o, in a reservoir may be calculated from the Hobson equation modified as follows:

$$z_o = 2\gamma \left[\frac{1}{r_t} - \frac{1}{r_p} \right] / g(\rho_w - \rho_o),$$

where γ is the interfacial tension between oil and water, r_t is the radius of pore throats in the barrier rock, r_p is the radius of pores in the reservoir rock, g is acceleration of gravity, and ρ_w and ρ_o are the densities of water and oil, respectively, under subsurface conditions. To apply the equation, pore sizes must be estimated from mean effective grain sizes of the reservoir and barrier rocks. Effective grain size, D_e, in centimeters can be approximated from core analysis data by means of an empirical permeability equation from which

$$D_e = [1.89 \ kn^{-5.1}]^{1/2},$$

where n is porosity in percent and k is permeability in millidarcys. Then pore and throat sizes may be estimated as functions of mean effective grain size as based on theoretical packings of grains. The oil-column equation assumes hydrostatic conditions, and additional column, Δz_o, may be trapped if hydrodynamic flow occurs down the dip from barrier to reservoir facies, or

$$\Delta z_o = \frac{\rho_w}{\rho_w - \rho_o} \frac{dh}{dx} x_o,$$

where dh/dx is the potentiometric gradient and x_o is the width of the oil accumulation.

Calculations of oil columns in stratigraphic fields show that the equations give values which are in fair to good agreement with observed oil columns; that porous and permeable, very fine-grained sandstones and siltstones are commonly effective barriers to oil migration; and that the recognition of such barriers can be important in exploration for stratigraphic traps.

INTRODUCTION

The role of capillary pressure in trapping oil is well understood, and the theoretical relations between oil and water in rock pores give an insight to the requirements for oil migration and trapping in porous sandstones. Nevertheless, these relations commonly have not been applied to actual accumulations, primarily because the values for several rock and fluid properties needed for quantitative solutions are not readily estimated for subsurface conditions.

It is possible to measure capillary pressures in the laboratory on samples of reservoir and barrier rock and, from these measurements, to calculate pore sizes and heights of oil column that may be present in the subsurface. Such measurements, however, are not made commonly, and when they are, the results apply only to the specific rocks measured and are not otherwise applicable for predicting new oil accumulations. Clearly, a more universal method is needed in exploration to evaluate the effectiveness of trapping conditions.

One approach is to use the measurements of standard core analysis to estimate pore sizes by means of an empirical equation. Such an approach has obvious advantages in that there are many more core analyses than capillary pressure measurements. Consequently, the method can be applied more widely in exploration and development because the basic data are relatively abundant.

When core analyses are available for both reservoir and barrier rock, the expected oil column may be calculated. As applied to simple stratigraphic traps, where the traps are formed by facies change, the results of calculations can be especially revealing. A rock that retains significant porosity and permeability may form an effective barrier to oil migration, thereby trapping a sizable accumulation of oil in a more porous and permeable facies downdip. The trap facies may produce water on test and still form an effective barrier.

The implications of such a condition are particularly important in prospecting for stratigraphic traps. What seems to be a permeable and water-bearing facies actually may be a barrier to oil migration. The recognition of such barriers requires not only a knowledge of reservoir morphology but also an understanding of fluid relations within the rocks.

[1] Manuscript received, July 22, 1974; accepted, October 18, 1974.

[2] Department of Geology, Texas A&M University. Summaries of this paper were presented at the annual meeting of the Gulf Coast Association of Geological Societies, October 1972, and at the annual meeting of the AAPG, April 1974. Parts of an earlier paper (Berg, 1972) are reprinted here by permission of the Gulf Coast Association of Geological Societies. Oil field data were supplied in part by John Ahlen, Stewart Chuber, J. C. Harms, J. A. Hartman, J. M. Henton, and S. E. Drum; their assistance is gratefully acknowledged. Douglas Von Gonton provided helpful advice on fluid-density corrections. The manuscript was read by A. F. Gangi, Department of Geophysics, Texas A&M, and was greatly improved by his suggestions.

Capillary-Pressure Applications

The importance of capillary pressure long has been recognized in the primary and secondary migration of oil. In one of the earliest descriptions of oil migration, Munn (1909, p. 524-528) clearly stated the role of capillarity in the expulsion of oil from small pores of fine-grained rocks and its accumulation in the larger pores of reservoir rocks, with its migration aided by the "hydraulic" pressure of flowing water. This idea was placed on a firm theoretical base by Hubbert (1953), who showed that the magnitude of capillary pressures caused expulsion of oil from fine pores and its accumulation in large pores, whereas the surrounding water may be free to move according to hydrodynamic gradients.

Experiments to confirm the effects of capillary pressures have not been numerous or highly successful. Except for some simple, early experiments (for example, Cook, 1923) laboratory demonstrations have not been conducted to confirm the geologic role of capillary pressures. In fact, a recent experiment resulted in questioning the importance of capillary pressure in oil accumulation (Cartmill and Dickey, 1970).

A knowledge of capillary effects was established early in the field of petroleum engineering (Leverett, 1941) and has been widely applied in the measurement of capillary pressures, their relation to oil saturation, to pore size of reservoir rocks, and to heights of oil columns as, for example, by Habermann (1960). These principles also have been successfully applied to the understanding of both primary and secondary recovery of oil.

Geologic applications, however, have been limited. A practical classification of carbonate rocks has been based on pore size because of the relation among pores, capillary pressures, and other physical properties (Archie, 1952); and the pore geometries of diverse carbonate reservoirs and barrier rocks have been explained by measured capillary pressures (Stout, 1964). The conditions for oil migration in carbonate rocks have also been examined by considering pore sizes and capillary pressures (Aschenbrenner and Achauer, 1960), and the variable effectiveness of faults as barriers to migration has been explained on the basis of capillary properties of sandstones in contact across fault zones (Smith, 1966).

Other than these examples, geologic explanations largely have accepted the phenomenon of capillarity without attempting to apply the principle to the solution of specific problems.

A notable exception is the derivation by Hobson (1954, p. 73) of an equation for the height of oil necessary for secondary migration. Although he gave only a hypothetical example of its use, the fact that the equation is based on rock and fluid properties that might be estimated for subsurface conditions suggests that it also can be applied to the trapping of oil for the solution of practical problems in exploration.

Consequently, Hobson's equation was modified and applied to a simple stratigraphic trap (Berg, 1972), and its apparent success in predicting an oil column indicated that the equation may be more widely applicable. This paper attempts to apply the equation over a wider range of subsurface conditions.

Capillary-Pressure Equation

All capillary phenomena arise from the fact that when two immiscible fluids are in contact, molecular attractions between similar molecules in each fluid are greater than the attractions between the different molecules of the two fluids. There is, then, at the boundary between the fluids a region having properties different from those in the fluids farther from the boundary. Ordinarily this region contains only a small part of the matter of the system, and its thickness can be neglected. Molecular attraction is greater on the side of the more dense fluid, and the surface of contact is drawn into a curvature which is convex toward the more dense fluid. When an immiscible fluid is completely immersed in another fluid, it assumes a spherical shape of minimum surface area.

The force which acts on the contact surface is called surface tension in the case of a gas in contact with a liquid, or interfacial tension in the case of two liquids. Surface or interfacial tension is thought of as a force which acts in the plane of the surface, and the result of this force is to produce a pressure difference across the contact surface. This pressure difference is called capillary pressure.

The explanation for capillary pressure is best given in terms of energy change or work. When a drop of oil is immersed in water, the oil assumes a spherical shape, and work is accomplished. The work, W, done by interfacial tension, γ, results in a change of surface area, dA_o, of the drop, or

$$W = \gamma dA_o. \qquad (1)$$

Work also is done by fluid pressures to produce a change in volume of the drop. Pressure in the oil, p_o, acts normal to the contact surface and tends to produce a change in volume of the oil, dV_o. Pressure in the water, p_w, also acts normal to the surface but in the opposite direction to that of the oil pressure, and water pressure also tends to produce a change in volume so that the work done by fluid pressures is

$$W = p_o dV_o - p_w dV_o. \qquad (2)$$

At equilibrium the work done by interfacial tension is equal to work done by fluid pressures, or

$$\gamma dA_o = p_o dV_o - p_w dV_o; \qquad (3)$$

and rearranging terms,

$$p_c = \gamma dA_o / dV_o, \qquad (4)$$

where p_c is capillary pressure and is equal to the pressure difference $(p_o - p_w)$. Because the oil is spherical, and has a radius, r, its area is $A_o = 4\pi r^2$, and $dA_o = 8\pi r dr$; its volume is $V_o = (4/3)\pi r^3$ and $dV_o = 4\pi r^2 dr$. Substituting these expressions for dA_o and dV_o in equation (4) gives

$$p_c = 8\pi r dr \gamma / 4 \pi r^2 dr, \qquad (5)$$

which reduces to

$$p_c = 2\gamma / r. \qquad (6)$$

Thus, the pressure difference is proportional to the interfacial tension between the oil and water, and inversely proportional to the radius of curvature of the drop. If the drop is not spherical, an element of surface can be characterized by two principal radii, r_1, and r_2, measured in planes at right angles to each other and normal to the surface, and the curvature of the surface is $(1/r_1) + (1/r_2)$. Then for a nonspherical surface equation (6) becomes

$$p_c = \gamma \left(\frac{1}{r_1} + \frac{1}{r_2} \right). \qquad (7)$$

This is the Laplace equation for pressure difference across the contact between two immiscible fluids.

Equation (6) may be applied to the subsurface if it is assumed that the rock is water-wet, as is usually the case, and that the fluid boundary is approximately a spherical surface.

The term capillary pressure also is used for a rock property that may be measured in the laboratory by injecting a rock sample with a nonwetting fluid, such as mercury, and noting the pressure required for fluid injection. The injection pressure commonly is plotted as a function of the volume of fluid injected. In this case the measured injection pressure or "capillary" pressure, p_c, is

$$p_c = 2\gamma \cos\theta / r, \qquad (8)$$

where θ is the contact angle between the fluid boundary and the solid surface, as measured through the more dense fluid, and r is a radius of curvature determined by pore size of the rock.

The mean injection pressure that causes displacement of a major part of the wetting fluid is called the displacement pressure. In the case of injecting a wetting fluid, such as water, the contact angle, θ, is zero, cos θ equals one, and equation (8) reduces to equation (6).

There are, then, two distinct meanings for the term "capillary pressure:" (1) the pressure difference across the interface between two immiscible fluids, and (2) the injection pressure measured for a specific rock. These two concepts might better be distinguished by adopting two different terms, *interfacial pressure* for the fluid property and *injection pressure* for the rock property. In applying the concepts of capillary pressures in the subsurface, it is the interfacial pressure that is responsible for the trapping of oil and gas, but it is the injection pressure that must be predicted for a specific rock in order to determine the amount of oil or gas that can be trapped.

MIGRATION OF OIL

Oil must migrate through water-filled pores in order to accumulate in significant quantities in a reservoir rock, and migration is promoted by buoyant forces but is inhibited by the capillary pressures which must be overcome for an oil globule to pass from a rock pore through an adjacent pore throat.

The migration problem is described best by considering the forces that oppose the buoyant rise of oil in a hydrostatic environment (Hobson, 1954, p. 69-73). If a spherical globule of oil is at rest in a pore, the radius of the globule is approximately equal to the pore radius, r_p (Fig. 1A). The capillary pressure, p_p, is given by the Laplace equation,

$$p_p = 2\gamma / r_p, \qquad (9)$$

where γ is the interfacial tension between the oil and water. It is assumed that the buoyant force is not sufficient to distort the globule and force it into the adjacent pore throat of radius, r_t, and the globule is in equilibrium with surrounding water.

Now suppose that the globule is distorted so that its upper end is halfway through the pore throat (Fig. 1B). The capillary pressure at the upper end is

$$p_t = 2\gamma / r_t, \qquad (10)$$

and as the throat radius is smaller than the pore radius, the capillary pressure is greater in the throat than in the pore. So

$$2\gamma / r_t > 2\gamma / r_p, \qquad (11)$$

and the pressure gradient is still assumed to be great enough to oppose further movement through the pore throat. In other words, the net capillary-pressure gradient, ∇p_c, is downward in opposition to buoyancy, and

$$\nabla p_c = (p_t - p_p)/z = 2\gamma\left(\frac{1}{r_t} - \frac{1}{r_p}\right), \qquad (12)$$

where z is the vertical height of the globule (Fig. 1B).

If the globule is distorted so that it is halfway through the pore throat (Fig. 1C), the radii are equal at the upper and lower ends, the capillary pressures are equal, and the globule can easily move upward by buoyancy.

If the globule is distorted further so that it attains a position more than halfway through the pore throat (Fig. 1D), the radius at its upper end is now r_p and is larger than the radius at its lower end, which is now r_t. Therefore, the capillary pressure at the upper end is less than at the lower end, and the gradient of capillary pressure is in the same direction as the buoyant force. The globule now can move upward rapidly from the narrow throat into the larger pore above.

Thus, the migration problem can be simply stated. The buoyant force acting on a globule must be large enough to overcome the capillary-pressure gradient within the globule caused by the greater pressure in the pore throat than in the pores. The conditions necessary for migration can be determined by considering the globule with its upper end in the pore throat (Fig. 1B). The buoyant force acting on the distorted globule is caused by the hydrostatic pressure gradient, ∇p_h, between the upper and lower ends, and this gradient is

$$\nabla p_h = g(\rho_w - \rho_o), \qquad (13)$$

where ρ_w is density of the water and ρ_o is density of the oil. If the globule is at rest, then the buoyant-pressure gradient (13) is equal to the capillary-pressure gradient (12) and, rearranging terms,

$$zg(\rho_w - \rho_o) = 2\gamma\left[\frac{1}{r_t} - \frac{1}{r_p}\right]. \qquad (14)$$

For any given set of conditions the fluid properties of density and interfacial tension remain constant, as do the pore and throat diameters and, of course, the acceleration due to gravity. Therefore, if the forces are to become unbalanced and the globule is to move, the vertical height is the only value that can change. This may be accomplished by the addition of other smaller droplets of oil which move into the pore, accrete on the globule and increase its length so that

$$zg(\rho_w - \rho_o) > 2\gamma\left(\frac{1}{r_t} - \frac{1}{r_p}\right). \qquad (15)$$

When the length of the globule is increased so that the buoyant pressure, represented by the left-

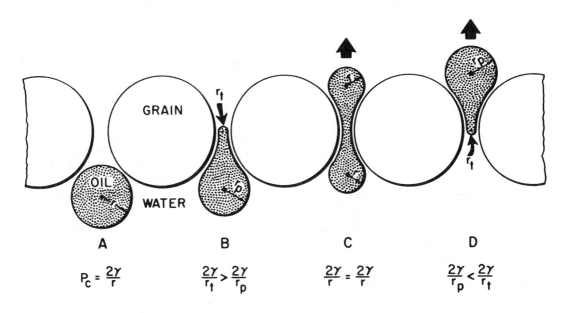

WHERE γ = INTERFACIAL TENSION

FIG. 1—Diagram of secondary migration of oil globule in water-wet, clastic rock **A** globule at rest in pore void, **B** globule entering pore throat, **C** globule half way through pore throat, and **D** globule passing through throat into next void.

hand terms of the equation, exceeds the capillary-pressure difference, represented by the right-hand terms, then the globule will move through the pore throat.

The vertical height that must be attained for the buoyant force to equal the internal-pressure gradient is called the critical height, z_c, for migration (Hobson, 1954, p. 73). Any increase in this height will cause the oil to move upward. The expression for critical height in terms of rock and fluid parameters is given by solving equation (15) for height, z_c, or

$$z_c = 2\gamma \left(\frac{1}{r_t} - \frac{1}{r_p} \right) / g(\rho_w - \rho_o). \qquad (16)$$

To find the value of critical height for oil migration, equation (16) may be solved for an idealized aquifer. Consider a well-sorted, fine-grained sandstone that has a porosity of 26 percent. The natural aggregate may approximate a rhombohedral packing of uniform spheres in which pore sizes are 0.154D, 0.225D, and 0.414D, where D is sphere diameter (Graton and Fraser, 1935). The migration path through the aggregate is by means of alternate pores of diameter 0.225D and 0.414D that are connected by pore throats of diameter 0.154D (Fig. 2). Along this path the larger pore radii are

$$r_p = (1/2)(0.414D), \qquad (17)$$

and pore-throat radii are

$$r_t = (1/2)(0.154D), \qquad (18)$$

where D is the diameter of grains. These radii determine the maximum capillary-pressure difference that will exist in a continuous oil phase which is composed of connected globules that fill the pores.

This method of approximating pore and throat sizes appears to be valid when compared to experimental data. For laboratory measurements Hubbert (1953, p. 1977) gave the equation for minimum capillary pressure, or displacement pressure, in the form of equation (8), as

$$p_c = \frac{C\gamma\cos\theta}{D}, \qquad (19)$$

where C is a constant of proportionality, θ is the contact angle of the oil-water interface at the grain surface, and D is the mean grain diameter of the rock. For water-wet rock, θ is zero and cos θ equals 1. The value of C is said to be about 16 from measurements by Purcell (1949) and unpublished work by R. H. Nanz.

In equation (16) for water-wet rock, the value for minimum capillary pressure is

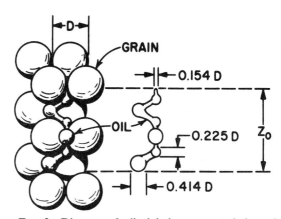

FIG. 2—Diagram of oil globules connected through pores in rhombohedral packing of uniform, spherical grains where D is grain diameter and porosity is 26 percent.

$$p_c = 2\gamma \left(\frac{1}{r_t} - \frac{1}{r_p} \right), \qquad (20)$$

or where pore and throat sizes are functions of mean grain diameter, D, then

$$p_c = \frac{4\gamma}{D} \left[\frac{1}{0.154} - \frac{1}{0.414} \right] = \frac{16.3\gamma}{D}, \qquad (21)$$

which gives 16.3 as an ideal value for the constant, C, a value quite close to the experimental value given above.

The fluid properties also must be known or estimated for subsurface conditions. The interfacial tension can be taken as 35 dynes/cm, and fluid-density contrasts may range from 0.1 to 0.3 gm/cc for oil and water, or approach 1.0 g/cc for gas and water. Then solving equation (16) for subsurface conditions shows that the critical height is many times larger than the dimensions of pores or, in other words, the vertical length of the oil in continuous phase is such that oil migration probably takes place in the form of "stringers" rather than "globules." For example, to migrate upward through a fine-grained sandstone with mean grain size of 0.2 mm, a stringer of low gravity oil ($\Delta\rho = 0.1$) must be about 300 cm or 10 ft in vertical length. This length of oil would have sufficient buoyancy to overcome the capillary pressure in pore throats of the rock. On the other hand, a stringer of gas ($\Delta\rho = 1.0$) need only be about 30 cm or 1 ft in vertical length to overcome the capillary pressure in the pore throats because of its greater buoyancy.

TRAPPING OF OIL

Equation (16) also may be applied to trapping of oil by a change in grain size. Consider the

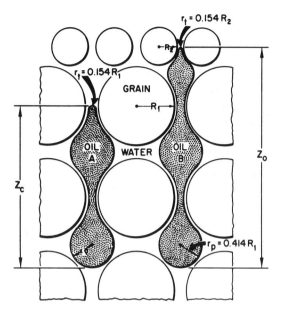

FIG. 3—Oil stringer of critical height, z_c, migrating upward through water-wet grains of sandstone, and height of oil, z_o, trapped by decrease in grain size. Reprinted from Berg (1972, Fig. 1) with permission of the Gulf Coast Association of Geological Societies.

stringer of oil described above migrating upward in a fine-grained sandstone until it encounters a sandstone in which mean grain size is decreased by one-half (Fig. 3). To overcome the higher capillary pressure in smaller pore throats, the critical height must increase to about 760 cm or 25 ft in vertical length for its buoyancy to increase sufficiently for continued migration upward. In other words, the critical height of the stringer now represents the height of oil column, z_o, that can be trapped by a decrease in mean grain size.

To calculate the height of oil column, equation (16) can be used, but the pore sizes must be redefined. The radius of pore throats, r_t, is that radius in the finer grained sandstone, whereas the larger radius of pores, r_p, is that radius in the coarser grained sandstone.

The height of oil column also may be calculated for subsurface conditions by use of equation (16). Consider again a reservoir rock composed of well-sorted, fine-grained sandstone that has a porosity of 26 percent. If the grains have a diameter, $D_p = 0.2$ mm, the height of oil that can be trapped within the rock is the critical height, z_c, as calculated above (Fig. 4). If the oil in this rock migrates upward until it encounters a change in grain size, the height of oil in the reservoir can be calculated if the grain size of the barrier rock is known.

For example, if the barrier rock is coarse siltstone with a grain size, $D_t = 0.05$ mm (Fig. 4), a

column of 55 ft can be held in the reservoir for a low-gravity oil ($\Delta\rho = 0.1$), but a column of only about 5 ft of gas can be held ($\Delta\rho = 1.0$). If the barrier is fine siltstone with a grain size, $D_t = 0.01$ mm, a column of 300 ft can be held for a low-gravity oil and 30 ft for gas. At finer grain sizes of the barrier, the columns are even greater, but the calculations are probably not reliable in this range because the analogy of the barrier rock as a systematic packing of uniform spheres is not realistic. Finer grains such as clay particles can be expected to pack more tightly and form smaller pores and, consequently, oil columns of greater height may be trapped.

Nevertheless, these calculations show that a moderate change in grain size can form an effective barrier to oil migration and, furthermore, a barrier facies may be porous and permeable and yet trap a significant oil column.

HYDRODYNAMIC EFFECTS

The hydrodynamic flow of water affects the height of oil that can be trapped by capillarity alone in stratigraphic traps (Hubbert, 1953, p. 2000-2001). Therefore, the calculation of oil column based only on hydrostatic conditions does not apply if hydrodynamic flow is present.

Consider a stringer of oil in a uniform aquifer past which there is a downdip flow of water (Fig. 5). Flow is parallel with the boundaries of the

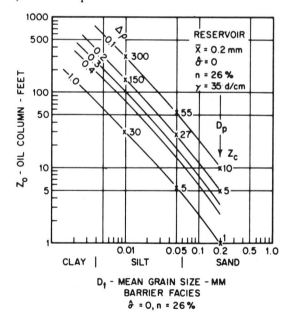

FIG. 4—Height of oil column, z_o, that can be trapped by barrier rock of mean grain size, D_t, in a reservoir rock of grain size, $D_p = 0.2$ mm where both rocks are composed of uniform spherical grains in rhombohedral packing and porosity, n, is 26 percent. Interfacial tension, γ, is assumed to be 35 dynes/cm.

aquifer, and the stringer is aligned parallel with the direction of flow and at rest. The pressure gradient along the stringer of oil increases the amount of oil held by capillarity alone, and the height of the oil stringer can be shown to be proportional to the hydrodynamic gradient within the aquifer.

There is a pressure difference, $\Delta p = p_2 - p_1$, between the upper and lower ends of the stringer because of the difference in heights of the water column, or

$$\Delta p = \rho_w g \Delta z_o - \rho_w g \Delta h, \qquad (22)$$

where Δh is the difference in hydrostatic heads, h_1 and h_2, and z_o is the elevation difference between the ends of the stringer (Fig. 5).

Neglecting the capillary pressure difference within the stringer, the pressure difference, Δp, also will be nearly equal to the pressure within the oil stringer because of the height of the oil column, Δz_o, or

$$\Delta p = \rho_o g \Delta z_o. \qquad (23)$$

Equating these expressions for pressure differences gives

$$\rho_o g \Delta z_o = \rho_w g \Delta z_o - \rho_w g \Delta h, \qquad (24)$$

which reduces to

$$\Delta z_o = \frac{\rho_w}{\rho_w - \rho_o} \Delta h, \qquad (25)$$

where Δz_o is the height of the oil column that is held in place by the downdip flow of the water. Equation (22) also may be expressed in terms of gradients as

$$\frac{dz_o}{dx} = \frac{\rho_w}{\rho_w - \rho_o} \frac{dh}{dx} = \tan\theta, \qquad (26)$$

where θ is the angle of inclination of the oil stringer. This equation (26) is identical to the expression for inclination of the oil-water contact under hydrodynamic conditions (Hubbert, 1953, p. 1987), but the derivation is given in simplified form (Willis, 1961, p. 13).

Thus, the additional oil column trapped by hydrodynamic flow may be calculated from an observed potentiometric gradient or from an observed tilt of an oil-water contact, or

$$z_o = \frac{\rho_w}{\rho_w - \rho_o} \frac{dh}{dx} x_o, \qquad (27)$$

where x_o is the horizontal width of the oil accumulation.

If flow of water is in an updip direction, the hydrostatic head, h_2, is greater than the head, h_1, and equation (27) will have the same form but

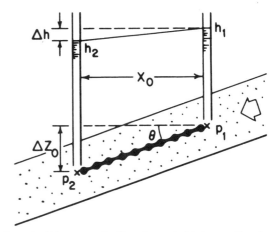

FIG. 5—Diagram of oil stringer held in aquifer by downdip flow of water.

will be of opposite sign, meaning that z_o is the loss of oil column resulting from flow.

The total oil column in a stratigraphic trap, z_{ot}, may be expressed by combining the equation for the capillary oil column (16) and the hydrodynamic oil column (27), or

$$z_{ot} = \frac{2\gamma \left[\frac{1}{r_t} - \frac{1}{r_p} \right]}{g \ (\rho_w - \rho_o)} \pm \left[\frac{\rho_w}{\rho_w - \rho_o} \right] \frac{dh}{dx} x_o, \qquad (28)$$

where the sign for the hydrodynamic column is determined by the direction of flow relative to the permeability barrier.

SUBSURFACE APPLICATION

If capillary pressures are effective in trapping oil, equation (16) should apply to the calculation of oil columns in the subsurface under hydrostatic conditions. The problem in subsurface applications, however, is that the rock and fluid properties must be estimated for subsurface conditions.

Fluid densities are determined most easily. Salinity of reservoir water may be known, but when water analysis is not available, salinity can be calculated from electric logs according to standard methods (Schlumberger Limited, 1969, p. 63-66). The estimate of sodium chloride salinity then may be converted to relative density for the solution (Fig. 6). Corrections for reservoir temperature, pressure, and dissolved gas can be made, but at moderate depths the density correction is generally small (McKetta and Wehe, 1962, p. 20-22). Therefore, the correction can be approximated for average subsurface conditions (Fig. 6) with only negligible error.

The API gravity of the oil generally is known, and its relative density (specific gravity) at the

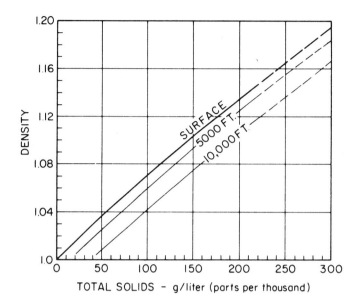

FIG. 6—Relative density of subsurface waters as function of total dissolved solids in grams per liter (parts per thousand) based on NaCl solutions and projected on data of Reistle and Lane (1928). Approximate corrections for subsurface density based on surface temperature of 70°F, temperature gradient of 1.5°F/100 ft, and pressure gradient of 0.465 psi/ft of depth.

surface, γ_o, can be determined by the relation

$$\gamma_o = 141.5/ \text{°API} + 131.5. \qquad (29)$$

For an estimate of oil density in the subsurface, the formation-volume factor, B_o, can be calculated by means of well-known correlation methods as, for example, by the empirical relation (Standing, 1962, p. 24)

$$B_o = 0.972 + 0.000147 \, F^{1.175}, \qquad (30)$$

where the correlation factor, $F = R_s(\gamma_g/\gamma_o)^{0.5} + 1.25T$, and R_s is the gas-oil ratio in cu ft/bbl, γ_g is relative density of produced gas, γ_o is relative density of the oil, and T is temperature in degrees F. The subsurface oil density, ρ_o, then is estimated as

$$\rho_o = (\gamma_o + 2.17 \times 10^{-4} R_s \, \gamma_g)/ B_o. \qquad (31)$$

Equations (29), (30), and (31) are satisfactory in most cases of oil-column calculation because API gravity and gas-oil ratio commonly are known. The relative density of gas, however, is not usually available but can be estimated as 0.7 for average conditions. The correlation equation (30) holds for most oils that have relatively small amounts of dissolved gas, say up to 1,500 cu ft/bbl. With larger amounts of gas or with gas-condensate systems, the total formation-volume factor should be used to calculate subsurface density (Standing, 1962, p. 32-34).

Interfacial tension between oil and water is more difficult to determine, but some measurements at higher pressures and temperatures have been made (Hocott, 1938). At surface conditions most oils have interfacial tensions against water of about 25 dynes/cm, and increase of temperature and pressure tends to increase interfacial tension (Fig. 7). In general, medium-gravity oils have interfacial tensions of 30 to 35 dynes/cm against their connate waters at pressures of about 1,000 psi or more. The surface tension of methane against water decreases from about 73 dynes/cm at surface conditions to about 35 dynes/cm at 150°F and pressures about 3,000 psi. At higher pressures and temperatures the tensions of both gas and oil approach 30 dynes/cm.

The approximation of pore sizes is most hazardous in the absence of grain-size or capillary-pressure measurements, but one approach is to use the more abundant measurements of core analysis to estimate the dominant grain size of a sandstone which, in turn, determines the dominant pore size. An empirical equation has been derived for permeability as a function of porosity (Berg, 1970). This equation applies to well-sorted sandstones of 30 percent or more porosity and can be given in the form

$$k = 5.3 \times 10^{-3} n^{5.1} (P_{90})^2, \qquad (32)$$

where k is permeability in millidarcys, n is porosity in percent and P_{90} is the ninetieth-percentile grain size in millimeters as determined by cumulative weight-percent analysis. In other words, the ninetieth percentile is the grain size that controls the dominant pore size for well-sorted quartzose sandstones.

In estimating pore sizes, it is not required that a correct percentile measure be determined, but only that an effective grain size which determines pore size be estimated. Therefore, equation (32) can be stated in more general terms and rearranged as

$$D_e = (1.89 \, kn^{-5.1})^{1/2}, \qquad (33)$$

where D_e is effective mean grain size in centimeters. When porosity and permeability are known for a given sandstone, as from core analysis, equation (33) may be solved for effective grain size. The use of equation (33) assumes that it applies to tighter packings of less than 30 percent porosity as well as to looser packings. This is a reasonable assumption for quartzose sandstone, because porosity in these rocks commonly is reduced by pressure solution and redeposition of dissolved silica as grain overgrowths, a process that can be visualized as a systematic decrease of pore size with decrease in porosity (Rittenhouse, 1971).

When grain size has been calculated, pore size may be estimated by considering the size of pores as a function of grain size in systematic packings of uniform spheres, as described previously (Fig. 2). This method of estimating pore sizes from porosity and permeability gives only a crude approximation of dominant pore size for natural sandstones. Nevertheless, such an empirical procedure may be attempted in the absence of more precise data.

FIELD EXAMPLES

Marine-Bar Sandstones

The calculation for height of oil column should yield values that approximate natural conditions of oil accumulation if the equation (16) is soundly based. A favorable situation in which to test the calculation would be a simple stratigraphic trap; that is, one in which accumulation is caused by facies change from a more permeable reservoir facies to a less permeable barrier facies. Such an oil accumulation is present at the Milbur field, Burleson County, Texas, where a lower Wilcox sandstone forms a simple lenticular reservoir at shallow depth (Chuber, 1972).

Lower Wilcox sandstones at Milbur were formed as littoral marine bars deposited in a generally regressive sequence. The sandstone has a

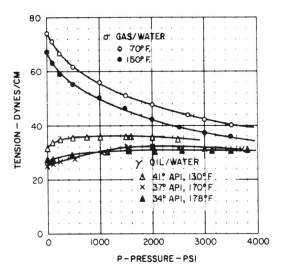

Fig. 7—Surface and interfacial tensions as functions of temperature and pressure (from Hocott, 1938).

maximum thickness of nearly 30 ft, and oil production is obtained from the updip edge of the bar where it grades laterally to a thinner lagoonal facies (Fig. 8). Although normal faults are present in the field, oil accumulation is controlled by the facies change, and the oil column is 60 ft at the northeast end of the field and may be as much as 75 ft at the southwest end (Chuber, 1972, Fig. 2).

The properties of the reservoir and barrier facies are known from the Clark Cotton 1 well in the nearby Burmil field (Fig. 9). Here the lower Burmil sandstone is a littoral marine-bar sandstone with an average porosity of 32.4 percent and average permeability of 900 md. These values are assumed to be representative also of the Milbur sandstone, which is of similar origin.

The overlying sandstone at Burmil represents the lagoonal facies equivalent to the Milbur reservoir sandstones, and within this unit average porosity and permeability range from 24 percent and 153 md to 19 percent and 25 md, respectively. In other words, these values also can be assumed to represent maximum and minimum properties for the barrier facies at Milbur. Although permeabilities of barrier facies are greatly reduced as compared to reservoir sandstone, still the lagoonal facies retains the ability to produce fluid, but in this case, only water.

The rock and fluid properties for the Milbur field are given in Table 1, and from these values the oil column may be calculated. The higher values of porosity and permeability for the barrier facies give a minimum oil column of 53 ft, and the lower values give a maximum column of 64 ft. The latter value agrees well with the actual oil column of 60 to 75 ft, and it may be concluded

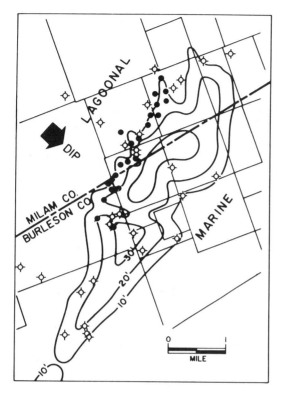

FIG. 8—Isopach map of reservoir sandstone, Milbur field, Milam and Burleson Counties, Texas (from Chuber, 1972, Fig. 2). Reprinted from Berg (1972, Fig. 2) with permission of Gulf Coast Association of Geological Societies.

that the less permeable lagoonal facies is an effective barrier to oil migration, although it has significant porosity and permeability.

Fluvial Sandstone

A fluvial, reservoir sandstone offers an interesting example of capillary trapping because the barrier facies not only yields water adjacent to the channel reservoir but also produces oil updip from the channel sandstone. This field, therefore, demonstrates that a reservoir sandstone itself can form an effective barrier to oil migration from a more porous and permeable sandstone.

Lower Cretaceous "J" sandstones of the Dakota Group produce oil from a variety of stratigraphic traps in the Denver basin. One type of stratigraphic trap is a narrow, channel sandstone that is present in several small fields in western Nebraska (Harms, 1966), and the Lane field is a typical example (Fig. 10).

Structure on top of the "J" sandstone has the form of a low-relief structural closure on the east with an adjacent nose downdip on the northwest. Two units in the "J" sandstone, called the "J_1" and "J_2", produce oil on the structural closure,

but increasing amounts of water are produced with oil to a datum of -430 ft. On the nose in the northwest a more porous and permeable channel sandstone cuts the regional "J" sandstone. The channel sandstone produces water-free oil to a datum of -440 ft and has an oil-water contact estimated at -450 ft. Therefore, significantly better oil production is present in the more porous and permeable channel sandstone to levels considerably below oil and water production in the regional "J" sandstone.

The relation between the channel reservoir and the regional "J" sandstone is shown by electric logs from selected wells (Fig. 11). The channel sandstone is fine grained, quartzose, cross-bedded and has an average porosity of 21 percent and average permeability of 533 md. In contrast, the regional "J_1" and "J_2" sandstones are finer grained and have lower permeabilities, but the "J_2" sandstone is very fine grained, well sorted and cross-bedded, quite similar to the channel sandstone.

An oil column of at least 43 ft is trapped in the channel sandstone by the less permeable "J_1" and "J_2" sandstones. Although the "J_2" sandstone, in particular, appears to be an unlikely barrier facies, capillary-pressure measurements show that the sandstone is capable of trapping 30 to 40 ft of oil in the channel sandstone (Harms, 1966, p. 2143). Assuming no structural closure on the channel reservoir, a total of 43 ft of oil is trapped by the upper "J_1" sandstone, and 23 ft of oil is trapped by the lower "J_2" sandstone. There is no evidence that hydrodynamic effects contribute to the trapping of oil (Harms, 1966, p. 2141).

The calculated oil columns are not completely in agreement with the observed oil column (Table 1). The "J_2" sandstone should trap 30 ft of oil, which compares well with the oil column of 23 ft

CLARK NO. I COTTON

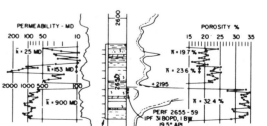

FIG. 9—Electric log and core analysis of lower Wilcox sandstones, Clark Cotton No. 1, Burmil field, Burleson County, Texas. Reprinted from Berg (1972, Fig. 3) with permission of Gulf Coast Association of Geological Societies.

TABLE 1. BASIC DATA AND OIL-COLUMN CALCULATIONS FOR SELECTED STRATIGRAPHIC OIL FIELDS

Field	Milbur, Texas[a]			Lane, Nebraska[b]			Main Pass Block 35[c] Offshore Louisiana			Paduca, New[d] Mexico	
	Reservoir	Barrier (min)	(max)	Reservoir	Barrier "J1"	"J2"	Reservoir	Barrier (upper)	(lower)	Reservoir	Barrier
Formation	Lower Wilcox			"J" Sandstone			Miocene "G2"			Bell Canyon	
Porosity, %	32	20	24	21	18	24	34	26	29	24	20
Permeability, md	900	25	153	533	65	77	3000+	75	170	25	5
Depth, ft	2750			4640			6700			4630	
Temperature, °F	100			120			140			100	
Pressure, psi	1000			1040			3000			2035	
Oil Gravity, °API	23			37			29			41	
Gas/Oil Ratio, ft^3/bbl	trace			70			420			270	
Oil Density (ρ_o) g/cm	0.91			0.81			0.77			0.76	
Gas Gravity (assumed)				0.70			0.70			0.70	
Gas Density (ρ_g) g/cm^3							0.17				
Water Salinity, ppm	8000			100,000			145,000			324,000	
Water Density, (ρ_w) g/cm^3	1.00			1.06			1.09			1.20	
Interfacial Tension (γ) dynes/cm	30			35			35			35	
Gas Column, ft							20				
Oil Column, ft	60 to 75			43			57			120+	
CALCULATED VALUES											
Grain Size, (D$_e$) cm x10^{-3}	6.0	3.5	5.5	13.5	6.9	3.6	9.5	2.9	3.3	2.1	1.5
Gas Column, ft							10				
Oil Column, ft		64 or	53		14	30		29 or	25		100 + 34

[a] In part from Chuber (1972).

[b] In part from Harms (1966, and personal communication).

[c] In part from Hartman (1972, and personal communication).

[d] From Scott (1967) and Jenkins (1961).

against it, but the "J$_1$" sandstone should trap only 14 ft as compared with the total column of 43 ft. The "J$_1$" sandstone has only one thin permeable sandstone in an otherwise dominantly shale section (Fig. 11), and the average permeability for the calculation was taken from this sandstone. Apparently, this sandstone makes the "J$_1$" section an inefficient barrier to oil migration, and it may be concluded that a significant amount of oil may have leaked from the channel reservoir to a position farther updip. This suggestion is supported by the fact that oil is produced from the "J$_1$" sandstone on adjacent structural closure (Fig. 10).

These calculations seem to indicate that a significant oil column can be trapped in more porous and permeable sandstone even when the barrier facies is not completely effective and is itself an oil reservoir.

Fluvio-Deltaic Sandstone

Fluvio-deltaic sandstones commonly form excellent reservoirs in stratigraphic traps where there is an abrupt lateral change from more permeable channel sandstones to less permeable natural-levee and overbank deposits. A typical example is the Miocene "G$_2$" sandstone in the Main Pass Block 35 field, offshore Louisiana (Hartman, 1972).

Structure on the "G$_2$" sandstone shows that production is on a classic, Gulf Coast fold associated with a down-to-the-basin normal fault (Fig. 12). Oil and gas in the "G$_2$" sandstone, however, are produced largely from a narrow channel that is up to 3,000 ft wide and has a southwest trend across the fold. Maximum thickness of the channel section may be greater than 80 ft (Hartman, 1972, Fig. 4, p. 556), and the total gas column is 41 ft and total oil column is 57 ft.

The relation between the channel sandstone and associated deposits is shown by the correlation of electric logs from wells at the top of the structure, a producing well and an adjacent dry hole outside the channel (Fig. 13). The most porous and permeable channel sandstone is 40 ft thick and contains gas in the upper 20 ft and oil below. It consists of well-sorted, fine-grained, quartzose sandstone that has an average permeability of more than 3 darcys (Hartman, 1972, p. 557). This sandstone is completely enclosed by very fine-grained sandstones, siltstones, and shales that probably were formed as natural-levee and overbank deposits lateral to the fluvial channel and as fine-grained fill in the main "G$_2$" channel. These sandstones in the upper "G$_2$" have an average porosity of 26 percent and average permeability of 75 md, and in the lower "G$_2$," an

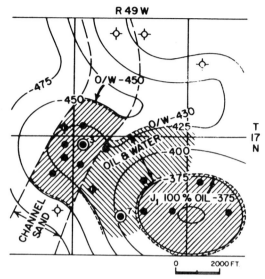

FIG. 10—Structure of "J" sandstone, Lane field, Morrill County, Nebraska. Contour interval 25 ft. Logs of numbered wells (circled) are shown in Figure 11.

average porosity of 29 percent and average permeability of 170 md (Hartman, 1972, p. 557, and personal commun.). These values are based on a limited number of sidewall cores, but they are probably representative of the overbank sections.

Gas in the channel sandstone is trapped by fine-grained sandstones and shales of the upper "G_2," and the oil is in the channel sandstone adjacent to the lower "G_2" sandstones. There is no evidence for any other barrier between the closely spaced channel and nonchannel wells.

Calculations show that only moderate columns of oil and gas can be trapped by the indicated permeability contrasts (Table 1). The upper "G_2" sandstones may trap 10 ft of gas, whereas the actual gas column is at least 20 ft; and the lower "G_2" sandstones outside of the channel can support an oil column on the order of 25 ft, which is less than half the actual column of 57 ft. One explanation for the lack of agreement between the actual and calculated columns may be in the varied nature of the channel deposits, because the electric logs show significant amounts of shale and siltstone above, and lateral to, the oil and gas sands (Fig. 13). These relations suggest that the barrier facies may be somewhat finer grained, on the average, and forms a more effective seal than is indicated by the given values of porosity and permeability. An alternative explanation might be that at least two sandstone units are present within the channel section and that each might hold a column of about 25 ft for an apparent total oil column of 50 ft.

Another explanation for the difference between the calculated and observed columns may be that the "G_2" sandstones outside the channel are not barrier facies at the given values of porosity and permeability and that these sandstones may be reservoirs also. This explanation is supported by the fact that four oil wells have been completed in sandstones outside the channel in the north part of the field (Fig. 12), and each of these wells has produced several hundred thousand barrels of oil. The overbank facies may be more widely productive of oil than was indicated by the previous in-

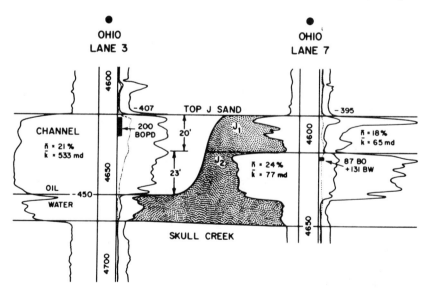

FIG. 11—Correlation of selected electric logs showing properties of reservoir and trap facies in the "J" sandstone, Lane field, Nebraska. Location of wells shown in Figure 10.

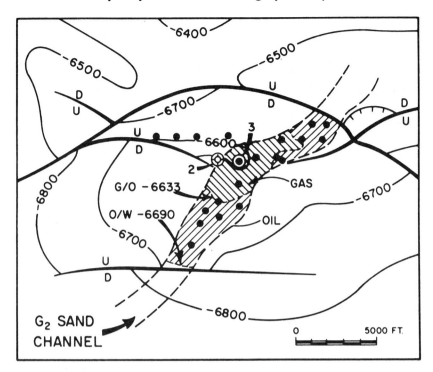

FIG. 12—Structure on "G$_2$" sandstone, Main Pass Block 35 field, offshore Louisiana, showing wells that produce from "G$_2$" sandstone. Contour interval 100 ft (adapted from Hartman, 1972, Fig. 3, p. 555).

terpretation of only channel-sandstone reservoirs.

In this case, the oil-column calculations seem to suggest that the so-called barrier facies contains reservoir sandstone, and it may be concluded that oil-column calculations could be helpful in similar fields during development drilling.

Hydrodynamic-Stratigraphic Traps

Oil-column calculations based on hydrostatic conditions may be greatly in error if hydrodynamic flow has caused a tilt of the oil-water contact. Such an example is the hydrodynamic-stratigraphic trap in Guadalupian Bell Canyon sandstone at the Paduca field in Delaware basin, southeastern New Mexico.

Structure at Paduca field is a gentle homocline that dips eastward at a rate of 100 ft/mi (Fig. 14). The Bell Canyon sandstones are of turbidite origin, and the reservoir facies is a massive, very fine-grained, channel sandstone that is about 1.5 mi wide and is enclosed by a barrier facies of thin-bedded, finer grained siltstone of turbidite-fan origin (Payne, 1973). The channel sandstone has a maximum thickness of about 30 ft and an average net pay thickness of 20 ft (Scott, 1967). The top of the producing zone is about 25 ft below the top of the mapped structure.

Production of oil is accompanied by significant amounts of water around the margins of the field,

and water production can be explained in part by decreased relative permeability to oil near the channel edges. The variable nature of oil and water production also has been attributed to the oil in individual sandstone lenses (Grauten, 1965). In either case, if Paduca field were a simple stratigraphic trap under hydrostatic conditions, the expected oil column might be on the order of the maximum thickness of the channel sandstone, or somewhat greater, because the barrier facies has significant permeability, but oil production extends over a much greater height. The Bell Canyon sandstone is known to have a potentiometric gradient generally parallel with dip (McNeal, 1965, Fig. 7, p. 316), so hydrodynamic flow can be expected to have an influence on the height of oil column at Paduca.

The exact configuration of the oil-water contact is difficult to determine without a detailed study, but reported completion figures suggest a tilted oil-water contact. The maximum oil column in the north half of the field is approximately 120 ft, from an elevation of $-1,170$ ft to $-1,290$ ft as mapped at the top of the reservoir sandstone. The oil-water contact in the south half of the field is not well defined, but significant water cuts are present as high as $-1,200$ ft and oil production occurs at least to $-1,250$ ft for wells about 1 mi apart, which suggests a tilt of the oil-water con-

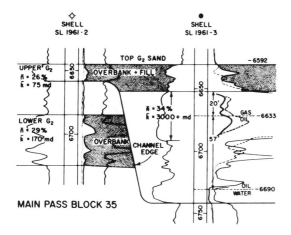

FIG. 13—Correlation of electric logs showing properties of reservoir and trap facies in the "G₂" sandstone, Main Pass Block 35 field. Location of wells shown in Figure 12.

tact of perhaps 50 ft per mi or about 80 ft across the field. Therefore, it is estimated that of the total oil column of 120 ft, about 30 to 40 ft may be due to capillary trapping, and an additional 80 to 90 ft is the result of hydrodynamic tilt (Fig. 15).

Reservoir properties of Bell Canyon sandstones show an average porosity of 24 percent and average permeability of 25 md (Scott, 1967). A minimum value of permeability as the limit for economic oil production has been measured as 5 md, which corresponds to an average porosity of 20 percent (Jenkins, 1961, Fig. 1, p. 1231). These values, then, represent the minimum contrast of porosity and permeability between the reservoir and trap facies. On the basis of these values (Table 1), the minimum height of oil column that can be trapped under hydrostatic conditions is calculated to be 34 ft.

Available pressure data and a published potentiometric map (McNeal, 1965, Fig. 7, p. 316) suggest a hydrodynamic gradient of 20 to 30 ft/mi in this part of the Delaware basin. Assuming an average value of 25 ft per mi, the tilt of the oil-water contact should be about 68 ft per mi, or a total of about 100 ft across the 1.5-mi width of Paduca field as measured in the direction of dip. This tilt corresponds to the additional oil that can be trapped by hydrodynamic flow, as given by equation (27).

The calculated values for hydrostatic and hydrodynamic trapping total 134 ft as compared with the observed oil column of 120 ft or more. The calculated hydrostatic column is 34 ft as compared with the estimated hydrostatic column of 30 to 40 ft, and the calculated hydrodynamic column is 100 ft which corresponds to the estimated hydrodynamic column of 80 to 90 ft. The

close agreement of observed and calculated columns strengthens the idea that the calculations may be valid for both hydrostatic and hydrodynamic factors in oil accumulations.

Fault Traps

Capillary trapping is not only effective in stratigraphic accumulations but applies to other types of traps as well. Of particular interest are those in which reservoir sandstone is in fault contact with water-bearing sandstone. In some cases, the fault zone itself may not be detected on electric logs as an impermeable horizon, but the interpretation of a fault leads to the conclusion that such a zone is present and forms an effective seal. This conclusion obviously is justified in many cases, but there are others in which the aquifer can be suspected of providing the trap, with or without the aid of a fault zone.

The Gulf Coast Tertiary province offers many examples of fault traps in which a sand-to-sand contact is present. A typical example is in a well from the Weeks Island field, Louisiana, in which

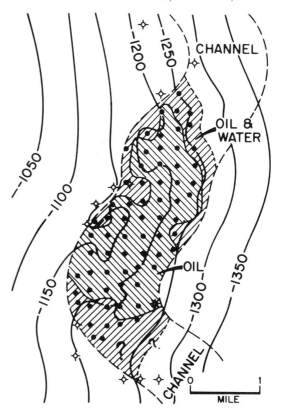

FIG. 14—Structure on Bell Canyon sandstone, Paduca field, New Mexico, showing outline of turbidite channel, reservoir sandstone, and areas of reported oil and water production that suggest possible tilted oil-water contact. Contour interval 50 ft. Adapted from Scott (1967).

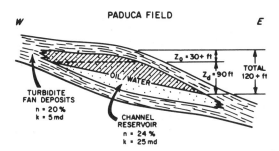

FIG. 15—Diagrammatic cross section of Paduca field showing capillary and hydrodynamic oil columns that total 120 ft or more.

the electric log shows an oil zone at a depth of 12,656 ft below a thick water-bearing sand (Fig. 16). A fault is interpreted as separating the oil and water zones, and the fault is confirmed by structural interpretation of the field.

Commonly in many such examples, however, the adjacent water-bearing sandstone is of a different character from the reservoir sandstone. In the illustrated log (Fig. 16) both the spontaneous-potential and gamma-ray curves show variations that suggest more abundant interbedded siltstone and shale above the reservoir. In other words, the adjacent sandstone quite likely has lower porosity and permeability as compared to the reservoir sandstone, and regardless of the effectiveness of the possible fault zone as a seal, the adjacent sandstone can be suspected as contributing significantly to the oil column across the fault.

The variability of fault zones as effective seals has been discussed by Smith (1966), but his examples were presented schematically. Unfortunately, no examples of faulted sandstones are known for which core analyses are available to assist in calculating the oil columns that might be held by sand-to-sand contacts. However, the possibility of sandstone barriers across faults should be kept in mind for evaluation and development of faulted structures.

Discussion

Field examples appear to show that capillary pressure is an effective mechanism for trapping oil, even where the barrier facies has significant porosity and permeability. Calculations of oil columns, however, are approximations, and their accuracy is limited not only by the basic assumptions but also by the lack of precision in determining rock and fluid properties.

The calculation of oil column by means of equation (16) assumes hydrostatic conditions, and a deficiency in the calculated height of oil may indicate a condition of hydrodynamic flow. For example, a potentiometric gradient of only 10

ft/mi and an oil-water density contrast of 0.2 gm/cc can result in a tilt of the oil-water contact of 50 ft/mi, and this amount of tilt is the height of oil that can be trapped by flow in addition to that trapped by capillary pressure. Therefore, a low potentiometric gradient could add significantly to the oil column in simple stratigraphic traps in areas of low dip.

The detection of low potentiometric gradients depends on accurate pressure measurements, and a freshwater gradient of 10 ft/mi would require the detection of a pressure gradient on the order of 4 psi/mi, a degree of precision that is highly unlikely with the present practice and frequency of drill-stem testing.

The presence of tilted oil-water contacts is often difficult to determine. Oil-field measurements probably are not accurate to better than ± 10 ft under best conditions, and oil-water transition zones in low-permeability reservoirs and lack of

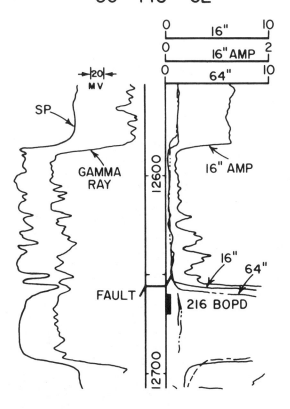

FIG. 16—Electric log from Weeks Island field, Iberia Parish, Louisiana, showing interpretation of fault between reservoir and overlying aquifer (from Shaw, 1962).

downdip well control complicate the determination of the level for 100 percent water production.

The estimation of effective grain and pore size in the reservoir and barrier facies introduces another approximation, but in this case the errors are not believed to be serious and are probably within the limits of oil-field measurements.

The permeability equation (30), from which an effective grain size can be derived, has been tested on a variety of sandstones that range from 40-percent porosity to less than 20-percent porosity, well beyond the limits for which the equation was derived, and calculated values of permeability are commonly within the range of ± 50-percent error. This error may result largely from the porosity term rather than from the grain size term in the equation. Nevertheless, the calculation of effective grain size depends on the square root of permeability, and the maximum error thereby is reduced to about ± 25 percent. In most cases a smaller error can be expected, and calculated values for oil column may be well within the range of error for subsurface measurements.

A more practical limitation for subsurface applications may be the lack of cores in updip dry holes from which the grain size of the barrier facies may be calculated. However, this difficulty commonly can be overcome by the fact that the barrier facies may be present in the same core as the reservoir facies. The Milbur field calculation is a good example because both trap and reservoir facies were assumed to be represented in the nearby cored well (Fig. 9).

For the most accurate calculations of oil column, fluid densities must be reduced to subsurface conditions. In general, the fluid-density contrast increases because of the greater volume increase of oil at higher temperatures as compared with water. The use of proper values in equation (30) may result in an error of only 1 percent (Standing, 1962, p. 24), but failure to account for dissolved gas can result in significant error. For example, if an average oil has a relative density of 0.8 and a gas-oil ratio of 500 cu ft/bbl, then according to equations (30) and (31) its subsurface density would be estimated at 0.69 at a temperature 150°F, whereas if dissolved gas were neglected, its subsurface density would be estimated as 0.77 for an error of 0.08 g/cc. With an observed oil column of 50 ft and fresh water, the calculated oil column might be 67 ft by using the greater oil density, for an error of +34 percent.

In the field examples given, the amounts of dissolved gas are relatively small, and errors in estimating subsurface densities are probably not significant. Under other conditions, however, error may be larger as, for example, in correcting density for a highly gassy oil or gas condensate system.

Assumptions regarding the values of surface tension or interfacial tension may introduce error. The choice between 30 or 35 dynes/cm for oil will result in only ± 15 percent error, and oil-column calculations will be within the limits of oilfield measurements. However, there have been few measurements of interfacial tensions of crude oils against their connate waters, and some recent measurements of pure hydrocarbon compounds against distilled water have resulted in values different from those indicated by previous measurements (Fig. 7). For example, surface tension for 100-percent methane against distilled water was about 50 dynes/cm (Jennings and Newman, 1971) as compared with 35 dynes/cm for the older measurements in the same range of temperature and pressure. Interfacial tension for n-decane against water also was high, but mixtures of equal amounts of methane and n-decane gave lower values in the same range as crude oils. In addition, interfacial tensions for fresh water and hydrocarbon liquids from C^3 through C^8 are on the order of 50 dynes/cm at higher temperatures and pressures, whereas benzene against water has an interfacial tension on the order of 30 dynes/cm (Hassan et al, 1953). Although there is some doubt about what the proper values of interfacial tension should be for oils with their dissolved gas, the more recent measurements indicate that the functions shown (Fig. 7) may be satisfactory for crude oil but may be somewhat low for light oils and gas.

In attempting to apply capillary-pressure relations to the subsurface, some misleading statements may be found in publications that deal with geologic applications. For example, it is well known that increase in temperature alone decreases interfacial tension, and failure to account for the effect of pressure, as shown by Hocott (1938), may result in assuming too low a value for interfacial tensions.

Another source of possible confusion is in discussions of capillary pressure as illustrated by the capillary rise of water. These explanations employ an equation which includes the cosine of the fluid-contact angle, θ, as measured from the solid surface through the more dense fluid, and use of the contact angle implies that its value must be known before capillary-pressure equations can be applied to the subsurface (Levorsen and Berry, 1967, p. 552). On the contrary, the contact angle can be ignored under subsurface conditions because in most cases the reservoir rock is water-wet, oil does not touch the mineral grains, and the contact angle is zero.

Capillary-pressure phenomena are subject to experimental confirmation, and laboratory experiments can be designed to test the validity of the oil-column equation. At least one such experiment has been conducted to test the hypothesis that oil migrates through aquifers as fine, disperse emulsions, and that oil accumulates where pore size is reduced (Cartmill and Dickey, 1970). The conclusion, however, was that the screening of oil did not occur as a result of capillary effects, but that other mechanisms, possibly electrostatic forces, were important in causing the oil droplets to accumulate in clusters around small pores. The results of this experiment are not entirely convincing, primarily because pore size was not measured and cannot be compared to the size of oil droplets. It is clear that other experiments are needed before the capillary-pressure mechanism for oil entrapment is rejected.

The calculation of hydrostatic oil column gives a maximum height of oil that can be trapped by the grain-size change between the reservoir and barrier facies, assuming that values selected for the calculation are representative of subsurface conditions. It is possible for a shorter oil column to be present than is indicated by the calculations because other conditions for origin and migration of hydrocarbons may not have been favorable for the generation of oil in sufficient amounts to fill the reservoir. This is not true, however, for the field examples described previously, nor, in fact, have calculations in other fields given values for possible oil columns that greatly exceed observed columns. It may be concluded that the calculation of excessive oil columns most likely indicates that improper values for porosity and permeability have been assumed and that the barrier facies is more permeable or "leakier," than is suggested by available core analyses.

When all the variables are considered, it is probably the presence of low potentiometric gradients that can introduce the most serious discrepancies between calculated and observed oil columns. It is interesting to note that the calculated, hydrostatic oil columns in field examples are generally less than the actual oil columns, and this common deficiency might be attributed to the approximate nature of the calculations. One is tempted to speculate, however, that additional columns are trapped by low hydrodynamic gradients, even though flow has not been demonstrated.

Conclusion

The calculation of oil columns by means of the capillary-pressure equation (16) appears to give values that are in accord with those observed in stratigraphic traps. These calculations require that certain properties of the rocks and fluids be

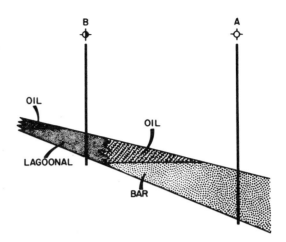

FIG. 17—Diagrammatic cross section of littoral marine sandstone before discovery of stratigraphic trap.

estimated for subsurface conditions; and, although such properties can only be approximated, the agreement between calculated and observed values is surprisingly good. This agreement suggests that the capillary-pressure mechanism for petroleum migration and trapping is a valid hypothesis and that its application to subsurface problems may yield results of practical value.

For example, consider a lenticular sandstone such as that in the Milbur field (Fig. 17). Before the field was discovered, a downdip hole, well A, penetrated 25 ft of porous and highly permeable bar sandstone. A second test, well B, was drilled updip and encountered the lagoonal facies, a thin sandstone of reduced porosity and permeability, still capable of fluid production but containing water, not oil.

At this point an oil accumulation might be anticipated farther updip at the pinchout, but the reservoir thickness and properties are not encouraging for a significant field to exist. It is not apparent that a sizable oil accumulation is present between the two dry holes. If the porosity and permeability are known, or can be estimated, for the sandstones in the dry holes, a simple calculation for height of oil column would suggest a favorable location between the dry holes and result in a field discovery.

Millions of barrels of oil probably are present in simple stratigraphic traps, but these reservoirs will be difficult to find because there is no structural evidence for their existence. A knowledge of fluid properties, especially capillary and hydrodynamic pressures, combined with interpretations of depositional environments, will be a significant aid in exploration.

References Cited

Archie, G. E., 1952, Classification of carbonate reservoir rocks and petrophysical considerations: AAPG Bull., v. 36, p. 278-298.

Aschenbrenner, B. C., and C. W. Achauer, 1960, Minimum conditions for migration of oil in water-wet carbonate rocks: AAPG Bull., v. 44, p. 235-243.

Berg, R. R., 1970, Method for determining permeability from reservoir rock properties: Gulf Coast Assoc. Geol. Socs. Trans., v. 20, p. 303-317.

———— 1972, Oil column calculations in stratigraphic traps: Gulf Coast Assoc. Geol. Socs. Trans., v. 22, p. 63-66.

Cartmill, J. C., and P. A. Dickey, 1970, Flow of a disperse emulsion of crude oil in water through porous media: AAPG Bull., v. 54, p. 2438-2443.

Chuber, S., 1972, Milbur (Wilcox) field, Milam and Burleson counties, Texas, in R. E. King, ed., Stratigraphic oil and gas fields: AAPG Mem. 16, p. 399-405.

Cook, C. W., 1923, Study of capillary relationships of oil and water: Econ. Geology, v. 18, p. 167-172.

Graton, L. C., and H. J. Fraser, 1935, Systematic packing of spheres with particular relation to porosity and permeability: Jour. Geology, v. 43, p. 785-909.

Grauten, W. F., 1965, Fluid relationships in Delaware Mountain Sandstone, in A. Young and J. E. Galley, eds., Fluids in subsurface environments: AAPG Mem. 4, p. 294-307.

Habermann, B., 1960, A study of the capillary pressure-hydrodynamic relationship to oil accumulation in stratigraphic traps: Canadian Mining Metallurgy Bull., v. 53, p. 811-817.

Harms, J. C., 1966, Stratigraphic traps in a valley fill, western Nebraska: AAPG Bull., v. 50, p. 2119-2149.

Hartman, J. A., 1972, "G_2" channel sandstone, Main Pass Block 35 field, offshore Louisiana: AAPG Bull., v. 56, p. 554-558.

Hassan, M. E., R. F. Nielsen, and J. C. Calhoun, 1953, Effect of pressure and temperature on oil-water interfacial tension for a series of hydrocarbons: Petroleum AIME Trans., v. 198, p. 299-306.

Hobson, D. G., 1954, Some fundamentals of petroleum geology: London, Oxford Univ. Press, 139 p.

Hocott, C. R., 1938, Interfacial tension between water and oil under reservoir conditions up to pressures of 3,800 psia and temperatures of 180°F.: AIME Trans., v. 132, p. 184-190.

Hubbert, M. K., 1953, Entrapment of petroleum under hydrodynamic conditions: AAPG Bull., v. 37, p. 1954-2026.

Jenkins, R. E., 1961, Characteristics of the Delaware Formation: Jour. Petroleum Technology, v. 13, p. 1230-1236.

Jennings, H. Y., Jr., and G. H. Newman, 1971, The effect of temperature and pressure on the interfacial tension of water against methane-normal decane mixtures: Soc. Petroleum Engineers Jour. Trans., v. 251, p. 171-175.

Leverett, M. C., 1941, Capillary behavior in porous solids: AIME Trans., v. 142, p. 152-169.

Levorsen, A. I., and F. A. F. Berry, 1967, Geology of petroleum, 2d ed: San Francisco, Freeman, 724 p.

McKetta, J. J., and A. H. Wehe, 1962, Hydrocarbon-water and formation water correlations, in T. C. Frick and R. W. Taylor, eds., Petroleum production handbook: New York, McGraw-Hill, Chap. 22, p. 1-26.

McNeal, R. P., 1965, Hydrodynamics of the Permian basin, in A. Young and J. E. Galley, eds., Fluids in subsurface environments: AAPG Mem. 4, p. 308-326.

Munn, M. J., 1909, The anticlinal and hydraulic theories of oil and gas accumulation: Econ. Geology, v. 4, p. 509-529.

Payne, M. W., 1973, Basinal sandstone facies in the Delaware Mountain Group, West Texas and southeast New Mexico: PhD thesis, Texas A&M Univ., 150 p.

Purcell, W. R., 1949, Capillary pressures—their measurement using mercury and the calculation of permeability therefrom: AIME Trans., v. 186, p. 39-46.

Reistle, C. E., and E. C. Lane, 1928, A system of analysis for oil-field waters: U.S. Bur. Mines Tech. Paper 432, 14 p.

Rittenhouse, G., 1971, Pore-space reduction by solution and cementation: AAPG Bull., v. 55, p. 80-91.

Schlumberger Limited, 1969, Log interpretation principles: New York, Schlumberger, 100 p.

Scott, R. J., 1967, Paduca Delaware, in E. E. Kinney, ed., Oil and gas fields of southeastern New Mexico: Roswell Geol. Soc. Symposium, p. 144-145.

Shaw, W. S., ed., 1962, Electric logs of south Louisiana: New Orleans Geol. Soc., p. X-1.

Smith, D. A., 1966, Theoretical consideration of sealing and non-sealing faults: AAPG Bull., v. 50, p. 363-374.

Standing, M. B., 1962, Oil system correlations, in T. C. Frick and R. W. Taylor, eds., Petroleum production handbook: New York, McGraw-Hill, Chap. 19, p. 1-42.

Stout, J. L., 1964, Pore geometry as related to carbonate stratigraphic traps: AAPG Bull., v. 48, p. 329-337.

Willis, D. G., 1961, Entrapment of petroleum, in G. B. Moody, ed., Petroleum exploration handbook: New York, McGraw-Hill, Chap. 6, p. 6-1-6-68.

The American Association of Petroleum Geologists
V. 68, No. 11 (November 1984), P. 1752-1763, 25 Figs., 1 Table

Evaluating Seals for Hydrocarbon Accumulations[1]

MARLAN W. DOWNEY[2]

ABSTRACT

Seals are an important and commonly overlooked component in the evaluation of a potential hydrocarbon accumulation. Effective seals for hydrocarbon accumulations are typically thick, laterally continuous, ductile rocks with high capillary entry pressures. Seals need to be evaluated at two differing scales: a "micro" scale and a "mega" or prospect scale. Quantitative "micro" data measured on hand specimens of seal rock are difficult to extrapolate a billion-fold to the scale of the sealing surface for a hydrocarbon accumulation. Fortunately, each class of exploration prospects has a different set of seal problems. Geologic work can be focused on the characteristic seal problems that plague classes of prospects. Anticlines have relatively little seal risk, because any layer serving as a top seal will also be a lateral seal. Stratigraphic traps and faulted prospects have substantial seal risks. When attention is focused toward the expected sealing surface of a potential hydrocarbon accumulation, it is possible to assess the relative risk that a seal is present. Improvements in assessing seal risk for an exploration prospect directly affect the estimation of exploration success.

PURPOSE AND SCOPE

Modern petroleum geology dates from the recognition that hydrocarbon accumulations require hydrocarbons to be generated from organic-rich source rocks, to be contained in porous and permeable reservoirs, and to be entrapped by a natural seal.

Recognition and mapping of the geometric form of the trap have been major facets of study by explorationists for over 70 years. Properties of reservoirs in which oil and gas are contained have been widely and intensively studied. Enormous progress has been made in the past two decades in techniques for identifying where and how oil and gas are generated and how they migrate, segregate, and transform. In contrast, surprisingly little has been published on the recognition and evaluation of the properties of the physical boundaries or "seals" that allow for the entrapment of hydrocarbons. As long ago as 1860, Henry D. Rogers was able to point out the importance of a hydro-

carbon seal in his investigation of oil springs by noting that the "imperviousness of the argillaceous strata, except where they are fissured...serves to hold down the elastic volatile products." (Quoted by Dott and Reynolds, 1969.)

In this article, I use the term "seal" to refer to a layered lithologic unit capable of impeding hydrocarbon movement, and "sealing surface" to describe the three-dimensional entrapping surface confining a hydrocarbon accumulation.

In 1980, the Research Committee of AAPG sponsored a conference on "Seals for Hydrocarbons" in an attempt to refocus attention toward understanding seals. That conference fostered a useful exchange of knowledge, and resulted in a collection of abstracts (Downey and Schowalter, 1980).

This article is intended as a personal essay on the evaluation of the relative risk that various seals can entrap hydrocarbon accumulations. It is a primer for explorationists, and should serve as a guide to the diverse literature.

As a business venture, exploration is greatly dependent on intelligent risk analysis. The proper risk assessment made on an exploration prospect or play is the product of the individual risks seen for reservoir, structure, charge, and seal. In assessing the proper risk to be placed on an exploration prospect, it is useful to remember that it is as important to have a seal as to have a reservoir, as necessary to have a seal as a trapping configuration, as vital to have a seal as to have access to hydrocarbon charge. Evaluation of an exploration prospect demands consideration of the probability of occurrence of an effective sealing surface.

MICRO PROPERTIES OF SEALS

Capillary Properties

Seals for hydrocarbons need to be analyzed and described within two different (micro and macro) scales. Fundamentally, the quality of a seal, at a given time, is determined by the minimum pressure required to displace connate water from pores or fractures in the seal, thereby allowing leakage.

Capillary pressure (Pd) of a water-filled rock is a function of hydrocarbon-water interfacial tension (γ), wettability (θ), and radius of largest pore throats (R), according to the relationship $Pd = 2\gamma \cos\theta/R$ (Purcell, 1949) (Figure 1).

Capillary pressure (sealing capacity) of the rock increases as (1) the throat radius of the largest connected pores decreases, (2) the wettability decreases, and (3) as the hydrocarbon-water interfacial tension increases.

Capillary forces of a seal act to confine hydrocarbons within an accumulation. The buoyancy forces of the

[1]Manuscript received, November 7, 1983; accepted, April 6, 1984.
[2]Pecten International Co., 1250 Woodbranch Park Drive, Houston, Texas 77079.

I express my thanks to H. G. Williams of Shell Development Co. for his helpful review of the manuscript, to John McCormick of Pecten International Co. for his effort in preparing the illustrations, to Pearly Bellvin and Freda Jones of Pecten International Co. for their pains in redrafting the article, to Elizabeth Engelke of Pecten International Co. for her efforts in coordinating and editing, and to Aphrodite Mamoulides of Shell Development Co. for her library research. I appreciate the advice and review of Shell Oil Co. and the permission to publish provided by Pecten International Co.

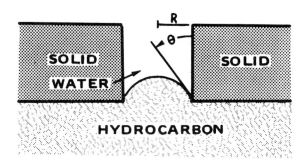

Figure 1—Capillary displacement pressure is dependent on three parameters; radius of largest connected pore throats (R), wettability (θ), and hydrocarbon-water interfacial tension (γ).

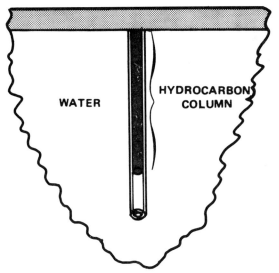

Figure 2—Buoyancy pressure of hydrocarbon column acts to attempt to force hydrocarbons through larger pores in seal.

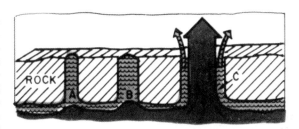

Figure 3—Rock pore throats A and B are constricted (have high entry pressures); pore throat C is sufficiently large that buoyancy pressure of hydrocarbon column can displace pore water and pass through pore throat.

hydrocarbon column of a static accumulation are given by the product of the hydrocarbon column height and the difference in density between the hydrocarbon and the reservoir pore water (Figure 2). These hydrocarbon buoyancy forces must be matched or exceeded by the capillary entry pressure that characterizes the pore structure of the seal (Figure 3).

One can measure in a laboratory the displacement pressure necessary to force a given hydrocarbon mixture through a given rock under specified conditions of temperature and pressure; such measurements provide quantitative data about the capacity of that seal to entrap those hydrocarbons. Indeed, the displacement pressure of sandstone seals can be estimated from grain size and sorting data, using the method of Berg (1980). Quantitative data are very valuable and, where properly used (Schowalter, 1979), can provide an important starting point for assessment of seal capacity. Unfortunately, such "micro" data taken from a rock sample have limited use when attempts are made to extrapolate rock sample data to the entire "macro" sealing surface bounding an accumulation (Downey, 1981).

The difficulty of extrapolating precise measurements made on, for instance, a piece of core 4 in. (10 cm) in diameter to the entire entrapping surface area can be understood by a simple example. Assuming a domal closure area of 6,400 acres (2,590 ha.), a core sample of the top seal would provide a ratio between area of seal sample and area of top seal of 1 to about 3.5 billion. What is the probability that the "micro" properties characterizing the core are invariant over the entire domal sealing surface? Such large extrapolations of data are commonly necessary in geologic work, but it is important in assessing seal properties to remember that *average* values are nearly meaningless in determining probability of seal for a hydrocarbon accumulation. In looking at sealing surfaces, we are basically concerned with the properties of the "weakest" point of the sealing surface; the measured values from a random core sample, unfortunately, have little relevance to the problem of determining the weakest leak point of the seal. Just as little comfort can be taken from a guarantee that your parachute will (on the average) open, explorationists are not really interested in the average properties of an enclosing, sealing surface.

Diffusion Losses Through Seals

As a gross geologic simplification, diffusion of hydrocarbons through seals is dependent mainly on (1) hydrocarbon type, (2) the characteristics of the water-filled pore network of the contacting seal, and (3) time available for diffusion. Accordingly, we can describe some nonrigorous generally applicable guidelines useful to geologists in assessing the risk of significant diffusive loss of hydrocarbons through seals.

If the reservoired hydrocarbon is oil, little or no diffusive leakage through a seal is likely. If the reservoired hydrocarbon is methane, and is sealed by rock layers lacking connected pore water (salt, anhydrite, or gas hydrates), no diffusive loss should be expected. However, if methane is sealed by a porous, water-bearing shale, substantial diffusive losses can occur over significant geologic time (Leythaeuser et al, 1982).

To summarize, exploration worries about diffusive loss of a hydrocarbon accumulation through a seal can be neglected except for exploration prospects where the expected hydrocarbon is methane, where the seal has porous water-filled pore space, and where long-term retention in the trap is required.

MACRO CHARACTERISTICS OF SEALS

Lithology

Any lithology can serve as a seal for a hydrocarbon accumulation. The only requirement is that minimum displacement pressure of the lithologic unit comprising the sealing surface be greater than the buoyancy pressure of the hydrocarbon column in the accumulation. As a practicality, however, the overwhelming majority of effective seals are evaporites, fine-grained clastics, and organic-rich rocks. These lithologies are commonly seen as seals because they typically have high entry pressures, are laterally continuous, maintain stability of lithology over large areas, are relatively ductile, and are a significant portion of the fill of sedimentary basins. Statistical analyses of the top seals of apparently unfaulted structural traps by Nederlof and Mohler (1981) indicated that lithology is the most important factor correlating to a good seal, with thickness and depth to seal having positive, but lesser, correlations to hydrocarbon column heights.

Ductility

The folding and faulting that accompany the formation of many traps put significant strain on the sealing surfaces of accumulations. Brittle lithologies develop fractures, whereas ductile lithologies tend to flow plastically under deformation (Table 1). Carbonate mudstones may have very high entry pressures, but under conditions of deformation, they may fracture much more readily than salts, anhydrite, clay shales, and organic-rich rock. Ductility, of course, is a rock property that varies with pressure and temperature (burial depth) as well as with lithology. The evaporite rock group make extraordinarily good ductile seals under overburdens of several thousand feet, but can be quite brittle at shallow depths.

Table 1. Seal Lithologies Arranged by Ductility*

Salt
Anhydrite
Kerogen-rich shales
Clay shales
Silty shales
Carbonate mudstones
Cherts

*Most ductile lithologies at top of column.

Highly organic-rich rocks are fine-grained rocks containing deformable layers of kerogen; such rocks commonly have a plastic behavior during folding and the flowage of the soft kerogen layers makes for very high displacement pressures in the relict pores.

In the overthrust provinces of the world, where deformation and fracturing are expected to be most intense, ductility becomes very important when assessing sealing layers for accumulations. Lithologies having a very fine pore structure and a ductile matrix can retain sealing properties even under severe deformation. A recent analysis of the characteristics of the world's 25 largest gas fields indicates that all of those in thrust provinces depend on evaporite seals. Indeed, in reviewing the world's 176 giant gas fields, almost all depended on shale or evaporite seals (Grunau, 1981).

Thickness

A few inches of ordinary clay shale are theoretically adequate to trap very large column heights of hydrocarbons. A clay shale with particle size of 10^{-4} mm would be expected to have a capillary entry pressure of about 600 psi (Hubbert, 1953), theoretically capable of holding back an oil column of, for instance, 3,000 ft (915 m). Unfortunately, there is a low probability that a zone only a few inches thick could be continuous, unbroken, unbreached, and maintain stable lithic character over a sizable accumulation. The benefits of a very thick seal are that it provides many layers of contingent sealing beds and a larger probability that a sealing surface will actually be distributed over an entire prospect. A thick seal is important and beneficial, but does not linearly influence the amount of hydrocarbon column that can be held by a top seal.

Where traps are created by fault offset of reservoirs, thickness of top seal can be very important. In such places, the top seal can be offset to become a lateral seal, and seal thickness can relate directly to column height of entrapped hydrocarbons.

Stability

Stratigraphic layers identified as having the capillary properties of seals need to be studied to see whether those layers are lithologically uniform throughout the areal distribution of the stratigraphic unit. Identifiable stratigraphic units may vary greatly in their capillary properties with only modest changes in lithology.

A stratigraphic cross section of the potential seal unit, utilizing electric log character and lithology, is an excellent start toward establishing that the seal unit is stable over the area of interest.

SIGNIFICANCE OF REGIONAL SEALS

Most basins contain major widespread sealing layers. These regional seals are characterized by broad extent, significant thickness, laterally stable character, and generally by relatively ductile lithologies. Where these seals are found above mature source rocks and reservoirs, they control, in large measure, the regional emplacement of hydrocarbons. A recent publication on the oil fields of the Arabian Gulf (Murris, 1980) provides excellent examples of this interdependence. Two major source rocks are present in the area, the Upper Jurassic Hanifa and the Aptian Shuaiba. The Upper Jurassic Hith Anhydrite and the Albian Nahr Umr Shale are the two principal seals. Hydrocarbon accumulations are concentrated in reser-

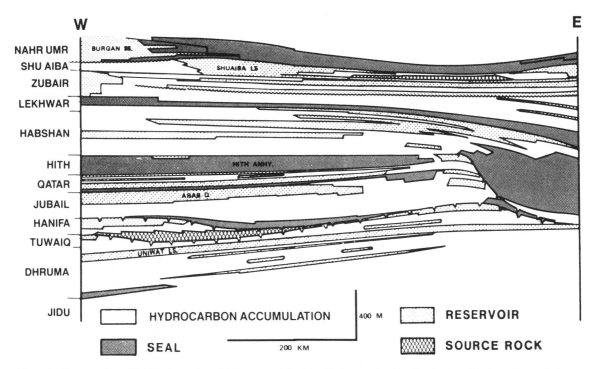

Figure 4—Cross section of Middle Jurassic to Albian, central Arabian Gulf area, showing distribution of hydrocarbons relative to seals. Adapted from Murris (1980).

voirs under these regional seals. In areas where the seals are absent or disrupted by faulting, the hydrocarbons have migrated upward to be trapped by secondary seals (Figure 4).

Major roofing seals act to confine migrating hydrocarbons to particular stratigraphic units. Regional evaluation of the exploration potential of a basin should start with (1) determination of stratigraphic position and areal distribution of mature source rocks, (2) identification of the regional seals for migrating hydrocarbons, (3) an analysis of trapping conditions focused on the areas under the regional seals and updip from the mature source rocks, and (4) an examination of the distribution of hydrocarbon shows and production.

ASSESSMENT OF SEAL RISK IN EXPLORATION

Different seal problems are characteristically associated with each of the major types of hydrocarbon traps. In each accumulation type, specific classes of seal problems tend to recur, and the wise explorationist focuses analysis on the expected problem areas. Proper review of seismic and well log data will help to assure a correct seal risk analysis of individual prospects.

Anticline

Domal closures.—A simple anticlinal closure has comparatively little seal risk. Convex upward folding of typical layered sedimentary sequences provides multiple seal interfaces. The lateral seal in a domal closure is the same as the top seal; that is, the sealing surface is defined by the attitude of the top seal (Figure 5).

In rare instances, the sedimentary sequence is nearly completely composed of porous and permeable rocks; in such areas, stratigraphic and fault traps are highly unlikely, and even anticlinal closures may have hydrocarbon retention problems.

In strongly folded settings, considerable stress may be placed on the top seal of possible accumulations. The major seal risk in domal closure is that of creating open fractures during folding. Development of open extension fractures in the earth is inhibited by gravitationally induced confining pressure. Development of open extension fractures is normally expected to be limited to the upper few hundred meters, perhaps few kilometers, of the earth. Therefore, domal folds at shallow depths will have a risk of crestal open fractures. In addition, under the spe-

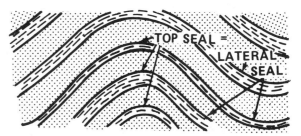

Figure 5—In a domal closure, any seal unit provides both top and lateral sealing surfaces.

55

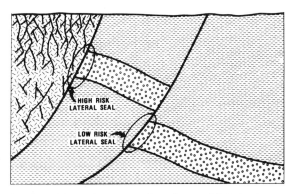

Figure 6—Schematic cross section of potential fault closure traps; indicated trap against shallow basement block will have substantial lateral seal risk.

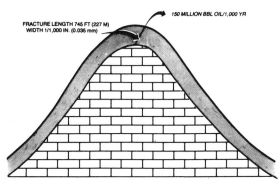

Figure 7—Sample calculation of migration loss through fracture. Assumptions: fracture width = 1/1,000 in. (0.035 mm); length = 745 ft (227 m); migration rate = radial flow capacity of reef with drainage radius of 745 ft (227 m); potential energy = oil-water density difference of 0.3 g/cm³ and water level change of 1,200 ft (365 m) in 1,000 ft (305 m); limestone permeability = 10 md; net pay = 500 ft (152 m).

cial circumstance of geopressured layers where pore pressures approach overburden weight, open extension fractures could form at any depth in anomalously stressed areas (Secor, 1965).

The earth forces that fold rocks into trapping configurations tend to stress and crack the folded units. Anticlinal closures in thrust belts need excellent top seals. Fracturing will be likely, unless the top seal is a lithology capable of deforming plastically.

Exploration along steeply faulted mountain fronts often indicates potential traps of reservoirs against basement blocks. Traps that depend on shallow brittle basement for lateral seals are very risky (Figure 6).

Even narrow open fractures have extraordinary transmissive properties to fluids. To emphasize the importance of top seal integrity and absence of open extension fractures, a sample calculation from data by Muskat (1949) shows that in a modeled case, a single 1/1,000 in. (0.035 mm) open fracture overlying a 500-ft (150-m) oil column would leak at the rate of 150 million bbl/1,000 yr (Figure 7)! It is interesting to note that the limiting parameter to the leak rate, in the model chosen, is the permeability of the reservoir. Such sample calculations are mainly useful in reminding us of the enormous transmissive powers of open fractures and the importance of unfractured seal boundaries.

Fault Traps

Faulted structures.—The term "fault trap" is somewhat of a misnomer; faults do not trap. Faults can place porous reservoirs adjacent to seals and form traps. This trivial-sounding distinction deserves attention because of the general tendency of explorationists to consider that the mapping of a fault trace is sufficient to define a sealing surface.

Entirely different seal risk is attached to the portion of an exploration prospect with simple domal closure (requiring only a top seal) versus fault closure (dependent on seal juxtaposing laterally to reservoir). The portion of the structure depending on fault closure has two seal requirements: (1) top seal and (2) presence of a lateral seal across the fault from the reservoir.

Fault closures require not only a top seal, stable in lithology over the trap area, but require lateral juxtaposition of the entire hydrocarbon-bearing reservoir against a sealing lithology. Thick hydrocarbon columns become almost impossible to seal if the trapping configuration depends on multiple faults with throws that vary along the individual faults. Beware of interpreters who map "fault closure" without demonstrating what the fault is juxtaposing!

Hydrocarbons are not distributed arbitrarily or randomly on complexly faulted structures. Their distribution follows very simple physical principles, and preferential hydrocarbon distribution can be predicted, given sufficient subsurface data. An "Allan" fault plane map (Allan, 1980) is a useful conceptual tool for analyzing the probable conjunction of seals and reservoirs in a trap configuration. Strata representative of the rock section in the area are categorized as seal or nonseal. These seal and/or nonseal layers are mapped in their proper structural relationships and their traces projected onto the boundary fault planes. An "Allan" fault plane map superposes seal and nonseal bed traces from the footwall against the hanging wall for each fault plane bounding an exploration prospect. When fault plane traces of seal and reservoir are considered in structural context, they demonstrate where a given reservoir might be in a trapping configuration (Figure 8).

Analysis of cross sections of faulted prospects can provide a preliminary test of the likelihood of a sealing surface. If there may be permeable zones contacting the fault in a trapping element, the probability of a sealing surface becomes strongly dependent on the dip attitude of the beds in the trapping block (Figure 9).

Fault planes as seals.—Fault planes are normally inconsequential to migrating fluids, and generally are of significance as sealing surfaces only because they may juxtapose rocks of differing capillary properties and fluid pressures (Smith, 1966).

The fault plane itself offers open passage to migrating fluids only under special conditions. The most significant

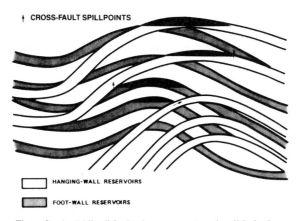

‡ CROSS-FAULT SPILLPOINTS

☐ HANGING-WALL RESERVOIRS

▨ FOOT-WALL RESERVOIRS

Figure 8—An "Allan" fault-plane map categorizes lithologies as seal or nonseal and superimposes the lithostratigraphic units of the hanging wall onto the footwall.

of those special conditions is the circumstance of shallow, near-surface faulting in an overall tensional regional stress field. In such cases, field observations and theory (Secor, 1965) agree that the fault plane may act as an open transmissive fracture. Faulted shallow structures will have substantial risk that the fault plane may leak.

In special circumstances, where thick, undercompacted clay shales are interspersed between reservoirs, clay smears can be emplaced along a fault plane (Figure 10). Such fault-plane shale smears are common small-scale phenomena where thick, soft shales are present in the stratigraphic section. These clay smears have been reported in east Texas outcrops (Smith, 1980), have been studied in coal mines in Germany, and have been inferred from log interpretation of fault zones in Nigeria (Weber et al, 1978). Thin clay smears have been invoked to prevent fluid migration between some adjoining sandstone reservoirs in the Gulf Coast Tertiary (Smith, 1980).

The special circumstances governing the emplacement of these fault-plane clay smears indicate that they are relatively rare in providing traps for significant hydrocarbon accumulations. Important exceptions are found in the Tertiary sediments of the Mississippi and Niger Deltas.

Stratigraphic Accumulations

Stratigraphic traps for hydrocarbons typically require a top seal and bottom seal, as well as an updip lateral seal. This tripartite requirement adds substantial additional risk to the likelihood that a typical stratigraphic prospect is enclosed by seals. Stratigraphic prospects inherently have high seal risks.

Facies changes.—A stratigraphic trap can be caused by the lateral facies change of a reservoir to a seal in an updip direction. Such reservoir facies changes are quite common, but commercial accumulations are relatively rare. The difficulty attending such facies traps is that the reservoir must change completely and abruptly updip to sealing rock. It is not enough that the updip lithology be a good seal (on the average); it must totally lack zones of low-entry, pressure-connected porosity. In assessing stratigraphic traps created by updip facies change, the rate of change from reservoir to seal is an important and mappable parameter. If the facies transition is not abrupt, the hydrocarbons will be largely emplaced in an updip waste zone that is "not quite reservoir, hardly seal" (Schowalter and Hess, 1982) (Figures 1-13).

Where is the seal for a hydrocarbon accumulation? The true seal lacks connected filaments of hydrocarbons (Schowalter and Hess, 1982). Many so-called seals for accumulations are really noncommercial reservoirs, or are reservoirs whose pore structure and hydrocarbon saturations do not allow production of water-free hydrocarbons (Figures 14, 15).

"Hydrocarbon reservoir" is a commercial term, whose definition depends on the economics of the moment. "Seal" is a technical term, and does not vary with the price of oil.

Mapping of the structural configuration of the updip limit of commercial thickness of reservoirs is of little use in exploration for stratigraphic traps. The structural configuration of the sealing surface is the definitive mapping boundary (Figure 16).

Reefs.—Reefs are a favorable type of stratigraphic trap because they may form their own structural closure. The actual entrapping closure of a pinnacle reef is that which is defined by the attitude of the first seal above the reef reservoir. A common error in reef prospect interpretations con-

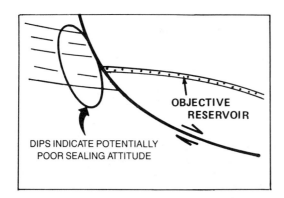

OBJECTIVE RESERVOIR

DIPS INDICATE POTENTIALLY POOR SEALING ATTITUDE

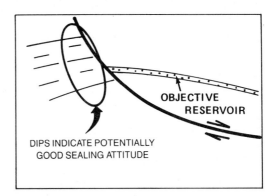

OBJECTIVE RESERVOIR

DIPS INDICATE POTENTIALLY GOOD SEALING ATTITUDE

Figure 9—Dip attitudes of sediments in trapping block provide information as to likelihood of a sealing surface.

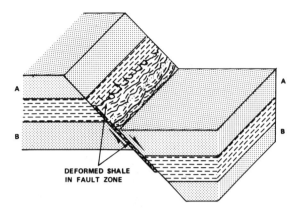

Figure 10—Dragging of undercompacted clays into fault plane can locally emplace sealing material along a fault. Modified from Smith (1980).

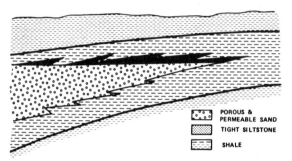

Figure 11—If porous and permeable sands change abruptly into shales, hydrocarbon column will all be in reservoir quality rock.

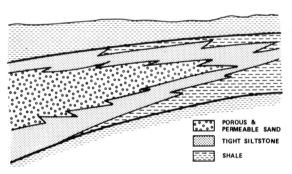

Figure 12—Generally, porous and permeable sandstones grade laterally into nonpermeable siltstones before grading into shales.

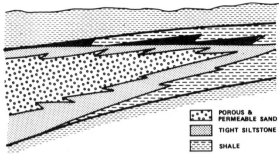

Figure 13—If rate of change from reservoir to seal is gradual, a limited hydrocarbon column is restricted to nonreservoir "waste-zone" facies.

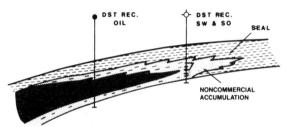

Figure 14—Although updip well did not flow oil, it has tested part of the accumulation, rather than the seal.

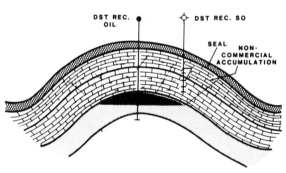

Figure 15—Tight limestones in this fold are not seals for oil field, but are noncommercial parts of accumulation.

sists of mapping the maximum relief available (generally the uppermost "reef" surface) without regard to the likelihood of that surface being a sealing surface (Figure 17).

Seismic-stratigraphic analysis of reef prospects is very helpful. Dipping acoustical events that impinge on a reef are danger signals. They indicate acoustical inhomogeneity of the potential seal, and may indicate avenues for leakage (Figure 18). Although Figures 17 and 18 are reef examples, it is apparent that examination of seismic data can provide excellent guides to the proper seal risk to be assigned to many types of prospects. Structural and lithologic information from seismic sections can provide significant insights as to likely seal risks.

Shelf-margin reef trends form important hydrocarbon traps. Simple drape closure of the top seal can provide low-risk domal entrapment. Updip facies changes of the reef reservoir to tight back-reef facies are frequently observed, but such facies changes impose a substantial seal risk for containment of thick hydrocarbon columns. Where back-reef facies are expected to be a component of the sealing surface, the back-reef lithology must be uniformly a seal.

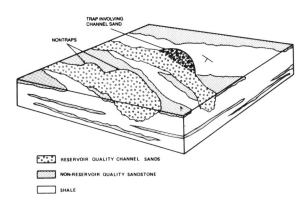

RESERVOIR QUALITY CHANNEL SANDS

NON-RESERVOIR QUALITY SANDSTONE

SHALE

Figure 16—Mapping updip reentrants of porous channel sandstones does not necessarily constitute mapping of trap configurations.

Sandstone channels.—Channel sandstones form an interesting class of stratigraphic traps. Channel sandstones generally have relatively sharp lateral boundaries, and the most prevalent seal problem is that created by the lithology of the incised lateral sealing surface (Figure 19). Seal risk for channel sandstones is heavily dependent on the capillary properties and homogeneity of the adjacent

lithostratigraphic units (Harms, 1966). If channel sandstones are enclosed in relatively homogeneous impermeable strata, each updip channel extension becomes a potential trap. But if the channels are incised into highly variable strata, true traps are uncommon. For example, early drilling in the Cretaceous of the southern Denver basin of Colorado suggested numerous opportunities for exploring for channel sandstones. Logs of a first series of wildcats demonstrated top seals and reservoirs with oil shows (Figure 20). It was easy to visualize that additional drilling would locate numerous sizable channel sandstone stratigraphic traps. Substantial effort went into mapping the size, character, and distribution of the thicker channel sandstones. Results of additional drilling indicated, however, that the seal layers (top as well as lateral) lacked continuity (Figure 21). Rather than hunting for thick porous channel sandstones, it became obvious that the exploration search should be directed toward areas where top and lateral seals could be expected.

Clay-filled channels.—Significant hydrocarbon accumulations can occur in reservoirs truncated by valley erosion and sealed by back-filling clays. Recent exploration in the Sacramento Valley of California has resulted in the discovery of several gas fields created by sand layers being truncated by canyon erosion and being laterally sealed by impermeable valley fill (Garcia, 1981) (Figure 22).

A similar type of configuration commonly characterizes the Pennsylvanian Minnelusa sandstone fields of the eastern flank of the Powder River basin (Van West, 1972). The

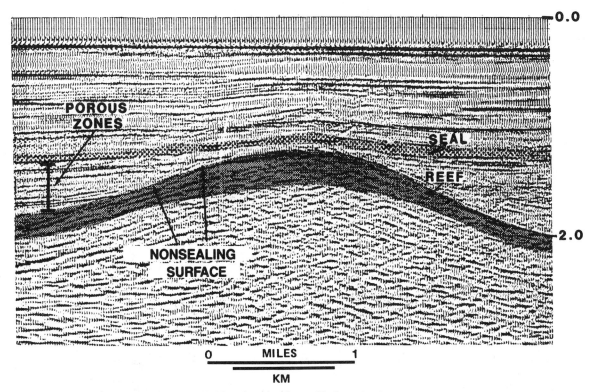

Figure 17—Porous and permeable stratigraphic units onlapping reef indicate that flanks of reef probably lack a sealing surface.

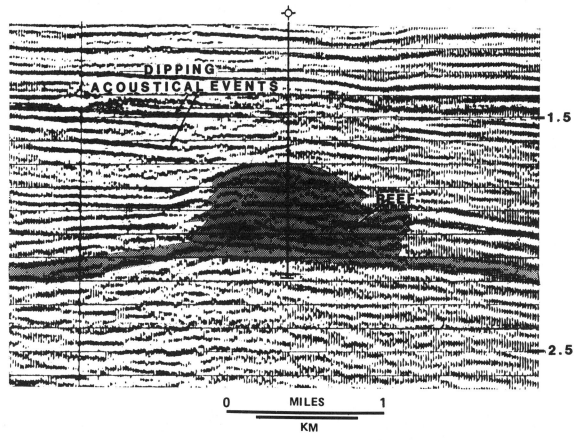

Figure 18—Dipping acoustical events impinging on reef hint that uppermost reef surface is unlikely to maintain itself as uniform sealing surface.

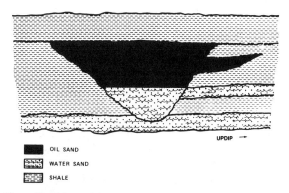

Figure 19—Schematic cross section of channel sandstone incised into varying lithologic units; hydrocarbon column is limited by lateral seals.

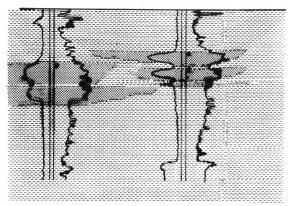

Figure 20—Early, widely spaced drilling demonstrated oil staining in upper portions of channel sandstones.

Minnelusa Formation was locally incised by Permian erosion, and the resulting valleys were filled by red shales of the Permian Opeche Formation. Most of the 200 million bbl of oil recoverable from the Minnelusa is trapped against seals provided by shale-filled paleo-valleys.

Novel Seals

Novel and unusual seals for hydrocarbon accumulations are sometimes present. These novel seals basically reflect

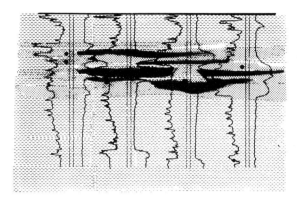

Figure 21—In-fill drilling demonstrates such numerous cut-and-fill channel sequences that little lateral seal lithology could be demonstrated.

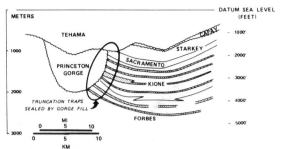

Figure 22—Reservoir terminations are caused by truncation by gorge; seal for reservoirs is provided by impermeable gorge fill. After Garcia (1981).

dynamic flow regimes (change in effective capillary entry pressure) or alterations in the pore throat size distribution of a rock by plugging of pores.

Hydrodynamic influences on seals.—Pioneer work by Hubbert (1953) has allowed a quantitative understanding of the basis for fluid movement in the subsurface. An important corollary to Hubbert's work demonstrates that the effective capillary pressure barrier of a seal is strongly altered by the fluid potential gradient caused by water movement across a seal/reservoir interface. Where water flows downdip from seal to hydrocarbon reservoir, the effective seal entry pressure is increased. If the water flow is from reservoir updip to seal, the effective entry pressure is decreased and the capacity of the seal to impede hydrocarbons is lessened.

On a worldwide basis, hydrodynamics exerts relatively modest effect on the seal risk for most hydrocarbon accumulations. However, in those basins where significant hydrodynamic flow is demonstrated, the magnitude and vector of the hydrodynamic flow are vitally important considerations in assessing seal risk for reservoirs in which hydrocarbon traps are expected.

In the Powder River basin of Wyoming, for example, studies of Recluse Muddy and Kitty Muddy fields by Berg (1975) indicate that downdip hydrodynamic flow greatly enhances the effective entry pressure of the very leaky lateral seals trapping the hydrocarbon column in these fields.

Where geopressured impermeable sediments are juxtaposed to normally pressured permeable reservoirs, the trapping capacity of the sediments is greatly enhanced.

Hubbert's work provided the theoretical basis for such "pressure seals." Weber et al (1978) have noted that trapping of hydrocarbons on the downthrown side of a growth fault in Nigeria is remarkably improved if reservoirs are juxtaposed against overpressured sediments across the fault. These observations have special significance to prospect evaluation in areas where geopressure distribution in the subsurface can be determined from analysis of seismic data (Pennebaker, 1968), or from electric logs (Hottman and Johnson, 1965). Proper analysis of the distribution of pressures in various fault blocks provides a much improved assessment of trapping risk for the exploration prospects (Figure 23).

Within fault blocks, pressure gradients usually parallel lithostratigraphic boundaries; adjacent fault blocks can have the pressure gradients strongly displaced (Figure 24).

In the Gulf Coast of the United States, soft Tertiary shales are commonly dragged up to form a complexly faulted sheath around a salt dome. This shale "sheath" is generally geopressured (Stuart, 1970) and greatly enhances the trapping capacity to the adjoining sandstone reservoirs. Such geopressuring of the sheath makes Gulf

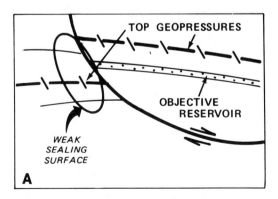

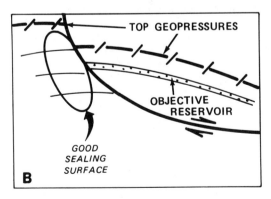

Figure 23—In example B, effective entry pressure of sealing block is increased, rendering it capable of laterally sealing a very large hydrocarbon column.

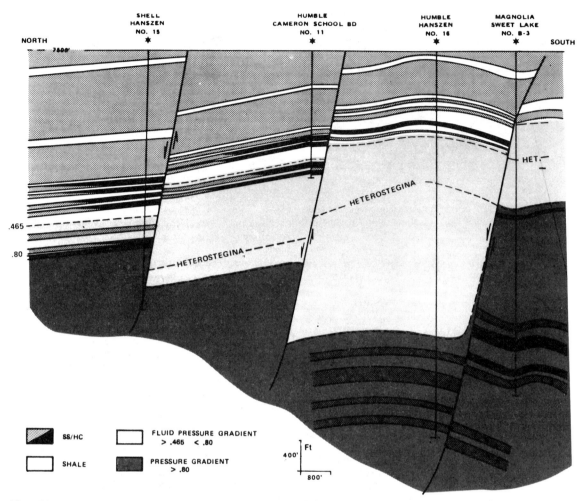

Figure 24—Cross section on north flank of Chalkley field, Louisiana, illustrates pronounced offsets of pressure gradients in adjacent fault blocks. Adapted from Stuart (1970).

of Mexico salt domes outstanding loci of hydrocarbon traps. In contrast, where salt domes have pushed upward through hard or brittle rocks and lack geopressured sheath, lateral sealing capacity of reservoir against faulted sediments around the salt may be lacking or much diminished.

Diagenetic seals.—Chemical and mechanical diagenesis of reservoirs can alter and plug pore space. As a generality, diagenetic alteration of pore space in and peripheral to a reservoir mainly alters the gross flow properties within a reservoir. The distribution of these pore-blocking cements can profoundly modify and impede reservoir fluid movement, but it is very rarely significant in creating a continuous sealing surface for an entire commercial hydrocarbon accumulation.

In rare places, diagenesis of a marginal portion of a reservoir has been reported to create a local seal. Schmidt and Almon (1980) have pointed to the Recinar Cardium "A" pool in Canada as an example of a field in which the seals have all formed by diagenetic cementation of a marginal sand facies.

Tar seals.—The Temblor Sandstone production in the East Coalinga field of the San Joaquin Valley of California has been cited as an example of a tar-sealed accumulation (Wilhelm, 1945). The productive sandstone interval is heavily tar-impregnated a short distance updip at the outcrop. Literature data appear equivocal as to whether the tar-impregnated outcrops represent a seal for the accumulation or merely record the slow transit of hydrocarbons.

Russian authors (Vinogradov et al, 1981) have proposed that seven recently discovered oil fields in Devonian rocks in the Volga-Ural region are trapped on a monocline by tar seals. The authors point out that the oil-water contacts are inclined, that present-day structural closure is lacking, and that neither facies changes nor faulting can account for the trapping.

Gas hydrates.—Gas hydrates are solid crystalline precipitates of gas and water. The gas hydrate is a metastable ice-like material that contains the gas molecules in a lattice of water molecules. The formation of the solid gas hydrate in the rock pores depends on the salinity of the pore water, pressure, temperature, and the gas composition (Makogon, 1981). Conditions for the formation of gas hydrates require that gas and water interact under pressure and at low temperatures. From a geologic point of view, these conditions can be satisfied in two natural settings: (1) in shallow onshore reservoirs in permafrost areas, and (2) in shallowly buried reservoirs underlying deep, cold waters. In an onshore setting, gas hydrates are reported from as deep as 2,000 m (6,560 ft) in Yakutsk, Siberia, where 0°C (32°F) occurred at 1,400 m (4,600 ft) drilling depth. Gas hydrates were reported 600 m (1,970 ft) below the sea floor in cores recovered from deep-water drilling on the Blake Plateau by the *Glomar Challenger* (Stoll et al, 1971).

Hydrocarbon gases migrating updip in porous beds can precipitate pore-filling hydrates where these hydrocarbons enter a specific pressure-temperature zone in the earth. As the migrating gas moves into the proper pressure-temperature range, the solid hydrate precipitates and plugs the rock pores. Gas hydrate plugging pores makes an unusual seal because it represents a relatively sudden phase transformation from gas to solid. Large quantities of gas are known to be self-trapped in the pore-filling hydrates; additionally, these pore-filling hydrates could seal a dipping reservoir to trap downdip gas and oil. Conceptually, any reservoir bed containing migrating gas could create its own updip seal as the reservoir bed intersects the zone of hydrate stability (Figure 25).

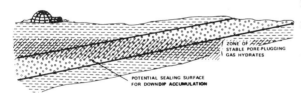

Figure 25—Gas hydrate plugging of reservoir pore space can create an intraformational seal, trapping additional hydrocarbons downdip.

SUMMARY AND CONCLUSIONS

Evaluation of the technical risk associated with a given exploration prospect requires an assessment of the likelihood of a sealing surface being present. An understanding of the basic requirements for a sealing surface to a prospect can contribute to the proper judgment on the likelihood of occurrence of a seal. Different classes of traps have specific types of sealing surface risks; for each class of potential traps, there are concepts and geologic techniques available to improve seal risk assessment. Understanding seal risk allows a more intelligent choice between alternative exploration prospects.

REFERENCES

Allan, U. S., 1980, A model for the migration and entrapment of hydrocarbons (abs.): AAPG Research Conference on Seals for Hydrocarbons, Keystone, Colorado, September 14-17, unpublished book of abstracts, 16 p.

Berg, R. R., 1975, Capillary pressure in stratigraphic traps: AAPG Bulletin, v. 59, p. 939-956.

———— 1980, Calculation of seal capacity from porosity and permeability data (abs.): AAPG Research Conference on Seals for Hydrocarbons, Keystone, Colorado, September 14-17, unpublished book of abstracts, 16 p.

Dott, R. H., and M. J. Reynolds, 1969, Sourcebook for petroleum geology: AAPG Memoir 5, 471 p.

Downey, M. W., and T. T. Schowalter, conveners, 1980, Seals for hydrocarbons: AAPG Research Conference on Seals for Hydrocarbons, Keystone, Colorado, September 14-17, unpublished book of abstracts, 16 p.

———— 1981, Evaluating seal risk in prospect analysis: Society of Exploration Geophysicists 51st International Meeting Technical Papers, v. 3, p. 1439-1470.

Garcia, R., 1981, Depositional systems and their relation to gas accumulation in Sacramento Valley, California: AAPG Bulletin, v. 65, p. 653-673.

Grunau, H. R., 1981, Worldwide review of seals for major accumulations of natural gas (abs.): AAPG Bulletin, v. 65, p. 933.

Harms, J. C., 1966, Stratigraphic traps in a valley fill, western Nebraska: AAPG Bulletin, v. 50, p. 2119-2149.

Hottman, C. E., and R. K. Johnson, 1965, Estimation of formation pressures from log-derived shale properties: Journal of Petroleum Technology, June, p. 717-722.

Hubbert, M. K., 1953, Entrapment of petroleum under hydrodynamic conditions: AAPG Bulletin, v. 37, p. 1954-2026.

Leythaeuser, D., R. G. Schaefer, and A. Yukler, 1982, Role of diffusion in primary migration of hydrocarbons: AAPG Bulletin, v. 66, p. 408-429.

Makogon, Y. F., 1981, Hydrates of natural gas: Tulsa, Oklahoma, Penn-Well Books, 237 p.

Murris, R. J., 1980, Middle East: stratigraphic evolution and oil habitat: AAPG Bulletin, v. 64, p. 597-618.

Muskat, M., 1949, Physical principles of oil production: New York, McGraw-Hill Book Co., section 6.3, p. 241-250.

Nederlof, M. H., and H. P. Mohler, 1981, Quantitative investigation of trapping effect of unfaulted caprock (abs.): AAPG Bulletin, v. 65, p. 964.

Pennebaker, E. S., Jr., 1968, An engineering interpretation of seismic data: Society of Petroleum Engineers of American Institute of Mining and Metallurgical Engineers SPE Paper 2165, 12 p.

Purcell, W. R., 1949, Capillary pressures—their measurement using mercury and the calculation of permeability therefrom: Petroleum Transactions, American Institute of Mining Engineers, v. 186, p. 39-48.

Schmidt, V. K., and W. Almon, 1980, Development of diagenetic seals in carbonates and sandstones (abs.): AAPG Research Conference on Seals for Hydrocarbons, Keystone, Colorado, Sept. 14-17, unpublished book of abstracts, 16 p.

Schowalter, T. T., 1979, Mechanics of secondary hydrocarbon migration and entrapment: AAPG Bulletin, v. 63, p. 723-760.

———— and P. D. Hess, 1982, Interpretation of subsurface hydrocarbon shows: AAPG Bulletin, v. 66, p. 1302-1327.

Secor, D. T., Jr., 1965, Role of fluid pressure in jointing: American Journal of Science, v. 263, p. 633-646.

Smith, D. A., 1966, Theoretical considerations of sealing and non-sealing faults: AAPG Bulletin, v. 50, p. 363-374.

———— 1980, Sealing and nonsealing faults in Louisiana Gulf Coast salt basin: AAPG Bulletin, v. 64, p. 145-172.

Stoll, D., J. Ewing, and J. Bryan, 1971, Anomalous wave velocities in sediments with gas hydrates: Geophysical Research, v. 76, p. 10.

Stuart, C. A., 1970, Geopressures: Proceedings of the Second Symposium on Abnormal Subsurface Pressure, Louisiana State University, Baton Rouge, Louisiana, Supplement, 121 p.

Van West, F. P., 1972, Trapping mechanisms of Minnelusa oil accumulations, northeastern Powder River basin, Wyoming: Mountain Geologist, v. 9, p. 3-20.

Vinogradov, L. S., I. S. Aver'yanov, and I. S. Nigmati, 1981, Catagenetically sealed oil pools of the Volga-Ural region: Neftegazovaya Geologiya i Geofizika, no. 12, p. 15-17; 1983, English translation *in* Petroleum Geology, v. 19, no. 6, p. 266-268.

Weber, K. S., G. Mandl, W. F. Pilaar, F. Lehner, and R. G. Precious, 1978, Growth fault structures: 10th Offshore Technology Conference, Paper 3356, p. 2643-2653.

Wilhelm, O., 1945, Classification of petroleum reservoirs: AAPG Bulletin, v. 29, p. 1537-1580.

The American Association of Petroleum Geologists Bulletin
V. 63, No. 5 (May 1979), P. 723-760, 35 Figs., 3 Tables

Mechanics of Secondary Hydrocarbon Migration and Entrapment[1]

TIM T. SCHOWALTER[2]

Abstract The mechanics of secondary hydrocarbon migration and entrapment are well-understood physical processes that can be dealt with quantitatively in hydrocarbon exploration. The main driving force for secondary migration of hydrocarbons is buoyancy. If the densities of the hydrocarbon phase and the water phase are known, then the magnitude of the buoyant force can be determined for any hydrocarbon column in the subsurface. Hydrocarbon and water densities vary significantly. Subsurface oil densities range from 0.5 to 1.0 g/cc; subsurface water densities range from 1.0 to 1.2 g/cc. When a hydrodynamic condition exists in the subsurface, the buoyant force of any hydrocarbon column will be different from that in the hydrostatic case. This effect can be quantified if the potentiometric gradient and dip of the formation are known.

The main resistant force to secondary hydrocarbon migration is capillary pressure. The factors determining the magnitude of the resistant force are the radius of the pore throats of the rock, hydrocarbon-water interfacial tension, and wettability. For cylindrical pores, the resistant force can be quantified by the simple relation: $Pd = (2\gamma \cos \Theta)/R$, where Pd is the hydrocarbon-water displacement pressure or the resistant force, γ is interfacial tension, $\cos \Theta$ is the wettability term, and R is radius of the largest connected pore throats. Radius of the largest connected pore throats can be measured indirectly by mercury capillary techniques using cores or drill cuttings. Subsurface hydrocarbon-water interfacial tensions range from 5 to 35 dynes/cm for oil-water systems and from 70 to 30 dynes/cm for gas-water systems. Migrating hydrocarbon slugs are thought to encounter water-wet rocks. The contact angle of hydrocarbon and water against the solid rock surface as measured through the water phase, Θ, is thus assumed to be 0°, and the wettability term, $\cos \Theta$, is assumed to be 1.

A thorough understanding of these principles can aid both qualitatively and quantitatively in the exploration and development of petroleum reserves.

INTRODUCTION

Primary migration is here defined as the movement of hydrocarbons (oil and natural gas) from mature organic-rich source rocks to an escape point where the oil and gas collect as droplets or stringers of continuous-phase liquid hydrocarbon and secondary migration can occur. The escape point from the source rock can be any point where hydrocarbons can begin to migrate as continuous-phase fluid through water-saturated porosity. The escape point then could be anywhere the source rock is adjacent to a reservoir rock, an open fault plane, or an open fracture. Secondary migration is the movement of hydrocarbons as a single continuous-phase fluid through water-saturated rocks, faults, or fractures and the concen-

tration of the fluid in trapped accumulations of oil and gas. Numerous mechanisms for primary migration have been proposed. The main proposed mechanisms for secondary migration are buoyancy and hydrodynamics.

The mechanisms of primary hydrocarbon migration and the timing of hydrocarbon expulsion have been debated by petroleum geologists since the beginning of the science. Mechanisms proposed for primary hydrocarbon migration include: solution in water, diffusion through water, dispersed droplets, soap micelles, continuous-phase migration through the water-saturated pores, and others. Early workers generally favored early expulsion of hydrocarbons with the water phase of compacting sediments. Recent geochemical evidence, as summarized by Cordell (1972), suggests that oil is formed at depths where the petroleum source rocks have lost most of their pore fluids by compaction. On the basis of these conclusions, Dickey (1975) suggested a case for primary migration of oil as a continuous-phase globule through the pores of the source rock. This concept was documented in part by Roof and Rutherford (1958) who suggested that continuous-phase oil migration from source rock to reservoir is required to explain the chemistry of known

[1]Manuscript received, April 4, 1978; accepted, October 16, 1978. An earlier version of this paper appeared in *Wyoming Geological Association Earth Science Bulletin*, v. 9, no. 4, p. 1-43.

[2]Kirkwood Oil and Gas, Casper, Wyoming 82602.

This paper is based on work done at Shell Development Research in Houston during 1972-74. I thank Shell Development Co. for permission to publish this paper. Special thanks are extended to Bob Purcell, Higby Williams, Paul Hess, and Ben Swanson for their help in formulating and carrying out the project, and to my supervisors, Larry Meckel and Garland Spaight, with credit for some of the figures to R. E. Tenny and John Howell.

Article Identification Number
0149-1423/79/B004-0001$03.00/0

oil accumulations. Gas accumulations, however, can be explained by either continuous-phase primary migration or by discontinuous molecular-scale movement of gas dissolved in water (Roof and Rutherford, 1958). Price (1976) offered still another expulsion concept. He postulated molecular solution at high temperature, upward movement with compaction fluids, and exsolution at shallower depths in low-temperature saline waters.

Regardless of the correct answer or combination of answers to the question of time and mechanism of primary hydrocarbon migration, secondary migration through reservoir carrier beds is the necessary next step for the formation of a commercial oil or gas accumulation. A thorough understanding of the mechanics of secondary hydrocarbon migration and entrapment is useful in the exploration for oil and gas. Knowledge in this area of exploration can be critical in tracing hydrocarbon migration routes, interpreting hydrocarbon shows, predicting vertical and lateral seal capacity, exploiting discovered fields, and in the general understanding of the distribution of hydrocarbons in the subsurface. The importance of understanding the mechanics of secondary migration and entrapment, particularly in the exploration for subtle stratigraphic traps, is illustrated by McNeal (1961), Harms (1966), Smith (1966), Stone and Hoeger (1973), Berg (1975), and by numerous papers by the Petroleum Research Corp. These articles provide an excellent starting point for a sound understanding of the principles involved. However, none of these papers adequately discuss the range of variables involved in secondary migration and how to cope with them. Nor do they discuss fully the quantitative and qualitative exploration implication of these principles. A thorough review of these principles is presented here with input of new research where appropriate.

MECHANICS OF SECONDARY HYDROCARBON MIGRATION AND ENTRAPMENT

If an oil droplet were expelled from a source rock whose boundary was the seafloor, oil would rise through seawater as a continuous-phase droplet because oil is less dense than water and the two fluids are immiscible. The rate of rise would depend on the density difference (buoyancy) between the oil and the water phase. The main driving force then for the upward movement of oil through sea water is buoyancy. Buoyancy is also the main driving force for oil or gas migrating through water-saturated rocks in the subsurface. In the subsurface, where oil must migrate through the pores of rock, there exists a resistant force to the migration of hydrocarbons that was not present in the simple example. The factors that determine the magnitude of this resistant force are (1) the radius of the pore throats of the rock, and (2) the hydrocarbon-water interfacial tension, and (3) wettability. These factors, in combination, are generally called "capillary pressure." Capillary pressure has been defined as the pressure difference between the oil phase and the water phase across a curved oil-water interface (Leverett, 1941). Berg (1975) pointed out that capillary pressure between oil and water in rock pores is responsible for trapping oil and gas in the subsurface. A more thorough discussion of capillary pressure than is presented here is contained in Berg's paper.

To begin our discussion of the mechanics of secondary migration and entrapment and the variables involved, we look at an oil accumulation in a reservoir under static conditions.

Driving Forces in Secondary Migration

Under hydrostatic conditions, buoyancy is the main driving force for continuous-phase secondary hydrocarbon migration. When two immiscible fluids (hydrocarbon and water) occur in a rock, a buoyant force is created owing to the density difference between the hydrocarbon phase and the water phase. The greater the density difference, the greater the buoyant force for a given length hydrocarbon column (always measured vertically). For a static continuous hydrocarbon column, the buoyant force increases vertically upward through the column. Figure 1 illustrates the buoyant force for a stratigraphically trapped static oil column in a porous reservoir sandstone. As illustrated on the right of the figure, the reservoir sandstone is confined vertically by a caprock shale seal and seat seal, the oil is trapped laterally by siltstone, and an oil-water contact is present downdip in the homogeneous reservoir sandstone. On the left of the figure the pressure due to the weight of the column of oil (density 0.77 g/cc) and the pressure due to an equal column of water (density 1.00 g/cc) are plotted on the horizontal axis; the vertical axis is the height in feet above the free water level. The free water level is the level at which water would stand in a large open hole. In terms of buoyancy this can also be defined as the point of zero buoyant force. The 100% water level is the vertical position above which the reservoir rock has a water saturation less than 100%.

When the pressure of a static fluid is plotted against depth, each fluid will have a particular

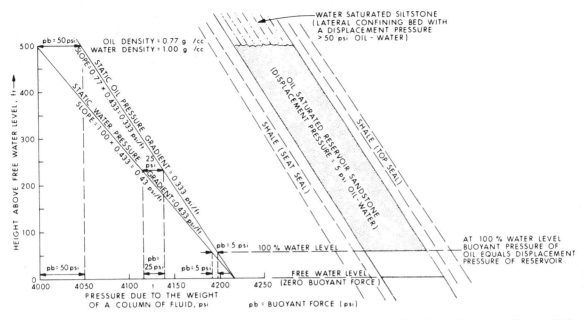

FIG. 1—Buoyant force in oil reservoir under static conditions (after Petroleum Research Corp., 1960; Smith, 1966).

slope depending on the density of the fluid. The slope or static fluid pressure gradient in psi/ft for any fluid can be calculated by multiplying the density in g/cc by 0.433.

For the example in Figure 1, the static fluid pressure gradient for the water phase would be 0.433 (4.33 × 0.1); the static fluid pressure gradient for the oil phase would be 0.333 (0.433 × 0.77).

The pressure decrease with height above the free water level or the static fluid pressure gradient (as plotted on the left side of Fig. 1) is greater in the denser water phase (0.433 psi/ft) than in the oil phase (0.333 psi/ft). The difference in pressure between the water phase and the oil phase at any point above the free water level is the buoyant force at that point. The buoyancy gradient, or the rate of buoyant pressure increase with height above the free water level, can be calculated by subtracting the oil pressure gradient (0.333 psi/ft) from the water pressure gradient (0.433 psi/ft). For the oil and water in Figure 1, the buoyancy gradient then is 0.1 psi/ft. With these conditions, a 100-ft oil column would produce a driving force of 10 psi at the top of the column and a 500-ft oil column would have a buoyant force of 50 psi as illustrated in Figure 1.

An analogy for the upward buoyant or driving force of a static oil column is the upward force generated by a wooden two-by-four vertically trapped in a tank of water. The longer the two-

by-four, the greater the buoyant force at the top of the board. In Figure 1, if the length of the vertical column of oil were increased, the buoyant force at the top of the oil column would be increased. Also, in our two-by-four example, the lower the density of the two-by-four, the greater the buoyant force for a given length of board. If the density of the oil were decreased or if the density of the water were increased for a given length hydrocarbon column, the buoyant force would be greater than the 50-psi illustration in Figure 1 for a 500-ft oil column.

Subsurface densities of hydrocarbon and water phases are important, then, in determining buoyant driving forces in secondary migration and entrapment of hydrocarbons. Subsurface water densities generally range from 1.0 to 1.2 g/cc, resulting in static water pressure gradients of 0.433 to 0.52 psi/ft. Subsurface oil densities vary from approximately 0.5 to 1.0 psi/ft, resulting in static oil pressure gradients of 0.22 to 0.43. Oil-water buoyancy gradients, using these densities, can range from zero to 0.3 psi/ft. Oil-water buoyancy gradients for the subsurface oil and water densities usually encountered are generally on the order of 0.1 psi/ft. However, the range of oil and water densities that are encountered in the subsurface suggests that there are vast differences in the ability of oil in different oil-water systems to migrate through a given reservoir rock or to be trapped by a given seal.

Gas densities range from as low as 0.00073 g/cc for methane at atmospheric pressure to approximately 0.5 g/cc for typical natural gas mixtures at high pressures (5,000 to 10,000 psi). Static pressure gradients for naturally occurring gas in the subsurface range from less than 0.001 to more than 0.22 psi/ft. The buoyancy gradient for gas-water systems in the subsurface can range from approximately 0.2 psi/ft to 0.5 psi/ft. The migration and entrapment of natural gas in a continuous phase in the subsurface then would vary greatly depending on the gas-water system in question. Gas-water systems generally have higher driving force than oil-water systems.

To quantify the buoyant force for a given hydrocarbon-water system the density of the water phase and of the hydrocarbon phase must be determined. To be useful in exploration these values must be obtainable from information generally available to the petroleum explorationist.

The three main variables affecting subsurface water density are: pressure, temperature, and the amount and kinds of dissolved solids. Figure 2 provides a means of estimating subsurface water densities considering the mentioned variables. In situations where the dominant negative ion is chloride, the chloride-ion concentration scale can be used. For waters that contain appreciable amounts of negative ions other than chloride, the upper scale for total dissolved solids should be used. The chlorinity or the total dissolved solids are generally available in exploration settings as are appropriate temperature and pressure information. Direct measurement of water densities can also be used but should be converted to subsurface temperature and pressure.

The density of oil in the subsurface is dependent on composition of the oil and dissolved gases, temperature, and pressure. Oil or condensate subsurface density can be estimated with workable accuracy if the stock tank API gravity and the solution gas-oil ratio in standard cubic feet/stock tank barrel (SCF/STB) are known (Fig. 3). Direct measurements of oil and its associated gas recombined at subsurface temperature and pressure are sometimes made by petroleum engineers. When these pressure-volume-temperature (PVT) values are available, they provide the most reliable estimates of subsurface oil densities.

The density of a gas in the subsurface is a func-

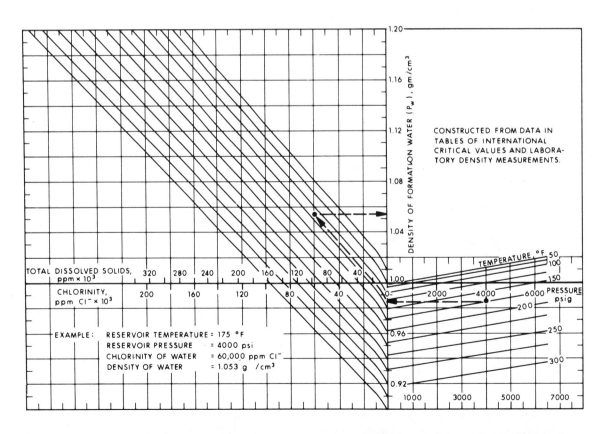

FIG. 2—Nomograph to determine density of formation water at subsurface conditions (after R. E. Tenny).

tion of the ratio of its mass to volume. The mass of a given amount of gas is related to the apparent molecular weight of the gas. The volume occupied by the gas is related to the pressure, temperature, and the apparent average molecular weight. The deviation in the behavior of a gas mixture from that postulated by the ideal gas law is related to the gas and subsurface conditions through a compressibility factor Z.

The equation used to determine the density of a gas in the subsurface is:

$$\rho g = 1.485 \times 10^{-3} \frac{mp}{ZT},$$

where ρg = subsurface density of gas (g/cc); m = apparent average molecular weight; p = absolute subsurface pressure (lb/sq in.); Z = compressibility factor; and T = absolute subsurface temperature (Rankine). '

If the apparent molecular weight (which can be estimated from gas composition), subsurface temperature, and pressure are known, the gas density can be estimated by using Figures 4 to 6. The following procedure can be used to determine ρg: (1) determine the apparent molecular weight of the gas mixture by calculating the percentage and molecular weight of each component in the gas mixture (e.g., the molecular weight for methane, CH_4, is 16, as carbon has a molecular weight of 12 and hydrogen a molecular weight of 1); (2) read the pseudo-reduced temperature and pressure from Figure 4; (3) determine a compressibility factor, Z, from Figure 5; (4) determine subsurface gas density by use of Figure 6. An example is shown on each figure using a gas with an apparent molecular weight of 23, a subsurface temperature of 200°F, and pressure of 2,600 psi.

Effects of Hydrodynamics on Driving Forces

The importance of hydrodynamics with regard to oil entrapment in structural traps has been discussed in detail by Hubbert (1953). Numerous other authors have since documented the effects of hydrodynamics on structural oil reservoirs throughout the world. In thinking of the effects of hydrodynamics on secondary migration and primarily stratigraphic-type entrapment of hydrocarbons, we must consider how a hydrodynamic condition would effect the buoyant driving force of a hydrocarbon filament in the subsurface. Hydrodynamic conditions in the subsurface change the buoyant force, and therefore the migration potential, for a hydrocarbon column of a given height. Buoyancy, as has been defined for a static

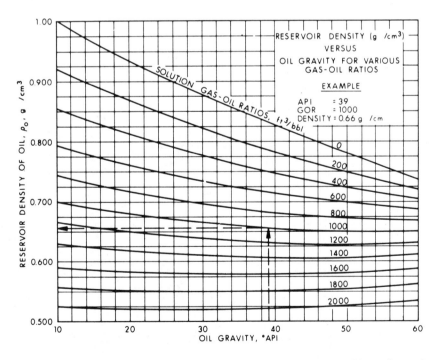

FIG. 3—Nomograph to determine subsurface oil density from API gravity and gas-oil ratio (after R. E. Tenny).

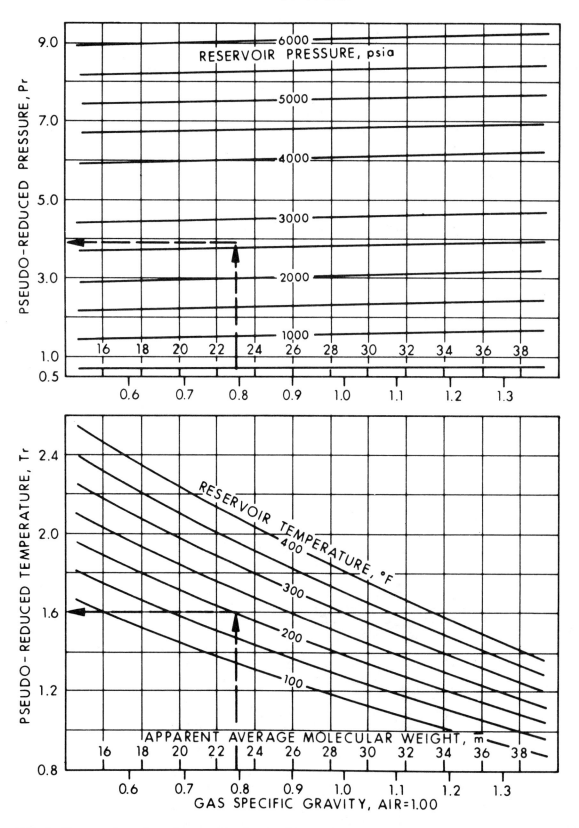

FIG. 4—Nomograph to determine pseudo-reduced pressure and temperature from apparent molecular weight, reservoir pressure, and temperature (after R. E. Tenny).

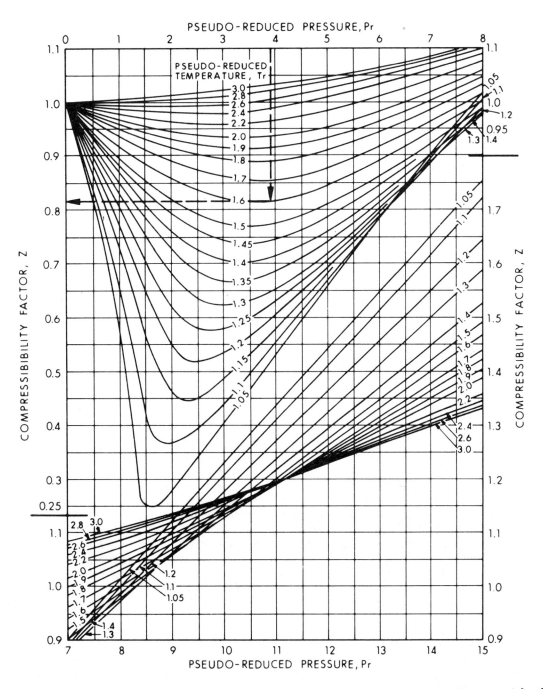

FIG. 5—Nomograph to determine compressibility factor, z, at pseudo-reduced temperature and pressure (after R. E. Tenny).

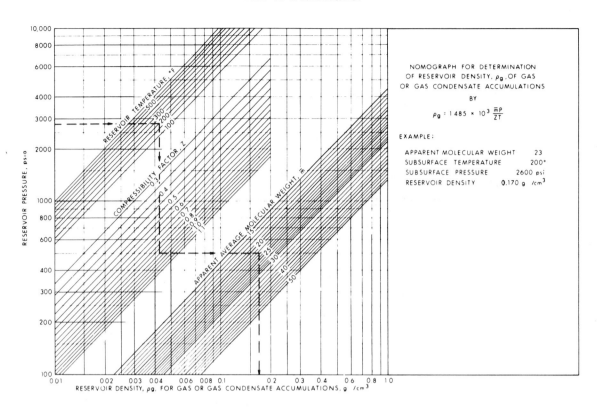

FIG. 6—Nomograph to determine reservoir density of gas condensate (after R. E. Tenny).

oil filament, is the pressure in the water phase minus the pressure in the oil phase at a given height above the free water level. When a hydrodynamic condition exists, the pressure in the water phase (and therefore the buoyant force) at any point will be different from that for hydrostatic conditions. Figure 7 (left side) illustrates the pressure difference in the water phase of an aquifer for an artesian gravity-type hydrodynamic condition for both updip and downdip flow. A hydrodynamic condition will also affect the water pressure–depth plot for a reservoir (Fig. 7, right side). Relative to hydrostatic conditions, downdip flow increases the slope of the pressure-depth plot; conversely, updip flow decreases the slope (Fig. 7, right side).

The pressure-depth plot (Fig. 1) was used to study the buoyant pressure for a given hydrocarbon column under hydrostatic conditions. In Figures 7 and 8, this same type of graph is used to show how the buoyant pressure of a given oil column will be different for hydrodynamic conditions. With upward water flow through a reservoir, the pressure difference between the water phase and the oil phase at the top of a given trapped oil column will be greater than the pressure difference for the same height oil column in the hydrostatic case (Fig. 8). When downward water flow occurs in a reservoir, the pressure difference between the water phase and the oil phase at the top of a given oil column is less than the hydrostatic case (Fig. 8) for the same height oil column.

From Figure 8 we can see that downdip flow reduces buoyancy or migration potential, and updip flow increases buoyancy or migration potential for any given oil filament in the subsurface. Transferring this observation to lateral seal capacity in the stratigraphic entrapment of hydrocarbons, downdip flow increases the seal capacity of a given lateral confining bed along a migration path by reducing the buoyant pressure of any hydrocarbon filament through a reservoir. Updip flow would effectively reduce lateral seal capacity in a given zone because the buoyant force for a given hydrocarbon filament would be increased from the hydrostatic. In the exploration for subtle stratigraphic traps, we can readily see the importance of hydrodynamics on the entrapment of hydrocarbons. The positive effect of a downdip hydrodynamic condition in increasing lateral seal capacity and trapping commercial volumes of hydrocarbons has been documented by several authors. This downdip flow or energy potential can be the result of either gravity-type (artesian or

Secondary Hydrocarbon Migration and Entrapment

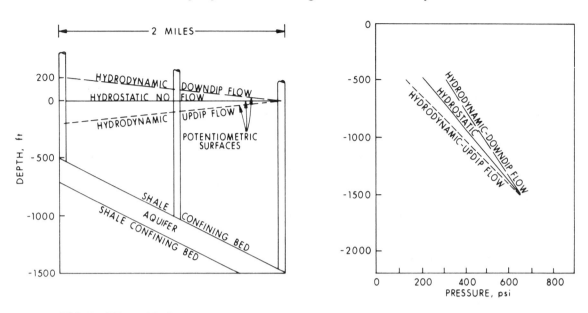

FIG. 7—Effect of hydrodynamics on pressure-depth plot of water phase in artesian condition.

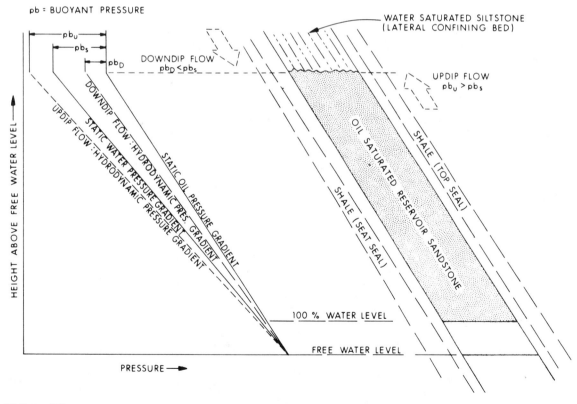

FIG. 8—Effect of hydrodynamics on buoyant force in oil reservoir for constant hydrocarbon column height (after Petroleum Research Corp.).

confined) hydrodynamic flow or geopressure-type (dewatering) hydrodynamic flow. Case histories of gravity-type (artesian) downdip flow affecting stratigraphic entrapment were discussed by Berry (1958). Hill et al (1961), McNeal (1961, 1965), and Stone and Hoeger (1973). The effects of geopressure (dewatering) hydrodynamic conditions were discussed by Meyers (1968) with examples from the Gulf Coast of the United States where hydrostatically pressured blocks are faulted down against geopressured fault blocks, creating fault traps.

It is clear that attempts to assess secondary hydrocarbon migration and entrapment in a given area must incorporate the effects of hydrodynamics. Berg (1975) derived a formula for determining the effect of updip or downdip flow on buoyancy and/or seal capacity. A nomograph (Fig. 9) has been prepared to provide a quick method of quantitatively assessing the effects of hydrodynamics on buoyancy or seal capacity. The data required for this estimate are the mapped potentiometric gradient (ft/mi) of the reservoir in question, the dip of the reservoir bed (ft/mi) and the density of the hydrocarbon phase. The density of the water phase is assumed to be 1.0 g/cc for simplification.

To read the nomograph, divide the mapped potentiometric gradient (ft/mi) by the dip of the reservoir (ft/mi). Enter the nomograph for that value and read across to the known hydrocarbon density, then down to the percent effect on trap capacity or buoyancy. For example, a structural dip of 500 ft/mi, potentiometric gradient of 50 ft/mi, and an oil density of 0.7 g/cc would have a 50% effect on buoyancy or lateral seal capacity. For these conditions, if the flow was in the downdip direction, the buoyant force of any oil filament would be reduced by 50%. The effect on lateral seal capacity for any facies change along the reservoir would conversely be increased by 50%. Updip flow would reduce seal capacity by 50%, as the buoyant force of any oil column would be increased by 50%.

Attempts to quantify hydrodynamic effects on stratigraphic entrapment by the use of this nomograph or Berg's formula must be made with caution. First, the construction of potentiometric

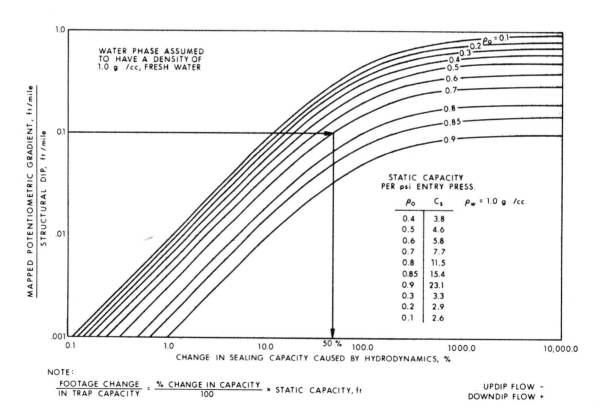

FIG. 9—Nomograph to estimate percent effect on seal capacity by hydrodynamics; assumes water density of 1.0 g/cc (after Higby Williams).

maps is not always accurate because of lack of usable pressure data, structural and stratigraphic complications, etc. Second, the approach in this paper and in Berg's avoids the effects of the flow of water around an existing oil accumulation due to low relative permeability to water within the oil-saturated reservoir and the change of the potentiometric slope across permeability facies changes within a reservoir. Another factor to consider in exploration applications is that the positive effect of increased lateral seal capacity in a particular rock unit will not trap a larger volume of oil than for the hydrostatic case unless secondary migration continues after initiation of the hydrodynamic condition. Also, the initiation of updip water flow will not necessarily cause the updip lateral seal of an existing stratigraphically trapped hydrocarbon column to leak if the size of the accumulation is already limited by spill around the flanks of the stratigraphic trap and therefore not at critical seal capacity.

Resistant Forces to Secondary Migration

In a previous example we discussed how a filament of oil released at the seafloor would rise through seawater because of the force of buoyancy. If the same filament of oil or gas is required to move through a water-saturated porous rock we have introduced a resistant force to hydrocarbon movement. For the hydrocarbon filament or globule to move through a rock, work is required to squeeze the hydrocarbon filament through the pores of the rock. In more technical terms, the surface area of the hydrocarbon filament must be increased to the point that it will pass through the previously water-saturated pore throats of the rock. The magnitude of this resistant force in any hydrocarbon-water-rock system then is determined by the radius of the pore throats of the rock; the hydrocarbon-water interfacial tension (surface energy); and wettability as expressed by the contact angle of hydrocarbon and water against the solid pore walls as measured through the water phase. This resistant force to migration is generally termed "capillary pressure."

For a simplified example, visualize a hydrocarbon filament trying to move upward through a water-saturated cylindrical pore (Fig. 10). The variables of the resistant force to hydrocarbon movement can be expressed by a simple equation (Purcell, 1949):

$$Pd = \frac{2\gamma \cos \Theta}{R} ,$$

where Pd = hydrocarbon-water displacement pressure (dynes/cm^2); γ = interfacial tension (dynes/cm); Θ = wettability, expressed by the

contact angle of hydrocarbon and water against the solid (degrees); and R = radius of largest connected pore throats (cm). The displacement pressure is that force required to displace water from the cylindrical pore and force the oil filament through the pore. This resistant force to migration is analogous to injection pressure as defined by Berg (1975, p. 941).

A change in any of the three variables in this formula will change the displacement pressure or resistant force to secondary migration (Fig. 10). The smaller the radius of the cylinder, the greater the displacement pressure. The greater the hydrocarbon-water interfacial tension, the greater the displacement pressure. The smaller the contact angle of hydrocarbon and water against the cylinder wall, the greater the displacement pressure.

For water-saturated porous rocks rather than cylindrical pores, Smith (1966) defined the displacement or breakthrough pressure as the minimum pressure required to establish a connected hydrocarbon filament through the largest interconnected water-saturated pore throats of the rock. When a continuous hydrocarbon filament has been established through the pores of the rock, secondary hydrocarbon migration can occur. If the displacement pressure for any hydrocarbon-water-rock system can be determined, the vertical hydrocarbon column necessary to migrate hydrocarbons through this rock can be calculated. The displacement pressure for any hydrocarbon-water-rock system then could be of importance in subsurface petroleum exploration, as the magnitude of this value would determine the sealing capacity for a caprock seal, the trap-

THE HYROCARBON-WATER DISPLACEMENT PRESSURE OF A ROCK IS A FUNCTION OF HYDROCARBON-WATER INTERFACIAL TENSION, WETTABILITY, & RADIUS OF THE PORE THROAT.

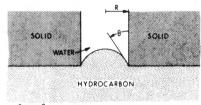

$$pd = \frac{2\gamma \cos \theta}{R}$$

WHERE pd = DISPLACEMENT PRESSURE
    γ = OIL-WATER INTERFACIAL TENSION
    θ = CONTACT ANGLE OF OIL AND WATER AGAINST THE SOLID
    R = RADIUS OF THE PORE THROAT

        AS γ INCREASES pd INCREASES
        AS θ DECREASES pd INCREASES
        AS R DECREASES pd INCREASES

FIG. 10—Resistant forces in secondary hydrocarbon migration (Purcell, 1949).

ping capacity for a lateral facies change or fault, or the minimum vertical hydrocarbon column needed to explain an oil show in a given rock.

In determining the displacement or breakthrough pressure for a given hydrocarbon-water-rock system in the subsurface, the hydrocarbon-water-interfacial tension, wettability, and radius of the largest connected pore throats must be measured or estimated. The range of these variables and methods of estimating subsurface values for these variables will be discussed.

Interfacial Tension

Interfacial tension can be defined as the work required to enlarge by unit area the interface between two immiscible fluids (e.g., oil and water). Interfacial tension is the result of the difference between the mutual attraction of like molecules within each fluid and the attraction of dissimilar molecules across the interface of the fluids.

Oil-water interfacial tension varies as a function of the chemical composition of the oil, amount and type of surface-active agents, types and quantities of gas in solution, pH of the water, temperature, and pressure. At atmospheric pressure and 70°F, interfacial tension of crude oils and associated formation water for 34 Texas oil reservoirs of different ages ranged from 13.6 to 34.3 dynes/cm, with a mean of 21 dynes/cm (Livingston, 1938). Oil-water interfacial tension generally tends to decrease with increasing API gravity and decreasing viscosity (Livingston, 1938).

With increasing temperature, oil-water interfacial tension generally decreases. For pure benzene-water and decane-water systems, interfacial tension decreases between 0.03 to 0.08 dynes/cm/°F (Michaels and Hauser, 1950) depending on the pressure. McCaffery (1972) found a decrease of interfacial tension of 0.03 dynes/cm/°F for a pure dodecane-water-system and 0.09 dynes/cm/°F for a pure octane and water system between 100 and 250°F. Natural crude oil and formation water interfacial tension decreases between 0.1 and 0.2 dynes/cm/°F according to Livingston (1938). Hocott (1938) documented a decrease in interfacial tension of approximately 0.1 to 0.15 dynes/cm/°F for natural crude oils between temperatures of 130 and 170°F. The preceding research documents the effect of increasing temperature on oil-water interfacial tension. The effect is complex, but the general trend is for oil-water interfacial tension to decrease as temperature increases. Extrapolation of the results for pure systems and crude oil and for formation water suggests that, for exploration purposes, an oil-water interfacial tension decrease of approximately 0.1 dynes/cm/°F appears to be a reasonable assumption.

The effect of increasing pressure on oil-water interfacial tension is also complex. For pure benzene and water, interfacial tension decreases approximately 0.3 dyne/cm per 100 psi pressure change; for decane and water interfacial tension increases with increasing pressure (Michaels and Hauser, 1950). Dodecane-water and octane-water interfacial tensions also increase slightly with increasing pressure (McCaffery, 1972). Crude oil-formation water interfacial tension tends to increase only 10 to 20% from atmospheric to saturation pressure and then to decrease slightly with increasing pressure (Hocott, 1938). Kusakov et al (1954) found, however, that at pressures above approximately 1,500 psi, continued increase in pressure had no effect on interfacial tension for crude-formation water systems. The data presented here suggest that for pure laboratory systems, increase in pressure can cause oil-water interfacial tension to increase or decrease. For crude oil-formation water systems, the effect of increasing pressure appears to increase interfacial tension slightly and then have little or no effect at pressures above 1,500 psi. In summary, then, the effect of pressure on crude oil-formation water interfacial tension appears small enough that it can be considered negligible.

In attempting to quantify oil-water-rock displacement pressure, a value for oil-water interfacial tension in the subsurface must be measured or estimated. Sophisticated laboratory equipment can measure oil-water interfacial tension at reservoir temperature and pressure. If this equipment is not available, interfacial tension can generally be measured at atmospheric conditions in most chemical laboratories. The results of atmospheric interfacial tension measurements must be extrapolated to subsurface temperature and pressure. If no laboratory data are available for the oil-water system in question, then an estimate must be made. Livingston's mean value for 34 Texas crude oils of 21 dynes/cm at 70°F is the best value for medium-density crude oils (30 to 40° API). A value of approximately 15 dynes/cm may be appropriate for higher gravity crude oils (greater than 40° API) with 30 dynes/cm being a reasonable approximation for low-gravity crude oils (less than 30° API). These estimates or measurements at atmospheric temperature (70°F) must be extrapolated to reservoir temperature. It is suggested that the oil-water interfacial tension value at 70°F be decreased 0.1 dynes/cm/°F temperature increase above 70°F. A nomograph (Fig. 11) has been prepared to estimate oil-water interfacial tension at reservoir temperature that assumes this linear decrease. Interfacial tension values at very high temperature and pressure are unknown and the nomograph lines do not extend below 5 dynes/cm. A recent paper, however, by Cartmill (1976) suggested that oil-water interfacial tension may continue to decrease at high temperature and pressure and eventually become zero. He postulated that this reduction of interfacial tension at high temperature and pressure may be a mechanism for primary migration of oil from source rocks to carrier beds and reservoirs.

From inspection of the displacement pressure equation (Fig. 10) a change in the oil-water interfacial tension will directly affect the displacement pressure for a given oil-water-rock system. From the data presented, subsurface oil-water interfacial tension can range from 5 to 35 dynes/cm. Therefore, the variation of oil-water interfacial tension could affect the displacement pressure of a given rock seven-fold. This effect is obviously very significant in attempting to quantify secondary migration. For example, the seal capacity of a lateral facies change or caprock seal could change by a factor of seven simply by changing the oil-water system present in the subsurface.

Gas-water interfacial tension—Methane gas–formation water interfacial tension at atmospheric temperature and pressure is approximately 70 dynes/cm. Gas-water interfacial tension varies with the amount of surface-active agents in the water, the amount of heavy hydrocarbons in solution in the gas, temperature, and pressure. Gas-water interfacial tension decreases 5 to 10 dynes/cm/1,000-psi pressure increase depending on the temperature (Hocutt, 1938; Hough et al, 1951). Gas-water interfacial tension decreases with increasing temperature from 0.1 to 1.0 dynes/cm/°F depending on the pressure (Hough et al, 1951). The effects of temperature and pressure on methane-water systems (from Hough et al, 1951) have been combined in a nomograph (Fig. 12) to estimate methane-water interfacial tension at any given subsurface temperature and pressure. Estimates from this chart should be sufficiently accurate for exploration application of gas-water interfacial tension to gas-water-rock

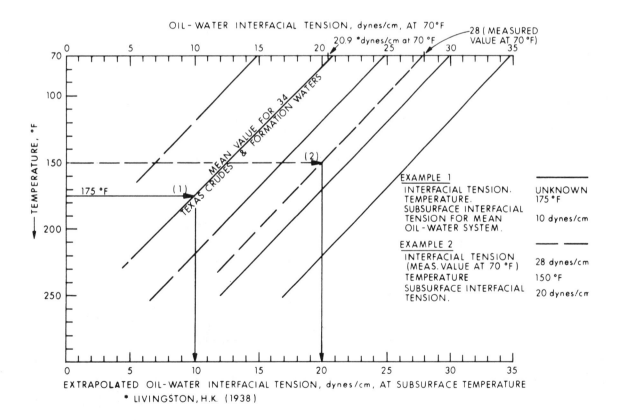

FIG. 11—Nomograph to estimate oil-water interfacial tension at reservoir temperature. Nomograph assumes decrease of 0.1 dynes/cm/°F temperature increase.

displacement pressures. Excessive amounts of ethane, propane, and other heavy gases in the gas phase will decrease interfacial tension from that of the pure methane-water systems as shown in the nomograph.

From Figure 12 it can be seen that methane-water interfacial tensions start as high as 70 dynes/cm at 75° and decrease to approximately 30 dynes/cm at high reservoir temperature and pressure. In contrast, the mean oil-water interfacial tension for 34 Texas crude oils and formation waters was 21 dynes/cm at 70°F (Livingston, 1938). As previously documented, oil-water interfacial tension tends to decrease with increasing subsurface temperature, reducing subsurface oil-water interfacial tension to roughly 10 to 20 dynes/cm. Gas-water interfacial tensions then are generally higher than oil-water interfacial tensions for both surface and subsurface conditions. A gas-water displacement pressure would then be greater than oil-water displacement pressure for the same rock. The high gas-water interfacial tension as compared to oil-water interfacial tension significantly reduces the migration potential of

gas through water-saturated rocks in the subsurface. The potential magnitude of this effect is discussed later with appropriate examples.

Wettability

Wettability can be defined as the work necessary to separate a wetting fluid from a solid. In the subsurface we would generally consider water the wetting fluid and the solid would be grains of quartz in a sandstone, calcite in a limestone, etc. The adhesive force or attraction of the wetting fluid to the solid in any oil-water-rock system is the result of the combined interfacial energy of the oil-water, oil-rock, and water-rock surfaces. Wettability is generally expressed mathematically by the contact angle of the oil-water interface against the rock or pore wall as measured through the water phase. For rock-fluid systems with contact angles between 0 and 90°, the rocks are generally considered water-wet; for contact angles greater than 90°, the rocks are considered oil-wet. Water-wet rocks would imbibe water preferentially to oil. Oil-wet rocks or oil-wet surfaces would imbibe oil preferentially to water. Al-

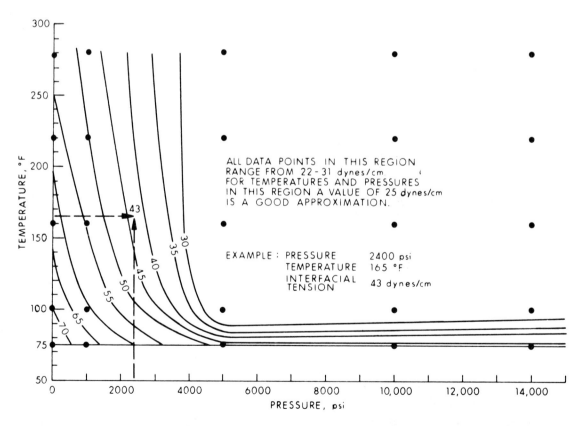

FIG. 12—Nomograph to estimate methane-water interfacial tension at different temperatures and pressures (black-circle experimental data points and extrapolated curves from Hough et al, 1951).

though a contact angle of 90° has generally been considered the breakover point to an oil-wet surface, Morrow et al (1973) stated that a contact angle of greater than 140° in dolomite laboratory packs was necessary for oil to be imbibed.

Water-laid sedimentary rocks are generally considered to be preferentially water-wet owing to the strong attraction of water to rock surfaces and the initial exposure of pore surfaces to water rather than hydrocarbons during sedimentation and early diagenesis. Water is thought by many workers to be a perfect wetting fluid and a thin film of water would coat all grain surfaces. If this is the situation, the contact angle for oil-water-rock systems would be zero. The wettability term in the displacement pressure equation would then be unity, as the cosine of zero is one. If water is not a perfect wetting fluid and the oil-water contact angle is greater than zero, the displacement pressure should theoretically decrease for that oil-water rock system. L. J. M. Smits (1971, personal commun.) has done experimental work on identical size bead packs which suggests that displacement pressures are only slightly affected by changing the oil-water-solid contact angle from 0 to 85°. Similar results were obtained by Morrow et al (1973) on displacement pressure tests in dolomite packs with contact angles ranging from 0 to 140°. These data and the general assumption that most rocks are preferentially water-wet suggest that the wettability term in the displacement pressure equation can be considered unity.

If the rocks are partially oil-wet, then the wettability term can be significant in reducing displacement pressure from that for the water-wet case. In the subsurface, rocks are seldom completely oil-wet but are fractionally oil-wet, that is, some of the grain surfaces are oil-wet and some are water-wet. According to Salathiel (1972), this would most likely occur in reservoir rocks where oil has been trapped and the grain surfaces in the larger pores would be exposed to the surface-active molecules in the oil phase and form an oil film or coating on the grain, making it preferentially oil-wet. The pore surfaces at the smaller pores or in the corners of the larger pores that are not saturated with oil would remain water-wet. Fatt and Klikoff (1959) have determined that when a rock is partially oil-wet there is a reduction in the oil-water displacement pressure for that oil-water-rock system. They suggested that the degree of fractional wettability needed to significantly reduce displacement pressure from that for the water-wet case is greater than 25% oil-wet grain surfaces.

Salathiel (1972) suggested that surface films of oil can produce fractional wettability in oil reser-

voirs. This has been further documented by Treiber et al (1972) who suggested that in most of the Amoco reservoirs studied, oil wets the rock more strongly than water. Other rocks that might develop grain surfaces that are partially oil-wet are rocks with large quantities of organic material such as source rocks which could adsorb oil surface-active agents. Rocks rich in iron minerals could also be partially oil-wet, as iron can preferentially adsorb surface-active material from crude oils. However, most sedimentary rocks would not contain enough iron minerals to have a significant effect on the overall wettability of the rock.

In summary, oil reservoirs and rocks rich in organic matter such as source rock would be the main exception to the water-wet case in the subsurface. The exploration application of hydrocarbon-water-rock displacement pressure values is generally directed at seal potential of various caprocks, the lateral seal capacity at facies changes in stratigraphic traps, and the migration potential of hydrocarbons through reservoir carrier beds. The likelihood of oil-wet rocks being present in these situations is considered remote. Therefore, it is generally recommended that the wettability term in the displacement pressure equation be considered unity in the quantitative application of displacement pressure values.

Radius of Pore Throats

The third critical factor in estimating the displacement pressure of a given water-rock-system is the radius of the largest connected pore throats in the rock. By inspection of the displacement pressure equation, the smaller the radius of the connected pore throats in a rock the greater the displacement pressure. The displacement pressure for a reservoir-quality sandstone would be significantly less than that of a fine-grained shale. Specific measurements of pore-size distribution are necessary to quantify secondary migration and entrapment.

Methods for estimating the radius of largest connected pore throats are numerous and varied. Pore-throat size and distribution can be measured visually in thin sections (Aschenbrenner and Achauer, 1960) or from scanning electron microscope photos. Pore geometry and pore-size distribution can also be measured by studying pore casts of leached carbonate rocks (Wardlaw, 1976). These direct measurement procedures have problems in that they generally only measure one plane of the rock and not the three-dimensional relations of one pore to another. Another problem with these methods is that they cannot be used effectively on nonreservoir rocks, which have pore throats too small to measure visually. These rocks are often of interest in hydrocarbon exploration, as they control hydrocarbon trapping. Other methods must be used for these fine-pored rocks.

Berg (1975) provided an empirical, mathematical formula for estimating pore throats for sandstones. Estimates from Berg's formula require that the porosity, permeability, and ideally the grain-size distribution of the sandstone be known. Porosity and permeability data are often available from core analyses, and therefore this approach may be useful in many instances. Berg discussed several examples where he used this method to estimate pore-throat size. However, Berg's method gives only a crude approximation of dominant pore-throat sizes for natural sandstones.

Visual or empirical estimates of pore-throat size as discussed in the preceding section are difficult to make and probably of limited value. A better approach would be to measure the displacement pressure directly. This can be done in the laboratory by injecting a nonwetting fluid into a rock under progressively increasing pressure and measuring the pressure at which a connected filament of nonwetting fluid extends across the sample. This technique would be analogous to the secondary migration of hydrocarbons through a water-saturated rock. Tests of this type are called capillary-pressure tests. Petroleum laboratories have run capillary-pressure tests for years on reservoir core samples. If the injection of the nonwetting fluid is continued incrementally beyond the pressure needed to establish a connected filament of nonwetting fluid across the sample, then the entire capillary properties or pore-size distribution of the rock can be determined.

Laboratory capillary pressure tests on rock samples can use almost any kind of fluid for the wetting and nonwetting phases. Oil or gas can be used for the nonwetting fluid and water for the wetting fluid. Although tests with these fluids would obviously be the best for petroleum exploration applications, they are difficult and time consuming. Purcell (1949) developed and demonstrated the validity and expediency of measuring rock capillary properties by mercury injection. Mercury capillary tests are now standard procedure for most private and commercial laboratories. Results from these tests can provide valuable exploration and production exploitation data. Results and application of capillary pressure test data have been reported in the literature by numerous authors (Stout, 1964; Harms, 1966; Smith, 1966; Roehl, 1967).

Mercury Capillary Pressure Tests

A brief discussion of mercury capillary pressure tests is warranted before proceeding. A perm-plug-type core sample or large sample cuttings are placed in a calibrated pressure chamber. Irregular shaped samples can be used in a mercury test because the volume of the sample is accurately measured during the test. Mercury (nonwetting phase) is introduced into the cell and completely surrounds the sample. Mercury then is forced into the sample by incrementally increasing the pressure on the mercury. The cumulative volume of mercury injected at each pressure is a measure of the nonwetting-phase saturation. This procedure is continued until the injection pressure reaches some predetermined value (usually 1,500 psi for normal laboratory equipment). The curve in Figure 13, a plot of mercury pressure versus volume of mercury injected, expressed as percent pore volume occupied, is the result of this process.

Mercury capillary pressure curves such as those in Figure 13 can be used to estimate displacement pressures, irreducible water saturations, the thickness of the hydrocarbon-water transition zone, and permeability. The irreducible water saturation is that percent of the pore space that hydrocarbons cannot penetrate and is often called ineffective porosity. This porosity is an important property of reservoir-rock petrophysics.

In a static oil reservoir as illustrated in Figure 1, the oil saturation as a percent of oil space will increase upward through the oil column as the forces of buoyancy overcome the forces of capillary pressure. As oil saturation increases, the ability of oil to flow to the well bore increases to the point where water-free oil production occurs. The interval from water production at the base of the oil-saturated reservoir to water-free oil production higher in the reservoir is termed the "oil-water transition zone." The thickness of the oil-water transition zone will depend on the capillary properties of the rock and the fluid properties of the system. This relation has been illustrated in Figure 14 (after Arps, 1964). Arps also discussed application of these principles in evaluating tilted oil-water contacts and the problem of minimum structural or stratigraphic closure required for water-free production in a petroleum reservoir.

Displacement pressure, which is critical in estimating hydrocarbon seal capacity, has been previously defined as that pressure required to form a continuous filament of nonwetting fluid through the largest connected pore throats of the rock. Purcell (1949) and Thomas et al (1967) have discussed the use of mercury capillary pressure

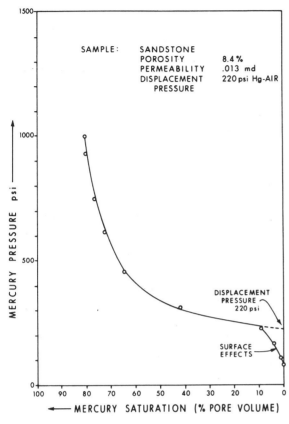

FIG. 13—Typical mercury capillary-pressure curve.

curves in estimating rock permeability.

The significance of all these capillary rock properties has been discussed in detail by Aufricht and Koepf (1957), Arps (1964), Stout (1964), and numerous other authors. The conversion of mercury pressure information to hydrocarbon-water pressure is discussed in detail later in the paper.

The validity of mercury tests to estimate various rock parameters, displacement pressure, irreducible water saturation, hydrocarbon-transition zones, and permeability (Purcell, 1949) is a function of the scale of heterogeneity of the rock. If most of the pores of the rock are small in comparison to the size of the test sample, then the results should be quite good. If, for example, the rock in question is known to be a vuggy carbonate rock or a fractured sandstone, where the very important larger pores of the rock cannot be adequately sampled, then the validity of the results should be poor. Therefore, all geologic knowledge available to a particular problem should be applied in choosing samples for mercury tests and in applying the results.

Displacement Pressures

Displacement pressure is one of the principal subjects of this paper, as it is the pressure which will determine the minimum buoyant pressure

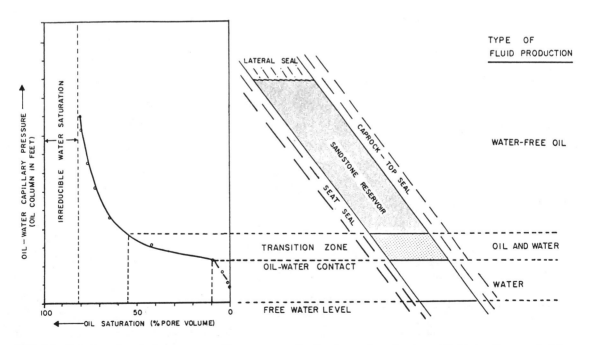

FIG. 14—Relation of typical mercury capillary curve to distribution and production of fluids in oil reservoir (after Arps, 1964).

needed for secondary migration. A reasonably accurate estimate of displacement pressure for various rock samples is then critical to quantifying of secondary hydrocarbon migration principles for exploration purposes.

For migration to occur a continuous hydrocarbon filament must extend through the interconnected pores of a water-saturated rock. In estimating displacement pressures from capillary pressure curves, it has been assumed that a continuous nonwetting filament would occur somewhere on the capillary plateau. This approach seems quite adequate where the capillary plateau is nearly flat as illustrated in Figure 15. Note that the pressure difference between saturations of 10 and 50% is quite small and, regardless of the minimum nonwetting saturation, the chance for error in estimating displacement pressure is minimal. However, for rocks with steep capillary plateaus or no plateau as illustrated in Figure 16, the displacement pressure cannot be accurately estimated without knowing the critical nonwetting-phase saturation needed to form a continuous nonwetting filament through the rock. This saturation is analogous to the critical gas saturations required

for gas breakthrough in depletion-type reservoirs containing a spreading oil. Critical saturations needed for migration have been reported by Rudd and Pandey (1973) to be generally less than 10% for shales and carbonate rocks. Additional direct measurements of critical saturation were needed to determine how accurately displacement pressures could be estimated for various rock types from readily available standard mercury capillary pressure curves.

Laboratory Tests of Displacement Pressure

Direct measurements of displacement pressure and critical saturation at breakthrough were conducted with two sets of equipment. A nitrogen-water system was used where nitrogen is displaced through water-filled rock samples under a confining pressure. The nitrogen pressure is increased in increments against one end of the rock sample, and the amount of effluent water at the other end is monitored. A constant and higher flow of effluent occurs at that nitrogen pressure when a nitrogen filament is continuous across the length of the sample. A high pressure (5,000 psi) mercury apparatus was also used where the form-

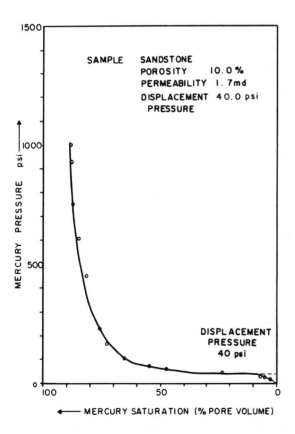

FIG. 15—Capillary-pressure curve with flat plateau.

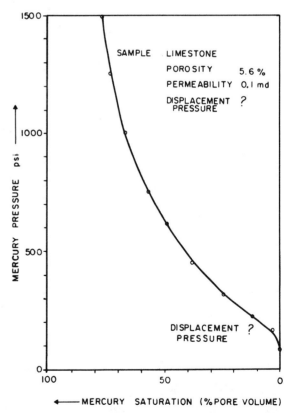

FIG. 16—Capillary-pressure curve with no plateau.

ation of a continuous thread of mercury across the length of the sample is detected by electrical conduction. The mercury system was superior to the nitrogen system because it was significantly faster and because breakthrough was distinct and instantaneously determined.

Test Results

Sandstones, shales, and chalks were used in the breakthrough studies. Four samples were tested with the nitrogen-water system. After completing these nitrogen-water tests, the samples were cleaned and standard mercury capillary pressure tests were run on the same samples. The measured nitrogen displacement or breakthrough pressures were converted to mercury capillary pressure values by using a conversion factor of 5X (Purcell, 1949) to compare the results to other samples tested with mercury equipment. Five samples were tested with the high-pressure mercury cell. Results from both techniques are reported in Table 1.

The nonwetting phase saturation needed to establish a connected filament across the length of the samples ranged from 4.5 to 17% of the rock pore volume. The average saturation for all the samples tested was 10%. The capillary curve for each sample and the percent saturation at breakthrough is illustrated in Figure 17 by the large "X" on the capillary curve.

From inspection of the capillary pressure curves in Figure 17, it is obvious that a wide spectrum of pore-size distribution was tested in the nine samples. The critical saturation for this varied rock sampling, however, has a relatively restricted range, 4.5 to 17%. From this sampling then, it would appear that migration can occur in most rock types at a nonwetting phase saturation of approximately 10% of the rock pore volume. These data suggest that displacement pressures could be estimated from standard mercury capillary pressure curves by determining the mercury pressure on the capillary curve at 10% mercury saturation. Sophisticated equipment as used in these experiments would not be necessary to get workable values for displacement pressure for any given rock.

The determination from this study that secondary hydrocarbon migration can occur at hydrocarbon saturations of around 10% can be applied in exploration. Saturations as low as 10% may be difficult to detect as a subsurface show in normal drilling operations. However, hydrocarbon shows with only 10% saturation may provide important exploration information in identifying hydrocarbon-transition zones in trapped accumulations and in defining hydrocarbon migration paths. Another interesting aspect of these data can be applied to bright-spot geophysics. Flowers (1976) demonstrated that a small percentage of free gas in a reservoir, too small to affect the resistivity measurements on bore hole logs, should produce a strong velocity change and hence a bright-spot amplitude anomaly. From the data presented here we can infer that the minimum saturation need for migration is approximately 10%. Gas saturation values of 5 to 10% are enough to cause bright-spot anomalies. These gas accumulations, then, probably represent locally generated gas bubbles that have not formed the connected gas

Table 1. Results of Capillary-Breakthrough Experiments

Sample [1]	Lithology	ϕ (%)	K_{air} (md)	K_{water} (md)	Nitrogen-water P_d (psi)	Equivalent Hg P_d (psi)	Measured Hg P_d (psi)	Estimated[2] Hg P_d (psi)	Pore Volume Saturation (%)
Outcrop Pecos Sandstone	Sandstone	18.9	0.12	0.06	40	200	--	245	8.9
Sandstone	Rerun of same sample	18.9	0.12	0.06	39	195	--	245	4.5
10079 h	Chalk	22.1	22.6	1.0	18	90	--	100	8.0
10079 v	Chalk	20.	30.9	0.35	22	110	--	110	11.6
3216.5 h	Sandstone calcite cement	1.0	--	--	--	--	2700	2500	16.0
8150 h	Sandstone	9.14	0.5	--	--	--	220	220	7.5
8150 v	Sandstone	9.05	0.5	--	--	--	230	220	13.6
Altamont	Sandstone	4.96	<1.0	--	--	--	550	500	17.0
11538 h	Argillaceous limestone	0.60	<0.01	--	--	--	>4500	>4500	--
14493 h	Silty shale	2.25	<0.01	--	--	--	2600	3200	6.6

P_d = displacement pressure
1. h or v after sample number refers to horizontal or vertical plug.
2. capillary pressure at mercury saturation of 10%.

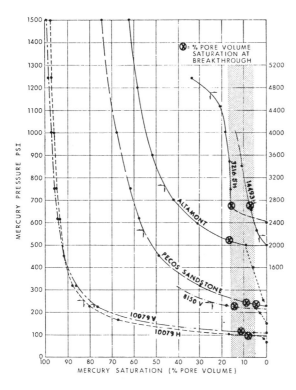

FIG. 17—Capillary-pressure curves and breakthrough saturations of samples tested.

filament needed to migrate and form a commercial deposit. Accumulation of this type could occur in off-structure positions as pointed out by Flowers (1976).

Capillary Properties of Drill Cuttings

In practice, exploration application of data from mercury capillary pressure tests has been considered limited to situations where regular-shaped core samples were available. However, Purcell (1949) in his original paper stated that capillary properties of irregular-shaped rock chips and drill-cutting-size samples can be measured accurately and without difficulty by mercury capillary pressure equipment. He measured the capillary properties of two reservoir sandstones that were broken into drill-cutting-size chips. Permeability estimated from capillary pressure data by a devised formula shows good agreement between data derived from rock chips and those derived from core samples.

Additional tests have been completed on four rock samples of different lithologies to evaluate further the reliability of mercury capillary properties derived from drill-cutting-size samples. For these tests three sandstones and one chalk were used. Four adjacent perm plugs were cut from each sample and numbered one through four. The four perm plugs from each sample were then measured for porosity and permeability to determine the heterogeneity in adjacent samples (Table 2). The numbered plugs for each sample were broken into various size rock chips (Fig. 18). Standard mercury capillary pressure tests were made on each plug or group of rock chips. The capillary pressure curves derived from each sample are shown in Figures 19 through 22.

These curves suggest that there is generally good agreement between data derived from full size core chips and those derived from rock chips of various sizes. Detailed examination, however, suggests that the capillary plateau appears to decrease slightly with the decreasing size of the rock chips. The irreducible water saturation seems to increase with a decrease in the size of the rock chips. The capillary pressure at 10% saturation is listed in Table 2 for comparison of displacement-pressure estimates from various size rock chips.

These data suggest that capillary properties of irregular-shaped rock chips as small as drill-cutting-size samples can be measured with workable accuracy with standard mercury capillary pressure equipment. The smaller the sample, however, the more likely the capillary plateau and the displacement pressure estimated at 10% saturation are to be less than that measured from a full-size perm plug. The rock types in which these techniques would be applicable would be those rocks that have a scale of heterogeneity smaller than the rock chips used.

In conclusion, capillary properties provide useful information in exploration or production studies, and usable data can be obtained from full-diameter cores, side-wall cores, or drill cuttings.

Conversion of Mercury Data to Hydrocarbon-Water Data

Quantitative application of mercury capillary pressure data to subsurface conditions requires the conversion of mercury capillary pressure values to subsurface hydrocarbon-water capillary pressure values. This conversion factor can be accomplished by using the following equation (Purcell, 1949):

$$(Pc)hw = \frac{\gamma hw \cdot \cos \Theta hw}{\gamma ma \cdot \cos \Theta ma} \cdot (Pc)ma,$$

where Pchw = capillary pressure for hydrocarbon water system, γhw = interfacial tension of hydrocarbon and water in dyne/cm, Θhw = contact angle of hydrocarbon and water (wettability), γma = interfacial tension of mercury plus air (surface tension energy), and Θma = contact angle of mercury and air against the rock.

Secondary Hydrocarbon Migration and Entrapment

The variability of subsurface hydrocarbon-water interfacial tension and methods of estimating these values have been discussed in the previous sections. As previously discussed, the contact angle of hydrocarbon-water systems is generally considered to be zero and the cos Θhw becomes unity. The interfacial tension of mercury and air is 480 dynes/cm at laboratory conditions. The contact angle between mercury and a solid is 40°, making the cos Θm equal to 0.776.

Subsurface values for hydrocarbon-water capillary pressure can be calculated by estimating the subsurface hydrocarbon-water interfacial tension and plugging it into the equation. A simple graphic solution to determine the conversion factor, from mercury to oil water or gas water, is provided in Figure 23. Once a conversion value has been estimated from Figure 23, this value is then multiplied by the mercury capillary pressure value in question. For example: (1) subsurface oil-water interfacial tension 21 dynes/cm; (2) mercury air to hydrocarbon-water conversion factor 0.055 (Fig. 23); (3) mercury displacement pressure 200 psi; (4) oil-water displacement pressure = 200 × 0.055 = 11 psi.

Calculations of Hydrocarbon Column Heights

It has been suggested that in quantifying secondary hydrocarbon migration and entrapment the calculation of vertical hydrocarbon volume a given rock pore system can seal or trap would be important in the exploration process. This can be accomplished by using the equation of Smith (1966):

$$H = \frac{PdB - PdR}{(pw - ph) \times 0.433},$$

where H = maximum vertical hydrocarbon column in feet above the 100% water level (oil-water contact) that can be sealed; PdB = subsurface hydrocarbon-water displacement pressure (psi) of the boundary bed; PdR = subsurface hydrocarbon-water displacement pressure (psi) of the reservoir rock; pw = subsurface density (g/cc) of water; ph = subsurface density (g/cc) of hydrocarbon; 0.433 = a unit's conversion factor.

The variables in this formula and methods used in determining the appropriate values have been discussed in previous sections. The only variable not directly plugged into this formula is hydrodynamics. A simple nomograph (Fig. 9) to estimate the percent effect on seal capacity can be used to quantify the effects of hydrodynamics. The percent effect on seal capacity from the nomograph can be multiplied by the results of the equation and added or subtracted to the original value depending on whether the hydrodynamic flow is up-dip or downdip.

One variable in this formula may be modified depending on the desired results. The value of PdR or displacement pressure if used in the formula will give the vertical hydrocarbon column to the 100% water level (Fig. 1). The explorationist may wish to know the vertical height to the point of water-free oil production rather than the 100% water level. This can be done by determining the subsurface hydrocarbon-water capillary pressure for water-free oil production for the reservoir

Table 2. Results of Rock Chip Capillary-Pressure Tests

	Sample	Lithology	ø%	K md	Hg P_d* (psi)
6405	-1	Interbedded sand and shale	7.0	0.64	1350
	-2		5.4	0.36	1350
	-3		6.8	0.56	1150
	-4		6.2	0.22	800
20454	-1	Silica-cemented sandstone	5.4	0.12	240
	-2		5.4	0.12	220
	-3		5.6	0.16	190
	-4		5.6	0.11	170
9587	-1	Chalk	18.5	2.32	170
	-2		21.4	2.00	130
	-3		17.3	.46	160
	-4		24.5	5.66	80
Pecos Sandstone	-1	Sandstone	17.9	0.32	250
	-2		18.2	0.42	240
	-3		17.9	0.31	240
	-4		18.1	0.53	190

* At 10% Hg saturation.

SAMPLE 6405

SAMPLE 20454

SAMPLE 9587

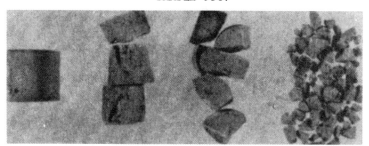

PECOS SANDSTONE

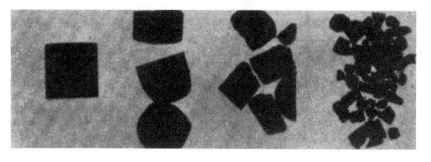

PLUG NUMBER   1   2   3   4

SCALE
⊢―――⊣
1 inch

FIG. 18—Photographs of samples tested to determine capillary properties of drill cuttings.

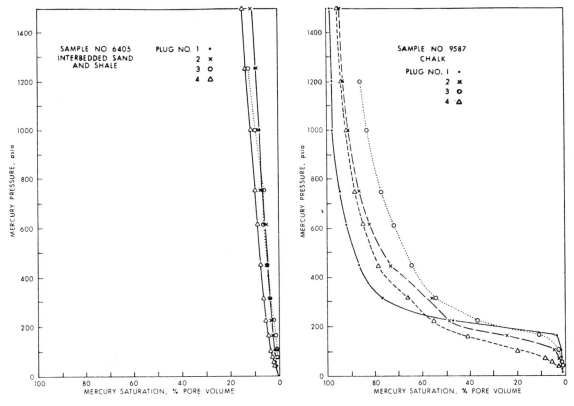

FIG. 19—Capillary-pressure curves for interbedded sand and shale, sample 6405 (Fig. 18).

FIG. 21—Capillary-pressure curves for chalk, sample 9587 (Fig. 18).

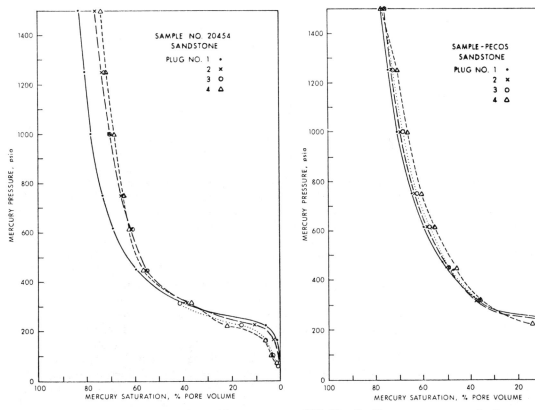

FIG. 20—Capillary-pressure curves for sandstone, sample 20454 (Fig. 18).

FIG. 22—Capillary-pressure curves for Pecos sandstone (Fig. 18).

rock in question (Arps, 1964; Fig. 14).

A complete sample calculation of the potential hydrocarbon seal capacity of a given rock is helpful in illustrating this process.

Sample Calculation

The following properties are given:

Mercury-air displacement pressure	220 psi
Oil-water interfacial tension in subsurface	unknown
Subsurface depth	8,000 ft
Subsurface pressure	4,000 psi
Subsurface temperature	175°F
Estimate of subsurface oil-water interfacial tension (Fig. 11)	10 dyne/cm
Mercury-air to oil-water conversion factor (Fig. 23)	0.025
Oil-water displacement pressure	5.5 psi
Water composition	60,000 ppm Cl⁻
Subsurface water density (Fig. 2)	1.05 g/cc
Oil characteristics	39° API GOR 1,000:1
Subsurface oil density (Fig. 3)	0.68 g/cc
Oil-water displacement pressure of reservoir rock	1.0 psi
Downdip hydrodynamic flow	
Potentiometric gradient	50 ft/mi
Dip	500 ft/mi
Percent effect on lateral trap capacity (Fig. 9)	50%

The calculation then, is

$$H(ft) = \frac{PdB - PdR}{(pw-ph) \times 0.443} = \frac{5.5 - 1}{(1.05-0.68)\,0.433} = 28 \text{ ft.}$$

The hydrodynamic effect = 28 ft × 0.50 = + 14 ft.

The total seal capacity = 28 ft + 14 ft = 42-ft oil column.

The previous sections of this paper discuss the variables involved in secondary migration and entrapment and how estimates of values for these variables can be made with information generally available in petroleum exploration. To show the importance of these values in the calculation of seal capacity, maximum and minimum values for each critical variable were substituted in the sample calculation with all other values held constant (Table 3). This table shows that a 220-psi mercury displacement pressure rock could seal from a minimum of 12.5 ft to a maximum of 124 ft of oil column depending on the value of the variables used. For gas-water systems the same 220-psi rock could seal a gas column of from 31 to 95 ft.

Table 3 illustrates the importance of the critical nongeologic parameters in quantifying secondary hydrocarbon migration and entrapment. The table also shows that a given rock can seal a larger

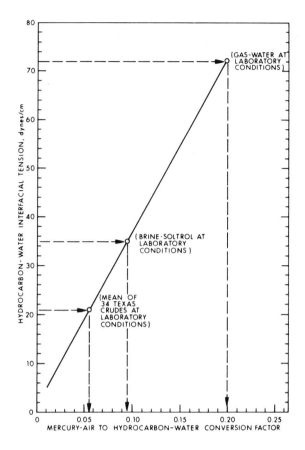

FIG. 23—Nomograph to determine mercury-air to hydrocarbon water conversion factor.

gas column than an oil column. The reason for this unexpected relation is that the high interfacial tension of the gas-water system compared to oil-water systems counteracts the higher buoyant pressure generated by gas-water systems.

The thrust of the first half of this paper has been to discuss how to determine the minimum hydrocarbon column required to migrate through a given rock. If the displacement pressure for the rock in question can be measured and the subsurface conditions for the test case are known, calculations can be made following the outlined procedure. This hydrocarbon column height can be useful in determining minimum requirements for migration in reservoir rocks and to estimate trap capacities of exploratory prospects by quantifying caprock and lateral-seal capacity. The same procedures can be used to quantify other aspects of secondary hydrocarbon migration. Hydrocarbon shows in any rock can be interpreted quantitatively if capillary properties and oil saturations are known. The capillary pressure of the rock at the saturation in question can be related to a

Table 3. Effects of Variables on Calculation of Seal Capacity

Critical Variables in Seal Calculation	Low Values	Calculations of Seal Capacity (ft)	Text Example Values	Calculations of Seal Capacity (ft)	High Values	Calculations of Seal Capacity (ft)
OIL-WATER SYSTEM						
Interfacial tension	5.0	12.5	10	28	35	124
Water density	1.0	32	1.05	28	1.2	20
Oil density	0.5	19	0.68	28	0.9	69
GAS-WATER SYSTEM						
Interfacial tension	25	31	50	66	70	95
Water density	1.0	74	1.1	66	1.2	60
Gas density	0.01	61	0.1	66	0.3	82

Assumptions   (1)   Seal P_d = 220 psi (mercury system)
                (2)   Reservoir P_d = 1 psi (hydrocarbon-water system)
                (3)   Hydrostatic conditions
                (4)   Water-wet rocks.

Calculations: Each critical variable in the seal-capacity calculation was changed independently from the text example calculation to illustrate the maximum and minimum effect of each variable in determining the seal capacity for a given rock with a mercury injection pressure of 220 psi.

maximum hydrocarbon column that must be associated with the hydrocarbon show. This approach is analogous to estimating seal capacity as illustrated in the given example. Data of this type can be used to estimate the oil-water contact in developing oil or gas fields and in quantitatively interpreting hydrocarbon shows in near-miss wildcat wells. Detailed examples of these techniques are discussed in the following sections.

Seal Capacity

Seal capacity estimates of various rock types can be useful at several different levels of exploration activity. In a virgin basin the identification or regional caprock seals to migrating hydrocarbons can be important in migration and reservoir-change studies. When a structural prospect has been identified, the caprock seal capacity of the formation immediately overlying the hydrocarbon-charged reservoir is important in determining the producibility of the prospect. In stratigraphic traps the lateral-seal capacity of the rock type updip from a charged reservoir will determine the vertical hydrocarbon column the lateral facies change can trap.

The prediction of caprock seals to migrating hydrocarbons on a prospect or regional scale should be based on all available geologic data. Lithologically, the perfect caprock seal would have very small pores (to trap a large hydrocarbon column), and be very ductile so that it would not yield by brittle fracturing. Stratigraphically, the perfect seal would be thick and laterally continuous across the basin or area in question. Regional salt beds and marine clay shales fit these

criteria and are generally considered as regional caprock seals to continuous-phase hydrocarbon migration. Where these more obvious caprock seals are not present, prediction of local or regional caprock seals requires additional data and the quantitative application of the principles of secondary hydrocarbon migration.

Estimation of caprock seal capacity in exploration settings requires two types of information. The first piece of information needed is the capillary properties of the caprock in question. If these data are known, the hydrocarbon column that the pore system of the rock can seal or trap can be calculated. The next type of information needed is some estimate of the mechanical properties of the rock (e.g., brittleness) and the structural setting of the rock layer in question. Rock mechanical properties can be measured directly in the laboratory or estimated empirically from published data. The structure setting can be determined by subsurface mapping and seismic sections. These data are necessary to determine whether the rock layer in question is likely to fail by brittle fracturing. If brittle fracture is dominant in the rock layer, it will not be an effective caprock seal, even if the pore system of the rock can seal a large hydrocarbon column. In the simplified problem discussed in the next paragraph all the rock layers in the example will be considered ductile and only the pore system of the rocks in question is considered in predicting seal capacity.

A simplified problem is presented in Figure 24 to serve as an example in caprock prediction. An anticlinal prospect has been identified as an exploration target. A good quality reservoir (bed A) that is thought to be charged with hydrocarbons is overlain by two distinct rock layers (beds B and C) which are not reservoir-quality rock. If bed B is a seal to migrating hydrocarbons, then the reservoir (bed A) will be filled with hydrocarbons before the spillpoint of the trap is reached and a commercial accumulation should be present. If bed C is a hydrocarbon seal and bed B is not, the trap would spill hydrocarbons updip before the reservoir quality rock of bed A is saturated with oil. Migrating hydrocarbons would be trapped in this second situation but the accumulation would be noncommercial because bed B, which is saturated with hydrocarbons, cannot produce at economic rates.

In attempting to solve this problem (Fig. 24), the seal capacity of the pore systems of rock types in beds B and C must be estimated. If rock samples are available, this can be done by direct measurement of displacement pressure as previously discussed. Calculations of vertical hydrocarbon seal capacity of each rock type in the appropriate subsurface environment can be made. If the pore network of the rock type in bed B has a low seal capacity, or if it is a brittle rock, there is the possibility that the defined prospect may be a noncommercial trapped hydrocarbon accumulation. This type of information should then be considered in the exploration-decision process.

The example just described considers the situation where two distinct lithologies of considerable thickness overlie a potential reservoir rock. In settings where the rocks overlying the potential reservoir are thin bedded, the chance of one ductile bed with a high displacement pressure acting as a seal is quite good. As pointed out by Hill et al (1961), in an oil column trapped by simple anticlinal closure, the buoyant force is directed vertically upward and perpendicular to the bedding. If the first thin bed overlying the reservoir has a low displacement pressure, and the bed immediately above it has a high displacement pressure that can act as a seal for the accumulation, then oil will be trapped in the commercial reservoir rock (Fig. 25).

Stratigraphic traps—For stratigraphic traps we have the added problem of lateral-seal capacity in addition to caprock-seal capacity discussed for structural traps. Stratigraphic traps as a general class include all trapped hydrocarbon accumulations that are formed by a displacement pressure barrier along a reservoir carrier bed. Any lateral termination of a vertically sealed reservoir-quality rock charged with migrating hydrocarbons would then be a commercial stratigraphic trap. This definition would include reservoir-rock lateral terminations owing to depositional facies changes, diagenetic facies changes, faults, unconformities, etc. Stratigraphic traps include all traps except simple anticlinal closure and tilted oil-water contacts on structural terraces which can form traps that would not hold hydrocarbons in the hydrostatic case as pointed out by Hubbert (1953).

A simplified stratigraphic trap has been diagramed in Figure 26 to compare and contrast structural and stratigraphic traps. For stratigraphic traps, Hill et al (1961) pointed out that the buoyant force of the oil column is directed updip parallel with bedding rather than perpendicular to bedding as in the structural trap (Fig. 25). In contrast to the structural trap, the lowest displacement pressure bed at the lateral termination of the reservoir will determine the stratigraphic-trap capacity. The prediction of trap capacity for

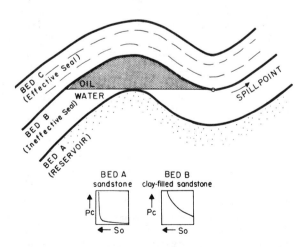

FIG. 24—Structural trap where commercial production is limited by caprock-seal capacity.

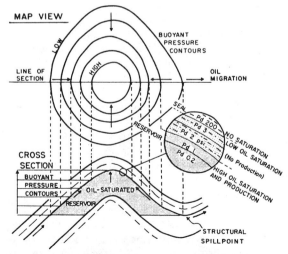

FIG. 25—Structural trap (after Hill et al, 1961).

stratigraphic traps must consider both the seal capacity of the rocks above and below the reservoir and the seal capacity of rocks laterally equivalent to the reservoir. As shown in Figure 26, thin continuous beds with low displacement pressure can be the controlling factor in stratigraphic-trap capacity.

In quantification of lateral-seal capacity, it is then important to know the displacement pressure of the rock at the updip termination of the reservoir, and the vertical and lateral continuity of the potential lateral seal. If a particular facies has been mapped as a potential lateral seal and rock samples are available, quantitative estimates of seal capacity can be made by running mercury capillary pressure tests and making the calculations for the hydrostatic or hydrodynamic case, whichever is appropriate (Fig. 26). In sampling a potential lateral seal, numerous samples should be taken vertically across the zone that is laterally equivalent to the reservoir. The rock with the lowest displacement pressure will act as the controlling lateral seal depending on its lateral continuity both updip and downdip. Berg (1975) has documented several cases where quantitative attempts at lateral-seal-capacity estimates have proved to be quite accurate.

Quantitative Hydrocarbon Show Interpretation

Another situation where attempts to quantify secondary hydrocarbon migration and entrapment can be useful in exploration is in the inter-

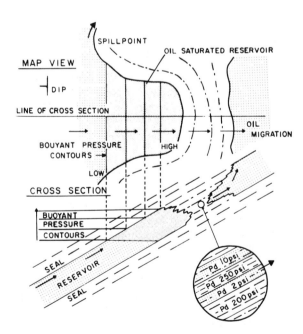

FIG. 26—Stratigraphic trap (after Hill et al, 1961).

pretation of hydrocarbon shows. In a laterally continuous reservoir rock that is charged with hydrocarbons we would generally expect to encounter two types of subsurface hydrocarbon shows. Type one would be a continuous-phase hydrocarbon occurrence that is associated with a trapped hydrocarbon accumulation of finite size. The other type would be a residual hydrocarbon stain along a migration path. In a very simplified approach to show interpretation, we can consider a flow of oil or gas while drilling, drill-stem testing, or production testing, as an indication of a trapped accumulation of hydrocarbons, because oil or gas along a migration path would be at residual saturation with no permeability to hydrocarbons. If we can determine that a given show is associated with a trapped accumulation of hydrocarbons we can estimate the probable areal extent of the accumulation by quantitatively applying the principles of secondary hydrocarbon migration and entrapment.

There are several situations where an explorationist can estimate the extent of a given accumulation. Let us examine what could be done when an exploratory well was drilled in the center of a structural or stratigraphic trap (Fig. 27). Once the well in Figure 27 is completed, the next step is to develop the field. One key question during development is where is the producible oil-water contact or how far downdip can wells be drilled before excessive water production will be encountered. Assuming uniform reservoir rock, this can be estimated by applying the mechanics of secondary migration and entrapment. Two approaches can be used to estimate the depth to water-free oil production. If the saturation of the reservoir rock is accurately known from log calculations and the capillary properties of the reservoir rock are known, calculations of the oil column required to account for the buoyant pressure required to reach that given saturation can be made. The procedure involved here is exactly as discussed in the calculation of seal capacity except the capillary pressure at reservoir saturation is used instead of the displacement pressure of the reservoir. Another approach that can be used if a continuous core is available through the reservoir is to run capillary pressure tests on rock samples where oil-saturated rocks are adjacent to water-saturated rocks. By comparing the oil column needed to saturate stained and unstained samples the oil column in the reservoir can be estimated. For example, an oil-stained rock that has a displacement pressure equivalent to a 30-ft oil column may be immediately overlain by a rock with a displacement pressure equivalent to 40 ft. The oil column present downdip from this sample

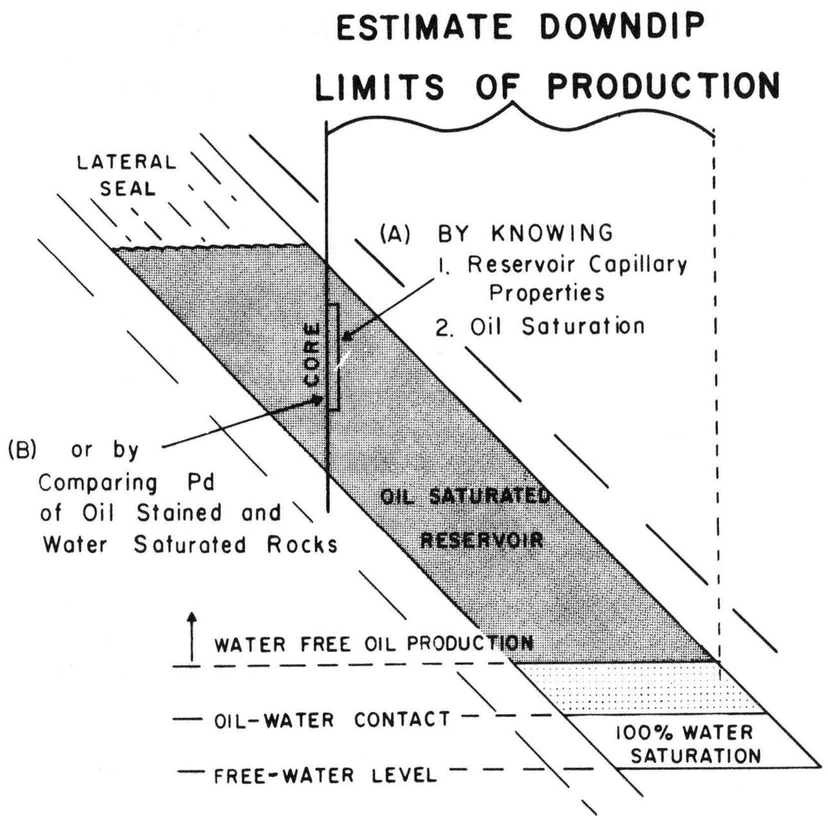

ESTIMATE DOWNDIP
LIMITS OF PRODUCTION

FIG. 27—Illustration of method for estimating downdip limits of production in stratigraphic trap.

would be greater than 30 ft but less than 40 ft. In complex stratigraphic traps, dry holes with oil shows in noncommercial reservoir rock can be drilled in the middle of a commercial oil accumulation. If a well of this type were drilled as an initial exploratory test, the extent of accumulation in the downdip direction could be estimated by the method just described.

Field development can also be aided by an understanding of capillary properties, particularly if there is a strong variation in the capillary pressure of different facies within a producing zone or between different producing beds with a common oil-water contact. Figure 28 illustrates the potential variation in the productive oil-water contact where there are two facies with widely different capillary properties crossing the crest of a closed structure. Figure 29 illustrates the possible variation of the producible oil-water contact in a structure trap with two producing beds that have widely different capillary properties. This diagram assumes a common free-water level and communication between the different producing beds. Quantitative estimates of producible oil-water

contacts can be made during field development (if the capillary properties of different facies or producing zones are known) by following the procedures outlined previously.

Another situation where quantitative show interpretation could be helpful is in the updip portion of subtle stratigraphic traps. Stratigraphic traps with a gradual updip change from reservoir-quality rock to the updip seal, will have a zone where the oil cannot be produced economically because of poor quality reservoir rock and/or low oil saturation. This zone, diagramed in Figure 30, has been termed the waste zone by Bob Dunham of Shell (personal commun., 1973), as the oil in this zone is wasted and cannot be produced. The recognition of waste zones is critical to the exploration for stratigraphic traps if we are to improve our oil-finding techniques. If the oil-stained zone in this well can be determined to be part of a trapped accumulation of hydrocarbons, then calculations can be made to determine the extent of the accumulation downdip. The approach would be the same as that described in the previous section. Calculations can be made by using capillary

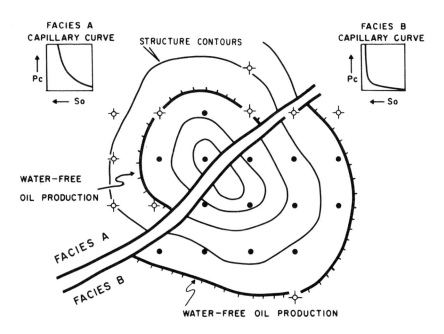

FIG. 28—Facies effects on water-free oil production (after John Howell).

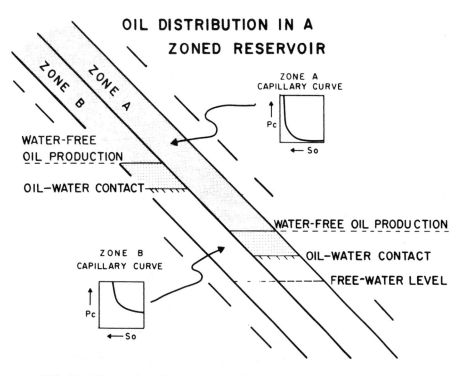

FIG. 29—Effects of capillary properties in zoned reservoir (after John Howell).

properties and oil saturation of a given oil-saturated rock or by comparing the displacement pressure of oil-stained and unstained samples.

The third situation where quantitative show interpretation can assist in exploration is where an exploratory well is drilled in the oil-water transition zone of a commercial reservoir (Fig. 31). Wells in this position test oil with uneconomically high water cuts. The obvious direction for an offset well is updip to get a higher oil saturation due to increased buoyant ·forces. The question is at what height above the first well will water-free oil production or oil production with low water cuts be encountered.

This question can be answered if the saturation and capillary properties of the reservoir rock are known. The height above the well in question that is required for commercial water-free oil production can be calculated following the same procedure previously described. The concept behind this approach has been discussed by Arps (1964). The question could be critical in situations where the height needed for commercial production is greater than the elevation to be gained by an additional test in the case of a structural accumulation. In the case of a stratigraphic trap this question can be important if the height needed above the transition-zone well is greater than the eleva-

tion available as defined by dry holes located in the lateral seal of the accumulation as illustrated in the example (Fig. 31).

MIGRATION AND ENTRAPMENT MODEL

The mechanical principles of secondary migration can logically be applied to developing a model for secondary hydrocarbon migration and entrapment. In summary, these principles state that if the driving force (buoyancy) of a continuous-phase hydrocarbon accumulation exceeds the retarding forces (displacement pressure) of a rock acting as a barrier to migration, oil or gas will displace water from the confining pore throats and migrate as a continuous filament through the largest connected pore throats of the rock.

To develop a workable secondary-migration model, a simplified geologic environment can be used as an illustration. Consider a laterally continuous homogeneous reservoir rock overlain by a high-displacement-pressure caprock seal and underlain by a hydrocarbon source rock. Oil or gas expelled from the source rock will begin to accumulate at the source rock–reservoir boundary. The method of primary migration is not inferred here but these principles can be applied whenever the expelled oil or gas occur as a continuous phase in the rock on a scale from droplets to

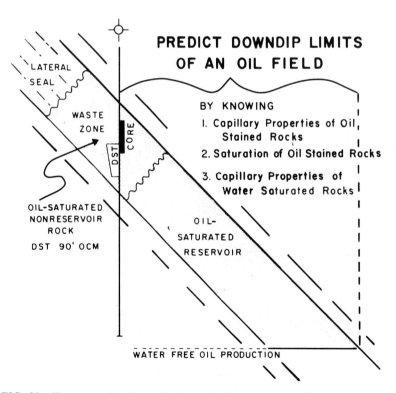

FIG. 30—How to predict downdip limits of oil accumulation from near-miss show.

larger connected filaments. As oil or gas accumulates at the source rock–reservoir boundary, the buoyant force of a continuous oil or gas filament will eventually exceed the displacement pressure of the reservoir rock and the hydrocarbon phase will then migrate vertically upward through the reservoir rock until it encounters the overlying caprock seal. The vertical oil or gas column necessary to migrate vertically upward through the reservoir rock will depend on the density of the hydrocarbon and water phases, the size of the largest connected pore throats of the reservoir, the interfacial tension, and the wettability of the hydrocarbon-water-rock system. These variables have been discussed in detail with methods to quantify the vertical hydrocarbon column needed for migration.

Using average values of oil and water densities, interfacial tension, and pore throat sizes measured from thin sections, Aschenbrenner and Achauer (1960) calculated that it takes a continuous vertical oil filament of 7½ ft to migrate vertically upward through the average reservoir carbonate rock. For a water-wet medium-grained sandstone they calculate that the vertical oil column needed for migration would be approximately 1 ft using average densities and interfacial tensions. Direct measurements of displacement pressure for 23 sandstone reservoirs and six carbonate reservoirs suggest that critical vertical oil columns needed for migration range from 1 to 10 ft for sandstones and 3 to 5 ft for the carbonate reservoirs. These calculations have assumed water-wet rocks, oil-water interfacial tension of 30 dynes/cm, hydrostatic conditions, and a buoyancy gradient of 0.1 psi/ft. Both these studies suggest that the continuous-phase vertical oil column needed for oil to migrate through average reservoir rocks at subsurface condition ranges from roughly 1 to 10 ft. Although these numbers would vary for gas and also for oil as the densities of the fluids, the interfacial tension, wettability, and hydrodynamic conditions vary, they can be used as workable numbers in constructing a migration model.

In the model the oil or gas would migrate vertically upward through the reservoir until it reached the reservoir seal boundary where it would spread out along this interface. Now an additional volume of oil must accumulate to migrate laterally updip along the reservoir seal boundary. The lateral length of a continuous oil or gas filament required to reach the critical vertical oil or gas column will depend on the dip of the beds. The steeper the dip the shorter the length of the hydrocarbon filament needed to obtain the

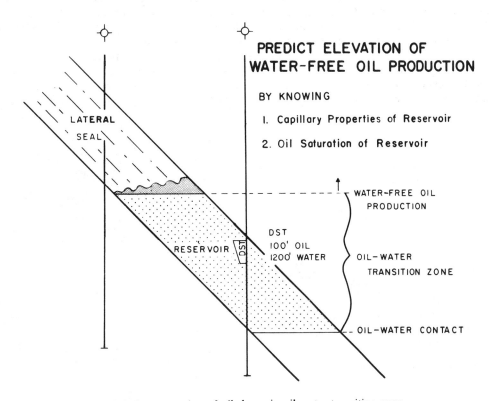

FIG. 31—Interpretation of oil shows in oil-water transition zone.

critical vertical hydrocarbon column height required for updip migration. Aschenbrenner and Achauer (1960) made a graph to determine the minimum length of hydrocarbon filament required at various dips to obtain the 7½-ft vertical oil column to migrate through their average carbonate reservoir. This additional volume of oil is obtained by the continual addition of oil being expelled from the source rock and migrating vertically upward through the reservoir. When the critical length of hydrocarbon column is obtained, oil will migrate laterally updip through the reservoir.

As migration occurs laterally updip through the reservoir the oil saturation could be as low as 10%, as this is the minimum saturation needed to migrate across the length of 1-in. permeability plugs tested in the laboratory. The hydrocarbon filament will migrate through only the upper few feet of the carrier system and the remaining reservoir section will be barren of hydrocarbons (Fig. 32). At the base of the migrating hydrocarbon filament small isolated droplets will be left behind as residual oil as it migrates upward. These shows of oil can be called "migration-path shows" and can provide important exploration information. The amount of oil left behind will depend on the initial saturation. The greater the initial saturation, the greater the residual saturation. Residual saturations along migration paths are thought to be on the order of 20% or less as hydrocarbon saturations during migration range from 10 to 30%. These residual droplets of oil are permanently trapped by capillary forces. The soluble portion of this residual oil can be dissolved in the surrounding water phase and dispersed by diffusion. Enough residual oil should be left behind as residual saturation along any migration path to create an oil show in samples or cores. In a uniform reservoir this migration residual stain should be located immediately below the caprock seal and only the upper few feet of reservoir should have any detectable oil show. Oil migration paths may be difficult to detect in drilling for this reason. Migration of gas as a separate phase through a reservoir may leave no residual saturation as a separate phase because of the high solubility of gas which may permit all the capillary-trapped gas to dissolve and dissipate by diffusion.

The hydrocarbon filament will migrate laterally updip perpendicular to strike in a tortuous manner, seeking the path of least work by moving through the rocks with the largest connected pore throats or lowest displacement pressure. This tortuous movement, if considered from map view, will leave some rocks with a residual oil saturation where migration has occurred and rocks immediately adjacent will be completely barren of oil. This fact should be considered when evaluating the likelihood of an 8-in. drill hole encountering a migration path in a potential reservoir carrier bed. This relation has been pictured in Figure 33.

As the migrating filament loses oil in the form of residual oil or gas at the base of the filament, the length of the hydrocarbon filament will be shortened and the buoyant force will be reduced. Eventually, the buoyant force of the filament will be reduced to the point that it will no longer be able to overcome the capillary resistant force of the pores of the reservoir carrier bed. Migration will cease at this point until another hydrocarbon

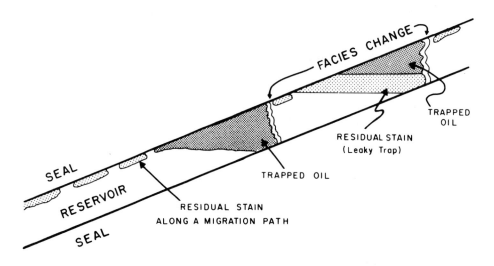

FIG. 32—Cross-section view of migration path.

filament migrates updip to the stalled filament and then migration will continue. This continuous pulsating process will continue as long as oil is being added downdip. This can be accomplished by continual generation of oil in the source rock or by addition of oil due to remigration. Oil or gas can continue to migrate laterally updip or vertically through any rock section so long as the buoyant force of the hydrocarbon column is greater than the resistant force of the carrier bed. Therefore, there are no physical limits to the distance oil or gas can migrate laterally or vertically in a given geologic situation.

Expanding the simplified model to the scale of a petroleum basin, we can envision oil being expelled at points of maturity within the basin and migrating updip, perpendicular to strike, through the reservoir carrier bed. The migrating front of oil or gas can be concentrated in areally small zones or migration paths by structural anomalies such as anticlinal axes plunging into the basin or by facies variations within the reservoir carrier bed. Oil or gas will be trapped along a migration path whenever a closed anticlinal trap or a displacement pressure barrier is present within the reservoir carrier bed. These traps can be of any size. For structure-type traps, size will depend on the size of the anticlinal feature and the vertical-seal capacity of the caprock. For stratigraphic traps the size will depend on the lateral-seal capacity of the displacement pressure barrier and the size and geometry of the displacement pressure barrier. Oil or gas will continue to migrate laterally updip into a trap along a migration path until the trap is full. In the simplified model with a reservoir carrier bed overlain by a high-displacement-pressure caprock seal, any structural trap (Fig. 25) will fill to its geometric spillpoint and then oil will spill updip and continue to migrate laterally updip through the reservoir carrier bed. As oil or gas continues to migrate updip into the trap, oil will spill out of the trap and migrate updip in a continual process. If the vertical oil or gas column that can be contained by the caprock seal in a structural trap is less than the oil or gas column at the spillpoint of the trap, oil or gas will leak vertically through the caprock seal and will not spill updip.

For stratigraphic traps (Fig. 26) the trap may fill to the point that oil or gas can spill around the displacement-pressure barrier somewhere along the strike of the reservoir carrier bed. This is analogous to a structural trap filling to its geometric spillpoint and spilling oil updip. Another possibility for the stratigraphic trap is that, as the trap is filling, the buoyant force of the hydrocarbon column could exceed the resistant force of the dis-

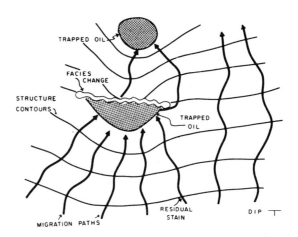

FIG. 33—Map view of migration path.

placement pressure and oil could leak laterally updip through the displacement-pressure barrier and continue to migrate updip through the reservoir carrier bed. Stratigraphic traps can then be considered both to spill oil or gas updip or leak oil or gas updip through the displacement-pressure barrier or lateral seal.

Oil or gas accumulations along a migration path are permanently trapped as long as geologic conditions remain constant. If there is a change in any parameter that is critical to the entrapment of a certain volume of oil or gas, then remigration as a continuous phase will occur. Such things as a change in dip, hydrodynamic conditions, densities of the hydrocarbon or water phase, sealing capacity of the caprock or lateral seal could cause remigration of oil or gas as a continuous-phase fluid out of the trap. If geologic conditions remain constant the oil or gas will remain permanently trapped and there will be no gradual leakage of bulk-phase hydrocarbons out of the trap. Hydrocarbons can, however, escape from the trap, but not as continuous droplets or filaments. If the trapped hydrocarbons are soluble they can be dissolved in the water phase within the reservoir and dissipated by diffusion from the trap or be swept away in solution in a moving-water phase. Oil molecules are generally quite insoluble and loss of oil from a trap by solution is probably minimal, except in the case of shallow reservoirs in an active hydrodynamic setting. Gas, particularly methane, is quite soluble in formation waters and gas loss by solution and diffusion could be significant in the case of trapped hydrocarbon gas. Gas in solution can diffuse through any porous water-saturated rock and this type of gas loss and migration from reservoirs, migration paths, and source rocks may account for the high-

97

amounts of gas in solution in formation waters in some petroleum basins.

In the migration model developed we have suggested that oil migrating into a trap will be permanently trapped as long as geologic conditions at the time of entrapment remain constant. This implies that the displacement or breakthrough pressure of a caprock seal in a structural trap or a lateral seal in a stratigraphic trap is independent of time and will not gradually leak droplets or filaments of continuous-phase oil or gas. Thomas et al (1967), in their study of threshold displacement pressures required to store natural gas in the subsurface, agree that threshold displacement pressures are independent of time. In their experiments they subjected two different water-saturated rock samples to gas pressure less than their threshold pressures for 3 to 10 days and observed no movement of water from the cores over this period of time. This laboratory work correlates with theoretical work that states that no continuous-phase migration will occur unless the buoyant pressure in the oil or gas column is greater than the resistant force of the confining seal or barrier.

In the migration model it has also been discussed that when the buoyant pressure of the oil or gas column exceeds the displacement pressure of the confining porous-rock barrier, oil or gas will displace water from the confining pore throats and migrate as a continuous oil or gas filament through the pore throats of the rock. The next question in further defining the model is how much oil or gas will leak through the constricting pore throat or throats before water will move back into the throat and snap off or collapse the oil or gas filament. How much oil or gas will escape from a trap when leakage through a seal occurs? Will the barrier allow the whole accumulation to migrate updip or will it leak one drop at a time?

For a migrating hydrocarbon filament to be snapped off, water must be able to flow into the confining pore throat and collapse the oil or gas filament. The confining pore throat would then be filled with water and the barrier to migration would, in effect, be resealed. Roof (1970) has calculated that for snap-off to occur in circular pores the leading edge of the oil or gas interface must extend past the confining pore throat for a distance of at least seven times the radius of the pore throat. For snap-off to occur in his model the pore would have to be large in relation to the size of pore throat. Roof then modeled migration through a stack of doughnut-shaped pores and determined that snap-off would not occur as oil or gas migrated through this series of pores. On

the scale of pores then it does not appear that traps or barriers would leak oil or gas one drop at a time.

How much oil or gas would leak through a barrier to migration before snap-off occurred and the rock resealed? Petroleum Research Corp. (1959) has determined that for the oil or gas filament to collapse, capillary pressure must be reduced to between one-fourth and one-half the pore-entry pressure. The reduction of capillary pressure required before water can be imbibed back into the rock and collapse the hydrocarbon filament is documented by the hysteresis effect during capillary injection and withdrawal (Pickell et al, 1966). In their studies mercury was injected into rock samples and the mercury saturation increased with increasing pressure. However, when the capillary pressure was reduced, no air was imbibed back into the sample until the pressure was reduced significantly below the entry pressure.

These data suggest that for snap-off or collapse in an oil or gas filament migrating through a rock to occur, the capillary pressure must be reduced to approximately one-half of the displacement pressure. The capillary pressure between the hydrocarbon and water phase could be reduced in our migration model by one-half simply by having approximately one-half of the oil or gas filament migrate through a displacement-pressure barrier. As the filament of oil or gas migrates through the displacement pressure barrier, snap-off would occur whenever capillary pressure or buoyant pressure was reduced enough that water could flow into the critical pore throats and collapse the oil or gas filament. This is a simplification of a complex phenomenon but, from the standpoint of developing a migration model, we can assume that when the buoyant force of an oil or gas filament exceeds the displacement pressure of a barrier along a migration path, a large part of the trapped oil or gas filament will migrate or leak through the barrier before collapse or snap-off. When snap-off occurs the barrier has been resealed and migration for the oil or gas filament downdip from the barrier will be halted. For simplicity, let us assume that for intergranular and intercrystalline porosity the amount of oil or gas allowed to leak through the displacement-pressure barrier will be approximately one-half of the trapped oil or gas column. The exception to this assumption would be for vugular porosity types which, as pointed out by Roof (1970), would snap off or collapse after only a few drops had migrated through the controlling pore throat.

From the standpoint of an explorationist, a trap along a migration path that has leaked oil or gas updip will reseal after approximately one-half of the trapped hydrocarbon column has migrated updip. The next logical question in our migration model then is: when a barrier has resealed, what is its displacement or threshold pressure? This problem has been investigated by Thomas et al (1967). Their gas-breakthrough experiments suggest that a rock can be resealed by water moving back into the rock. With sufficient time for water to move back into the rock, they suggest that a rock can reseal at or near its original displacement or threshold pressure. Theoretical calculations considering a single confining pore throat also suggest that once the oil or gas filament has collapsed and water has been imbibed into the critical pore throat, the displacement pressure of the pore throat would be the same as the original displacement pressure before leakage. Therefore, in our migration model, we can assume that stratigraphic traps that leak oil or gas updip through a lateral seal can reseal and refill to their original capacity if the resealed barrier holds.

Differential Entrapment

The migration model we have developed has taken us to the point of the first major trap along a reservoir carrier bed or migration path. As oil or gas continues to migrate updip beyond the first major trap, what will happen and how will oil and gas be distributed in long-range migration? Gussow (1954) was the first to deal with this problem. Gussow discussed how oil and gas will be expelled from a source, coalesce to form continuous slugs in a reservoir carrier bed, and migrate updip perpendicular to structure, honoring the lowest displacement-pressure rocks in its path. The slugs will join along major structural noses or permeability barriers to form "rivers" of oil. This updip movement of hydrocarbons as streams in carrier beds to the final condition of entrapment is known as secondary migration.

Gussow (1954) stated that the lowest trap along a migration path will be filled first, then the trap structurally higher, and so on. When there is a series of structural traps that spill petroleum updip, there is the potential for differential entrapment of oil and gas when the two phases are present. As illustrated in Figure 34 and discussed by Gussow, if both oil and gas are present in a trap as separate phases, gas will occupy the upper part of the trap and oil the lower. As the accumulation continues to fill with oil and gas migrating into the trap, gas will rise to the top of the structure and when the structure is full to the spillpoint, oil will be spilled updip. As the trap continues to fill with gas, oil will be spilled updip until the trap will be completely filled with gas and can no longer trap oil. As illustrated in the diagram, this would result in oil filling the higher traps and gas the lower traps along a migration path. Gussow listed numerous examples of the differential entrapment of oil updip from gas along migration paths along which a series of structural traps are present. He suggested that this relation will hold except where there is strong downdip hydrodynamic flow. As discussed by Hubbert (1953), in a structural trap filled with two phases (oil and gas), under hydrodynamic conditions oil can be flushed out of the trap and gas left behind. This situation would produce gas updip from oil and is an exception to Gussow's case.

Let us now consider the situation where the traps along a migration path are a series of displacement-pressure barriers that will hold a certain hydrocarbon column and then leak hydrocarbons updip through the barrier before the trap is filled to its stratigraphic spillpoint. Petroleum

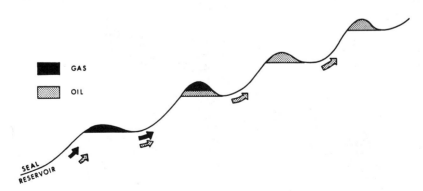

FIG. 34—Structural (spill) differential entrapment of oil and gas (after Gussow, 1954). For series of traps that spill updip, gas will be differentially entrapped downdip from oil.

Research Corp. (1960) discussed this situation in their report on differential entrapment. Figure 35 illustrates the effect of leak differential entrapment (compare to spill differential entrapment, Fig. 34). When oil and gas are present as separate phases in a stratigraphic trap, gas will be at the updip part of the trap and will be trying to break through the barrier. As the trap fills and the buoyant pressure increases, gas will leak out through the barrier first when the displacement pressure of the barrier is reached. As discussed previously, a large slug of gas and some oil will migrate through the barrier before it reseals. As migration continues, gas will eventually be the only phase migrating updip through the displacement pressure or permeability barriers along the migration path. This produces a situation where gas is differentially trapped updip from oil, which is exactly the opposite of spill differential entrapment.

When a migration path consists of both structural and stratigraphic traps, the distribution of hydrocarbons can become quite complex because of the opposite effect of leak and spill differential entrapment.

Other factors such as the depth and timing of oil and gas generation must also be considered when interpreting the distribution of oil and gas along migration paths. For example, shallow gas high up along a migration path that appears to be due to stratigraphic leak differential entrapment may actually be indigenous biogenic gas rather than gas that has migrated long distances and been differentially trapped updip from oil. Thermal generation of oil and gas will also pose prob-

lems in interpreting patterns of oil and gas distribution. Thermal generation models suggest that oil is generated first and expelled, then gas is generated and expelled. This sequence of generation and migration could cause gas to be distributed downdip from oil, which is analogous to structural spill differential entrapment. Another complication to consider is that, whenever the structural dip of a carrier bed is changed, remigration and further adjustments in the distribution of oil and gas will occur.

Differential entrapment of oil and gas along migration paths may cause dramatic chemical changes in oil composition. Hobson (1962) and Silverman (1965) gave detailed discussions of this phenomenon. In summary, they suggested that small but measurable changes occur as oil migrates as a single phase through reservoir carrier beds. These changes due to secondary migration of a single oil phase cannot explain the markedly different chemistry of some oils that are thought to be genetically related. These larger differences in composition of genetically related oils are speculated to be caused by phase separations of oil and gas during migration and the process was called separation-migration by Silverman (1965). For separation-migration to occur, two phases, oil and gas, must be present in the trap. The gas phase must escape, leaving the liquid phase behind. As the gas phase migrates updip to a lower pressure, retrograde condensation can occur and form a liquid and gas phase from the gas phase that was separated by migration. The liquid oil formed by this process will be compositionally distinct from the parent oil left behind. Silverman

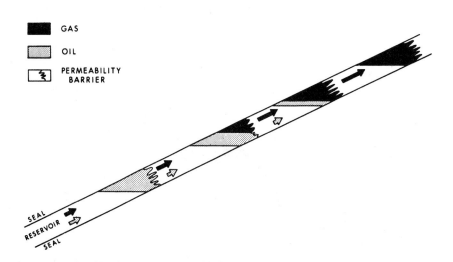

FIG. 35—Stratigraphic (leak) differential entrapment of oil and gas. For series of traps that leak updip, oil will be differentially entrapped downdip from gas.

suggested that the gas phase could be separated from the oil phase by fracturing the caprock so that only gas escapes from the trap. This type of separation is analogous to stratigraphic differential entrapment as discussed in the preceding section where a trap leaks through a displacement-pressure barrier. Gas will be at the updip portion of the trap and when the buoyant pressure of the oil and gas column exceeds the displacement pressure of the controlling barrier to migration, a large slug of gas will leak through the barrier and the barrier will then reseal. This would separate an oil and a gas phase and create the chemical changes discussed by Silverman.

The compositional changes during normal secondary migration and separation-migration are considered relatively unimportant to the explorationist because of the lack of predictive value.

CONCLUSIONS

The processes of secondary hydrocarbon migration and entrapment are well understood physical processes. The distribution of oil and gas in the subsurface can be examined in a logical quantitative fashion by following a few basic principles and using data generally available in petroleum exploration and development. A thorough understanding of these processes can be very useful in all phases of the search for oil and gas.

In the exploration for new oil and gas reserves these principals define critical factors needed to predict the location of entrapped oil or gas along a migration path. For structural traps the critical factors are the seal capacity of the reservoir caprock, the structural configuration at the base of the seal, and the tilt of the oil-water contact if a hydrodynamic condition is present. For stratigraphic traps the location, configuration, and seal capacity of a lateral barrier to oil or gas migration along a carrier bed are critical. The seal capacity of the barrier in terms of vertical hydrocarbon column will be affected by the density of the hydrocarbon and water phases, the hydrodynamic conditions in the carrier bed, the pore-throat sizes of the barrier, the interfacial tension of hydrocarbon-water phase and the wettability of the rock. The dip of the reservoir will not affect seal capacity but will affect the volume of hydrocarbons trapped.

Once a commercial field has been located, the principles of secondary migration and entrapment can be useful in field development. The updip and downdip limits of the accumulation can be calculated quantitatively from normally available well data and the information can be used in development drilling.

Wherever the processes of secondary migration and entrapment are used for prediction in the search for oil and gas, they should be used in conjunction with all available geologic information, as they cannot stand alone and provide meaningful data.

REFERENCES CITED

Arps, J. J., 1964, Engineering concepts useful in oil finding: AAPG Bull., v. 48, p. 157-165.

Aschenbrenner, B. C., and C. W. Achauer, 1960, Minimum conditions for migration of oil in water-wet carbonate rocks: AAPG Bull., v. 44, p. 235-243.

Aufricht, W. R., and E. H. Koepf, 1957, The interpretation of capillary pressure data from carbonate reservoirs: Jour. Petroleum Technology, v. 9, no. 10, p. 53-56.

Berg, R. R., 1975, Capillary pressure in stratigraphic traps: AAPG Bull., v. 59, p. 939-956.

Berry, F. A. F., 1958, Hydrodynamics and geochemistry of the Jurassic and Cretaceous Systems in the San Juan basin, N.W. New Mexico and S.W. Colorado: PhD thesis, Stanford Univ.

Cartmill, J. C., 1976, Obscure nature of petroleum migration and entrapment: AAPG Bull., v. 60, p. 1520-1530.

Cordell, R. J., 1972, Depths of oil origin and primary migration: a review and critique: AAPG Bull., v. 56, p. 2029-2067.

Dickey, P. A., 1975, Possible primary migration of oil from source rock in oil phase: AAPG Bull., v. 59, p. 337-345.

Fatt, I., and W. A. Klikoff, 1959, Effect of fractional wettability on multibase flow through porous media: AIME Petroleum Trans., v. 216, p. 71-77.

Flowers, B. S., 1976, Overview of exploration geophysics—recent breakthroughs and challenging new problems: AAPG Bull., v. 60, p. 3-11.

Gussow, W. C., 1954, Differential entrapment of oil and gas: a fundamental principle: AAPG Bull., v. 38, p. 816-853.

Harms, J. C., 1966, Stratigraphic traps in a valley fill, western Nebraska: AAPG Bull., v. 50, p. 2119-2149.

Hill, G. A., W. A. Colburn, and J. W. Knight, 1961, Reducing oil finding costs by use of hydrodynamic evaluations, in Economics of petroleum exploration, development and property evaluation: Englewood, N.J., Prentice-Hall, Inc., p. 38-69.

Hobson, G. D., 1962, Factors affecting oil and gas accumulations: Inst. Petroleum Jour., v. 48, no. 461, p. 165-168.

Hocott, C. R., 1938, Interfacial tension between water and oil under reservoir conditions: AIME Petroleum Trans., v. 32, p. 184-190.

Hough, E. W., M. J. Rzasa, and B. B. Wood, 1951, Interfacial tensions of reservoir pressures and temperatures, apparatus and the water-methane system: AIME Petroleum Trans., v. 192, p. 57-60.

Hubbert, M. K., 1953, Entrapment of petroleum under hydrodynamic conditions: AAPG Bull., v. 37, p. 1954-2026.

Kusakov, M. M., N. M. Lubman, and A. Yu. Koshevnik, 1954, Oil-water and oil-gas interfacial tensions under in-situ conditions: Neftyanoye Khozyaystvo, v. 32, no. 10, p. 62-69 (trans. by Assoc. Tech. Services, Inc.).

Leverett, M. C., 1941, Capillary behavior in porous solids: AIME Petroleum Trans., v. 142, p. 152-169.

Livingston, H. K., 1938, Surface and interfacial tension of oil-water systems in Texas oil sands: AIME Tech. Paper 1001.

McCaffery, F. G., 1972, Measurement of interfacial tensions and contact angles at high temperature and pressure: Jour. Canadian Petroleum Technology, v. 11, no. 3, p. 26-32.

McNeal, R. P., 1961, Hydrodynamic entrapment of oil and gas in Bisti field, San Juan County, New Mexico: AAPG Bull., v. 45, p. 315-329.

———— 1965, Hydrodynamics of the Permian basin, in Fluids in subsurface environments: AAPG Mem. 4, p. 308-326.

Meyers, J. D., 1968, Differential pressures, a trapping mechanism in Gulf Coast oil and gas fields: Gulf Coast Assoc. Geol. Socs. Trans., v. 18, p. 56-80.

Michaels, A. S., and E. A. Hauser, 1950, Interfacial properties of hydrocarbon-water systems: Jour. Phys. and Colloidal Chemistry, v. 55, p. 408-421.

Morrow, N. R., P. J. Cram, and F. G. McCaffery, 1973, Displacement studies in dolomite with wettability control by octonoic acid: AIME Petroleum Trans., v. 255, p. 221-232.

Petroleum Research Corp., 1959, Oil and gas migration through reservoir rocks: Denver, Colo., Research Rept. A-14.

———— 1960, Exploration applications of oil and gas differential entrapment: Denver, Colo., Research Rept. A-15.

Pickell, J. J., B. F. Swanson, and W. B. Hickman, 1966, Application of air-mercury and oil-air capillary pressure data in the study of pore structure and fluid distribution: SPE Jour., v. 6, no. 1, p. 55-61.

Price, L. C., 1976, Aqueous solubility of petroleum as applied to its origin and primary migration: AAPG Bull., v. 60, p. 213-244.

Purcell, W. R., 1949, Capillary pressure—their measurements using mercury and the calculation of permeability therefrom: AIME Petroleum Trans., v. 186, p. 39-48.

Roehl, P. O., 1967, Stony Mountain (Ordovician) and Interlake (Silurian) facies analogs of recent low-energy marine and subaerial carbonates, Bahamas: AAPG Bull., v. 51, p. 1979-2031.

Roof, J. G., 1970, Snap-off of oil droplets in water-wet pores: AIME Petroleum Trans., v. 249, p. 85-90.

———— and W. M. Rutherford, 1958, Rate of migration of petroleum by proposed mechanisms: AAPG Bull., v. 42, p. 963-980.

Rudd, N., and G. N. Pandey, 1973, Threshold pressure profiling by continuous injection: AIME-SPE Paper 4597, 7 p.

Salathiel, R. A., 1972, Oil recovery by surface film drainage in mixed-wettability rocks: AIME-SPE Paper 4104.

Silverman, S. R., 1965, Migration and segregation of oil and gas, in Fluids in subsurface environments: AAPG Mem. 4, p. 53-65.

Smith, D. A., 1966, Theoretical considerations of sealing and non-sealing faults: AAPG Bull., v. 50, p. 363-374.

Stone, D. S., and R. L. Hoeger, 1973, Importance of hydrodynamic factor in formation of Lower Cretaceous combination traps, Big Muddy–South Glenrock area, Wyoming: AAPG Bull., v. 57, p. 1714-1733.

Stout, J. L., 1964, Pore geometry as related to carbonate stratigraphic traps: AAPG Bull., v. 48, p. 329-337.

Thomas, L. K., D. L. Katz, and M. R. Tek, 1967, Threshold pressure phenomena in porous media: AIME-SPE Paper 1816, 12 p.

Treiber, L. E., D. L. Archer, and W. W. Owens, 1972, A laboratory evaluation of the wettability of fifty oil-producing reservoirs: AIME Petroleum Trans., v. 253, p. 531-540.

Wardlaw, N. C., 1976, Pore geometry of carbonate rocks as revealed by pore casts and capillary pressure: AAPG Bull., v. 60, p. 245-257.

Reprinted by permission of the World Petroleum Congresses and Applied Science Publishers, London, from *Proceedings of the Eighth World Petroleum Congress,* 1971, vol. 2, pp. 13-26 (ASPL).

MIGRATION, ACCUMULATION AND RETENTION OF PETROLEUM IN THE EARTH

Abstract

Increasingly numerous, varied and accurate analyses of petroleums and sedimentary rocks are providing a foundation for the development, testing and application of theoretical models of earth processes. Combining new experimental data with predictions derived from such models now permits better understanding of: (1) the relationship between petroleum source and reservoir rocks including the volume of source rock required for a given accumulation; (2) the relationship between isotopic compositions of the constituents of petroleum gases and their genesis; (3) mechanisms for transport of petroleum into a reservoir, loss from the reservoir and concomitant alterations in the petroleum.

Résumé

Des analyses de pétroles et de roches sedimentaires de plus en plus nombreuses, variées et précises fournissent une base au développement, l'essai et l'application de modèles théoriques de processus se passant à l'intérieur de la terre. L'utilisation combinée des données expérimentales nouvelles et des prédictions provenant de ces modèles permet maintenant une meilleure comprehension de: (1) la relation qui existe entre roche-mère et roche réservoir y compris le volume de roche-mère nécessaire pour une accumulation de pétrole donnée; (2) la relation entre la composition isotopique des constituants des hydrocarbures gazeux et leur genèse; (3) le mécanisme de la migration du pétrole dans le réservoir, la fuite à partir de ce réservoir et les modifications qui en résultent dans le pétrole.

1. INTRODUCTION

Petroleum accumulations are seldom found at the site of deposition of the source material and subsequent genesis. In fact, migration out of the source rock and accumulation in a trap is essential to formation of a petroleum accumulation of sufficient size to be of economic interest. Just as this mobility leads to accumulation, it also leads to leakage out of the reservoir, either to another trap or to the surface.

The chemical and isotopic composition of a petroleum accumulation depends upon the composition of the source material, the extent of conversion to petroleum and fractionation during migration. Qualitative descriptions of such processes are not adequate either for scientific understanding, for suggesting future studies or as a basis for more effective exploration methods. A quantitative approach can be achieved through mathematical models. In this paper, accumulation–depletion of reservoirs and alteration of the chemical and isotopic composition of the accumulations are modeled. Numerical calculations indicating the consequences in the earth are provided for each model.

by J. E. SMITH, J. G. ERDMAN and D. A. MORRIS
Phillips Petroleum Company,
Bartlesville, Oklahoma, U.S.A.

2. THE DYNAMICS OF SHALE COMPACTION AND THE EVOLUTION OF PORE FLUID PRESSURES

The major fluid in compacting fine grained rocks such as shales is water. The fluid pressure history in such rocks is dominated by water and by the rate at which it can flow out of the rock. Water flow and pressure gradients can serve to transport petroleum from the source rock to the reservoir and under favorable conditions contribute to preservation of the accumulation.

The following model[1] incorporates the assumption that the phenomenological equation for water movement is Darcy's law. The results are applicable to any rock type for which the framework compressibility is much greater than the compressibility of the contained water or the mineral grains. The model can be extended to take the compressibility of water into account. Although discussed in terms of shale, the model is applicable to all fine grained compressible rocks.

A simplification of the problem of shale compaction is obtained by recognizing that shales are tubular bodies with average lateral widths some hundreds or thousands of times greater than their average thickness. Further, the permeability of shales is typically several orders of magnitude smaller than that of most other lithologies in contact with the shale. For these reasons, water being squeezed out of shales would tend to take the shortest path to more permeable strata, which would generally be above and/or below the shale unit. Water would also be squeezed laterally into more permeable units but only a limited fraction of the tubular shale body would be close enough to the periphery to lose a high percentage of water laterally. The present exposition treats the upward and downward movement of water out of shales and should apply to the major portion of most shale bodies.

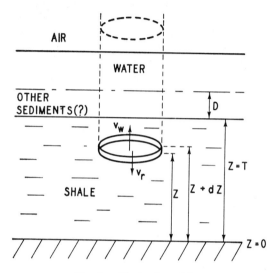

Fig. 1—Shale deposition.

The equations relating to compaction are obtained by considering the movement of solid grains and water through the upper and lower faces of a cylinder of unit cross section and infinitesimal thickness. The lower face of this cylinder (Fig. 1) is located a distance Z above the interface between the shale unit and the underlying lithologic unit. The upper face of this cylinder is located a distance $Z + dZ$ above the interface. Two continuity equations, one for the water and one for the solid grains, are obtained by considering that the difference between the mass of water, or grains, passing into the cylinder at Z minus the mass of water, or grains, flowing out of the cylinder at $Z + dZ$ in the same infinitesimal time increment, dt, is equal to the increase of the mass of the water, or grains, contained within the fixed cylinder in the same time increment. The resulting conditions on the porosity φ, the water density ρ_w, the water velocity v_w, the grain density ρ_r and the grain velocity v_r are:

$$-\frac{\partial \rho_w \varphi v_w}{\partial Z} = \frac{\partial \varphi \rho_w}{\partial t}, \tag{1}$$

$$-\frac{\partial \rho_r (1 - \varphi) v_r}{\partial Z} = \frac{\partial (1 - \varphi) \rho_r}{\partial t} \tag{2}$$

According to Darcy's law, the volume flow rate of water past grains, $(v_w - v_r)\varphi$, is proportional to the space gradient of the fluid potential, Ψ, and inversely proportional to the viscosity of the fluid, μ:

$$(v_w - v_r)\varphi = -\frac{K}{\mu}\frac{\partial \Psi}{\partial Z}, \tag{3}$$

$$\Psi = p - g \int_{Z=Z}^{T} \rho_w dZ - p_e \tag{4}$$

K is the permeability, g is the gravitational constant, p is the pore fluid pressure and p_e is the fluid pressure at $Z = T$ which is the interface between the overlying water and the sediments. The total vertical stress or overburden pressure, S, is given by

$$S = g \int_{Z=Z}^{Z=T} [\rho_w \varphi + \rho_r (1 - \varphi)] \, dZ + p_e. \tag{5}$$

The porosity of shales in flat-lying units which are not subjected to excess lateral compressive or tensional forces is determined by the frame pressure σ which is the difference between the total vertical stress and the fluid pressure.[2,3] This functional relationship may be obtained from experimentally determined shale density versus depth-of-burial relationships when the shales are normally pressured, i.e. $\Psi = 0$. The exponential dependence of porosity on depth, as determined by Athy,[4] was used to derive the following relationship which applies whether or not the fluid pressure is normal:

$$\sigma \equiv S - p = \frac{g(\rho_r - \rho_w)}{b}\left[\ln\frac{\varphi_T}{\varphi} - (\varphi_T - \varphi)\right]. \tag{6}$$

φ_T and b are empirical constants from the porosity-depth relationship. An additional term must be added to the right hand side of equation (6) if the shale unit is beneath other rock units.

When equations (1) and (2) are represented in integral form and the boundary conditions are taken into account, the variables v_w, v_r, Ψ, p and S can be eliminated among the above equations to obtain a single equation for the porosity:

$$\frac{1}{1 - \varphi}\int_{Z=0}^{Z}\frac{\partial \varphi}{\partial t} dZ = \frac{g(\rho_r - \rho_w)K(1 - \varphi)}{b\mu\varphi}\left\{\frac{\partial \varphi}{\partial Z} - b\varphi\right\}$$

+ time-dependent constant determined by boundary conditions. (7)

Permeability generally is a strong function of porosity. In obtaining numerical solutions to equation (7), the assumption is made that the permeability varies as the

TABLE I
SHALE PROPERTIES IMMEDIATELY AFTER DEPOSITION NORMALLY PRESSURED SUBSTRATUM

Fractional porosity	Normal porosity	Pore water pressure [bar]	Excess water pressure [bar]	Over-burden pressure [bar]	Frame pressure [bar]	Perme-ability [md × 10³]	Water velocity through rock [cm/10³ yr]	Water flow rate [cm³ water/cm² rock/10³ yr]	Rock velocity [cm/10³ yr]	Depth of burial [m]
(1)	(2)	(3)	(4)	(5)	(6)	(7)	(8)	(9)	(10)	(11)
0.4800	0.4800	0.0	0.0	0.0	0.0	11.0	1.57	0.75	−6.62	0
0.4671	0.4667	2.1	0.05	3.7	1.6	8.85	1.41	0.66	−5.97	20
0.4546	0.4539	4.1	0.09	7.4	3.2	7.12	1.25	0.57	−5.36	40
0.4424	0.4413	6.2	0.14	11.2	5.0	5.73	1.11	0.49	−4.80	60
0.4305	0.4291	8.3	0.20	15.0	6.7	4.61	0.96	0.42	−4.27	80
0.4190	0.4173	10.4	0.25	18.9	8.5	3.71	0.83	0.35	−3.78	100
0.4078	0.4058	12.5	0.31	22.8	10.3	2.99	0.69	0.28	−3.32	120
0.3968	0.3946	14.5	0.36	26.7	12.1	2.40	0.56	0.22	−2.89	140
0.3861	0.3837	16.6	0.41	30.6	14.0	1.93	0.43	0.17	−2.49	160
0.3757	0.3731	18.7	0.46	34.6	15.9	1.55	0.30	0.11	−2.11	180
0.3655	0.3628	20.7	0.49	38.6	17.9	1.25	0.17	0.06	−1.76	200
0.3555	0.3528	22.8	0.51	42.7	19.9	1.00	0.04	0.01	−1.43	220
0.3457	0.3430	24.8	0.51	46.8	22.0	0.80	−0.09	−0.03	−1.11	240
0.3360	0.3335	26.8	0.47	50.9	24.1	0.63	−0.21	−0.07	−0.81	260
0.3263	0.3243	28.7	0.39	55.0	26.3	0.50	−0.34	−0.11	−0.53	280
0.3167	0.3154	30.6	0.24	59.2	28.6	0.40	−0.48	−0.15	−0.26	300
0.3070	0.3067	32.4	0.00	63.4	31.0	0.31	−0.59	−0.18	0.0	320

TABLE II
SHALE PROPERTIES FOLLOWING DEPOSITION AND TIME LAPSE OF 100000 YEARS NORMALLY PRESSURED SUBSTRATUM

Fractional porosity	Normal porosity	Pore water pressure [bar]	Excess water pressure [bar]	Over-burden pressure [bar]	Frame pressure [bar]	Perme-ability [md × 10³]	Water velocity through rock [cm/10³ yr] × 10³	Water flow rate [cm³ water/cm² rock/10³ yr] × 10³	Rock velocity [cm/10³ yr] × 10³	Depth of burial [m]
(1)	(2)	(3)	(4)	(5)	(6)	(7)	(8)	(9)	(10)	(11)
0.4800	0.4800	0.0	0.0	0.0	0.0	11.0	22.2	10.7	−54.3	0
0.4673	0.4673	1.9	0.00	3.5	1.6	8.88	11.7	5.45	−54.1	19
0.4544	0.4544	4.0	0.00	7.3	3.3	7.10	6.12	2.78	−53.3	39
0.4419	0.4418	6.0	0.00	11.1	5.1	5.68	6.67	2.95	−51.8	59
0.4297	0.4296	8.0	0.00	14.9	6.9	4.54	7.01	3.01	−49.8	79
0.4178	0.4178	10.0	0.00	18.7	8.7	3.63	7.02	2.93	−47.0	99
0.4063	0.4062	12.1	0.00	22.6	10.5	2.90	6.67	2.71	−43.7	119
0.3951	0.3950	14.1	0.00	26.5	12.5	2.32	5.95	2.35	−39.7	139
0.3841	0.3841	16.1	0.00	30.5	14.4	1.85	4.88	1.88	−35.0	159
0.3735	0.3735	18.1	0.01	34.5	16.4	1.48	3.47	1.30	−29.9	179
0.3632	0.3632	20.2	0.01	38.5	18.4	1.18	1.77	0.64	−24.4	199
0.3532	0.3532	22.2	0.01	42.6	20.4	0.95	−0.16	−0.06	−18.7	219
0.3435	0.3434	24.2	0.01	46.7	22.5	0.76	−2.19	−0.75	−13.1	239
0.3340	0.3339	26.2	0.01	50.8	24.6	0.60	−4.16	−1.39	−8.04	259
0.3247	0.3247	28.3	0.00	55.0	26.7	0.48	−5.90	−1.92	−3.85	279
0.3158	0.3157	30.3	0.00	59.1	28.9	0.39	−7.09	−2.24	−1.02	299
0.3070	0.3070	32.3	0.00	63.3	31.0	0.31	−5.99	−1.84	0.0	319

TABLE III

SHALE PROPERTIES AFTER DEPOSITION OF 600 M OF OVERLYING SEDIMENTS NORMALLY PRESSURED SUBSTRATUM

Fractional porosity	Normal porosity	Pore water pressure [bar]	Excess water pressure [bar]	Over-burden pressure [bar]	Frame pressure [bar]	Perme-ability [md × 10⁵]	Water velocity through rock [cm/10³ yr]	Water flow rate [cm³ water/cm² rock/10³ yr]	Rock velocity [cm/10³ yr]	Depth [m]
(1)	(2)	(3)	(4)	(5)	(6)	(7)	(8)	(9)	(10)	(11)
0·1973	0·1973	60·7	0·0	129·5	68·7	0·90	1·08	0·21	−2·06	0
0·2048	0·1938	67·0	4·97	132·3	65·4	1·21	0·93	0·19	−1·89	13
0·2100	0·1895	72·9	9·26	136·0	63·1	1·48	0·782	0·16	−1·70	29
0·2134	0·1833	79·8	13·59	141·4	61·6	1·68	0·602	0·13	−1·45	53
0·2138	0·1772	85·3	16·84	146·8	61·5	1·71	0·455	0·10	−1·23	77
0·2122	0·1714	90·1	19·18	152·2	62·1	1·61	0·323	0·07	−1·03	101
0·2091	0·1657	94·2	20·85	157·7	63·5	1·43	0·199	0·04	−0·84	125
0·2044	0·1602	97·6	21·85	163·1	65·5	1·19	0·080	0·02	−0·66	149
0·1954	0·1532	100·8	21·83	170·4	69·6	0·83	−0·078	−0·02	−0·44	181
0·1892	0·1498	101·5	20·90	174·1	72·6	0·64	−0·160	−0·03	−0·33	197
0·1808	0·1465	101·1	18·80	177·8	76·7	0·45	−0·245	−0·04	−0·23	213
0·1682	0·1433	98·0	14·12	181·5	83·5	0·25	−0·339	−0·06	−0·12	229
0·1602	0·1417	95·2	10·55	183·4	88·1	0·17	−0·494	−0·08	−0·64	237
0·1420	0·1401	85·5	0·00	185·3	99·8	0·06	−0·471	−0·07	0·0	245

TABLE IV

SHALE PROPERTIES TWO MILLION YEARS AFTER BURIAL BENEATH 600 M NORMALLY PRESSURED SUBSTRATUM

Fractional porosity	Normal porosity	Pore water pressure [bar]	Excess water pressure [bar]	Over-burden pressure [bar]	Frame pressure [bar]	Perme-ability [md × 10⁶]	Water velocity through rock [cm/10³ yr] × 10²	Water flow rate [cm³ water/cm² rock/10³ yr] × 10²	Rock velocity [cm/10³ yr] × 10²	Depth [m]
(1)	(2)	(3)	(4)	(5)	(6)	(7)	(8)	(9)	(10)	(11)
0·1973	0·1973	60·7	0·0	129·5	68·7	8·97	4·74	0·94	−9·72	0
0·1937	0·1929	62·8	0·38	133·2	70·4	7·73	4·70	0·91	−9·64	16
0·1894	0·1875	65·3	0·89	137·8	72·5	6·46	4·54	0·86	−9·33	36
0·1853	0·1824	67·9	1·46	142·4	74·5	5·42	4·23	0·78	−8·79	56
0·1813	9·1773	70·5	2·05	147·0	76·5	4·56	3·73	0·68	−8·03	76
0·1774	0·1724	73·1	2·63	151·6	78·5	3·83	3·05	0·54	−7·08	96
0·1735	0·1677	75·7	3·15	156·3	80·6	3·21	2·22	0·39	−5·97	116
0·1694	0·1630	78·1	3·56	160·9	82·8	2·65	1·26	0·21	−4·75	136
0·1651	0·1585	80·3	3·76	165·6	85·3	2·16	0·22	0·04	−3·49	156
0·1604	0·1542	82·2	3·67	170·3	88·0	1·71	−0·09	−0·14	−2·26	176
0·1551	0·1499	83·8	3·15	175·0	91·2	1·31	−1·99	−0·29	−1·17	196
0·1490	0·1458	84·6	2·00	179·7	95·0	0·95	−2·72	−0·41	−0·034	216
0·1420	0·1417	84·6	0·00	184·4	99·8	0·64	−3·11	−0·44	−0·0	236

eighth power of the porosity.[5] The dependence of viscosity on salinity, temperature and pressure is also taken into account.

A finite difference technique was used to obtain a solution for a typical series of events providing an example of application of this model. The events are: (A) 320 m of shale are deposited at a rate of 0·1 m per thousand years, requiring 3·2 million years; (B) following the deposition of the shale, 100000 years elapse during which no further sedimentation occurs; (C) 600 m of normally pressured sediments are deposited over the shale at a rate of 0·5 m per thousand years, requiring 1·2 million years; (D) 2 million years elapse during which no further deposition occurs. Tables I to IV give profiles through the shale unit of the following properties at the end of periods A to D: (1) Fractional porosity of the shale; (2) Normal porosity of the shale, which is the value which would be approached after a long time lapse without further deposition, during which time the pore pressure would also approach normal; (3) The pore water pressure minus the fluid pressure at the top of the shale formation; (4) The excess water pressure. This is the pore water pressure minus the normal hydrostatic pressure for the subsurface depth in question; (5) The overburden pressure or total vertical stress, minus the pressure of the overlying water and atmosphere. This is the total weight, per unit area, of water-saturated rock overlying the depth in question; (6) The frame pressure. This is the difference between overburden pressure and pore water pressure and is the force per unit area acting to collapse the rock framework and reduce the porosity. This pressure is responsible for the slow expulsion of water from the low-permeability shale; (7) Permeability; (8) Velocity of water being squeezed out of the compacting shale, relative to the matrix material. After cessation of sedimentation, this flow will gradually decrease to zero as the shale becomes normally pressured; (9) Volume flow rate of water relative to the rock matrix; (10) Velocity of the compacting rock matrix relative to the base of the shale unit; (11) Depth from the top of the shale unit.

In this example it was assumed that the shale was underlain by a sandstone which remained normally pressured.

For contrast, another solution was obtained in which the shale is underlain by an impermeable base, for instance an evaporite sequence. The pressure profile following deposition of 600 m of normally pressured sediments over the shale was also calculated.

Two factors which influence the length of time for which high pressures persist are the thickness of the shale unit and the thickness of overlying normally

pressured sediments. The calculations described above were repeated for a shale unit originally 1000 m thick buried beneath 1300 m of sediments. The high pressures were found to persist for more than 100 million years following burial.

In these examples a portion of the shale unit next to a normally pressured sand tended to have a lower porosity and permeability, and to form a seal which slowed down further losses of water from the unit. The steep pressure gradients which develop in the shale favor the migration of petroleum into the sands.

Constants used in the above examples were $\varphi_T = 0·48$, $b = 0·0014/m$, and $\rho_r = 2·65$ g/cc. Permeability was related to porosity by the relationship $K = H\varphi^8$, H being determined so that $K = 10^{-5}$ md for $\varphi = 0·2$. The pore water was assumed to contain 5 per cent sodium chloride. Temperature was computed assuming 10°C for the sediment–water interface and a geothermal gradient of 0·03°C/m.

3. RELATIONSHIP BETWEEN PETROLEUM RESERVOIR ROCKS AND SOURCE ROCKS

A statistical relationship between the total ultimate recoverable oil in a sandstone and its average thickness is given by Curtis *et al.*[6] Curtis showed that the recoverable oil in a sandstone is proportional to the square of its average thickness (Fig. 2). The minor gas in the accumulations was included as an equivalent volume of oil.

The total in-place oil, V_0, in sandstone reservoirs may be related to the average sandstone thickness, Y, from Fig. 2 using an average recovery factor of 27%:[7]

$$V_0 = C Y^2. \qquad (8)$$

The dimensional dependence of equation (8) suggests that the area of contact between sandstones

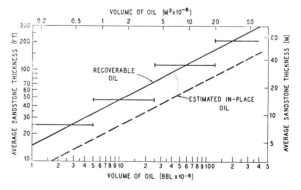

Fig. 2—Volumes of petroleum in sandstone reservoirs correlated with average thickness of sandstone.

107

and surrounding lithologies could be the factor controlling the amount of oil migrating into a sandstone.[1] This dependence can be rationalized on the basis of oil coming from low permeability rocks, such as shales, in contact with the sandstone. Shales have permeabilities typically three to six orders of magnitude smaller than sandstones.[8]

Two relationships between source rocks and sandstones are considered. Sandstones are represented as circular sheets of diameter D and thickness Y. One possible relationship between sandstone and source rock (Fig 3a) is where a contributing source rock is placed

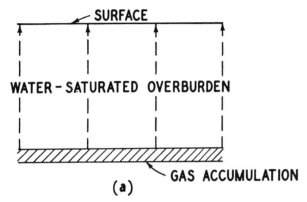

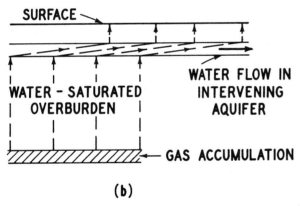

Fig. 3—Loss from gas accumulations to surface by (a) vertical diffusion only and (b) vertical diffusion with lateral transport by intervening aquifer.

laterally to the sandstone. The volume of oil yielded by the source rock is proportional to the product of source rock–sandstone contact area and the lateral width of contributing source rock, WDY, and to the average volume fraction of source rock yielded as oil, v:

$$V_0 = v(WDY). \qquad (9)$$

Equation (9) can be compared to equation (8) by factoring out Y^2:

$$V_0 = \left[vW\left(\frac{D}{Y}\right)\right]Y^2. \qquad (10)$$

The quantity in brackets must be independent of the sandstone average thickness and equal to C if it is to be consistent with equation (8). For this relation to be true D/Y must be independent of the sandstone thickness, statistically. Accordingly, the average lateral extent of productive sandstones should tend to increase in direct proportion to the thickness. Equations (8) and (10) together provide the following expression for the average width, W, of source rock contributing oil to sandstones which have proved to contain commercial accumulations:

$$W = \frac{C}{v(D/Y)}. \qquad (11)$$

A second arrangement of source rock and sandstone is shown in Fig. 3b. The source of thickness T may be below and/or above the sandstone, and oil migrates vertically into the sandstone. The volume of oil produced into the sandstone again is proportional to the volume of source rock, and is given by

$$V_0 = \left[\frac{\pi}{4}vT\left(\frac{D}{Y}\right)^2\right]Y^2. \qquad (12)$$

This result may be combined with equation (8) to give the source rock thickness, T, in terms of C, v and D/Y:

$$T = \frac{4C}{\pi v(D/Y)^2}. \qquad (13)$$

Both the average source rock width, equation (11), and the average source rock thickness, equation (13), now can be calculated by estimating the values of the constant C, the source rock yield v and the average ratio of sandstone width to thickness D/Y. The value of C is $2 \cdot 85 \times 10^4$ m^3 oil/m^2 (Fig. 2). An estimate of 10^3 was obtained for D/Y from a consideration of the statistics provided by Curtis *et al.*[6] and by Kaufman.[9] Estimates of v derived from work by Barbat,[10] McIver[11] and Hunt[12] were averaged to obtain $v \approx 1 \cdot 7 \times 10^{-3}$. These values of D/Y and v may differ from the correct average values by a factor of two. The order-of-magnitude estimates are

$W \approx 17$ km and $T \approx 21$ m. W is reduced to 5 km if source rock completely surrounding the sandstone periphery contributes oil.

A significant result of the above analyses is that the distance for primary migration of oil out of a source rock into a sandstone is two or three orders of magnitude greater for lateral migration than for vertical migration. This result is independent of the value of the fractional yield v of the source rock. In order for lateral source rocks to produce as much oil into a reservoir sandstone as source rocks above or below the sand, the horizontal permeability of the source rocks would have to exceed the vertical permeability by a factor of several hundred. In addition the bedding planes of the sandstone and source rock would have to be parallel to a corresponding degree of precision. An assessment of the relative contributions of vertical and horizontal migration must await the measurement of directional permeabilities.

4. LOSSES OF PETROLEUM FROM RESERVOIRS OVER GEOLOGIC TIME

Seepage of petroleum to the surface is a common occurrence in many petroliferous basins. A variety of micro methods have been developed for detection of petroleum hydrocarbons in surface soils and more recently in water offshore as an aid to exploration. Roof and Rutherford[13] and Sokolov *et al.*[14] have treated migration of petroleum hydrocarbons in the earth. The slowest mechanism for transport is diffusion except when the reservoir is overlain by rocks over-pressured relative to the reservoir.

The loss of light hydrocarbons from reservoirs by diffusion to the surface of the earth[1] is a major mechanism for depleting petroleum accumulations over geologic time. Fick's law describes the diffusion of gas components, through the pore waters of the rock overlying an accumulation:

$$\frac{\partial C_i}{\partial t} = D_i \frac{\partial^2 C_i}{\partial Z^2}. \qquad (14)$$

C_i is the concentration of gas component i in the pore waters at a distance Z above the accumulation. D_i is the diffusion coefficient for component i through the water-saturated rock and t is the time. The influences of lateral diffusion of water flow both vertically and laterally and of cap rocks are discussed following the discussion of steady-state and non-steady-state diffusion vertically through a uniform overburden.

For the models to be presented, it is assumed that the composition and size of the accumulations in question do not change in geologic time. This assumption requires that diffusional losses be replaced, e.g. from source rocks. The rate of loss determined for these models indicates how accumulations change in composition and size when losses are not replaced. Non-steady-state gas concentrations C_i in the pore water are given by[15]

$$C_i = C_i{}^0 \left(1 - \frac{Z}{T} \right)$$

$$- \frac{C_i{}^{02}}{\pi} \sum_{n=1}^{\infty} \frac{\sin(n\pi Z/T)}{n} e\left[-D_i \left(\frac{n\pi}{T} \right)^2 t \right]. \qquad (15)$$

$C_i{}^0$ is the concentration of component i in the pore water at the top of the accumulation ($Z = 0$). The overburden thickness is T. Numerical examples are concerned principally with the steady-state, for which the gas concentrations decrease linearly from $C_i{}^0$ near the accumulation to zero at the surface of the earth. The leading term in the above equation gives this time-independent solution.

Parameters relevant to gas production in the United States were usually adapted in the examples. An average depth to production of 1737 m and an average composition by mole per cent $CH_4 = 74.51$, $C_2H_6 = 5.56$, $C_3H_8 = 3.21$, n- and i-$C_4H_{10} = 1.57$, other components $= 15.15$ were estimated from Boone.[16] A reservoir temperature 70°C and pressure 172 atm were chosen to be consistent with the depth and a water salinity of 30000 ppm. Solubilities $C_i{}^0$ as mole fractions of the gas component in the salt water were estimated to be 12.9, 1.29, 0.605 and 0.205 × 10^{-4} for methane through the butanes.[17,18] These concentrations correspond to a total, for methane through the butanes, of 1.9 volumes of gas at standard conditions per volume of pore water. Average co-efficients of diffusion through water for the temperature range between the reservoir and surface are 2.70, 2.15, 1.74 and 1.40 × 10^{-5} cm²/sec for methane through the butanes.[19] The coefficients of diffusion through the overburden are uniformly reduced by a geometrical factor of 39.3, obtained by averaging data from Kartsev *et al.*[20] Waxman and Smits[21] have obtained similar values for this geometrical factor by conductivity measurements on very shaly sandstone cores. The statistics of Curtis *et al.*[6] show that a commercial accumulation of 10 × 10^9 scf or 2.8 × 10^8 m³ gas at standard conditions is close to the most frequently found size, and only a few per cent of accumulations contain more than 28 × 10^8 m³ gas. All computations are carried out for these two sizes of accumulation. The areas covered by these accumulations are estimated assuming a porosity of 15.7%,[22] a gas saturation of 80% and average sandstone reservoir thicknesses of 9.1 m and 12.2 m respectively.[6] A porosity of 15.7% is also adopted for the overburden.

The following results apply to the steady-state except where indicated.

The quantity of methane through butanes dissolved in pore water between the smaller reservoir and the surface exceeds the quantity in the reservoir by 60%. The dissolved gases exceed the quantity in the larger

TABLE V
LOSSES FROM RESERVOIRS BY DIFFUSION

Volume of Gas In Reservoir		
$[m^3 \times 1D^{-8}]$	2·8	28
Surface Area Above Reservoir [km²]	1·5	11
Component	*Total Losses* $[m^3/yr]$	
methane	2·84	21·3
ethane	0·23	1·70
propane	0·09	0·64
butanes	0·02	0·18
	3·2	23·8
Component	*Residence Time* $[yr \times 10^{-6}]$	
methane	74	99
ethane	70	93
propane	106	141
butanes	190	253

reservoir by 21%. The rate of loss of methane, ethane, propane, and the butanes at the surface would be 2·0, 0·16, 0·06 and 0·02 cm³/m²/yr for either reservoir. Whereas the relative molar composition $CH_4 : C_2H_6 : C_3H_8 : C_4H_{10}$ in the reservoirs was taken to be 100 : 7·5 : 4·3 : 2·1, these gases emerge at the surface in the ratio 100 : 8·0 : 3·0 : 0·82 after diffusing through the overburden. In the case of an actual reservoir both these sets of ratios would change with time since the composition and rate of supply of these gases from source rocks would tend to vary.

Table V shows the net losses from each reservoir per year and the residence times of the gas components in the reservoir. Although annual losses are small the reserves are rapidly depleted on a geologic time scale. The rates of loss may be compared to a marginally visible surface seep in water of only 15 gas bubbles per minute, with 0·15 cm³ per bubble. Such a seep would leak 1·2 m³ of gas per year.

For the diffusing gas component, the time lapses between leaving the accumulation and reaching the surface are 140, 170, 230 and 270 million years for methane, ethane, propane and the butanes, respectively.

The rate of diffusional loss of components from reservoir is much greater if the reservoirs are suddenly filled at some point in geologic time and the gas components begin to diffuse into the overlying pore waters which initially contain no dissolved gas [equation (15)]. For this non-steady-state condition only 2·7 million years are required for the pore water in the overburden to become an average of 10% saturated with methane. For the butanes which diffuse more slowly, 5·3 million years are required to attain the same degree of average saturation. These losses correspond to 16% of the component originally present in the smaller reservoir and 12% of the component originally present in the larger reservoir.

The initial losses from the reservoirs tend to saturate the pore waters nearest the reservoir and much more time is required for the diffusing gases to reach the surface of the earth, or for a steady-state to be attained. For both steady-state and non steady-state diffusion the rate of loss of gas from the reservoirs will be greater than has been computed if lateral as well as vertical diffusion is taken into account.

The steady-state velocity of the diffusional front is different for the several components, and also varies with depth as shown in Table VI. These diffusional velocities are of the same magnitude as the velocities of water movement through compacting shales, as has been shown in column 8 of Table I.

An intervening aquifer 3 m thick, with a lateral water velocity[23] of 1 m per year, will move the vertically diffusing gas an appreciable distance laterally. The maximum lateral displacements, shown in Table VI, were computed assuming that the rock fabric does not induce vertical mixing within the aquifer. Comparison of the columns for methane and butane indicates that fractionation occurs when mixed hydrocarbons diffuses into or out of strata containing moving water.

In vertical, water-filled joints and faults, the diffusion coefficients of the various gases would be increased by a factor of approximately 39. Since the voids would not be expected to extend unbroken from the accumulation to the surface, the diffusional losses of gases to the surface would be increased over faults and fractured zones but by a factor less than 39. There also would be corresponding decreases in the time required for the gas to reach the surface and in the lateral displacement effected by intervening aquifers.

It is likely that the diffusion coefficients for gases in some reservoir seals are much less than the uniform values that have been assumed for the entire overburden, although no measurements are known to exist. The qualitative effect of a seal may be anticipated by arbitrarily assuming that the 150 m of overburden immediately above the reservoirs is the reservoir seal, with the diffusion coefficients for gases

TABLE VI
TRANSPORT OF GASES DIFFUSING FROM RESERVOIRS

Subsurface depth [m]	*Velocity of vertical diffusion front* (cm/1000 yr)	
	methane	*butanes*
1740	0·8	0·4
870	1·6	0·8
174	8·0	4·1
Depth to Aquifer [m]	*Maximum lateral displacement* [km]	
	methane	*butanes*
174	38	73
870	190	360

reduced to 1/500 the coefficients in water. Porosity of the seal may be estimated as 4·5% if the formation factor for the seal is 500 and it is assumed that the formation factor varies inversely as the square of the porosity (Archie's relation). Coefficients of diffusion and porosity in the remaining 1590 m of overburden remain the same as previously assumed. With this modified steady-state model, the concentrations of the dissolved gases decrease linearly by 55% in diffusing from the reservoir up through the seal, and then decrease linearly through the remaining 1590 m of overburden to essentially zero at the surface. As compared to the previous model, (1) the quantities of gas held up between the reservoir and surface are decreased by 55%, (2) the rates of loss of gas to the surface are decreased by 50% and (3) the diffusional front velocities and the effects of lateral water flow are the same, in the interval between 1590 m and the surface.

5. INFLUENCE OF KINETICS ON CARBON ISOTOPIC COMPOSITION

Among the low molecular weight components of a petroleum, that is methane through the pentanes, the carbon isotopic compositions may differ by 40% or more of the total variation in nature. The major causes of this large variation are (1) migrational effects and (2) chemical kinetic effects stemming from the fact that $-C^{12}$ and $-C^{13}$ bonds break at slightly different rates in a given reaction. Both effects should be largest for methane and methane does show the greatest variation in the ratio C^{13}/C^{12} of any natural gas component.[24,25]

In regard to isotope effects accompanying losses by diffusion, the same concepts and equations discussed in the previous section generally apply. Discussion of actual examples is prevented primarily by the sparsity of data for isotopic effects on solubilities and diffusion coefficients in water. Investigators including Silverman,[26] Stahl[27] and Colombo *et al.*[28] have proposed both diffusional and chemical kinetic explanations for isotope distributions in gas and oil. In this section a model treating the influence of kinetics on carbon isotopic composition is discussed both as a basis for future evaluation of migrational effects and for better understanding of this important geochemical process.

Two situations will be considered: (1) A sample of gas has an isotopic composition representative of the cumulative gas produced from the organic source material and (2) a sample of gas has an isotopic composition representative of that last produced from the source, previously produced gas having been lost.

Assumptions providing the basis for the derivations to follow are:

(1) The precursor to a given petroleum molecular species initially contains C^{13} randomly distributed throughout the portion of the precursor which yields the species. This assumption may be relaxed to allow the α (alpha) carbon to have a different initial probability of being C^{13}, if experience shows this to be necessary.

(2) A given precursor produces only one petroleum molecular species.

(3) Interchange of C^{13} for C^{12}, or vice versa, within the precursor does not change the position in the precursor at which bond breakage occurs.

(4) The isotopic composition of that part of the precursor which does not become the product petroleum molecular species under consideration does not affect the ratio of the rates of breaking $-C^{12}$ and $-C^{13}$ bonds, and has a negligible effect on the fraction of any isotopic species of precursor remaining at any given time. The effect of a failure of the second part of this assumption would be relatively unimportant at the start of the reaction, and would become more important as the reaction goes to completion. Failure of either part of this assumption would be quantitatively unimportant if the abundance of the rarer isotope of the atom affecting the rate of bond breakage is of the order of 1% or less. This is the case when the reaction involves breaking a C–C, N–C or O–C bond, and secondary isotopic effects are negligible.

(5) The ratio of the rate constants for the isotopically different reactions leading to the production of any species is insensitive to temperature variations in the natural environment. The rate constants themselves may be sensitive functions of temperature.

The above assumptions suffice to derive all the formulas presented here.

A petroleum molecule C_nU is considered to be formed from a precursor $T-C_nU'$ as indicated by the following expression:

$$T-C_nU' \rightarrow T' + C_nU. \qquad (16)$$

It is not necessary to specify just what T, T′, U′ and U are, as the essential points are that (a) the petroleum molecule formed has *n* carbon atoms and (b) one of these carbon atoms is bonded to the group T– before the reaction. When the petroleum molecule is an alkane, U' is H_{2n+1} and U is H_{2n+2}. In cases where the petroleum molecule contains mC^{13} atoms and $(n-m)C^{12}$ atoms, the precursor will be written $T-(C_m^{13}C_{n-m}^{12}U')$. When it is necessary to show that a $-C^{13}$ bond is to be broken the precursor will be written $T-C_m^{13}C_{n-m}^{12}U'$, and when a $-C^{12}$ bond must be broken the precursor will be written $T-C_{n-m}^{12}C_m^{13}U'$. Both of these two precursors yield a petroleum molecule with mC^{13} and $(n-m)C^{12}$ atoms:

$$T-C_m^{13}C_{n-m}^{12}U' \rightarrow T' + C_m^{13}C_{n-m}^{12}U, \quad (17)$$

$$T-C_{n-m}^{12}C_m^{13}U' \rightarrow T' + C_m^{13}C_{n-m}^{12}U. \quad (18)$$

The rate of decrease in concentration of either type of precursor is given by

$$\frac{d}{dt}[T-C_m^{13}C_{n-m}^{12}U'] = -k_{13}[T-C_m^{13}C_{n-m}^{12}U']f, \quad (19)$$

$$\frac{d}{dt}[T-C_{n-m}^{12}C_m^{13}U'] = -k_{12}[T-C_{n-m}^{12}C_m^{13}U']f. \quad (20)$$

The symbol t is used for time. The factor f involves concentrations of reactants such as water or concentrations of catalytic sites, or other quantities which are the same for both isotopic reactions. k_{13} and k_{12} are the rate constants. The above two equations may be combined to eliminate f and give

$$\frac{d \ln [T-C_m^{13}C_{n-m}^{12}U']}{dt} = K\frac{d \ln [T-C_{n-m}^{12}C_m^{13}U']}{dt} \quad (21)$$

where

$$K \equiv \frac{k_{13}}{k_{12}}. \quad (22)$$

If the ratio of rate constants K is insensitive to temperature over the range of temperatures for which the reaction occurs in the earth, then equation (21) may be integrated to give

$$\frac{[T-C_m^{13}C_{n-m}^{12}U']}{[T-C_m^{13}C_{n-m}^{12}U']^0} = \left(\frac{[T-C_{n-m}^{12}C_m^{13}U']}{[T-C_{n-m}^{12}C_m^{13}U']^0}\right)^K. \quad (23)$$

The superscript ^{0} indicates the initial concentrations at $t = 0$. The quantity on the left of the above equation is the fraction of the starting material $T-C_m^{13}C_{n-m}^{12}U'$, remaining at time t, and the symbol X_{13} will be used for this quantity:

$$X_{13} \equiv \frac{[T-C_m^{13}C_{n-m}^{12}U']}{[T-C_m^{13}C_{n-m}^{12}U']^0}. \quad (24)$$

As long as $m \geq 1$, C_{13} is independent of m since the reaction rate is affected only by the carbon at the breaking bond. Similarly the fraction of starting material $T-C_{n-m}^{12}C_m^{13}U'$ remaining at time t is

$$X_{12} \equiv \frac{[T-C_{n-m}^{12}C_m^{13}U']}{[T-C_{n-m}^{12}C_m^{13}U']^0}. \quad (25)$$

Equation (23) may now be re-written

$$X_{13} = X_{12}^K. \quad (26)$$

The concentration of $C_m^{13}C_{n-m}^{12}U$ at any time is given by

$$[C_m^{13}C_{n-m}^{12}U] = ([T-C_m^{13}C_{n-m}^{12}U']^0 \\ - [T-C_m^{13}C_{n-m}^{12}U']) \\ + ([T-C_{n-m}^{12}C_m^{13}U']^0 \\ - [T-C_{n-m}^{12}C_m^{13}U']). \quad (27)$$

The initial concentrations of the isotopically distinct precursors may be related to the total initial concentration $[T-C_nU']^0$ of precursor under the assumption that C^{13} and C^{12} are initially randomly distributed in the labile portion of the precursor:

$$[T-C_m^{13}C_{n-m}^{12}U']^0 \\ = \frac{(n-1)!p_{12}^{n-m}p_{13}^m}{(n-m)!(m-1)!}[T-C_nU']^0, \quad (28)$$

$$[T-C_{n-m}^{12}C_m^{13}U']^0 \\ = \frac{(n-1)!p_{12}^{n-m}p_{13}^m}{(n-m-1)!m!}[T-C_nU']^0. \quad (29)$$

Here p_{12} is the initial probability of any carbon atom in the $-C_nU'$ group being C^{12}, and p_{13} is the initial probability of any carbon being C^{13}. By definition,

$$p_{12} + p_{13} = 1. \quad (30)$$

The numbers of gram atoms of C^{13} and C^{12} in the product petroleum molecules of the type C_nU are given by

$$[C^{13}]_n = \sum_{m=1}^{n} m[C_m^{13}C_{n-m}^{12}U], \quad (31)$$

$$[C^{12}]_n = \sum_{m=0}^{n-1} (n-m)[C_m^{13}C_{n-m}^{12}U]. \quad (32)$$

The subscript n is used to indicate the number of carbons in the produced molecule.

The ratio of $[C^{13}]_n$ to $[C^{12}]_n$ at any time may be determined with use of equations (23) through (32) to be

$$\frac{[C^{13}]_n}{[C^{12}]_n} = \left(\frac{p_{13}}{p_{12}}\right)$$

$$\left[1 - \frac{1 - (1 - X_{12}{}^K)/(1 - X_{12})}{n\{1 - p_{13}(n-1)/n\,(1 - [1 - X_{12}{}^K]/[1 - X_{12}])\}}\right]. \tag{33}$$

The quantity { } in the above result differs from 1 by a factor of 10^{-3} or less and may always be replaced by unity without any sensible loss of accuracy. Hence

$$\frac{[C^{13}]_n}{[C^{12}]_n} = \left(\frac{p_{13}}{p_{12}}\right)\left[1 - \frac{1 - (1 - X_{12}{}^K/1 - X_{12})}{n}\right]. \tag{34}$$

This expression gives the isotopic ratio $[C^{13}]_n/[C^{12}]^n$ for the cumulative product C_nU, in terms of (1) the

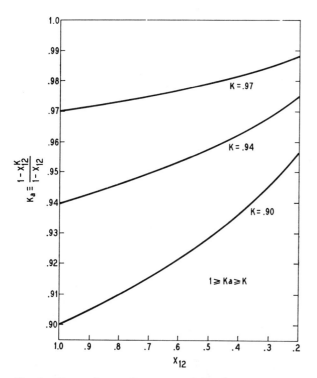

Fig. 4—*Dependence of apparent ratio of rate constants, K_a, on fraction of remaining reactants, X_{12}, and the actual ratio of rate constants, K.*

isotopic ratio for the initial starting material p_{13}/p_{12}, (2) the remaining fraction of reactants $\approx X_{12}$, (3) the number of carbons in the product molecule n and (4) the ratio of the rate constants for breaking $-C^{13}$ and $-C^{12}$ bonds, K.

The factor $(1 - X_{12}{}^K)/(1 - X_{12})$ in the above result will be called the apparent ratio of rate constants K_a. It is equal to K at the initiation of the reaction

producing C_nU and is always numerically between K and 1 (Fig. 4):

$$K_a \equiv \frac{1 - X_{12}{}^K}{1 - X_{12}}, \tag{35}$$

$$\frac{[C^{13}]_n}{[C^{12}]_n} = \frac{p_{13}}{p_{12}}\left[1 - \frac{1 - K_a}{n}\right]. \tag{36}$$

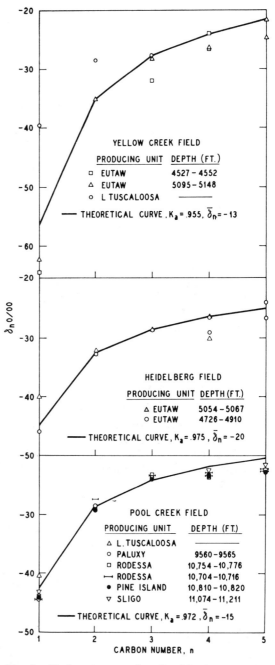

Fig. 5—*Carbon isotope data for Mississippi gases.*

Isotope ratios are usually compared to a standard, such as the PDB standard[30] and are reported in the "del" unit which is the parts per thousand fractional deviation from the standard:

$$\delta_n \equiv \left[\frac{([C^{13}]_n/[C^{12}]_n)}{([C^{13}]_{PDB}/[C^{12}]_{PDB})} - 1 \right] \times 10^3, \quad (37)$$

$$\delta \equiv \left[\frac{(p_{13}/p_{12})}{[C^{13}]_{PDB}/[C^{12}]_{PDB}} - 1 \right] \times 10^3. \quad (38)$$

Equation (36) can finally be re-written in terms of these δ values as

$$\delta_n = \bar{\delta} - [10^3 + \bar{\delta}]\left(\frac{1 - K_a}{n}\right). \quad (39)$$

Experimental values of δ_n versus n are given in Fig. 5 for methane ($n = 1$) through the pentanes ($n = 5$). The gases were produced from the Yellow Creek, Heidelberg and Pool Creek oil fields in Mississippi. For the individual fields, single values of K_a and $\bar{\delta}$ have been chosen so that equation (39) fits the data qualitatively. In employing equation (39) in this manner, it is being assumed that (1) the gases represent the cumulative production from the source, (2) the same fraction of the individual precursors for methane through the pentanes remains unreacted, (3) the ratio of isotopic rate constants K is the same for the precursors of methane through the pentanes and (4) the isotopic composition for the initial precursors, $\bar{\delta}$, is the same for methane through the pentanes. The only checks as to the plausibility of these assumptions are the values of $\bar{\delta}$ and K_a required to approximately fit the data. They are

$$-20 \leq \bar{\delta} \leq -13,$$

$$0.955 \leq K_a \leq 0.975.$$

The required values of $\bar{\delta}$ are within the range of possible organic precursors as shown in Fig. 6 (Silverman).[26] The range of K_a values suggests that $K < 0.955$. Stevenson *et al.*[31] and Melander[32] give examples for which the carbon isotope effect is of this magnitude, although the reactions are not necessarily similar.

An incremental quantity of gas produced from source materials between t and $t + dt$, at which time X_{12} is approximately the remaining fraction of source material, will have the isotope ratio

$$\frac{[C^{13}]_{nI}}{[C^{12}]_{nI}}$$

$$= \frac{p_{13}}{p_{12}}\left[1 - \frac{1 - KX_{12}^{K-1}}{n\{1 + [(n-1)/n]p_{13}[KX_{12}^{K-1} - 1]\}} \right]. \quad (40)$$

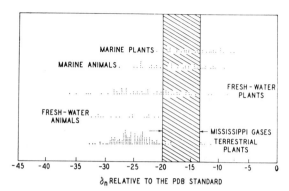

Fig. 6—*Comparison of carbon isotope ratios of modern fauna and flora with computed values for the precursors of the Mississippi gases.*

This result is obtained by considerations analogous to those used to derive equation (33), without any new assumptions. The factor { } equals 1 ± 10^{-3} as long as at least 10% of the precursors remain unreacted and will be replaced by unity. The subscript I indicates that the isotope ratio is for incremental gas production. The δ values for incremental gas production are:

$$\delta_{nI} = \bar{\delta} - (10^3 + \bar{\delta})\left[\frac{1 - KX_{12}^{K-1}}{n} \right]. \quad (41)$$

Fig. 7 shows δ_n and δ_{nI} as a function of n using $\bar{\delta} = -15$ and $K = 0.92$. From the figure it is seen

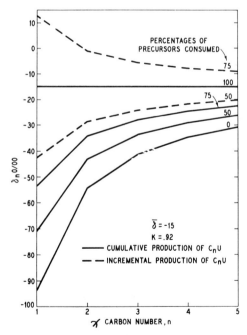

Fig. 7—*Theoretical isotopic compositions of the hydrocarbon gases.*

that when 75% of the source materials are used up, the methane has become isotopically heavier than the other hydrocarbon gases. This situation should be comparatively rare for commercial fields, since it would imply that most of the produced gases have escaped.

6. DISCUSSION AND CONCLUSIONS

For scientific understanding and full exploration benefits, genesis must be considered as well as migration–accumulation–depletion mechanisms and the changes in chemical and isotopic composition within the reservoir which accompany all of these processes.

The geochemical origin of petroleum and the chemical origin of low molecular weight hydrocarbon constituents have been discussed by Erdman.[33,34] Petroleum is conceived as deriving from the organic remains of plants and animals deposited either as particulate matter or as adsorbed molecules on the surfaces of fine grained minerals in an aquatic environment. The organic matter is preserved under these conditions partly by exclusion of oxygen and partly by the biostatic character of the older water diffusing upward through the sediment.

A limited number of the chemical constituents of petroleum, for example methane, n-heptane and n-paraffins above nonane are contained in the original biogenetic material. These petroleum constituents, mainly hydrocarbons, become dispersed in the large volume of water, either in solution or colloidly and are flushed out of the sediment or rock during the early stages of compaction. Only rarely does methane in natural gas have the very light isotopic composition characteristic of biogenetic methane.

The bulk of the organic matter retained in the sediment is either insoluble in water, or if soluble strongly polar hence adsorbed tenaciously on mineral surfaces. Recent researches by Eisma[35] indicate that the mineral surfaces catalyze reactions of the adsorbed organic molecules leading to the formation of the array of weakly polar compounds which comprise geochemically young petroleums. In general these processes are slow and are accelerated by increasing temperature with depth of burial.

Low molecular weight components, principally those of natural gas, will dissolve at least in part in the pore water and migrate as the water is squeezed out of the rock by compaction. If the productivity of the source rock is sufficient to nearly saturate the pore water and if it is not appreciably diluted during migration, the dissolved petroleum components will be released into a gas phase upon reaching a zone of decreased pressure. As shown in section 2 on compaction, the potential gradient may decrease either upward or downward, hence gas accumulation can be derived from source rocks either above or below the trap.

Higher molecular weight components will tend to accumulate as a separate phase in the interstices of the source rock, migration being restrained by the small passages between grains. Gas will also dissolve in this liquid oil. Again if productivity of the source rock is sufficient, a degree of compaction will be reached when the ratio of oil to water will be sufficiently high to permit the oil to move as a single phase. Again as shown by the compaction models, migration of oil phase may be either downward or upward into a trap.

A consequence of this concept of genesis–migration is that a source rock must possess a minimum productivity if it is to provide a petroleum accumulation. As indicated by the model development in section 3, declining productivity of source rock cannot be compensated indefinitely by increasing source rock volume.

The higher the productivity of a source rock, the lower the compaction necessary to permit migration under conditions favorable for accumulation. As a consequence of the kinetic effects on isotopic composition discussed in section 5, a young petroleum from a source rock of high productivity should contain methane, ethane, etc. much lighter in carbon isotopic composition than the higher molecular weight portion of the petroleum. The isotopic differences should decrease with decreasing source rock productivity. Continuing contributions from the source rock, chemical degradation of high molecular weight components of the oil in the reservoir, and diffusional losses from the reservoir will tend to decrease the difference in isotopic composition of the low molecular weight components.

On the basis of all considerations discussed, the best preserved accumulations will be found in traps overlain by a fine grained rock such as a shale which is over-pressured relative to the reservoir and which is itself the source rock for the petroleum accumulation.

Acknowledgements

Acknowledgement is made of the assistance of G. C. Dysinger who provided a computational technique and R. S. Scalan who helped with calculations.

REFERENCES

1. J. E. SMITH, submitted for publication in more complete form in J. International Ass. Math. Geol.
2. "Theoretical Soil Mechanics." K. TERZAGHI, New York, 1943, John Wiley.
3. M. K. HUBBERT and W. W. RUBEY, Bull. Geol. Soc. Am., 1959, **70**(2), 115–66.
4. L. F. ATHY, Bull. Am. Ass. Petrol. Geol., 1930, **14**(1), 1–22.
5. "Handbook of Well Log Analysis", S. J. PIRSON, New Jersey, 1963, Prentice-Hall.
6. "Geometry of Sandstone Bodies", Ed. J. A. PETERSON and J. C. OSMOND, Tulsa, 1961, Am. Ass. Petrol. Geol.
7. "Mechanics of Secondary Oil Recovery", C. R. SMITH, New York, 1966, Reinhold.
8. M. GONDOUIN and C. SCALA, Petrol. Trans. AIME, 1958, **213**, 170–9.
9. G. M. KAUFMAN, Paper presented at joint meeting of AAPG, SEPM, GAC, and MAC, Toronto, May 18–21, 1964.
10. "Habitat of Oil", ed. L. G. WEEKS, Tulsa, 1958, Am. Ass. Petrol. Geol.
11. "Seventh World Petroleum Congress Proceedings", Vol. 2, London, 1967, Elsevier.
12. J. M. HUNT, Geochim. et Cosmochim. Acta, 1961, **22**(1), 37–49.
13. J. G. ROOF and W. M. RUTHERFORD, Bull. Am. Ass. Petrol. Geol., 1958, **42**(5), 963–80.
14. "Sixth World Petroleum Congress Proceedings", Section I, Hamburg, 1963, Verein Zur Förderung Des 6. Welt-Erdöl-Kongresses.
15. "Mathematical Theory of the Conduction of Heat in Solids", p. 67, H. S. CARSLAW, New York, 1945, Dover.
16. "Helium-Bearing Natural Gases of the United States", W. J. BOONE, Jr., Washington, 1958, U.S. Govt. Printing Office.
17. "Handbook of Natural Gas Engineering", D. L. KATZ *et al.*, New York, 1959, McGraw-Hill.
18. T. A. MISHNINA, O. I. AVDEEVA and T. K. BOZHOVSKAYA, Mat. Vsesoyuz. Naucho-Issled. Geolog. Instituta, New Series, 1961, **46**, 93–110.
19. P. A. WITHERSPOON and D. N. SARAF, J. Phys. Chem., 1965, **69**(11), 3752–5.
20. "Geochemical Methods of Prospecting and Exploration for Petroleum and Natural Gas", A. A. KARTSEV *et al.*, Berkeley and Los Angeles, 1959, Univ. of California Press.
21. M. H. WAXMAN and L. J. M. SMITS, Soc. Petrol. Eng. J., 1968, **8**(2), 107–36.
22. W. D. VON GONTEN and R. L. WHITING, Soc. Petrol. Eng. of AIME, 1967, **240**, 266–72.
23. "Hydrology", O. E. MEINZER, New York, 1942, Dover.
24. W. M. SACKETT, Bull. Am. Ass. Petrol. Geol., 1968, **52**(5), 853–7.
25. F. GAZZARRINI, Riv. Combust., 1968, **20**(7–8), 377–88.
26. S. R. SILVERMAN, J. Amer. Oil Chem. Soc., 1967, **44**(12), 691–5.
27. W. J. STAHL, Erdöl Kohle Erdgas Petrochem., 1968, **21**(3), 514–8.
28. "Advances In Organic Geochemistry", ed. G. D. HOBSON and M. C. LOUIS, Guildford and London, 1966, Pergamon.
29. "An Introduction to Probability Theory and Its Applications", W. FELLER, New York, 1957, John Wiley.
30. H. CRAIG, Geochim. et Cosmochim. Acta, 1953, **12**(1, 2), 133–49.
31. D. P. STEVENSON *et al.*, J. Chem. Phys., 1948, **16**(10), 993–4.
32. "Isotope Effects on Reaction Rates", L. MELANDER, New York, 1960, Ronald Press.
33. "Fluids In Subsurface Environments", ed. A. YOUNG and J. E. GALLEY, Tulsa, 1965, Am. Ass. Petrol. Geol. publ.
34. "Seventh World Petroleum Congress Proceedings", Vol. 2, London, 1967, Elsevier.
35. J. W. JURG and E. EISMA, Science, 1964, **144**, 1451.

BULLETIN OF THE AMERICAN ASSOCIATION OF PETROLEUM GEOLOGISTS
VOL. 38, NO. 5 (MAY, 1954), PP. 816-853, 17 FIGS.

DIFFERENTIAL ENTRAPMENT OF OIL AND GAS:
A FUNDAMENTAL PRINCIPLE[1]

WILLIAM CARRUTHERS GUSSOW[2]
Calgary, Alberta, Canada

ABSTRACT

The principle of differential entrapment is outlined as an explanation of the so-called anomalous occurrences of oil and gas in contiguous structures. Selective trapping or "gas flushing" is responsible for the fact that some traps are gas fields and not oil fields, and explains why many apparently good structures are dry. The more important factors which modify accumulations of hydrocarbons are listed and discussed.

The Bonnie Glen-Wizard Lake reef trend of Alberta, Canada, is described as an example in which the accumulation of oil and gas is largely controlled by the principle of differential entrapment. Many other examples of actual geological occurrences that have been described in the literature are illustrated and reinterpreted. Many of the accumulations in the Rocky Mountain province of the United States are classic examples. Relationships in the Persian fields are mentioned briefly, and a new explanation of the synclinal occurrence of oil is presented.

New concepts introduced are: (1) hydrocarbons move along definite migration paths; (2) they migrate as streams or rivulets; (3) traps located along these migration paths will be filled with oil and/or gas; those not on the path of migration will remain loaded with salt water; (4) traps filled with oil are still effective gas traps but a trap filled with gas is not an effective trap for oil; (5) regional or distant migration is believed essential for commercial accumulations of hydrocarbons.

The greatest argument against the theory of a distant source and migration for long distances has been refuted, and a clear understanding of this important principle and its application should lead to many new-pool and new-field discoveries.

INTRODUCTION

In 1861, T. Sterry Hunt of the Geological Survey of Canada published his "Anticlinal Theory." This was later expanded to include all types of structural traps and is now commonly known as the "Structural Theory." Although nearly 40 years elapsed before it was generally accepted, it now forms one of the more important considerations in exploration for oil and gas.

During the past 50 years, many examples of apparently "anomalous" gas-oil relationships have been recorded and undoubtedly much evidence exists that has not found its way into print. The apparent "anomalous" relationships of oil and gas in interconnected or contiguous reservoirs have so far not had a satisfactory explanation. Most of the theories advanced depend on the presence or absence of source material, size of the drainage area, and movement of water which might flush the oil out of some traps. The existence of faulting has also been suggested as an important reason.

[1] Presented at the Alberta Society of Petroleum Geologists luncheon, Calgary, May 13, 1953. Manuscript received, January 3, 1954. Outlines of this paper have appeared in the Alberta Soc. Petrol. Geol. *News Letter* (June, 1953); *Canadian Oil and Gas Ind.*, Vol. 6, No. 6 (June, 1953); *Petrol. Explor. Digest*, Vol. 1, No. 7 (May, 1953), p. 5; *Oil and Gas Jour.*, Vol. 52, No. 18 (September 7, 1953), p. 168.

[2] Consulting geologist. Grateful acknowledgment is made to the following for pertinent comments and constructive criticism of the original brief outlines which were published in the trade journals: John G. Bartram, Parke A. Dickey, J. V. Howell, M. King Hubbert, G. S. Hume, F. H. Lahee, A. I. Levorsen, John L. Rich, E. W. Shaw. To A. J. Goodman, Joseph S. Irwin, E. H. Vallat, and Waldo W. Waring, many thanks are expressed for critical reading of the final manuscript. Particular acknowledgment is made to E. H. Vallat, Triad Oil Company Limited, and to Mid-Continent Pipelines Limited for permission to use basic data incorporated in Figure 4.

Multiple factors are believed to control the migration and accumulation of oil and gas. Many of these factors are fairly well understood and accepted, but it appears that a fundamental principle has been overlooked in the attempt to explain why some seemingly good structures are gas fields rather than oil fields, while others produce large quantities of salt water. This simple principle explains why gas fields may occur in a downdip position and produce little or no oil, while structures farther updip produce oil with little or no gas, and others in a still farther updip position may be water-bearing. In reality, this carries the structural theory to its logical conclusion.

It is with much trepidation that so simple a device is presented. Despite its simplicity, it is developed in considerable detail as a search of the literature reveals that it is a fundamental relationship that has been overlooked. More emphasis is placed on this one feature in the present paper but it is important not to lose sight of the many other factors that control the migration and accumulation of oil and gas.

STAGES OF ACCUMULATION OF HYDROCARBONS IN SIMPLE TRAP

The simple trap (Fig. 1) illustrates the various stages of accumulation of oil and gas in a simple trap under hydrostatic conditions. As oil and gas migrate updip, water is displaced from the trap, and the oil and gas segregate into two layers above the water table in accordance with the law of gravity (Fig. 1, Stage 1). If the supply of hydrocarbons is more than adequate to fill the trap to the spill point, oil and gas will continue to accumulate until the oil-water interface reaches the spill point. Thereafter a trap for more oil no longer exists (Fig. 1, Stage 2) and additional oil would cause oil to spill updip from the now "filled" trap. Gas, however, would continue to enter the trap, displacing more and more oil until, finally, all the oil has been displaced and flushed or "spilled" updip.

Once the gas-oil interface has reached the spill-point level, it would theoretically coincide with the oil-water interface and become a gas-water interface as the last oil escapes from the trap (Fig. 1, Stage 3). The end point has now been reached and this represents the final stage in this specific trap. Any additional gas would now cause gas to spill updip, while oil would continue to migrate updip past the gas-filled trap. Under hydrostatic conditions this would now be preserved for geologic time as a potential gas field, unless cut into by erosion or faulting, or affected by regional tilting. As the capacity of the trap to hold gas varies with the pressure or depth of burial, the gas will be compressed with increased load so that the trap would once more be able to hold more oil or gas. Conversely, if the load or depth of burial is reduced (by erosion), the gas expands, spilling more gas (or oil) updip. This might be referred to as remigration.

MIGRATION AND ACCUMULATION IN INTERCONNECTED RESERVOIRS

When the fundamental principle thus outlined is applied to a series of interconnected traps as illustrated in Figure 2, disregarding local or tributary migra-

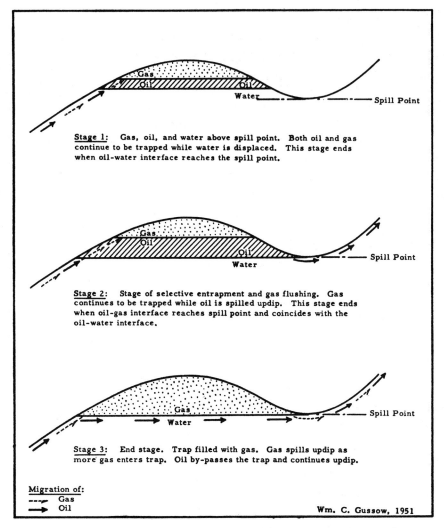

Fig. 1.—Accumulation of oil and gas in simple trap under hydrostatic conditions.

tion and the effect of solution gas, we find (Fig. 2-A) oil and gas accumulating only in the lowest trap (Trap I) until this is filled to the spill point: all other updip traps would be water-bearing until the lowest trap is filled. Only when oil begins to spill updip from Trap I will any oil accumulate in Trap II. Note that there is no primary gas cap in Trap II at this stage. Traps III and IV are water-bearing.

Further stages in the selective trapping of oil and gas are illustrated in Figure 2, B and C. Thus, Figure 2-B illustrates the conditions that might prevail when Trap I is completely filled with gas and all the oil has been flushed updip. Trap II might then be filled to the spill point with oil, and might be spilling oil updip into Trap III. Trap IV (and any other updip traps) would be filled with salt water.

Figure 2-C shows conditions at a still later stage. Trap I is filled with gas, and is spilling gas updip into Trap II; Trap II now has a gas cap, and is spilling

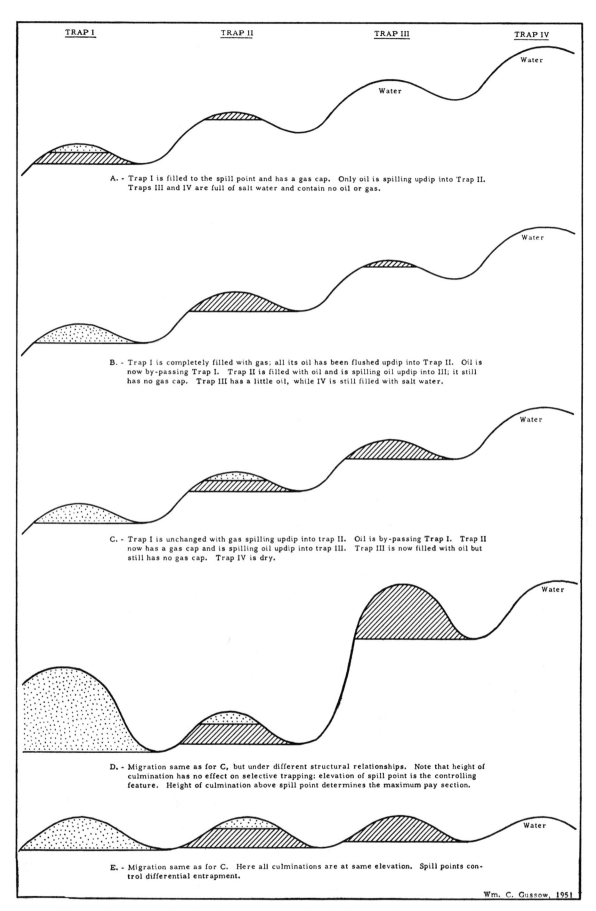

Water

A. - Trap I is filled to the spill point and has a gas cap. Only oil is spilling updip into Trap II. Traps III and IV are full of salt water and contain no oil or gas.

Water

B. - Trap I is completely filled with gas; all its oil has been flushed updip into Trap II. Oil is now by-passing Trap I. Trap II is filled with oil and is spilling oil updip into III; it still has no gas cap. Trap III has a little oil, while IV is still filled with salt water.

Water

C. - Trap I is unchanged with gas spilling updip into trap II. Oil is by-passing Trap I. Trap II now has a gas cap and is spilling oil updip into trap III. Trap III is now filled with oil but still has no gas cap. Trap IV is dry.

Water

D. - Migration same as for C, but under different structural relationships. Note that height of culmination has no effect on selective trapping: elevation of spill point is the controlling feature. Height of culmination above spill point determines the maximum pay section.

Water

E. - Migration same as for C. Here all culminations are at same elevation. Spill points control differential entrapment.

Wm. C. Gussow, 1951

FIG. 2.—Three stages in migration and accumulation of oil and gas in interconnected reservoirs. C, D, and E represent same stage in migration but under different structural relationships.

oil updip into Trap III; Trap III is more or less filled with oil but still has no primary gas cap; Trap IV is filled with salt water.

Figure 2, D and E, indicate more or less the same stage of migration as 2-C but under different structural relationships. Note that the height of the culmination of succeeding traps is not necessarily higher than in the previous trap. The spill-point level is the controlling feature of selective trapping. The height of the culmination above the spill point determines the maximum oil and/or gas column in the trap.

In a geanticlinal area where migration can go no further once it has reached the highest culmination, if gas continues to accumulate, gas will displace oil and water, backing them down the dip. This has given rise to the synclinal oil pools and so-called "dry sands" of the Appalachian region.

The traps illustrated in Figure 2 might represent a cross section through a series of parallel and contiguous anticlines or a longitudinal section (not necessarily a straight line—see Fig. 3) through a succession of culminations along one anticlinal axis, or they could represent a succession of culminations on a reef barrier or along a "shoestring" sand trend.

Only the simplest examples of the application of the principle of differential entrapment have been illustrated in Figure 2. There are many examples of combinations of structural traps to be considered but in every case the fundamental principles govern.

Many examples of this phenomenon have been described in the literature of the world and they have been passed up as "beyond explanation." The principle of differential entrapment would appear to explain some of the relationships. The present study has been handicapped because many of the earlier publications contain so little factual data that accurate interpretations can not always be made. Also, it would be desirable to have as accurately detailed information on dry structures as on those which are productive. Lacking such information, it is impossible to compare productive and non-productive structures. Frequently, discussions of the manner of accumulation of oil and gas have been based on observations on individual structures, rather than on an analysis of a composite picture throughout a region.

From the foregoing, it is obvious that *a trap filled with oil is still an effective gas trap but a trap filled with gas is not an effective oil trap*. It is, therefore, necessary to consider the migration and trapping of oil and gas separately, and not collectively as has been the tendency in the past.

The effect of solution gas in the oil has purposely been disregarded as this complication tends to confuse presentation of the underlying principle of differential entrapment. Actually, the oil is saturated with gas and, in migrating updip, gas is constantly being evolved with decrease in hydrostatic pressure. Thus, at the time of migration every oil-filled trap beyond the updip point of accumulation of primary gas will have a secondary gas cap of solution gas. The size of these will depend on the amount of oil that passed by a specific trap and will be proportional to the height of its oil-water interface above its downdip spill point. If the area of entrapment is subjected to deeper burial, most of these secondary gas caps soon disappear.

STATEMENT ON MIGRATION OF HYDROCARBONS FROM ORIGINAL DISPERSED STATE TO PRESENT POSITIONS OF ENTRAPMENT

As a background for the foregoing discussion on differential entrapment, an outline is given of the writer's views on migration of hydrocarbons in a water-saturated environment, from the initial dispersed state to the present position of entrapment and accumulation.

Regional or distant migration is believed to be essential for commercial accumulations to result.

Source beds are believed to be marine, clay and lime muds, in which hydrocarbons occur as minutely disseminated globules and may form any time after the material is deposited. With increased load due to deeper burial, more and more compaction of these muds results. At a critical load, which is a function of depth and the presence of suitable carrier beds, capillary pressure in the source beds is overcome and flush movement of hydrocarbons, out of the fine source beds and into the coarse carrier beds, begins. The rate of movement of the fluids is a function of pressure and permeability: thus, shale has a very low permeability while sandstone may be highly permeable. Shales are not impermeable in themselves; however, a shale-sandstone interface (or any coarse-fine interface) is a true impermeability barrier to hydrocarbons in a water-wet environment (from the coarse-textured side) on account of the relative capillary pressure in shale and in sandstone. The capillarity effect is to expel hydrocarbons out of the fine medium into the coarse medium at the interface. This is a one-way boundary for hydrocarbons but water can move freely in either direction.

In the ideal case, source beds are interbedded with porous sands or carrier beds, and the fluids in the source beds are expelled by "filter pressing." If a source bed is overlain and underlain by carrier beds, movement of the fluids will be both up and down; if a carrier bed lies only below, movement of the fluids will be down. All kinds of combinations and complications of this primary movement are visualized, but only the simplest examples are described.

If the carrier bed lies below the source bed, at first, minute globules of oil will appear (like perspiration) on the underside of the source bed. Gradually, these globules will grow in size and become droplets. Droplets will coalesce and become drops and, eventually, slugs. If the carrier bed lies above the source bed, when large enough, these drops or slugs will begin to migrate toward the top of the carrier bed where they will be trapped against the underside of an impermeability barrier. (Any hydrodynamic effect is disregarded for the moment.)

This primary movement of hydrocarbons in a finely dispersed state, from the source beds into the carrier beds, is known as *primary* migration. At first, water is the only fluid expelled. Then, at a critical depth of burial and resulting compaction, flush movement of hydrocarbons occurs. With further reduction of porosity expulsion of hydrocarbons continues.

If these impermeability barriers are horizontal surfaces, no further movement will result once the hydrocarbons reach the underside of the source beds or the top of the carrier beds. Slight irregularities may cause some local migration, but this is considered infinitesimal.

With differential compaction or regional isostatic subsidence in geosynclinal areas of deposition, a gradually increasing differential pressure due to regional dip or tilt will result. Lateral, updip movement of the hydrocarbons will begin only when the regional dip is great enough to overcome friction in the carrier beds. This has been illustrated by a smooth block on a highly polished table. The table is tilted gradually, and the block begins to slide only when the tilt is great enough to overcome the friction of the surface contact. Another way to illustrate this, is to turn a hose onto a flat plate of glass so that the surface becomes covered with droplets of water. When the surface is tilted by raising one edge, the droplets will all begin to "lean" down the slope. Then, as the tilt is increased, some of the larger droplets will begin to move slowly down the slope and, coalescing with other droplets, will begin to grow and form slugs which will eventually run off the sheet in rivulets.

The predominant tendency is for oil and gas to move updip, *the migration path being normal to the structure contours*—always seeking the highest spill point. At first the slugs of oil will begin to coalesce and form patches. These will move slowly and merge with other patches and, eventually, rivulets of oil will begin to stream up the dip. Irregularities and gentle undulations in the regional dip will cause rivulets to join and become small streams, and streams to unite as larger streams, tributaries, and finally rivers. All this takes place on the underside of an impermeability barrier—the converse of surface drainage.

The hydrocarbons will now be moving in a well defined migration path. This will have its local tributary streams, and from time to time will be joined by larger branches. The path of migration may be generally up the regional dip out of a basin and, on reaching the crest of an anticline, along it at right angles to the regional dip, spilling from one culmination to another. It is rarely, if ever, in a straight line (Fig. 3). The hydrocarbons will accumulate in traps along the migration path. When the lowest of these is filled, they will spill updip into the next highest trap and eventually the differential effect of oil and gas will cause a segregation or selective trapping of these two substances.

This lateral movement of hydrocarbons as streams in the carrier beds to the final positions of entrapment is known as *secondary* migration. It may follow closely after primary migration or tens of millions of years may elapse before any secondary or lateral migration occurs. Later spilling, due to further regional tilt and decantation, or to expansion of a gas cap, is known as *remigration*.

Under hydrodynamic conditions, the direction of movement will be the resultant of the dip and the hydraulic gradient. This might retard or augment the upward flow or even reverse it; or, if the gradient is at an angle to the dip, deflect it. It is even conceivable, under special circumstances, for oil to move downdip while gas moves updip.

Recently, the behavior of hydrocarbons under hydrodynamic conditions was masterfully demonstrated by Hubbert (1953), explaining tilted water tables, entrapment in unclosed structures, and the possible flushing of shallow oil and gas traps by a ground-water hydraulic gradient.

Figure 3 represents a subsurface contour map of an anticlinal trend along the western margin of a large supply basin, and demonstrates the migration path that would be followed by any oil or gas migrating updip out of the geosynclinal area on the east. The heavy line indicates the regional migration path. The lighter broken lines represent minor local tributaries.

When culmination D has filled to the spill point at saddle (1), oil and/or gas would stream updip into culmination C. Note that none would spill into Trap E; Traps A, B, and C would have to be filled to about the 1,030-foot contour before any oil would back up across saddle (2) and spill south into Trap E.

When Trap C is filled to approximately the 1,055-foot level, oil would spill across saddle (3) into Trap B, and B would have to be filled to approximately the 1,060-foot contour before any hydrocarbons would spill across saddle (4) into Trap A.

Culmination D is shown as being filled with gas. This would no longer be a trap for oil or more gas under existing conditions. Trap C is filled with oil and has a gas cap. Trap B is indicated as being filled to the spill point with oil (all primary gas is trapped downdip in Trap C). Traps A and E are dry or at best would have showings of hydrocarbons derived from local migration.

If a series of traps is filled with oil, any gas entering the lowest trap would form a gas cap and begin spilling oil updip. This would cause oil to spill updip, all along the line into a still higher water-filled trap. Eventually the lowest trap would become filled with gas and then gas would spill updip into the next trap, forming a gas cap in it. Conversely, if a series of traps is filled with gas, any oil migrating from below would by-pass all the gas-filled traps and end up in a water-filled (or partly gas-filled) trap at the upper end. Thus, it makes no difference which came first; the end result is the same.

When the supply basin is exhausted, secondary migration will cease and the reservoir fluids will come to rest so that, along the path of migration, the following general sequence will result. The deepest (downdip) traps will be filled to the spill point with gas. These will be followed by one or more traps containing oil and having a gas cap, and then by one or more traps filled with oil only (having no primary gas cap). When the supply of oil is exhausted, all remaining updip traps will be filled with salt water. The number of traps that will be filled with gas only or with oil only, will depend on the size of the individual traps and on the supply of oil and/or gas.

It should be remembered that oil and gas are closely associated, the amount of gas dissolved in the oil or the amount of oil becoming volatile varying with the temperature and pressure. Thus, if the area of entrapment is subjected to deeper burial, the gas in filled reservoirs having an oil column and a gas cap, will go into solution in the oil, increasing the gas-oil ratio. The size of the gas cap will

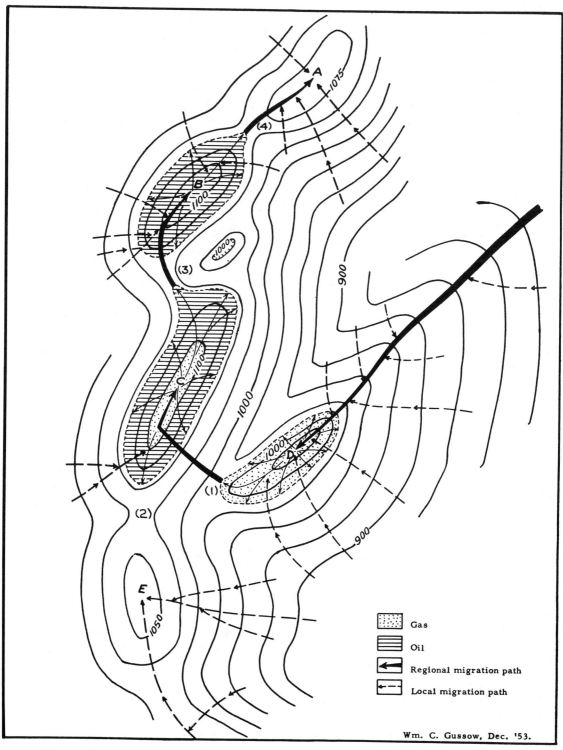

Wm. C. Gussow, Dec. '53.

FIG. 3.—Subsurface structure map showing migration path of hydrocarbons entering anticlinal trend along west margin of large supply basin. Contour interval, 25 feet.

also be further reduced due to compression of the gas remaining in the gas cap. (It is possible for the gas cap to disappear entirely.) As a result, the oil-water interface will no longer be at the spill point, making this an effective trap for oil again as well as for gas. Reservoirs filled with gas only will become effective traps again for oil and gas, but reservoirs filled with oil only will remain unchanged, so that the oil will now be undersaturated.

DIFFERENTIAL ENTRAPMENT IN BONNIE GLEN-WIZARD LAKE REEF TREND OF ALBERTA, CANADA

This great reef barrier of Upper Devonian age has not been described in print, although papers have been published by various authors on some of the individual accumulations. The trend has been traced more than 125 miles, extending 80 miles south-southwest from the Leduc field, and more than 50 miles north of Leduc. Figure 4 is a longitudinal section along the reef barrier showing the fluid content of the various traps.

It is visualized that oil and gas migrated updip out of the Rocky Mountain geosyncline along various tributary migration paths and, on entering the reef trend, migrated along it from south to north. Reef débris along the flanks of the reef trend forms an excellent "pipe line" or carrier for the migration of oil and gas along the barrier. At Leduc, the migration path appears to have split, some oil spilling updip as far as the St. Albert reef, but most of it spilling northeast into the Redwater field. It is also conceivable that this "funneling" effect (in Leduc and younger formations) accounts for the fabulous accumulations of hydrocarbons in the Athabaska oil sands, which are about 200 miles updip north-northeast. Certainly, exceptional conditions must have prevailed to account for such unique accumulations as the Athabaska oil sands, estimated at x hundred billion barrels, or even the Redwater field, which is estimated to have 1,400,000,000 barrels of oil in place.

All primary gas has been trapped in structures downdip from Wizard Lake. South Westerose and Rimbey-Homeglen, and any other reefs at the south, are filled with wet gas to the exclusion of oil, while Westerose and Bonnie Glen reefs have large gas caps. Wizard Lake and Glen Park are filled to the spill point with oil and have no gas caps. A gas cap reappears at Leduc-Woodbend. This is considered to be a secondary gas cap formed by gas that has come out of solution in spilling updip, 1,000 feet, from Glen Park into Leduc. All the gas which came out of solution was trapped at Leduc and most of the oil has been flushed updip. Acheson and St. Albert contain some oil, while Morinville and all the reefs on the north along this trend are loaded with salt water and contain no oil and no gas; only good showings of oil have been obtained in some wells.

A computation was made to see if there was enough oil in the Acheson, St. Albert, and Redwater reefs to account for the large secondary gas cap at Leduc-Woodbend. It was found that this was in material balance with the Wizard Lake gas-oil ratio. This seems to confirm that the Leduc-Woodbend gas cap originated

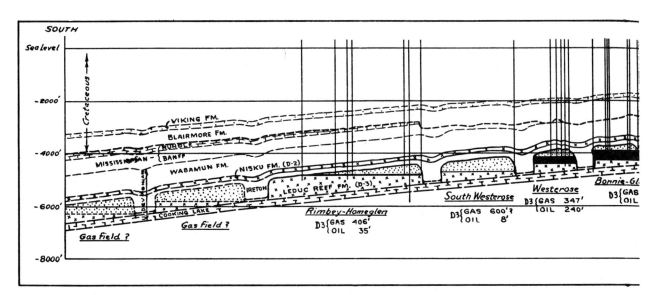

Fig. 4.—Bonnie Glen-Wizard Lake reef trend, Alberta, Canada, showing fluid content of

from solution gas which was originally in solution in the oil that spilled updip from the Wizard Lake reef.

The fact that Westerose and Bonnie Glen both have gas caps might be pointed out as an exception. This, however, may be explained by assuming that gas must have been spilling updip from South Westerose into both Westerose and Bonnie Glen, simultaneously, just as a braided stream splits and comes together again. The small oil columns at Rimbey-Homeglen and South Westerose may be explained by oil which has drained down out of the gas cap after the oil content was flushed updip. These reefs were once filled to the spill point with oil and had no primary gas caps. Then gas caps appeared and, finally, all the oil was flushed updip. Then, during geologic time, oil drained down out of the gas cap. Another possible explanation is that more oil was trapped in these reefs after the gas was compressed due to deeper burial. A combination of the two may be the answer.

The oil showings at Morinville are believed to represent local migration; it is firmly believed that commercial accumulation of oil can only result from regional migration.

A further observation, indicating that migration was from south to north is illustrated at Acheson where a small "bubble" of solution (?) gas is trapped downdip (−2,600 feet), while no gas cap occurs at the highest culmination at the extreme north end of this field (Fig. 5). The gas-oil interface was found at −2,638 feet and the spill point of this closure is at about −2,645 feet. More gas would have to accumulate in this small trap before any would spill north into the next culmination ($\frac{1}{2}$ mile north). The gas-oil interface would have to reach approximately −2,670 feet before any gas could spill into the highest culmination (−2,500 feet) at the north end of the field.

128

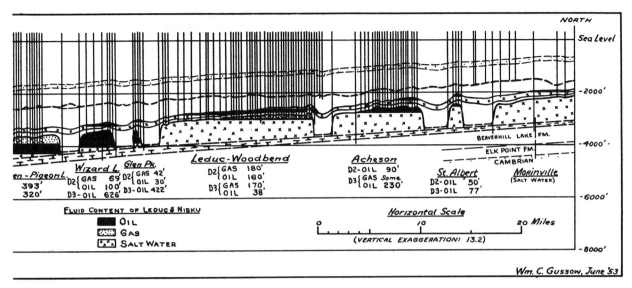

Leduc and Nisku traps. (Maximum pay sections are indicated for Leduc and Nisku reservoirs.)

The draping of the Nisku (D2) over the top of the Leduc reefs, due to differential compaction in the Ireton section, has caused the formation of shallow traps in the Nisku and younger formations. A good example is the accumulation of oil and gas in the Nisku in the Leduc-Woodbend field (Fig. 4) where there is a thick oil pay with a large gas cap. The oil-water interface is near the south end of the field. Updip, at Acheson and St. Albert, the Nisku contains oil without a gas cap. Farther updip, the Nisku is loaded with salt water and contains no oil or gas. Downdip at Wizard Lake and Glen Park both oil and gas are trapped in this zone. No oil pays have been reported in the Nisku south of Wizard Lake. Any of these traps should be filled with gas and were probably not reported, being uncommercial.

Differential entrapment seems to govern accumulation in the Nisku in a general way, giving rise to a sequence from south to north of traps filled with gas (?) to traps containing both oil and gas, to traps filled with oil and having no gas cap, to traps which are loaded with salt water. This general sequence is modified slightly by other factors not fully understood. While most of the hydrocarbons probably migrated along the reef trend, it is possible that there was some lateral tributary migration up the regional dip due to the shallow nature of these traps, and that the net result is a combination of these effects.

Other examples of this phenomenon in Alberta are the Stettler-Caprona-Big Valley, and the Duhamel-New Norway-Malmo reef trends.

EXAMPLES OF DIFFERENTIAL ENTRAPMENT IN UNITED STATES

Some of the best geological occurrences of differential effects are here discussed. These are based on a reconnaissance study of the published records, which was begun early in 1950 and carried on from time to time as conditions per-

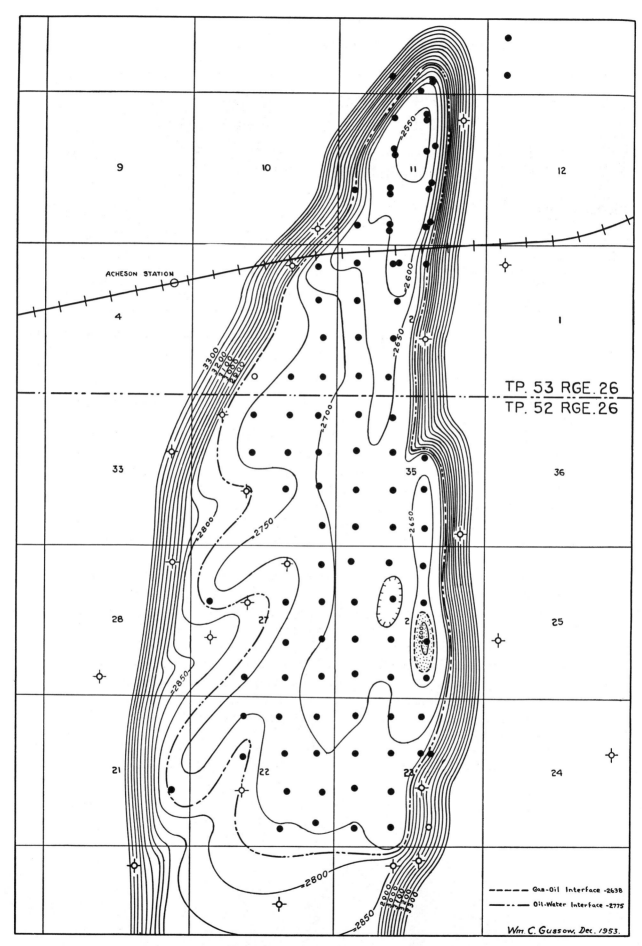

FIG. 5 (*See explanation on facing page.*)

130

mitted. Undoubtedly, many other (and better) examples exist but these have not been published.

Lost Soldier district, Wyoming.—J. S. Irwin (1929) described the relationships of several contiguous structures in south-central Wyoming. Wertz, Mahoney, West Ferris, Middle Ferris, and East Ferris are separate domes or culminations on a long anticlinal axis (Fig. 6). The Little Lost Soldier dome is contiguous to the Wertz dome, paralleling it on the west. The upper producing formations in each of these are the Sundance sand (Jurassic), Lakota and Dakota (lower Upper Cretaceous), and the Frontier sands (middle Upper Cretaceous). These produce oil to the exclusion of gas at Little Lost Soldier, and gas to the exclusion of oil at the Wertz dome, although their crests are only $2\frac{1}{2}$ miles apart. The cross section in Figure 7 shows the structural relationships of these two domes. The culmination of the Wertz dome is about 2,000 feet lower than that of the Lost Soldier dome. On the northeast side of the Wertz dome, and parallel with it, is the Bunker Hill dome (Fig. 6). The producing beds at Little Lost Soldier and Wertz were found to be water-bearing at Bunker Hill (Bartram, 1936).

Mahoney, West Ferris, and Middle Ferris domes are culminations on the Wertz-Mahoney-Ferris anticline, and are separated by low structural saddles. Their crests are about 1,000 feet higher than that of the Wertz dome. They yield gas to the practical exclusion of oil from the same formations at Wertz. The easternmost structure shown in Figure 6 is the East Ferris dome, the crest of which is about 600 feet higher than that at Middle Ferris, with an independent closure of 1,200 feet. Both oil and gas are trapped in this dome.

Twenty-five years ago Irwin attributed the apparent "anomalous" accumulation of hydrocarbons to the presence or absence of faults or their tightness. Thus, the absence of a gas cap at Little Lost Soldier was attributed to selective leakage of gas, whereas the occurrence of gas to the exclusion of oil at Wertz was attributed to the absence of faults or to their tightness if present.

Tillotson (1935, p. 320) expressed himself as follows:

An exceptional condition is the existence of oil on the Lost Soldier dome in the same formations that produce only gas at Wertz. The Wertz anticline is contiguous to Lost Soldier but 2,000 feet lower structurally. The reason for this phenomenon is not known . . . any attempt to explain the occurrence of oil in this structural closure and gas in the contiguous lower structure must be chiefly conjectural. Similar conditions exist elsewhere in Wyoming.

Dobbin (1947, p. 811) stated that the lack of gas in the Little Lost Soldier field "suggests strongly that there has been vertical migration of oil and leakage of gas." However, the remark that " . . . some of the conditions of oil accumula-

---

←⃖⃖⃖

FIG. 5.—Acheson area, Alberta, Canada. Structure contours on Leduc reef. Interval, 50 feet. Modified after J. W. Coveney and A. A. Brown, "Geology and Development History of the Acheson Field," *Bull. Canadian Inst. Min Met.* (1954, in press), Fig. 6. Gas is trapped downdip at −2,600 feet, while no gas occurs at the highest culmination (−2,500 feet) at the extreme north end.

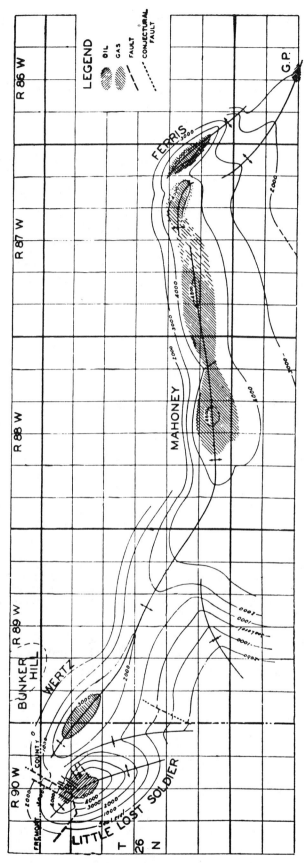

FIG. 6.—Structure of northern part of Lost Soldier district, Carbon, Sweetwater, and Fremont counties, Wyoming, contoured on top of Dakota sand. Datum, sea-level. Contour interval, 1,000 feet. After Irwin in *Structure of Typical American Oil Fields*, Vol. II (Amer. Assoc. Petrol. Geol., 1929), Fig. 9, p. 663.

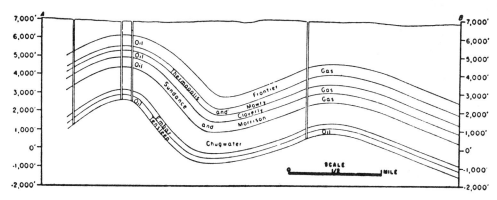

Fig. 7.—Lost Soldier and Wertz domes, Carbon and Sweetgrass counties, Wyoming. After C. E. Dobbin, "Exceptional Oil Fields in Rocky Mountain Region of United States," *Bull. Amer. Assoc. Petrol. Geol.*, Vol. 31, No. 5 (May, 1947), Fig. 11, p. 810. Gas is trapped downdip at Wertz while Little Lost Soldier produces oil to exclusion of gas.

tion in the region defy satisfactory explanation," indicates that he is not altogether satisfied with this explanation.

According to the principle of differential entrapment, no anomalous condition need exist.

It is suggested that oil and gas migrated northeastward out of the Washakie supply basin and entered the Wertz-Mahoney-Ferris anticline and accumulated in successive domes, spilling oil updip into the next highest structure and trapping gas in the lower structures, until finally oil was spilled updip from the Wertz dome into Little Lost Soldier, and into the East Ferris dome at the other end of the anticline. On account of the lower structural saddle between Wertz and Bunker Hill, the Bunker Hill dome remained full of salt water.

All of these accumulations have been modified by the regional hydraulic gradient so that the oil-water interfaces are tilted northwest.

Sage Creek and Winkleman fields, Wyoming (Dobbin, 1947, p. 813).—The Sage Creek and Winkleman fields are local domes on a long anticline on the west flank of the Wind River Basin. Oil is trapped downdip at Winkleman in a dome having 300 feet of closure, while the large Sage Creek dome with a minimum of 2,000 feet of closure and separated from Winkleman by a shallow saddle, is loaded with salt water and has only good showings of oil. Figure 8 shows the relationships of these two structures.

In describing these two fields, Dobbin states (1947, p. 813):

The Sage Creek and Winkleman fields have an exceptional relationship in that the restricted Winkleman dome is an excellent small oil field in formations that have produced only strong showings of oil in the large Sage Creek dome.

According to the principle of differential entrapment, it is suggested that oil was trapped downdip in the Winkleman dome and only enough spilled updip across the low structural saddle to give showings of oil in the large Sage Creek structure which is still full of salt water.

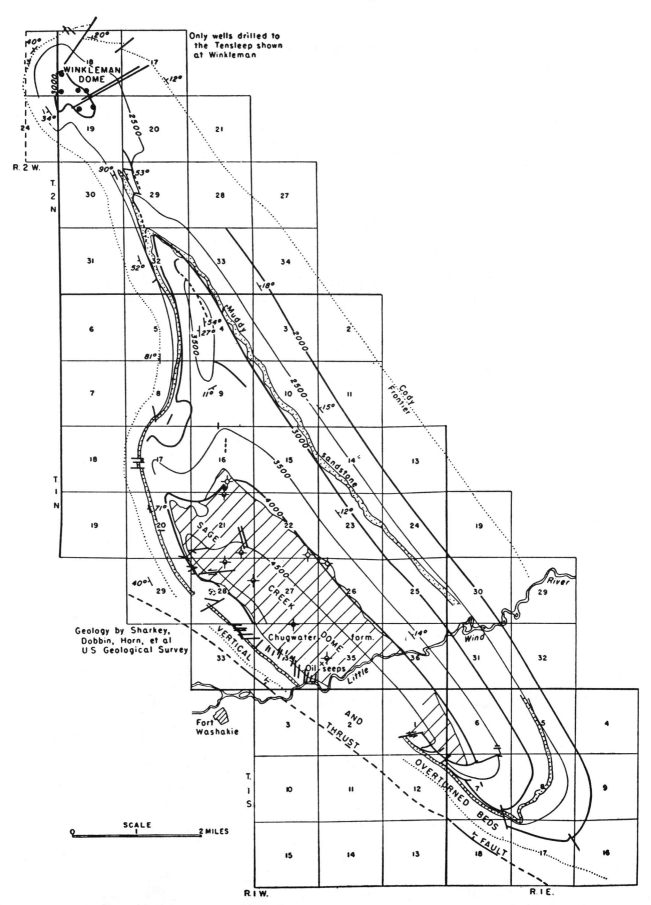

FIG. 8.—Sage Creek and Winkleman domes, Fremont County, Wyoming, contoured on Tensleep formation. After C. E. Dobbin, "Exceptional Oil Fields in Rocky Mountain Region of United States," *Bull. Amer. Assoc. Petrol. Geol.*, Vol. 31, No. 5 (May, 1947), Fig. 13, p. 814. Oil is trapped downdip at Winkleman while large Sage Creek dome is loaded with salt water.

Elk Basin oil and gas field, Wyoming and Montana.—John G. Bartram (1929, pp. 577–88; Pl. 1) described the accumulation of hydrocarbons in this structure, lying across the state line between Wyoming and Montana at the north end of the Big Horn Basin. The trap is a large anticline in formations of Upper Cretaceous age with about 4,000 feet of closure, and is broken by three sets of faults. In the Second Wall Creek sand, oil fills the crest of the structure and extends well down the sides, and in the words of Bartram:

> . . . there is a well-marked water level in the field [This is actually tilted 100 ft./mi. to the southwest].* In the main part of the field no gas wells have been found . . . but gas has been found in the fault blocks in the south end. In this locality . . . there is water below the gas, with evidently no oil between. . . .

It appears that gas has been trapped downdip and that oil only has been spilled updip into the main part of the structure as would be expected. The path of migration would be north out of the Big Horn Basin.

Salt Creek and Teapot domes, Natrona County, Wyoming (Beck, 1929).—These are two culminations on a long anticlinal axis (Fig. 9) on the southwest flank of the Powder River Basin. The Salt Creek field is a large dome having a structural closure of 1,600 feet, and produces oil from several horizons ranging from Upper Cretaceous to Pennsylvanian in age. The Teapot dome is separated from the larger Salt Creek dome by a low saddle and its culmination is 1,200 feet lower. Gas is trapped downdip in the Teapot dome as a gas cap whereas no gas occurs in the Salt Creek field. Due to the hydraulic gradient, the oil-water interfaces are tilted north at a low angle.

The Powder River anticline is located northwest of Salt Creek and is 16 miles long and 10 miles wide but, according to Ver Wiebe (1952, p. 332), "for some unexplained reason contains no oil or gas." These relationships can be explained by the principle of differential entrapment.

Garland and Byron fields, Wyoming (Dobbin, 1947, p. 811).—The Garland and Byron fields are two contiguous domes at the north end of the Big Horn Basin (Fig. 10). The Garland field has a closure of 2,600 feet and the Byron field has a closure of 1,500 feet but is about 2,000 feet lower structurally. Oil is trapped in the Tensleep in both structures but the Garland dome has a gas cap. The migration path was northeast out of the Big Horn Basin, all gas being trapped in the basinward trap (Garland), only oil spilling updip into Byron. (Note that the elevation of the culminations is not a controlling feature of accumulation.) Oil-water interfaces in both structures are tilted southwest at 240 feet per mile.

Northwest Lake Creek field, Hot Springs County, Wyoming.—This is located on the southwest flank of the Big Horn Basin, approximately 20 miles northeast of Thermopolis, on the northwest plunge of the Lake Creek anticline. Oil is trapped in the Tensleep sandstone, 500 feet down the plunge from the culmination of the anticline, in a trap which is a combination of structure and a hydraulic gradient.

* Part in brackets added by present writer.

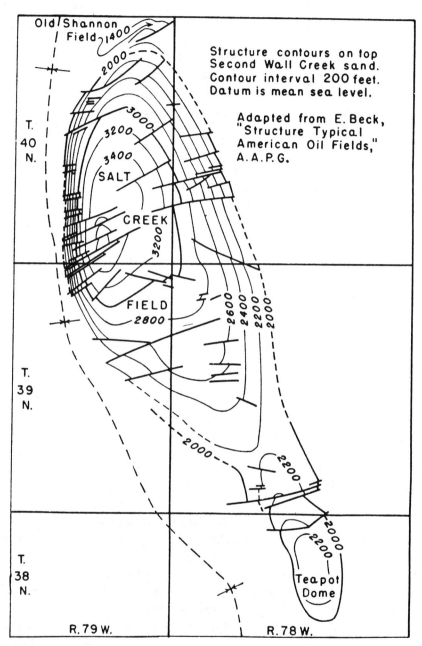

FIG. 9.—Salt Creek and Teapot dome oil fields, Natrona County, Wyoming. After C. E. Dobbin, "Exceptional Oil Fields in Rocky Mountain Region of United States," *Bull. Amer. Assoc. Petrol. Geol.*, Vol. 31, No. 5 (May, 1947), Fig. 14, p. 816. Gas is trapped downdip at Teapot as gas cap; no gas occurs in Salt Creek field.

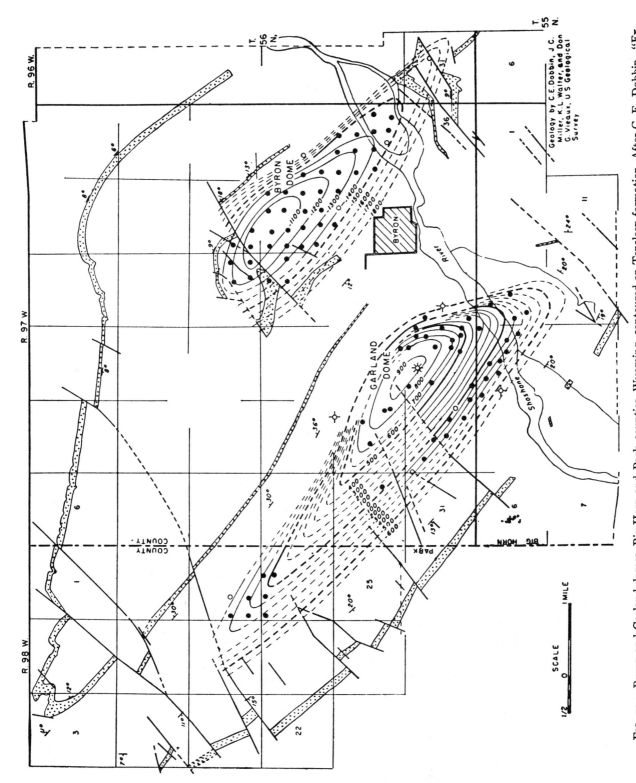

FIG. 10.—Byron and Garland domes, Big Horn and Park counties, Wyoming, contoured on Tensleep formation. After C. E. Dobbin, "Exceptional Oil Fields in Rocky Mountain Region of United States," *Bull. Amer. Assoc. Petrol. Geol.*, Vol. 31, No. 5 (May, 1947), Fig. 12, p. 812. Both structures produce oil from Tensleep but Garland has gas cap. Supply basin is southwest.

Green and Ziemer (1953, p. 178) visualize that the Northwest Lake Creek field once had oil *and gas* trapped at the culmination of the structure, and that

subsequent . . . flow forced the gas and much of the oil from the crest of the anticline. A small portion of the oil found a position of rest in the energy field at a small and local culmination along the anticlinal plunge forming Northwest Lake Creek field—a residual "blob" of a once-great oil field having been saved by a fortuitous series of events.

This, of course, could not be true. It is doubtful whether there was ever a gas cap in this field; certainly if there had been, all the oil would have been flushed downdip before any of the gas would have been flushed out of the trap. A more plausible explanation is that there never was any more oil in the Northwest Lake Creek field and that any gas must have been trapped downdip.

Big Horn Basin, Wyoming.—Frederick G. Clapp stated (1929, p. 679):

The question has frequently been raised why certain domes of the Rocky Mountain region—apparently as good structurally as some others that are eminently productive—yield no oil or gas commercially but only large volumes of water. In the region surrounding the Big Horn Mountains, for instance, little fluid except water is found in the belt of domes first removed from the mountains; yet many oil fields are found in domes of the second line.

In 1952, two gas condensate discoveries were made in the deeper part of the Big Horn Basin (Five-Mile, Fourteen-Mile), indicating that gas is trapped downdip from the known oil fields.

The absence of oil in the "belt of domes first removed from the mountains" has been attributed to the effect of artesian water, but this would not explain the presence of gas downdip. A hydraulic gradient, if effective, would cause oil to move downdip and leave gas in the updip traps.

A possible explanation appears to be the effect of differential trapping. This would explain gas in the deeper part of the basin while oil has been trapped in "domes of the second line," none having spilled updip into the "belt of domes first removed from the mountains."

The Oregon Basin anticline, Park County, Wyoming (Dobbin, 1947, Fig. 17, p. 819), on the west edge of the Big Horn Basin is a good example of this phenomenon. Oil is trapped in two large domes on the north-south Oregon Basin anticline, separated by a low structural saddle. A lower structural saddle separates the south dome of the Oregon Basin anticline from the Horse Center anticline farther west. This structure is dry, no oil having spilled across the lower saddle.

Another example is the Little Buffalo Basin field, Wyoming (Dobbin, 1947, Fig. 16, p. 818), at the southeast. This consists of two culminations on a larger dome, separated by a low structural saddle. Oil is trapped in the Embar and Tensleep on the east culmination, whereas in the west culmination these formations contain salt water.

In discussing the geology of the Baxter Basin gas fields, Sweetwater County, Wyoming, W. T. Nightingale (1935, p. 323) writes as follows.

Another problem of interest is the commercial gas production in the Sundance formation at the North Baxter Basin and its absence in the same formation at South Baxter Basin, although the latter is approximately 1,200 feet higher structurally than the former.

Both structures occur on the same major axis of folding, having a length of 90 miles and a width of 12 miles. Commercial gas and a small showing of oil were encountered downdip at North Baxter Basin in the Sundance and apparently none spilled updip into South Baxter Basin.

As stated by Clapp (1929, p. 676), "Some structurally perfect domes in northwest Colorado are barren owing to causes not fully established." It is suggested that differential entrapment is the cause.

Kevin-Sunburst dome, Toole County, Montana (W. F. Howell, 1929).—This is one of the northernmost fields in the United States. It is 15 miles south of the Canadian border on a large dome on the north-plunging end of the Sweetgrass arch. In 1929, W. F. Howell published a contour map (Fig. 11) showing the localization of oil in traps, 200 feet down the dip on the north flank of the main dome. There is no oil production at the apex of the dome. It would appear that oil migrated updip from the north and was all trapped in local closures on the flank of the main dome, and that there was not enough oil to spill updip into the main culmination.

Artesia field, Eddy County, New Mexico.—Morgan J. Davis (1929, p. 112) has provided us with an excellent description of the Artesia field, New Mexico. His structure map is reproduced in Figure 12. The structure is a northeast-trending anticline. Davis (p. 119) describes this as "an example of accumulation on the apex and flank of an anticline. Furthermore, the flank accumulation is on the side toward the regional dip." He continues:

there are a few anomalies that are prejudicial to this conception. The most productive parts of the field have been Sections 21 and 28. A glance at the contour map will show that this area lies lower structurally than the production in either Section 17 or Section 4. It is also a fact established by examination of samples, that the porosity of the producing zones in Section 28 is higher than that in other parts of the field. It is thought that porosity alone is the explanation for this irregularity. . . . It would seem that the situation of the porous spots on structure is purely a matter of chance since they are found on both the apex and the flank of the anticline.

And a little further (p. 120):

Oil and gas do not maintain the same ideal relationship as do the oil and water. Wells that make little more than gas are situated irregularly on the structure.

The effective traps on the southeast flank of this large anticline are a combination of structural terracing and a hydraulic gradient, although some slight closure is indicated. It seems plausible that differential entrapment is the fundamental principle of accumulation, accounting for the very small amount of oil at the culmination (Empire pool) while most of the oil and/or gas is trapped downdip.

San Juan Basin, New Mexico.—Nowels (1928; 1929) shows that Hogback, Rattlesnake, and Table Mesa, on the west side of the San Juan Basin in the northwestern corner of New Mexico, are confined to domes and produce from the Dakota sandstone. The oils in all three of these fields are high-gravity, being 74° A.P.I. for Rattlesnake, 63° for Hogback, and 58° for Table Mesa. The equally good non-productive structures of Chimney Rock, Tocito, and Beautiful Mountain contain water in the Dakota.

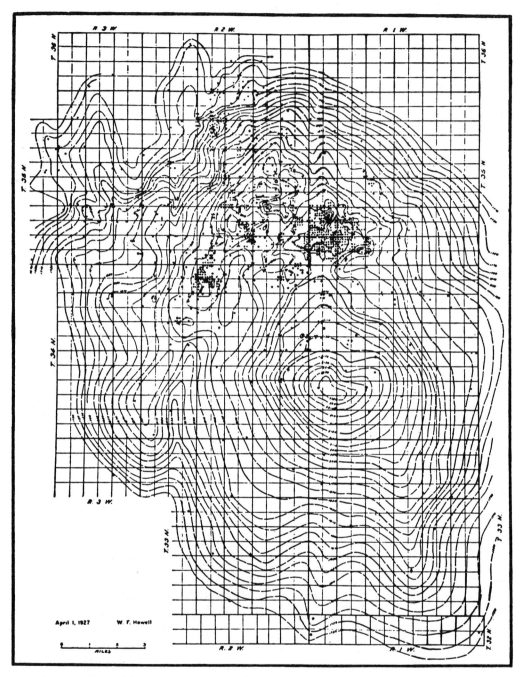

Fig. 11.—Subsurface structure of Kevin-Sunburst oil and gas field, Montana, contoured on Ellis-Madison contact. After W. F. Howell, "Kevin-Sunburst Field, Toole County, Montana," *Structure of Typical American Oil Fields*, Vol. II (Amer. Assoc. Petrol. Geol., 1929), Fig. 3, p. 261. Oil is trapped 200 feet downdip on north flank. No oil at apex.

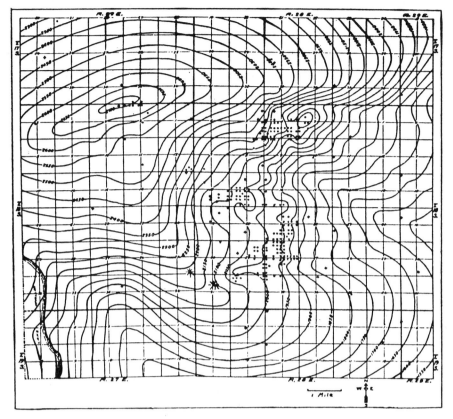

FIG. 12.—Artesia field, New Mexico. Subsurface structure contours on "Red Sand." Datum, sea-level. After Morgan J. Davis, in *Structure of Typical American Oil Fields*, Vol. I (Amer. Assoc. Petrol. Geol., 1929), Fig. 2, p. 118. Most of the oil and all of the gas are trapped downdip on east flank. Very little oil at culmination.

A detailed study might reveal that this is another example of differential entrapment.

Butler and Marion counties, Kansas.—C. R. Thomas' structure map (1929), contoured on the Ordovician, is reproduced in Figure 13. This shows five well developed domes close together, four of which are prolific producers, and one (Burns dome) is barren. Thomas visualizes oil migrating updip from the west, "where abundant source beds occur," and attributes the absence of production at Burns dome to the absence of source beds between Elbing and Burns domes.

Frank R. Clark (1934, p. 319) reproduces Thomas' structure map as evidence against lateral migration and in support of a local origin. He fails to recognize the principle of differential entrapment. According to this, oil would migrate updip from the west (any gas being trapped downdip) and enter the three western domes—Covert-Sellers, Peabody, and Elbing—more or less simultaneously. The spill point of this series of structures is between Covert-Sellers and Florence domes so that oil would begin to spill updip across this low saddle into the Florence dome. If this process had continued, oil would eventually have spilled

FIG. 13.—Subsurface structure on Ordovician producing zone, Butler and Marion counties, Kansas. Contour interval, 20 feet. Scale, one township approximately 6 miles square. After C. R. Thomas, "Flank Production of the Nemaha Mountains (Granite Ridge), Kansas," *Structure of Typical American Oil Fields*, Vol. I (Amer. Assoc. Petrol. Geol., 1929), Fig. 3, p. 65. Burns dome is loaded with salt water while Covert-Sellers, Peabody, Elbing, and Florence are prolific producers of oil.

updip from the Florence dome into the Burns dome. No oil could enter the Burns dome from the west, the saddle between Elbing and Burns domes being at a much lower elevation.

This is believed to be a good illustration demonstrating that local drainage areas—Burns, for example—are inadequate to provide commercial accumulations of oil and that migration over great distances is essential.

Geneseo uplift, Rice and Ellsworth counties, Kansas (S. K. Clark *et al.*, 1948). —The Lyons gas field and the Geneseo and Edwards oil fields are good examples of the effects of differential entrapment. All three are anticlinal accumulations in the Arbuckle limestone and occupy successive culminations on the Geneseo uplift (Fig. 14).

In describing the Lyons gas field, the authors state (Clark *et al.*, 1948, p. 240):

Structurally the field is identical in type with Geneseo and Edwards, but in this case the Arbuckle reservoir contains only gas, whereas the other two contain only oil. There is no obvious explanation for this phenomenon.

Actually, one small oil well was completed at the south edge of the Lyons gas field. All the gas has been trapped downdip in the Lyons gas field which has almost reached Stage 3 (Fig. 1). No doubt a small thin oil ring existed. At one time, the Lyons dome must have been filled with oil and was spilling oil updip into Geneseo and then from Geneseo into the Edwards dome. Eventually a gas cap appeared in the Lyons dome and, finally, as more and more gas was trapped, nearly all the oil was flushed updip into Geneseo and Edwards.

Burbank field, Osage County, Oklahoma.—In discussing accumulation in the Burbank field, Oklahoma, Sands (1929, pp. 220–29) writes as follows.

Gas also was present in large quantities in the extreme northern end of the field, which is the lowest part of it, the gas being found structurally below the oil. . . .

This was attributed to selective pore size, large enough to allow gas to accumulate but too fine-grained for oil or water to penetrate. It is possibly an example of differential entrapment.

Natural gas pools of southern Oklahoma.—C. W. Tomlinson (1935, p. 591), in discussing the natural gas pools of southern Oklahoma, notes that

the more important accumulations of oil-free natural gas . . . have been found . . . at depths exceeding 1,500 feet. . . . From this fact, together with the relative scarcity of gas at Tatums . . . , it seems a reasonable inference, that gas has been more successful than oil in escaping from the shallower and less perfectly sealed reservoirs.

The tilt of the oil-water interface in the Graham pool (p. 592) demonstrates that these accumulations have been modified by the hydraulic gradient of the district. However, it appears that the regional relationships must be explained by the effect of differential entrapment, causing gas to be trapped in the deeper parts of the basins.

Van field, Texas (Lahee, 1934, p. 415).—The Van field in Van Zandt County, Texas, occurs on a large dome cut by a major fault having a minimum displacement of 500 feet. On the upthrown (southeast) side, the Woodbine pay zone contains oil but little, if any, free gas, whereas on the downthrown side the Wood-

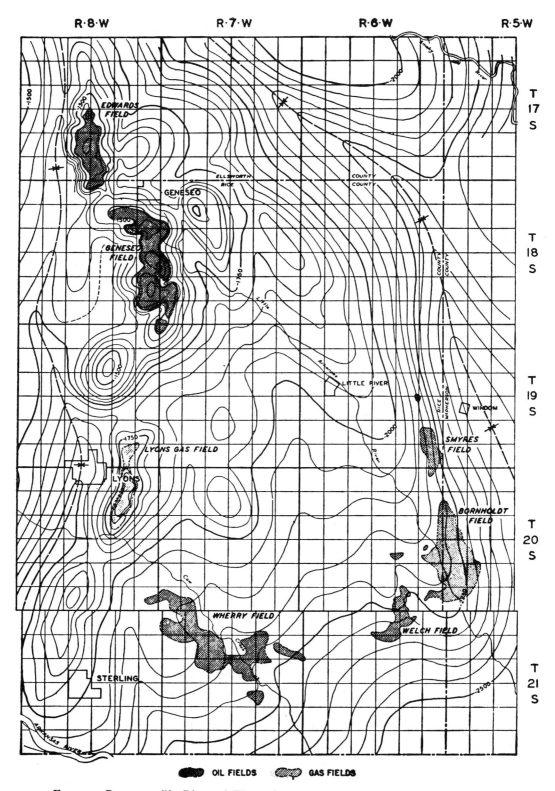

FIG. 14.—Geneseo uplift, Rice and Ellsworth counties, Kansas. Structure contours on top of Arbuckle limestone. Contour interval, 50 feet. After S. K. Clark, C. L. Arnett, and J. S. Royds, "Geneseo Uplift, Rice, Ellsworth, and McPherson Counties, Kansas," *Structure of Typical American Oil Fields*, Vol. III (Amer. Assoc. Petrol. Geol., 1948), Fig. 3, p. 229. Gas is trapped downdip at Lyons while Geneseo and Edwards produce only oil. Wherry, Welch, Bornholdt, and Smyres produce from basal Pennsylvanian and are stratigraphic-trap accumulations.

bine contains free gas as well as oil. The crest of the gas-bearing part of the structure is fully 300 feet lower than the crest of the dome in the upthrown block.

One explanation that has been given for this so-called "anomalous" occurrence of gas is that the faulting took place after the oil and gas had accumulated in the dome and that the original gas-bearing crest was downfaulted. In accordance with the principle of differential entrapment, no anomalous condition exists.

O'Hern oil field, Duval and Webb counties, Texas (Barnett, 1941).—The O'Hern field, 42 miles east of Laredo, Texas, is a combination pinch-out and structural-trap accumulation in the O'Hern sand of Eocene age. Barnett reproduces a structure map of the O'Hern field (Fig. 3, pp. 734–45), contoured on top of the O'Hern sand. This clearly shows one gas well, the Magnolia Petroleum Company's Brennan-Benavides No. 47, in the gas cap at the culmination of the structure. Two other gas wells, the Magnolia Petroleum Company's O'Hern-Seacord Nos. 3 and 4, were drilled into the O'Hern sand, approximately 150 feet downdip from the gas-cap well and near the downdip limits of the field. Although these two gas wells are shown by Barnett as being located on a nose of the main O'Hern oil field, they could easily be re-contoured as occurring on a separate culmination on the downdip flank of the O'Hern oil field, and separated from the main accumulation by a low saddle.

This revised interpretation would be in agreement with the principle of differential entrapment, most of the gas being trapped downdip on the flank of the structure, while a very small amount has spilled updip to form a one-well gas cap.

The oil-water contact is tilted eastward slightly by the hydraulic gradient. No anomalous conditions of accumulation are believed to exist.

Homer oil field, Louisiana (Spooner, 1929).—This field lies in Claiborne Parish, Louisiana, on a dome with nearly 1,100 feet of closure, and covering an area of more than 70 square miles. A major fault, with a downthrow of 500 feet on the south, roughly divides the dome into north and south halves. Oil in this field is produced from two sands of Upper Cretaceous age. The lower Oakes ("Blossom") sand, the top of which is 475 feet below the base of the Nacatoch, produces oil in the south (downdip) pool only. It contains salt water on the north (updip) side of the fault. This is in accordance with the principle of differential entrapment.

Natural gas in Gulf Coast salt-dome area, Texas and Louisiana.—L. P. Teas (1935, p. 683) states,

> There are only relatively small amounts of gas in most salt-dome fields in proportion to their oil. This relative scarcity of gas may be explained by the more fugitive character of gas which has permitted its escape into the atmosphere at different periods in the geologic history of the dome. . . . More recently discovered oil and gas fields that may be associated with very deeply buried salt masses show relatively much less deformation so that much larger deposits of gas occur.

Is this possibly due to the effect of differential entrapment?

East Coalinga Extension field, Fresno County, California.—A contour map of this field was published by L. S. Chambers (1943, p. 489) and is reproduced in

Figure 15. This indicates an oil field about 8 miles long, having two culminations, and a common oil-water interface. It is noted that gas is trapped in the northern culmination (−7,400 feet), while there is no gas cap in the southern culmination (−5,800 feet). Gas is thus trapped downdip, about 1,600 feet below the top of the

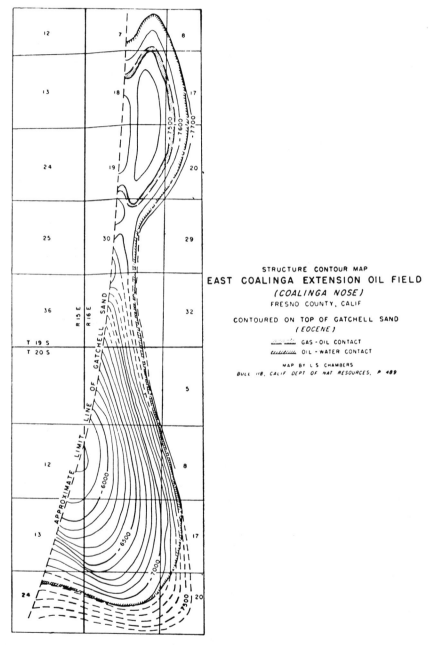

STRUCTURE CONTOUR MAP
EAST COALINGA EXTENSION OIL FIELD
(COALINGA NOSE)
FRESNO COUNTY, CALIF

CONTOURED ON TOP OF GATCHELL SAND
(EOCENE)

GAS - OIL CONTACT
OIL - WATER CONTACT

MAP BY L S CHAMBERS
BULL 118, CALIF DEPT OF NAT RESOURCES, P 489

FIG. 15.—East Coalinga Extension field, Fresno County, California. After L. S. Chambers, as reproduced by M. King Hubbert, "Entrapment of Petroleum under Hydrodynamic Conditions," *Bull. Amer. Assoc. Petrol. Geol.*, Vol. 37, No. 8 (Aug., 1953), Fig. 41, p. 2020. Gas is trapped downdip in northern culmination. No gas cap in southern culmination.

structure. The oil accumulation is modified by the hydraulic gradient so that the oil-water interface is tilted north at about 90 feet per mile.

Apparently, gas was trapped in East Coalinga Extension north dome while oil was spilled updip into the higher culmination (south dome). This relationship of oil and gas can not be explained by the hydraulic theory, but is in accordance with the principle of differential entrapment.

There are many other examples in the United States that suggest the principle of differential entrapment is the controlling feature, explaining the relationships of oil, gas, and water, in series of interconnected traps, however, sufficient data have not been published on these.

Examples that warrant further consideration are the relationships of: (1) production from the lower Glen Rose at Coke (gas), Pickton and New Hope (oil), Texas; (2) South Groesbeck, North Groesbeck, and Mexia (gas), and Wortham, Currie, North Currie, Richland, and Powell (oil), in the Mexia-Tehuacana fault zone, Texas; (3) Cat Creek, Montana; (4) Thornburg (gas), Iles (oil and gas), and Moffat (oil), in the Uinta Basin of northern Colorado; (5) Paint Creek uplift, Kentucky; (6) Elwood (oil) and La Goleta (gas), California; and (7) Big Piney (gas), Tip Top and LaBarge (oil), Wyoming.

Differential entrapment can explain why oil-producing structures are confined to the Pennsylvanian shelf area of eastern Oklahoma and eastern Kansas while in the McAlester Basin, all traps are filled with gas to the exclusion of oil. This view is contrary to the recent conclusions postulated by Weirich (1953).

DIFFERENTIAL ENTRAPMENT IN PERSIAN FIELDS

Much less information is available on petroliferous areas outside Canada and the United States. Hundreds of examples of differential entrapment must exist but too few data have been published on which to base an opinion regarding most foreign fields. Undoubtedly, accumulations in Venezuela, Mexico, Trinidad, Russia, Indonesia, Burma, and elsewhere will be found to exhibit similar relationships.

An attempt has been made to assemble data on the Persian fields (Lees, 1953; Baker and Henson, 1952; Ver Wiebe, 1952), which appear to illustrate the effect of differential entrapment rather well. The rich Persian oil fields of the "Central Area" occur in southwestern Iran, north of the northwest end of the Persian Gulf, in what are known as the foothills of the Persian mountain arc. They are on three main anticlinal trends, about 250 miles long and 5–6 miles across. These more or less parallel each other and have a regional plunge northwest, so that the northernmost culminations are the deepest. Those at the southern end are the highest and are commonly exposed as a mountain (kuh). Oil and gas are trapped in a series of culminations along these anticlines, in fields which average 16 miles in length by 4 miles in width with closures ranging from 2,000 feet to 7,000 feet or more.

Production is in part from the upper Asmari limestone. This is a thick transgressive limestone of lower Miocene age, which passes basinally into evaporites and shales. In many places, this Miocene reservoir is superimposed on Oligocene reef limestone (lower and middle Asmari) which may have contributed much of the oil. Free reservoir connection in the Asmari limestone has been proved over long distances. The cap rock is an evaporite series known as the lower Fars formation, also Miocene in age, consisting of cyclical alternations of anhydrite, limestone, and salt, which provide a plastic infrangible cover.

Figure 16, is a graphic compilation of information from the references cited. The sections are more or less true to scale horizontally and vertically, and show the relationships and fluid content of the respective culminations along three parallel anticlinal trends: the configuration, however, is diagrammatic. The three anticlinal trends are separated by deep synclinal troughs in which the Asmari limestone lies about 13,000 feet below sea-level. It is, therefore, most unlikely that lateral spilling has occurred between anticlinal trends.

The Lali, Masjid-i-Sulaiman anticlinal trend.—This trend lies nearest the mountains. and culminations along it are, from north to south: Lali, Masjid-i-Sulaiman, and Kuh-i-Asmari. In the upper Asmari, Lali has a large gas cap and an oil column of approximately 2,200 feet. Masjid-i-Sulaiman has a 2,200-foot oil column but no gas cap, while Kuh-i-Asmari, at the south end, is exposed as a mountain and is dry except for residual showings. Gas has apparently been trapped downdip at, or below, Lali, and little if any oil has spilled updip from Masjid-i-Sulaiman—the showings at Kuh-i-Asmari probably having resulted from local migration. On the other hand, the Asmari structure has been incised by erosion and lacks a proper seal so that if any oil had entered this structure by spilling updip from Masjid-i-Sulaiman, it would have escaped. These relationships are in accordance with the principle of differential entrapment.

Deeper drilling at Lali obtained oil in the Cretaceous, 1,600 feet below the Asmari limestone complex. The Cretaceous at Masjid-i-Sulaiman and at Kuh-i-Asmari is loaded with salt water. It would appear that Cretaceous oil is trapped downdip at Lali and that this structure is not filled to the spill point, with the result that Masjid-i-Sulaiman and Kuh-i-Asmari are dry. It would be difficult to explain this relationship by selective seepage.

Naft-i-Safid, Haft Kel anticlinal trend.—This axis lies southwest and somewhat *en échelon* to the Lali, Masjid-i-Sulaiman trend. Three culminations are shown, from north to south: Naft-i-Safid, Haft Kel, and Mamatain.

Naft-i-Safid has a 1,000-foot oil column and a 2,000-foot gas cap. Haft Kel, which is about 2,000 feet higher in elevation, has a 2,000-foot oil column and a very small gas cap. Mamatain contains about 200 feet of gas and heavy oil, and is largely filled with salt water. The gas cap at Haft Kel probably represents solution gas that has come out of the oil in migrating updip 2,000 feet in elevation. The small accumulation of hydrocarbons at Mamatain may be entirely the result of local migration. The Asmari accumulations along this anticlinal trend conform with the principle of differential entrapment.

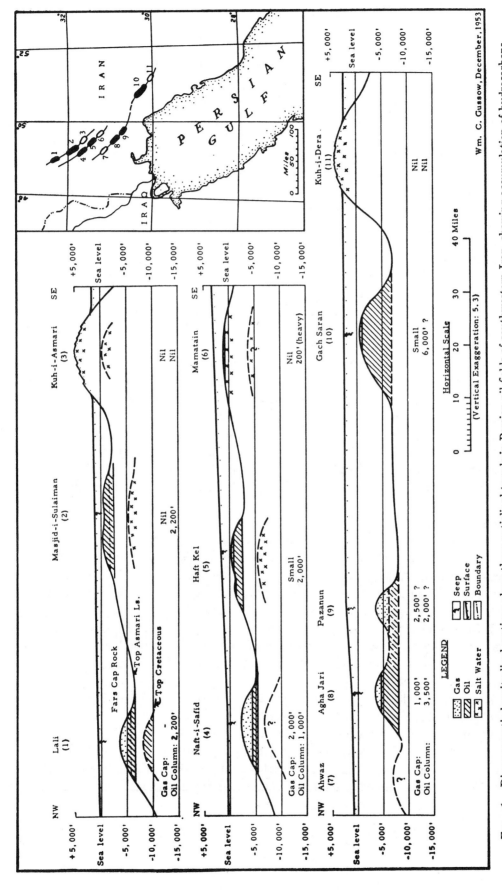

FIG. 16.—Diagrammatic longitudinal sections along three anticlinal trends in Persian oil fields of southwestern Iran, showing accumulation of hydrocarbons. Pay sections indicated for Asmari limestone reservoirs.

149

The Cretaceous has been tested only at Haft Kel where it was found to be loaded with salt water. According to the principle of differential entrapment the Cretaceous at Mamatain would be expected to be dry, whereas at Naft-i-Safid it could well be filled with oil, and probably there is no primary gas cap. A careful study of the probable migration path and spill-point levels in the Cretaceous, should eliminate much of the guess work in such a test.

Agha Jari, Gach Saran anticlinal trend.—Culminations along the western-most axis are, from north to south: Ahwaz, Agha Jari, Pazanun, Gach Saran, and Kuh-i-Dera. Hydrocarbon accumulations along this trend are somewhat irregular at first glance. Ahwaz is a newly discovered deep structure that has not been tested; Agha Jari has a 1,000-foot gas cap and an oil column reported to be more than 3,500 feet thick; Pazanun has a wet gas column of more than 2,000 feet and the presence or absence of an oil column has not been confirmed by drilling; Gach Saran is reported to have a small gas cap and a "very considerable oil column indeed"; Kuh-i-Dera is exposed as a mountain and is dry.

An analysis of the fluid relationships suggests that Pazanun will be found to have a common oil-water interface with Agha Jari and accordingly might have a maximum oil column of 2,500 feet: the elevation of the saddle (spill point) between these two culminations is the controlling factor. The oil column at Gach Saran could be as much as 7,000 feet, making this one of the largest known accumulations in the world. The migration path must have entered this anti-clinal trend from the southwest, at Pazanun. When this had filled with oil to the spill point, which is the saddle on the north, oil began to spill north into Agha Jari. When Agha Jari was filled both Agha Jari and Pazanun would have a common oil-water interface and as more and more oil accumulated this common interface would be backed down to the spill point south of Pazanun. Oil would now begin to spill south into Gach Saran. After many thousand feet of oil column had accumulated at Gach Saran, gas began to be trapped at Pazanun, causing more oil to be flushed updip into Gach Saran. When the gas-oil interface in Pazanun reached the spill-point level of the saddle on the north, a gas cap would appear at Agha Jari (as gas began to spill north out of Pazanun). More and more oil would spill south into Gach Saran as the gas cap formed in Agha Jari.

Information available suggests that the large, geosynclinal Persian Gulf area was the supply basin, and that the path of migration was northwest along the axis until it spilled northeast into the anticlinal trends, moving along them toward the southeast.

In the past, the apparent "anomalous" relationships in the Persian fields have largely been attributed to seepages. Thus it was argued that Masjid-i-Sulaiman, Haft Kel, Agha Jari, and Gach Saran owe their large oil columns and lack of gas to selective escape of their gas caps, permitting more oil to accumulate. In this connection, Lees (1953) states that "a complete explanation of all the circumstances cannot be offered." Baker and Henson (1952) write that "dry holes have been drilled in geologically favorable locations, but the easy explanation that oil was never formed in the vicinity is not satisfying in the face of abundant residual oil traces." As to the large gas cap at Pazanun, Lees (1953) believes this " . . . may be due to the greater thickness of the Fars, providing a more effective seal."

The present view is that no anomalous conditions exist, and it is suggested that the "residual" oil traces are the result of local migration. The fact that they are unproductive is explained by their structural position with respect to regional migration paths and by the effect of differential entrapment.

SYNCLINAL OIL POOLS

Two types of synclinal oil pools are known: those in blanket-type sands in which the oil accumulation is trapped far down the plunge of synclinal structures, and the "shoestring" sands, in which the oil lies in synclinal structures or in the downdip part of a lenticular sand body. Various theories have been advanced to account for the trapping of oil in synclines. This is generally attributed to the "absence of water" in the reservoir. So-called examples of this phenomenon have been investigated and to date no valid examples of water-free sands have been discovered.

Davis and Stephenson (1929) have listed many examples of synclinal oil pools in West Virginia; three examples (Griffithsville, Tanner Creek, and Granny's Creek fields) were described and illustrated. The Copley oil pool, also in West Virginia, is described by Reger (1929) as

an illustration of synclinal accumulation of oil in sands which carry no water. Oil occurs in the deepest part of the basin, at the foot of a descending axis, being almost surrounded by gas in the anticlines on either side and along the higher synclinal axis northeastward.

Although he mentions the possible "absence of water . . . due to displacement," he does not believe in this explanation and proceeds to describe accumulation in the synclinal areas on account of the absence of water, with gas migrating into the higher parts.

Heck (1941) describes another so-called "dry-sand" reservoir in the Gay-Spencer-Richardson trend. This is a "shoe-string" sand that has been folded across into a series of anticlines and synclines. Oil occurs in the synclinal areas. The Cabin Creek field, described by Wasson and Wasson (1929), is located in central West Virginia, 20 miles southeast of Charleston. This accumulation is in a "shoe-string" sand which parallels a plunging synclinal axis, oil being trapped downdip with a gas cap occupying the upper end.

In summarizing Volume II, *Structure of Typical American Oil Fields*, Clapp (p. 703) dwells on the subject of synclinal oil and remarks that "the cause of the absence of water from West Virginia sands has remained a mystery."

Figure 17 is a typical example of a synclinal accumulation of oil. In every case the anticlines or updip parts are filled with gas and the oil has been backed down into the synclines. Unquestionably, an oil-water interface occurs downdip,

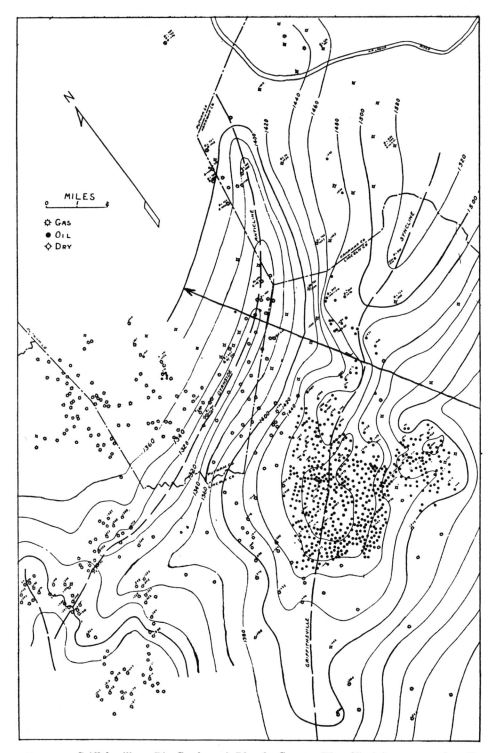

Fig. 17.—Griffithsville or Big Creek pool, Lincoln County, West Virginia, contoured on Berea sandstone. Contour interval, 20 feet. All contours below sea-level. After Ralph E. Davis and Eugene A. Stephenson, in *Structure of Typical American Oil Fields*, Vol. II (Amer. Assoc. Petrol. Geol., 1929), Fig. 1, p. 572. Oil and water have been backed downdip into synclinal troughs by large gas caps which fill anticlines and upper parts of synclines.

water having been backed farther down the synclinal trough. In some lenticular sands, the oil-water interface has been backed downdip right out of the sand so that it coincides with an impermeability barrier.

Oil must have occupied the anticlines at one time. Then, with the appearance of a gas cap, the gas began to back the oil (and water) downdip out of the anticlines, until to-day the anticlines and much of the synclines are filled with gas. Water is trapped in closed synclinal areas (Clendenin) so that a gas-water interface results, and the oil occurs down the plunge of the synclinal troughs above a water interface.

In every case, accumulation of oil and gas is in accordance with the structural theory of accumulation and the apparent absence of water in the sands is simply due to the fact that it has been displaced out of the anticlines and higher parts of the synclines by large quantities of gas. The absence of water in these sands seems no more puzzling than its absence in the gas cap of an anticlinal structure. There are thus no known exceptions to the structural theory.

CONCLUSION

In summary, three fundamental principles control the migration and accumulation of oil and gas.

1. The effect of gravity on inhomogeneous fluids causing oil and gas to migrate vertically upward in a water-saturated environment (potential due to difference in specific gravity of gas, oil, and water)
2. Impermeability barriers and structural control, causing deflection and lateral migration and eventual trapping of hydrocarbons
3. The effect of differential entrapment caused by a succession of traps

Some of the factors which modify the results of the fundamental principles are the following:

1. Hydrodynamic conditions in the strata, causing tilt of interfaces and slight spilling
2. Faulting and seepage, allowing insignificant escape of gas and/or oil
3. Depth of burial. (Capacity for a trap to hold gas is a function of pressure and increases with depth. Thus at 6,200 feet, the reservoir is capable of holding 200 times its volume of compressed gas)
4. Regional tilt, causing insignificant spilling or decantation of contents if reservoir filled to the spill point
5. The effect of temperature and pressure on the phase relationships of oil and gas, causing gas to go into solution in the oil or for solution gas to come out of solution forming a gas cap, *et cetera*

It should be noted that hydrodynamic conditions causing tilt of interfaces and flushing of shallow traps, are an admitted fact and have been adequately proved. Tilts up to a maximum of 850 feet per mile are known. The average is about 100 feet per mile, and while much greater tilts and anomalous effects are readily demonstrated in the laboratory, they have so far not been found in nature. If hydrodynamic conditions actually existed such as to cause flushing of oil from a reservoir, the result would be to cause the oil to migrate downdip and accumulate in a downdip position, leaving any gas caps in an updip position. This is exactly the reverse of the "anomalous" conditions found in nature.

Many oil and gas seeps which occur in nature substantiate the fact that faulting and fracturing do permit the escape of hydrocarbons. However, a great many seeps are the result of lack of closure or a good seal. It is rather remarkable what little effect later faulting and fracturing have on most accumulations. They do have a modifying effect but can not explain the so-called "anomalous" occurrences.

Regional tilt and resulting decantation have also been put forward to explain the "anomalous" occurrences. Without doubt this is a factor to be considered and there is no denying that it does occur. It is merely a matter of the degree to which it is effective. For example, the regional tilt in Alberta might effect accumulations in the Nisku (D2) but could not account for the effects seen in the Leduc (D3) reefs. (The regional tilt in Alberta in the vicinity of the reef trend would be the equivalent of sliding a thin dime under the edge of a full bowl of soup having a 5½-inch diameter base.)

Finally, some gas caps are undoubtedly the result of solution gas which has escaped from the oil with decrease in pressure as the oil migrates or spills updip into structurally higher traps, or as the reservoir is brought closer to the surface by erosion.

The greatest argument against the theory of a distant source and migration for long distances has been the so-called "exceptions" or "anomalous" occurrences, that did not fit the "anticlinal" or structural theory. For example, the " . . . absence of oil in many suitable reservoirs and effective traps enclosed by rocks of presumably the same physical character and visible organic content as those that enclose the oil-saturated rocks" (Frank R. Clark, 1934, p. 311). Thus, Clark errs in his assumption that " . . . every effective reservoir would be filled with oil" (p. 312), if a widely disseminated source and distant migration are postulated, rather the exception is the rule and is a condition which agrees with the facts observed in nature.

The principle of differential entrapment is based on the assumption that oil and gas migrate over great distances and that adequate source beds occur in a large downdip supply basin. No doubt source beds within an individual structure will contribute hydrocarbons, but local sources are not considered adequate to form commercial accumulations. This new concept would appear to refute one of the strongest arguments raised against distant migration or in support of a local origin. The principle of differential entrapment is, however, well substantiated.

More thought must be given to the path of migration in areas of accumulation. This can be determined and mapped with the conventional tools of exploration with the same degree of accuracy as exploration for structural traps.

REFERENCES

BAKER, N. E., AND HENSON, F. R. S., 1952, "Geological Conditions of Oil Occurrence in Middle East Fields," *Bull. Amer. Assoc. Petrol. Geol.*, Vol. 36, No. 10 (October), pp. 1885–1901.
BARNETT, D. G., 1941, "O'Hern Field, Duval and Webb Counties, Texas," *Stratigraphic Type Oil Fields*, pp. 722–49. A.A.P.G.
BARTRAM, JOHN G., 1929, "Elk Basin Oil and Gas Field, Park County, Wyoming, and Carbon

County, Montana," *Structure of Typical American Oil Fields*, Vol. II, pp. 577–88. A.A.P.G.

———, 1936, "Examples of Migration of Petroleum," *Bull. Amer. Assoc. Petrol. Geol.*, Vol. 20, No. 5 (May), p. 613.

BECK, ELFRED, 1929, "Salt Creek Oil Field, Natrona County, Wyoming," *Structure of Typical American Oil Fields*, Vol. II, pp. 589–603.

CHAMBERS, L. S., 1943, "Coalinga East Extension Area of the Coalinga Oil Fields," in "Geologic Formations and Economic Development of the Oil and Gas Fields of California," *California Dept. Nat. Resources Bull. 118*, Pt. 3, pp. 486–90.

CLAPP, FREDERICK G., 1929, "Rôle of Geologic Structure in the Accumulation of Petroleum," *Structure of Typical American Oil Fields*, Vol. II, pp. 667–716.

CLARK, FRANK R., 1934, "Origin and Accumulation of Oil," *Problems of Petroleum Geology*, pp. 309–35. A.A.P.G.

CLARK, STUART K., ARNETT, C. L., AND ROYDS, JAMES S., 1948, "Geneseo Uplift, Rice, Ellsworth, and McPherson Counties, Kansas," *Structure of Typical American Oil Fields*, Vol. III, pp. 225–48. A.A.P.G.

COVENEY, J. W., AND BROWN, A. A. (1954, in press), "Geology and Development History of the Acheson Field," *Bull. Canadian Inst. Min. Met.*, Fig. 6.

DAVIS, MORGAN J., 1929, "Artesia Field, Eddy County, New Mexico," *Structure of Typical American Oil Fields*, Vol. I, pp. 112–23. A.A.P.G.

DAVIS, RALPH E., AND STEPHENSON, EUGENE A., 1929, "Synclinal Oil Fields in Southern West Virginia," *ibid.*, Vol. II, pp. 571–76.

DOBBIN, C. E., 1947, "Exceptional Oil Fields in Rocky Mountain Region of United States," *Bull. Amer. Assoc. Petrol. Geol.*, Vol. 31, No. 5 (May), pp. 797–823.

GREEN, T. H., AND ZIEMER, C. W., 1953, "Tilted Water Table at Northwest Lake Creek Field," *Oil and Gas Jour.*, Vol. 52, No. 10 (July), p. 178.

HECK, E. T., 1941, "Gay-Spencer-Richardson Oil and Gas Trend, Jackson, Roane, and Calhoun Counties, West Virginia," *Stratigraphic Type Oil Fields*, pp. 807–27.

HOWELL, W. F., 1929, "Kevin-Sunburst Field, Toole County, Montana," *Structure of Typical American Oil Fields*, Vol. II, pp. 254–68.

HUBBERT, M. KING, 1953, "Entrapment of Petroleum under Hydrodynamic Conditions," *Bull. Amer. Assoc. Petrol. Geol.*, Vol. 37, No. 8 (August), pp. 1954–2026.

HUNT, T. STERRY, 1861, *Montreal Gazette* (March 1).

IRWIN, J. S., 1929, "Oil and Gas Fields of Lost Soldier District, Wyoming," *Structure of Typical American Oil Fields*, Vol. II, pp. 636–66.

LAHEE, FREDERIC H., 1934, "A Study of the Evidences for Lateral and Vertical Migration of Oil," *Problems of Petroleum Geology*, pp. 399–427.

LEES, G. M., 1953, "The Middle East" and "Persia," in "The World's Oil Fields—The Eastern Hemisphere," *The Science of Petroleum*, Vol. VI, Pt. 1.

NIGHTINGALE, W. T., 1935, "Geology of Baxter Basin Gas Fields, Sweetwater County, Wyoming," *Geology of Natural Gas*, pp. 323–39. A.A.P.G.

NOWELS, K. B., 1928, "Oil Production in Rattlesnake Field," *Oil and Gas Jour.* (June 7), pp. 106–07, 143–44.

———, 1929, "Development and Relation of Oil Accumulation to Structure in the Shiprock District of the Navajo Indian Reservation, New Mexico," *Bull. Amer. Assoc. Petrol. Geol.*, Vol. 13, No. 2 (February), pp. 117–51.

REGER, DAVID B., 1929, "Copley Oil Pool of West Virginia," *Structure of Typical American Oil Fields*, Vol. I, pp. 440–61.

SANDS, J. MELVILLE, 1929, "Burbank Field, Osage County, Oklahoma," *ibid.*, pp. 220–29.

SPOONER, W. C., 1929, "Homer Oil Field, Claiborne Parish, Louisiana," *ibid.*, Vol. II, pp. 196–228.

TEAS, L. P., 1935, "Natural Gas of Gulf Coast Salt-Dome Area," *Geology of Natural Gas*, pp. 683–740.

THOMAS, C. R., 1929, "Flank Production of the Nemaha Mountains (Granite Ridge), Kansas," *Structure of Typical American Oil Fields*, Vol. I, pp. 60–72.

TILLOTSON, ALLEN W., 1935, "Gas Fields of Lost Soldier District, Carbon and Sweetwater Counties, Wyoming," *Geology of Natural Gas*, pp. 305–22.

TOMLINSON, C. W., 1935, "Natural Gas Pools of Southern Oklahoma," *ibid.*, pp. 575–607.

VER WIEBE, W. A., 1952, *North American Petroleum*. Wichita, Kansas.

WASSON, THERON, AND WASSON, ISOBEL B., 1929, "Cabin Creek Field, West Virginia," *Structure of Typical American Oil Fields*, Vol. I, pp. 464–75.

WEIRICH, THOMAS EUGENE, 1953, "Shelf Principle of Oil Origin, Migration, and Accumulation," *Bull. Amer. Assoc. Petrol. Geol.*, Vol. 37, No. 8 (August), pp. 2027–45.

American Association of Petroleum Geologists Bulletin
V. 63, No. 4 (April 1979) pp. 608-620.

Differential Entrapment of Oil and Gas in Niagaran Pinnacle-Reef Belt of Northern Michigan[1]

DAN GILL[2]

Abstract The Niagaran pinnacle-reef belt in the northern part of the Michigan basin is about 170 mi (270 km long) and 10 to 20 mi (16 to 32 km) wide. Since 1969, 360 oil- and gas-producing reefs and 72 salt-plugged or otherwise barren, water-saturated reefs have been found in the belt. The reefs are of small areal extent (average 80 acres; 32 ha.), high relief (up to 600 ft; 180 m), and steep flanks (30 to 45°) and are effectively sealed by the lower Salina evaporite deposits. The reefs are hydraulically interconnected through the Lockport Formation, their common permeable substrate, which dips basinward at 70 to 140 ft/mi (13 to 26 m/km; 0.76 to 1.52°). Reef height, pay thickness, burial depth, reservoir pressure, hydrogen sulfide content, and extent of salt plugging increase progressively in a basinward direction across the belt, whereas oil gravity and degree of dolomitization increase systematically in the opposite direction.

The belt is distinctly partitioned in an updip direction into three parallel bands of gas-, oil-, and water-saturated zones. This zonation of reservoir fluids is in full accord with Gussow's classic theory on the differential entrapment of oil and gas and provides a textbook example of its applicability on a regional scale in a natural case history. Interruptions in the continuity of the water-saturated band are interpreted as indicating passageways through which hydrocarbons may have migrated farther updip into the carbonate-shelf platform bounding the pinnacle-reef belt on the north. This reasoning leads to the delineation of two additional favorable target areas for further exploration.

INTRODUCTION

Natural geologic environments inherently are fraught with inhomogeneities, anisotropisms, and a host of known and unknown local modifying factors. All of these combine to obscure the basic physical and chemical laws which control geologic processes. The severity of this situation naturally varies with the complexity of a phenomenon but, for general characterization, the assessment is probably valid for all branches of geology. Thus, most examples of a certain phenomenon may appear to depart in some way from the general physicochemical rules which are thought to govern it and examples which are in full accord with the rules are, paradoxically, the exceptions. The purpose of this study is to document such a rare occurrence in which one of the fundamental theories on the migration and entrapment of oil and gas is borne out in nature. This theory is W. C. Gussow's principle of differential entrapment of oil and gas (Gussow, 1954, 1968). The example is found in the Middle Silurian northern pinnacle-reef belt of the Michigan

basin. Here, differential entrapment can be demonstrated to have operated on a regional scale, affecting a migration front which encompassed a wide segment of the basin 170 mi (270 km) long. This textbook example provides strong substantiation for Gussow's theory. Furthermore, recognition of the fact that distribution of hydrocarbons in northern Michigan was controlled by the differential-entrapment mechanism allows postulation of other prospects in the area.

GENERAL GEOLOGIC BACKGROUND

The paleogeography and sedimentary succession of the Michigan basin and its surrounding areas during Niagaran (Middle Silurian) time were controlled by three main structural elements which experienced different degrees and styles of subsidence: (1) a relatively stable outer platform, coinciding with the structural arches surrounding the basin; (2) an unstable platform-to-basin transitional slope along which differential subsidence occurred; and (3) the relatively flat floor of the uniformly subsiding basin interior. These structural zones, in turn, gave rise to four main carbonate depositional environments which delineated continuous circular belts disposed concentrically around the center of the basin (Briggs and

[1]Manuscript received, February 24, 1978; accepted, October 2, 1978.

[2]Geological Survey of Israel, Jerusalem 95501, Israel.

Courteous cooperation by R. J. Burgess and the geologic staff of the Northern Michigan Exploration Co., Jackson, Michigan, is gratefully acknowledged. R. D. Dunn provided the water contact map (Fig. 3) and assisted in compiling data (Table 1) for the construction of the cross sections (Figs. 5, 6); W. Mantek provided valuable unpublished information for the construction of Figure 4. I also thank L. I. Briggs, U. Kafri, and O. Amit for helpful discussions. Sincere thanks are extended to the National Science Foundation for partial financial support under grant EAR76-17410 and to the Geological Survey of Israel for facilitating the completion of this study.

Article Identification Number
0149-1423/79/B003-0003$03.00/0

Briggs, 1974; Caughlin et al, 1976, Fig. 3). In a basinward direction these included (Fig. 1): (1) platform shelf carbonates; (2) a barrier reef belt or carbonate bank along the platform margin, with an associated narrow belt of fore-bank skeletal detritus basinward of it; (3) a pinnacle-reef belt, coinciding with the platform slope; and (4) a deeper water basinal environment. Detailed descriptions of carbonate facies in various geographic segments of the respective subsurface depositional belts can be found in Evans (1950), Alling and Briggs (1961), Pounder (1962), Burgess and Benson (1969), Sanford (1969), Brigham (1971), Huh (1973), Mantek (1973, 1976), Briggs and Briggs (1974), Meloy (1974), Mesolella et al (1974), Gill (1977a), Huh et al (1977), and Briggs et al (1978). Paleomagnetic (Ziegler et al, 1976) and paleobotanic data (Cramer, 1970) indicate

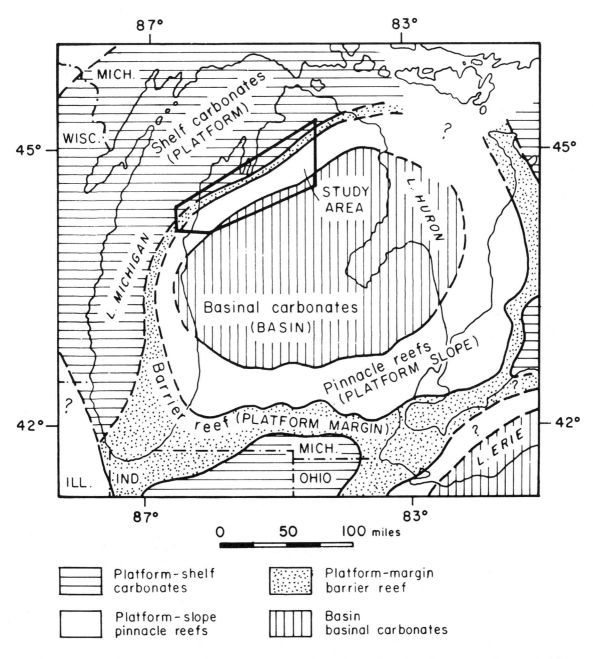

FIG. 1—Michigan basin Niagaran paleogeography and main carbonate depositional environments (modified from Briggs and Briggs, 1974; Lake Erie area data from Sparling, 1965; Sanford, 1969).

that during the Silurian, the Michigan basin was located at a latitude between about 10 and 15°S. This paleotropical geographic position might account for the fact that the Silurian was an extensive reef-building period throughout the Great Lakes area (Lowenstam, 1957).

Near the end of the Wenlockian Stage, the reef buildups along the platform margins eventually coalesced to form a continuous barrier which effectively barred the basin (Alling and Briggs, 1961). Augmented, perhaps, by a change to a more arid climate, this barring brought about a drastic change to an evaporitic sedimentologic regime. During the rest of the Silurian, the Michigan basin accumulated a great thickness of evaporites. In its central part, the evaporite cycles of the Salina Group attain a thickness of about 3,000 ft (900 m; Ells, 1967).

Figure 2 presents a schematic stratigraphic chart which serves to introduce the stratigraphic relations and terminology employed in the present study. Both formal and informal lithostratigraphic units are given on the chart. The inclusion of informal units is unavoidable because some of these names are widely used by stratigraphers working in the area. Although it is beyond the scope of this study to discuss or try to resolve

some of the outstanding stratigraphic problems concerning the studied interval, the following comments are pertinent.

1. The use of the name "Guelph Formation" for what commonly is referred to as the "Brown Niagaran" may be inappropriate (Gill, 1977a, p. 19-23; Shaver, 1974, p. 59-60). The Guelph, at least in part, is considered to be a reef facies of the Lockport Formation and thus has no precise stratigraphic level throughout the Great Lakes region (Huh et al, 1977). At present this stratigraphic name is applied without distinction to all rock bodies intermediate between the Lockport Formation (whose top cannot in all places be defined precisely) and the Salina Group. This interval encompasses different carbonate facies whose precise time correlation has been a subject of much debate. As commonly applied, it includes the higher part of the shelf-platform carbonate sequence, the shelf-margin barrier reef, the carbonate buildups of the pinnacle reefs, and a thin section deposited contemporaneously with the evolution of the pinnacles in interpinnacle and basinal areas. Quite clearly, a refinement of the stratigraphic terminology of this unit is in order.

2. The Cain Formation recently has been introduced as a formal lithostratigraphic name for a

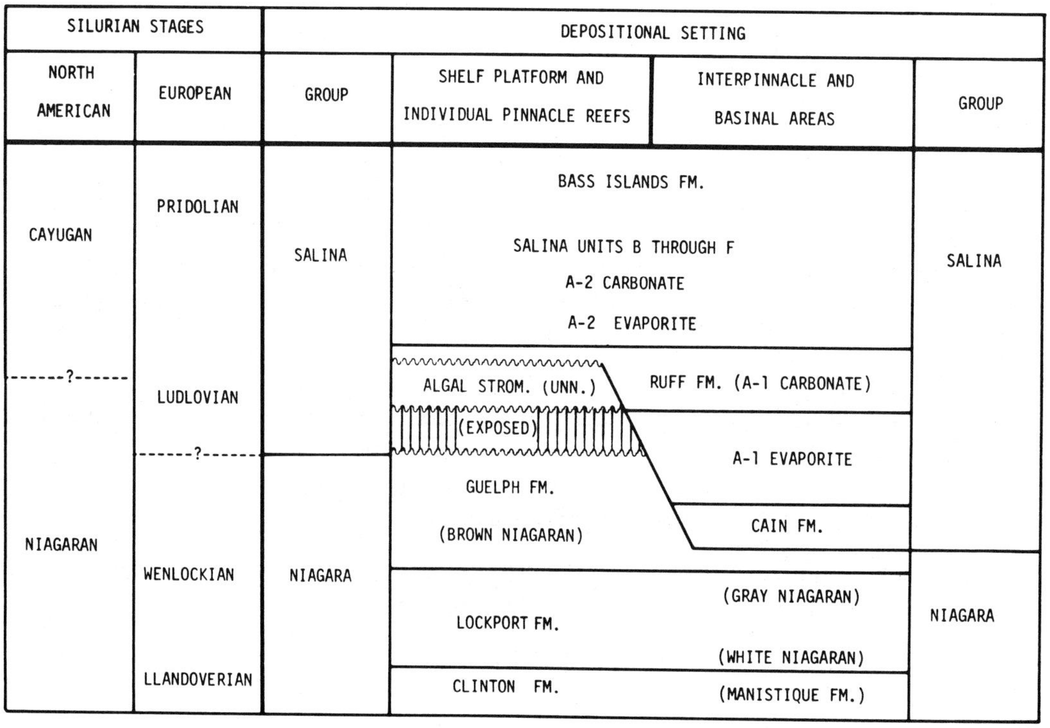

SILURIAN STAGES		DEPOSITIONAL SETTING			
NORTH AMERICAN	EUROPEAN	GROUP	SHELF PLATFORM AND INDIVIDUAL PINNACLE REEFS	INTERPINNACLE AND BASINAL AREAS	GROUP
CAYUGAN	PRIDOLIAN	SALINA	BASS ISLANDS FM. SALINA UNITS B THROUGH F A-2 CARBONATE A-2 EVAPORITE		SALINA
------?------	LUDLOVIAN		ALGAL STROM. (UNN.) (EXPOSED)	RUFF FM. (A-1 CARBONATE) A-1 EVAPORITE	
	------?------	NIAGARA	GUELPH FM. (BROWN NIAGARAN)	CAIN FM.	
NIAGARAN	WENLOCKIAN		LOCKPORT FM.	(GRAY NIAGARAN) (WHITE NIAGARAN)	NIAGARA
	LLANDOVERIAN		CLINTON FM.	(MANISTIQUE FM.)	

FIG. 2—Schematic representation of stratigraphic relations and nomenclature of Middle and Upper Silurian formations in subsurface of Michigan basin.

rock unit at the base of the Salina Group in inter-pinnacle and basinal areas (Gill et al, 1978; Briggs et al, 1979). The Ruff Formation recently has been introduced as a formal lithostratigraphic name for the Salina A-1 carbonate unit (Budros and Briggs, 1977).

3. The pinnacle reefs within the basin are capped by an algal stromatolite, which in places is up to 60 ft (18 m) thick (unnamed algal stromatolite in Fig. 2). Stratigraphically this unit belongs to the Salina Group. Informally, it has been referred to as the Belle River Mills stromatolite (Gill, 1977b). Most likely, it is correlative with the algal stromatolite which caps the Maumee reef in northern Ohio (the "Maumee algal stromatolite," Kahle, 1974, 1978). Such an algal stromatolite is not known to appear over the carbonate-shelf platforms. However, a thin interval of tidal-flat carbonate rocks at the equivalent stratigraphic position in the platform areas may be correlative to it. This unit may also be correlative with the Limberlost Dolomite of northern Ohio and Indiana (Droste and Shaver, 1976). Several episodes of subaerial exposure are recorded within the stromatolite in both the Maumee quarry (Kahle, 1978) and over pinnacle reefs (Huh et al, 1977). Its precise stratigraphic position is uncertain. It could be correlative with either the Cain Formation or the Ruff Formation.

4. Above the Ruff, the distinction of the various subunits of the Salina Group is clear only inside the Michigan basin. Above the shelf platform the equivalent section is very much reduced in thickness and the various Salina subunits are not well defined.

In the context of the petroleum geology of the pinnacle reefs, the following aspects of the Salina sequence are noteworthy. The restriction of the basin resulted, initially, in the prevalence of euxinic conditions in the basin interior (Schmalz, 1969). This stage is marked by the deposition of the Cain Formation (Gill et al, 1978; Briggs et al, 1979) which is about 25 ft (7.5 m) thick. It consists of a basal dark-gray to black argillaceous mudstone, followed by five cycles of calcite and anhydrite varvites alternating with halite. The carbonate components of the Cain are rich in carbonaceous bituminous matter. In the 11 samples of the basal argillaceous mudstone analyzed, total organic carbon was in the range of 0.07 to 0.70%, with a mean value of 0.30%. The formation is very similar to several other preevaporite sapropelic-bituminous carbonate rocks which have been inferred to constitute excellent source beds for hydrocarbons (Borchert and Muir, 1964, p. 232-236; Schmalz, 1969; Brongersma-Sanders, 1971, 1972). As evaporation proceeded and the

water gradually withdrew to the center of the basin, the circumferential carbonate banks and the pinnacle reefs became exposed to subaerial weathering processes. The resultant dolomitization and development of secondary porosity significantly improved the reservoir properties of the reefs. The reefs became progressively buried under evaporite cycles containing salt, anhydrite, and sabkha carbonate deposits. The A-1 carbonate unit, in the lower part of the Salina sequence, contains abundant organic matter of algal origin; its authigenic pyrite indicates the prevalence of reducing conditions within the sediment soon after its deposition. These attributes have prompted several authors to suggest the A-1 carbonate unit as a likely source rock for the Niagaran crudes (Sharma, 1966, p. 347; Ells, 1967, p. 17; Gill, 1973, 1976, 1977a, p. 119-122; Budros, 1974, p. 124-128). Finally, as far as impervious cover rocks are concerned, with the deposition of the A-2 evaporite, the reefs became effectively sealed by flanking and overlying evaporite deposits (Fig. 2).

NORTHERN PINNACLE-REEF BELT
General Characteristics and Systematic Geologic Gradients

The northern pinnacle-reef belt extends from Mason and Manistee Counties on Lake Michigan to Presque Isle and Alpena Counties on Lake Huron as a continuous arcuate belt which is about 170 mi (270 km) long and 10 to 20 mi (16 to 32 km) wide (Fig. 1). The reefs developed between the relatively stable platform and the uniformly subsiding basin interior. During the growth of the reefs this transitional belt underwent differential tectonic subsidence which tilted the reefs' substrate basinward. At present, the slope of this surface (the "Gray Niagaran" or the Lockport Formation) ranges between 70 and 140 ft/mi (13 and 26 m/km; 0.76 to 1.52°) across the belt. Slight and insignificant as this slope may seem, it nevertheless induced systematic gradients in a host of geologic parameters; these will be reviewed. The tectonic mechanism of this bending, whether vertical gravitational downwarping or a fault-controlled downstepping movement, is not well understood. Recently, Mantek (1976) has drawn attention to the fact that subgroups of reefs within the northern belt are distinctly aligned in lineaments which form a consistent en echelon pattern, oriented at about 15° to the trend's strike. He suggested that "these lineaments possibly represent very low relief en echelon tension fractures with associated minor normal faulting, resulting from a combination of basin subsidence and left lateral shearing at the commencement of reef growth during Niagaran time. Reef growth would tend to initiate along the favourable edges of the resulting very gentle terraces" (Mantek, 1976, p. 14).

The boundaries of the pinnacle-reef belt coincide with the changes in the slope of the "Gray Niagaran" surface. These changes are mirrored in the thickness of the overlying Salina Group sediments, particularly in the interval between the top of the A-2 carbonate and the base of the A-1 evaporite, and are readily detectable on seismic records. Within the belt this interval thickens basinward at a rate of 35 ft/mi (6.5 m/km). Basinward of the belt, beyond the 800-ft (240 m) isopach of this interval, the rate drops to 20 ft/mi or 4 m/km (Caughlin et al, 1976). The edge of the platform-margin barrier reef, defining the northern boundary of the belt, coincides with the 100-ft (30 m) isopach of the "Brown Niagaran" unit (Mantek, 1973, Fig. 9).

Exploration for reefs in the northern part of the lower peninsula started in 1966. The first discoveries in this area were reported in 1969 (Burgess and Benson, 1969). Most of the exploration activities concentrated in the western and central parts of the belt (Figs. 1, 3). Since then, new discoveries have been made at a staggering rate. By the end of 1976, more than 1,100 wells had been drilled in this area resulting in the discovery of 360 producing reefs. The pinnacle reefs are isolated carbonate buildups enclosed by salt and anhydrite deposits (Fig. 2). This juxtaposition between rocks of different petrophysical (density, seismic velocity) properties gives rise to lateral and vertical contrasts which are readily detectable by geophysical methods. The exploration in northern Michigan has been based almost entirely on the seismic reflection method, which, owing to recent advances in field and data processing techniques (Caughlin et al, 1976; McClintock, 1977), has been extremely effective. The success of seismic methods in locating reefs in northern Michigan has been on the order of 85% (Caughlin et al, 1976).

In June 1976 cumulative production from the northern trend reached 38.5 million bbl of oil and 0.2 Tcf of gas (Mantek, 1976). Estimated primary recoverable reserves are on the order of 300 to 400 million bbl of oil and 3 to 5 Tcf of gas (Mantek, 1976; Caughlin et al, 1976; Yelling and Tek, 1976).

On the basis of data from a large number of reefs, Mantek (1973, 1976) and Yelling and Tek (1976) compiled statistics on several reef properties. Reef heights above their common substrate level in the Lockport Formation ("Gray Niagaran") increase systematically across the trend in a

basinward direction from 300 ft (90 m) near the shelf edge to over 600 ft (180 m) at the basinward edge of the belt. A parallel trend exists in net-pay thicknesses which increase progressively basinward from 30 to 400 ft (9 to 120 m), as well as in the drilling depth to the reefs which ranges from 3,000 to 7,000 ft (900 to 2,100 m), depending on the geographic location within the belt. With increasing depth, there is a parallel increase in reservoir pressure, which has a gradient of 0.52 to 0.56 psi/ft across the belt. This gradient is expressed, in turn, in the level of the contact of hydrocarbon (oil or gas) and water within individual reefs. Figure 3 is a regional map of this contact. It is based on data from several hundred wells in which the contact was determined on laterologs (R. A. Dunn, personal commun., 1977). The slope of this surface changes slightly with lateral (east-west) position along the belt—100 ft/mi (19 m/km) in Manistee County; 150 ft/mi (29 m/km) in Grand Traverse County; 133 ft/mi (25.2 m/km) in Kalkaska County and 125 ft/mi (23.7 m/km) in Otsego County. Along any given profile this slope, which ranges from 1.08 to 1.63°, is slightly larger than the corresponding slope of the Lockport Formation. It should be emphasized that the surface shown in Figure 3 is discontinuous between individual pinnacle reefs. The map, nevertheless, provides a useful indirect means for predicting the height of the hydrocarbon-saturated column at any given location in the belt.

The reefs are true pinnacles of very small areal extent, and have very steep flanks ranging from 30 to 45°. The size of their basal part ranges from 30 to 850 acres (12 to 344 ha.) with an average of 80 acres (32 ha.). The number of producing wells per reef is, for the most part, one or two, with a maximum of 10 for the largest ones. Porosities range from 3 to 37% with an average of 6%, and water saturation is in the range of 5 to 25%. Reef size, net-pay thickness, porosity, and water saturation are approximately lognormal in distribution (Yelling and Tek, 1976).

Several additional geologic attributes display regular systematic trends of variations across the belt. Subsequent to their growth, the reefs experienced a complex diagenetic history which included subaerial exposure and consequent development of secondary karstic porosity and dolomitization. However, porosity-occluding processes of pore plugging by halite and anhydrite were operative during lower Salina Group deposition, when the reefs were gradually buried by evaporite deposits. The intensity of these processes varied in a regular fashion with position within the belt and, consequently, the resultant diagenetic changes display systematic gradients in a direction perpendicular to the belt's long axis. Salt and anhydrite plugging within the reefs increases basinward. Along the front of the belt facing the basin interior many reefs are barren of hydrocarbons owing to nearly complete pore plugging by evaporite minerals (Fig. 4). Inversely, the degree of dolomitization and the extent of secondary porosity development within the reefs increase gradually in an opposite direction toward the shelf edge. The outer reefs in the belt still consist of undolomitized limestone, those near the shelf edge are completely dolomitized, and those in the middle are partly dolomitized. The combined effect of these properties can be expected to be reflected in the "productivity" of the reefs, measured by the number of barrels of oil produced per unit drop in reservoir pressure. Raw productivity data for 98 fields have been compiled by Matzkanin (1976); productivity ranges from 10 to 4,500 bbl/psi and the figures do not display any perceptible systematic spatial pattern. It should be noted, however, that these raw data have not been standardized for the various production-enhancement treatments or for any other related factors. Until such standardization is performed, it will not be possible to assess whether the productivity of reefs does indeed correlate with their position within the belt.

Segregation of Reservoir Fluids

Of the 360 producing reefs discovered up to June 1976, 221 are oil fields and 139 are gas fields; 72 additional barren reefs, including both salt-plugged and water-saturated reefs, also have been found (Mantek, 1976). A map of the different types of fields in the belt is shown in Figure 4. The distribution of reservoir fluids follows a remarkably consistent pattern. Along its entire length, the belt is clearly partitioned into four longitudinal bands. From the basin interior toward the shelf, these are: barren salt-plugged reefs; outer gas-bearing reefs; intermediate oil-bearing reefs; marginal, barren, water-saturated reefs. The boundaries between the bands are rath-

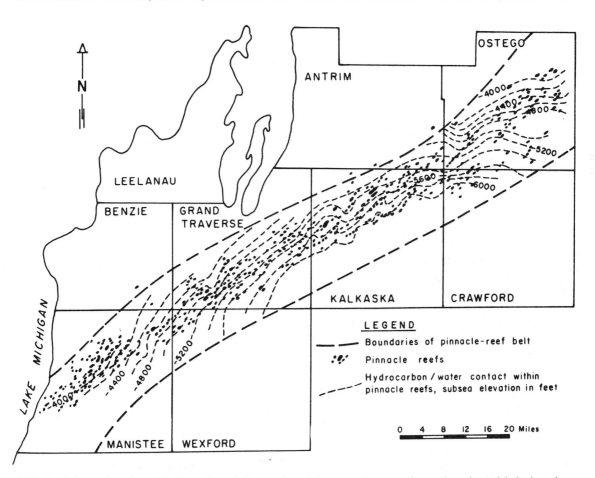

FIG. 3—Subsea elevation of hydrocarbon (oil or gas) and water contact, northern pinnacle-reef belt, based on information from several hundred boreholes in which water contact was determined from laterologs (R. A. Dunn, personal commun., 1977). Mapped contact exists physically only within pinnacle reefs.

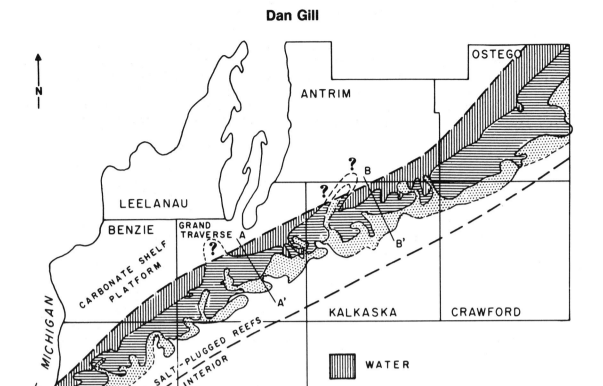

FIG. 4—Distribution of salt-plugged, gas-, oil-, and water-bearing reefs in northern pinnacle-reef belt of Michigan (based on information from 430 reefs obtained from Michigan Oil and Gas Assoc. Oil and Gas News, 1976; Reef Petroleum Corp., 1976; W. Mantek, personal commun., 1977). AA′ and BB′ mark locations of geologic cross sections shown in Figures 5 and 6.

er sinuous; the bands change width along their strike and are, in places, indented by irregular ingressions of adjacent bands. Two geologic traverses running perpendicular to the long axis of the belt are shown in Figures 5 and 6. The location of these cross sections is indicated in Figure 4. The identity of the boreholes appearing in the sections, along with additional pertinent data, are summarized in Table 1. The specific traverses shown in Figures 5 and 6 were deliberately selected to demonstrate the partitioning phenomena as best possible. Some profiles, when drawn strictly perpendicular to the regional strike, do show apparent local reversals in the segregation of fluids, for example, a differentiation in consecutive reefs into oil, gas, and back to oil in a downdip direction. Such diversions and irregularities, however, can be adequately accounted for in terms of the differential entrapment theory (see following) and thus, do not detract from the general applicability of the theory to the entire belt.

Associated with the partitioning are systematic gradients in two additional reservoir-fluid parameters. The oil becomes progressively lighter in a downdip direction. In API units, oil gravities increase from 43 near the shelf to 65 in the farthest oil fields, located between 7 and 9 mi (11 and 14 km) downdip from the shelf edge (Al-Muneef, 1973, Fig. 17). The hydrocarbons (both oil and gas) also become progressively more sour in the same direction, increasing from less than 18 ppm hydrogen sulfide near the shelf edge to over 340 ppm in some fields farthest downdip (Wilson, 1976).

DISCUSSION

The most plausible explanation to the fluid partitioning phenomenon in the northern pinnacle-reef belt of Michigan is offered by Gussow's "differential entrapment" principle (Gussow, 1954, 1968). By the same token, this Michigan basin case history provides one of the better natural

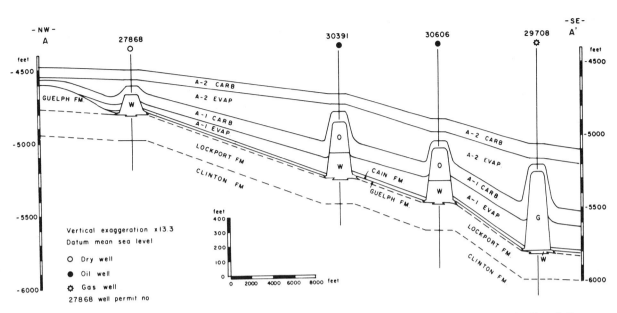

FIG. 5—Subsurface geologic cross-section AA′ (see Fig. 4 for location) across pinnacle-reef belt in Grand Traverse County, northern Michigan, demonstrating updip segregation of reservoir fluids in accordance with Gussow's differential-entrapment principle (*W*, water; *O*, oil; *G*, gas; draping of A-2 carbonate over reefs not shown; reef bases not to scale).

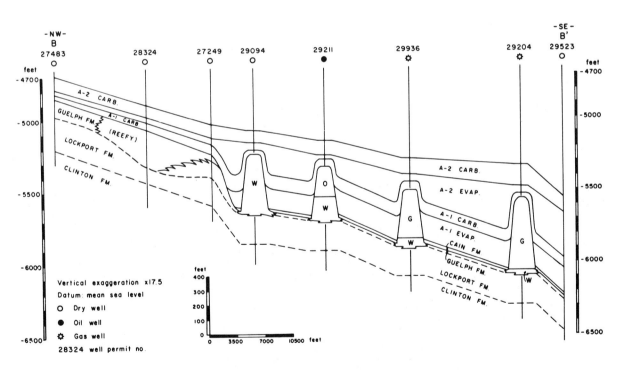

FIG. 6—Subsurface geologic cross section BB′ (see Fig. 4 for location) across pinnacle-reef belt in Kalkaska County, northern Michigan. Symbols are same as in Figure 5.

Table 1. Borehole Information Used to Construct Cross Sections (Figs. 5, 6)

Permit Number	Well Name	Location	K.B. Elev. (ft)	Geologic Setting	Production	Water Level (ft)	Depth Below Sea Level to Stratigraphic Tops (ft)						
							A-2 Carb.	A-2 Evap.	A-1 Carb.	A-1 Evap.	Cain Fm.	Guelph Fm.	Lockport Fm.
27868	N.M.E.C. 1-Goff	32-27N-10W	943	Pinnacle reef	Dry	-4,680	4,515	4,577	4,607	-	-	4,680	4,821
30391	Shell 1-20 State Paradise	20-26N-10W	925	Pinnacle reef	Oil	-5,100	4,703	4,775	4,821	-	-	4,893	5,283
30606	Shell 2-27 Edwards	27-26N-10W	934	Pinnacle reef	Oil	-5,300	4,878	4,972	5,034	-	-	5,083	5,460
29708	Shell 1-34 Widener	34-26N-10W	944	Pinnacle reef	Gas	-5,802	5,064	5,165	5,206	-	-	5,246	5,809
27483	McClure-Bailey 1	24-29N-7W	1,113	Shelf platform	Dry	-	4,697	4,790	4,821	-	-	4,850	4,995
28324	Shell 1-27 Elkins	27-29N-6W	1,320	Shelf-edge barrier reef	Dry	-	4,903	4,985	5,025	-	-	5,068	5,113
27249	N.M.E.C. 1-Adams	35-29N-6W	1,335	Shelf-edge slope	Dry	-	5,039	5,126	5,167	-	-	5,242	5,339
29094	1-23 State Rapid River "F"	23-28N-7W	1,108	Pinnacle reef	Dry	-5,315	5,084	5,166	5,214	-	-	5,241	5,657
29211	1-25 State Rapid River "E"	25-28N-7W	1,108	Pinnacle reef	Oil	-5,542	5,146	5,254	5,275	-	-	5,334	5,712
29936	1-31 Simpson Western Unit	31-28N-6W	1,204	Pinnacle reef	Gas	-5,832	5,279	5,393	5,444	-	-	5,506	5,890
29204	2-9 Weber	9-27N-6W	1,204	Pinnacle reef	Gas	-6,076	5,334	5,476	5,526	-	-	5,552	6,088
29523	1-22 State Excelsior	22-27N-6W	1,193	Basinal	Dry	-	5,546	5,663	5,965	6,047	6,255	6,265	6,273

examples of an updip partitioning of fluids of different gravities and serves to substantiate Gussow's classic theory. The impressive aspect of the Michigan example is its regional extent. Here, differential entrapment can be shown to have affected a migration front which entailed about one-fifth of the basin circumference.

It can be argued that the observed partitioning reflects an original differentiation within the source bed at the time primary migration began. In other words, closer to the shelf the source bed generated oil whereas away from the shelf and nearer the basin, the source bed expelled gas. Each reef drained only the source bed in its immediate vicinity and thus the original partitioning between oil and gas was preserved. Such an original differentiation could result from differences in depth of burial and consequent differences in pressure and temperature which would, in turn, lead to different degrees of maturation. It is, however, quite clear that across the narrow pinnacle-reef belt, the differences in overburden with reference to any given horizon are too small to generate such a differentiation. By invoking more than one type of source material and/or more than one source bed and by combining both short- and long-range primary and secondary migrations, one could conceivably propose several additional alternative explanations. However, except for Gussow's theory, all the alternative explanations examined by the writer were found to suffer from certain weaknesses, as they failed to account adequately for all the observations.

The structural and stratigraphic framework of the northern Michigan pinnacle reefs, entailing a series of evaporite-encased stratigraphic reef traps chained in an updip succession above a common regional permeable substrate, provides a rather ideal setup for differential entrapment to become operative. However, considering the high likelihood for the existence of modifying factors and the inherent irregularities which characterize natural environments, it is quite astounding that the more subtle predictions of Gussow's model are also borne out in this case.

In discussing the phenomenon of the gravitational stratification of oils of different gravities within a reservoir, Gussow (1968, p. 362) attributed it also to the differential-entrapment mechanism (and not to simple gravitational separation which, for chemically miscible fluids, is not possible). Carrying the argument further, he predicted that along a migration path between interconnected traps, oil gravities should increase updip. Such a gradient is indeed present in Michigan, as described previously. However, in his evaluation, Gussow implicitly presupposed that oils of differ-

ent gravities are physically separated from one another during migration. This supposition raises certain difficulties which require further elaboration. The guiding constraint lies in the presence of a continuous gradient of gravities, as opposed to a situation in which distinct families, or discrete portions of the gravity spectrum, are represented. The latter could conceivably result if different source beds were involved whereas, for the former to evolve, a progressive separation from a common source material is more plausible. Such a differentiation could conceivably evolve according to two different schemes.

1. The generation of oil is a continuous temperature, pressure, and time-dependent process during which, with increased overburden, progressively lighter hydrocarbons are formed. A gravitational differentiation is, thus, inherent in the oil-maturation process itself. The earlier formed heavy oils will initially saturate the downdip traps. With increased overburden, as progressively more paraffinic (lighter) oils migrate out of the source beds, differential entrapment will be set into operation. Heavier oils will be displaced updip and, eventually, an inverted updip gravity gradient will be formed. Such a differentiation will be more pronounced in a bowl-shaped basin in which the source bed forms a continuous horizon which, at any given time, is buried deeper at the basin center than along its shallower margin (e.g., Tissot et al, 1971). Furthermore, if the downdip traps are also concentrated along the basin margin, as in the Michigan basin, the first hydrocarbons to migrate into them will be the heavier ones drained from the shallow-buried source beds in their immediate proximity. This model implicitly assumes that when two oil phases of different specific gravities encounter each other, the lighter one will float over the heavier one rather than dissolve and mix with it. It also assumes a prolonged and continuous hydrocarbon generation and expulsion process. It may be debatable whether these assumptions are realistic.

2. The entire range of oil gravities now present across the belt formed more or less simultaneously in one phase. Separation by gravity was obtained during migration, the lighter phases traveling faster. Thus, by the time the heavier oils reached the downdip traps these were already occupied by the lighter fractions and the heavier oils had to migrate farther updip to find empty traps. In this admittedly oversimplified model, a fractionation by gravity during migration can lead to a spatially inverted gravity gradient, with the heaviest oils farthest updip, even without the more involved process of displacive remigration of heavier phases taking place. The key to this

model is the intuitive assertion that lighter hydrocarbons travel faster during migration. This relation is found to be valid for several flow processes which have been proposed to describe primary and secondary migration. These include capillary rise during primary migration and forces caused by fluid potential during secondary migration, as shown by Hubbert (1953); forces resulting from bouyancy; effective permeability and specific discharge in fluid flow according to Darcy's law (Craft and Hawkins, 1959, Chap. 7; Verruijt, 1970, Chap. 2); and flow due to diffusion (Glasstone, 1956, p. 85). In all of these processes, the impelling forces which set the fluid in motion, or the rate of flow as determined by the material properties of the medium, are inversely proportional to the fluid density or to other fluid properties, such as viscosity or molecular size, which vary concomitant with its density.

The observed gradient in the hydrogen sulfide content of the oil (decreasing updip) is also in accord with the differential entrapment formula of "lighter fractions trapped behind." However, because the hydrogen sulfide molecule is heavier than the prevalent methane gas, the existence of a similar gradient between the gas pools can be considered anomalous. The explanation for this apparent anomaly is hindered by our lack of knowledge as to the precise origin of the hydrogen sulfide. For example, it is quite possible that much of the hydrogen sulfide did not originate in the source bed but, instead, was generated within the reservoirs by postentrapment reduction of sulfate, either by anaerobic bacteria or by a reaction with already available hydrogen sulfide (Orr, 1974, p. 2314). Hydrogen sulfide content thus becomes a function of the amount of available sulfate. Because anhydrite, in both the reservoir pore space and in the enclosing A-1 evaporite, becomes more abundant basinward, the parallel increase in hydrogen sulfide may be linked to this trend.

As noted earlier, several formations, including the A-1 carbonate, the Cain Formation, and the Clinton have been proposed as likely source beds for the Niagaran crude oils. It is of interest to examine whether the interpretation framework set by the differential entrapment concept prescribes any preferences regarding this issue.

In northern Michigan, the A-1 carbonate covers the reefs and acts as an impervious seal. Hubbert (1953, p. 1978) has shown that unidirectional flow of hydrocarbons from an impervious bed into a reservoir can take place. Therefore, it is still possible that the A-1 carbonate could have contributed some hydrocarbons to the reefs. However, it is unlikely that this impervious formation

could have facilitated any large-scale regional migration. From the spatial relations of the rock units in question (Figs. 5, 6), it can be seen that the only aquifer through which both regional and interreef hydraulic communication could have taken place is the Lockport Formation (and overlying Guelph Formation, the thin interreef and basinal equivalent of the reef rocks). This formation most probably served as the regional carrier for long-distance migration and as local carrier for the interreef remigration of the spilled fluids. It therefore follows that the main source bed must have been in unimpeded hydraulic contiguity with the Lockport, and that this relation prevailed over large areas in which the Lockport dips south. The shaly facies of the Clinton Formation, one of the potential source rocks (Sharma, 1966, p. 347) which lies below the Lockport, is apparently confined to the southern part of the basin where the Lockport dips north. Thus, it appears that the Cain Formation, which overlies the Lockport and Guelph throughout the interior of the basin, remains the more likely source bed for most of the hydrocarbons in the reefs.

The most important consequences of recognizing the role of differential entrapment in northern Michigan are the implications which this theory embodies concerning further exploration in this area. The theory infers that as long as the updip boundary of gas and oil has not been encountered, favorable prospects may exist farther updip. The application of this simple concept to the delineation of additional prospective areas in northern Michigan is discussed next.

Owing to petrographic anisotropies, interception by traps, and other local modifying factors, the fluids migrating within the Lockport Formation did not form a homogeneous front that advanced uniformly everywhere across the belt. Rather, the flow network can be best likened to inverted, updip-converging, dendritic drainage systems. The sinuous irregular boundaries between the gas-, oil-, and water-saturated bands (Fig. 4) attest to this flow pattern. Along most of its length, the pinnacle-reef belt is bounded updip by a water-saturated zone. However, the continuity of this band is interrupted in two areas. These areas, designated by question marks in Figure 4, are zones through which hydrocarbons may have actually migrated into the carbonate-shelf platform. For the area in the northern part of Kalkaska County, this speculation is strengthened by the fact that natural gas, although apparently not in commercial quantities, has been discovered on the shelf-edge carbonate bank in Rapid River Township, Sec. 5, T28N, R7W (Michigan Oil and Gas Assoc. Oil and Gas News, 1976). If gas could have migrated that far up the platform slope, it is reasonable to assume that, given favorable traps, oil could still be found even farther north within the carbonate shelf. Favorable trapping conditions may indeed be present all along the shelf-edge Niagaran barrier-reef complex. This facies belt contains porous carbonate buildups which are sealed by the impervious A-1 carbonate and A-2 anhydrite units, thereby providing many potential stratigraphic traps. These prospects are not limited to the two areas shown in Figure 4. The continuity of the water-saturated band shown on the map might be only apparent, as it is based on interpolations between the few sporadic water-bearing reefs found thus far. Therefore, additional passages through which "tributaries" of migrating hydrocarbons may have entered the platform may be present elsewhere along the trend. It is, in a way, an ironic coincidence that the first discovery in northern Michigan (Onoway oil field, Presque Isle County, Sec. 29, T25N, R2E) should in fact have been made on the carbonate bank rather than in the pinnacle-reef belt. Onoway itself turned out to be of only marginal economic importance but this discovery stimulated the entire exploration program which eventually led to the discovery of Michigan's "north slope" petroleum province.

REFERENCES CITED

Alling, H. L., and L. I. Briggs, 1961, Stratigraphy of Upper Silurian Cayugan evaporites: AAPG Bull., v. 45, p. 515-547.

Al-Muneef, N. S., 1973, Productivity and characteristics of the Niagaran pinnacle reefs in northern Michigan: Master's thesis, Univ. Michigan, 45 p.

Borchert, H., and R. O. Muir, 1964, Salt deposits: London, D. Van Nostrand Co., 338 p.

Briggs, L. I., and D. Z. Briggs, 1974, Niagaran Salina relationships in the Michigan basin, in Silurian reef-evaporite relations: Michigan Basin Geol. Soc. Field Conf., p. 1-23.

———— et al, 1978, Stratigraphic facies of carbonate platform and basinal deposits, late Middle Silurian, Michigan basin, in Guidebook to field excursions, the north-central section: Geol. Soc. America, p. 117-131.

———— et al, 1979, Transition from open marine to evaporite deposition in the Silurian Michigan basin, in A. Nissenbaum, ed., Saline lakes and natural brines: New York, Elsevier Pub. Co., in press.

Brigham, R. J., 1971, Structural geology of southwestern Ontario and southeastern Michigan: Ontario Dept. Mines and Northern Affairs Petroleum Resources Sec., Paper 71-2, 110 p.

Brongersma-Sanders, M., 1971, Origin of major cyclicity of evaporites and bituminous rocks; an actualistic model: Marine Geology, v. 11, p. 123-144.

———— 1972, Hydrological conditions leading to the development of bituminous sediments in the preevaporite phase, in G. Richter-Bernburg, ed., Geology of saline deposits: Paris, UNESCO, p. 19-21.

Budros, R., 1974, The stratigraphy and petrogenesis of the Ruff Formation, Salina Group in southeast Michigan: Master's thesis, Univ. Michigan, 178 p.

———— and L. I. Briggs, 1977, Depositional environment of Ruff Formation (Upper Silurian) in southeastern Michigan, in Reef and evaporites—concepts and depositional models: AAPG Studies in Geology 5, p. 53-71.

Burgess, R. J., and A. L. Benson, 1969, Exploration for Niagaran reefs in northern Michigan: Ontario Petroleum Inst. 8th Ann. Conf. Tech. Paper 1; also 1969-70: Oil and Gas Jour., v. 67, no. 51, p. 80-83; v. 67, no. 52, p. 180-188; v. 68, no. 1, p. 122-127.

Caughlin, W. G., F. J. Lucia, and N. L. McIver, 1976, The detection and development of Silurian reefs in northern Michigan: Geophysics, v. 41, p. 646-658.

Craft, B. C., and M. F. Hawkins, 1959, Applied petroleum reservoir engineering: Englewood Cliffs, N. J., Prentice-Hall, Inc., 437 p.

Cramer, F. H., 1970, Middle Silurian continental movement estimated from phytoplankton-facies transgression: Earth and Planetary Sci. Letters, v. 10, p. 87-93.

Droste, J. B., and R. H. Shaver, 1976, The Limberlost Dolomite of Indiana: a key to the great Silurian facies in the southern Great Lakes area: Indiana Geol. Survey Occasional Paper 15, 21 p.

Ells, G. D., 1967, Michigan's Silurian oil and gas pools: Michigan Geol. Survey Rept. Inv. 2, 49 p.

Evans, C. S., 1950, Underground hunting in the Silurian of southwestern Ontario: Geol. Assoc. Canada Proc., v. 3, p. 55-85.

Gill, D., 1973, Stratigraphy facies, evolution and diagenesis of productive Niagaran Guelph reefs and Cayugan sabkha deposits, the Belle River Mills gas field, Michigan basin: PhD thesis, Univ. Michigan, 276 p.

———— 1976, Source-reservoir-seal relations in productive evaporite-encased reef reservoirs (abs.): AAPG Bull., v. 60, p. 674.

———— 1977a, The Belle River Mills gas field; productive Niagaran reefs encased by sabkha deposits, Michigan basin: Michigan Basin Geol. Soc. Spec. Paper 2, 187 p.

———— 1977b, Salina A-1 sabkha cycles and the Late Silurian paleogeography of the Michigan basin: Jour. Sed. Petrology, v. 47, p. 979-1017.

———— L. I. Briggs, and D. Z. Briggs, 1978, The Cain Formation; a transitional succession from open marine carbonates to evaporites in a deep water basin, Silurian, Michigan basin (abs): 10th IAS Internat. Cong. Sedimentology Abs., v. 1, p. 244-245.

Glasstone, S., 1956, The elements of physical chemistry, 15th printing: Princeton, N.J., D. Van Nostrand Co., 695 p.

Gussow, W. C., 1954, Differential entrapment of oil and gas: a fundamental principle: AAPG Bull., v. 38, p. 816-853.

———— 1968, Migration of reservoir fluids: Jour. Petroleum Technology, v. 20, p. 353-363.

Hubbert, M. K., 1953, Entrapment of petroleum under hydrodynamic conditions: AAPG Bull., v. 37, p. 1954-2026.

Huh, J. M., 1973, Geology and diagenesis of the Salina-Niagaran pinnacle reefs in the northern shelf of the Michigan basin: PhD thesis, Univ. Michigan, 253 p.

——— L. I. Briggs, and D. Gill, 1977, Depositional environments of pinnacle reefs, Niagara and Salina Groups, northern shelf, Michigan basin, in Reefs and evaporites—concepts and depositional models: AAPG Studies in Geology 5, p. 1-21.

Kahle, C. F., 1974, Nature and significance of Silurian rocks at Maumee quarry, Ohio, in Silurian reef-evaporite relationships: Michigan Basin Geol. Soc. Field Conf., p. 31-54.

——— 1978, Patch reef development and effects of repeated subaerial exposure in Silurian shelf carbonates, Maumee, Ohio, in R. V. Kesling, ed., Guidebook to field excursions: the north-central section: Geol. Soc. America, p. 63-115.

Lowenstam, H. A., 1957, Niagaran reefs of the Great Lakes areas: Geol. Soc. America Mem. 67, v. 1, p. 215-248.

Mantek, W., 1973, Niagaran pinnacle reefs in Michigan, in Geology and the environment: Michigan Basin Geol. Soc. Ann. Field Conf. Guidebook, p. 35-46.

——— 1976, Recent exploration activity in Michigan: Ontario Petroleum Inst., 15th Ann. Conf. Proc., 29 p.

Matzkanin, A. D., 1976, Subsurface pressure report on the northern Michigan reef development: Michigan Geol. Survey.

McClintock, P. L., 1977, Seismic data processing techniques in exploration for reefs, northern Michigan, in Reefs and evaporites—concepts and depositional models: AAPG Studies in Geology 5, p. 111-124.

Meloy, D. U., 1974, Depositional history of the Silurian northern carbonate bank of the Michigan basin: Master's thesis, Univ. Michigan, 78 p.

Mesolella, K. J., et al, 1974, Cyclic deposition of Silurian carbonates and evaporites in Michigan basin: AAPG Bull., v. 58, p. 34-62.

Michigan Oil and Gas Association Oil and Gas News, 1976, Oil field map section, year-end summary, 1976: v. 82, no. 52, p. 31.

Orr, W. L., 1974, Changes in sulfur content and isotopic ratios of sulfur during petroleum maturation—study of Big Horn basin Paleozoic oils: AAPG Bull., v. 58, p. 2295-2318.

Pounder, J. A., 1962, Guelph-Lockport Formations of southwestern Ontario: Ontario Petroleum Inst. 1st Ann. Conf. Proc., Sess. 5.

Reef Petroleum Corporation, 1976, North Michigan Niagaran exploration map: Michigan Oil and Gas Assoc. Oil and Gas News, v. 82, no. 52.

Sanford, B. V., 1969, Silurian of southwestern Ontario: Ontario Petroleum Inst. 8th Ann. Conf. Tech. Paper 5, 44 p.

Schmalz, R. G., 1969, Deep-water evaporite deposition, a genetic model: AAPG Bull., v. 53, p. 798-823.

Sharma, G. D., 1966, Geology of Peters reef, St. Clair County, Michigan: AAPG Bull., v. 50, p. 327-350.

Shaver, R. H., 1974, Structural evolution of northern Indiana during Silurian time, in R. V. Kesling, ed., Silurian reef evaporite relationships: Michigan Basin Geol. Soc. Field Conf., p. 55-77.

Sparling, D. R., 1965, Geology of Ottawa County, Ohio: PhD thesis, Ohio State Univ., 265 p.

Tissot, B., et al, 1971, Origin and evolution of hydrocarbons in early Toarcian shales, Paris basin, France: AAPG Bull., v. 55, p. 2177-2193.

Verruijt, A., 1970, Theory of groundwater flow: London, Macmillan and Co. Ltd., 190 p.

Wilson, G. A., 1976, Producing well locations containing hydrogen sulfide in the northern Michigan Niagaran reef fields: Michigan Geol. Survey, unpub. map.

Yelling, W. F., Jr., and M. R. Tek, 1976, Prospects for oil and gas from Silurian-Niagaran trend in Michigan: Univ. Michigan Inst. Sci. and Technology, 35 p.

Ziegler, A. M., et al, 1976, Silurian continental distribution, paleogeography, climatology and biogeography (abs.): 25th Internat. Geol. Cong., Sydney, Australia, Abs., v. 3, p. 729.

American Association of Petroleum Geologists Memoir 14,
Geology of Giant Petroleum Fields, copyright 1970,
pp. 50-90.

Geology of Beaverhill Lake Reefs, Swan Hills Area, Alberta[1]

C. R. HEMPHILL,[2] R. I. SMITH,[2] and F. SZABO[2]

Calgary, Alberta

Abstract The discovery in 1957 of oil in the remote Swan Hills region, 125 mi northwest of Edmonton, began a wave of exploration similar to that following the 1947 Leduc discovery which started the postwar oil boom in Western Canada. By the end of 1967 more than 1,800 wells had been drilled to explore and develop the Swan Hills region. Drilling has established in-place reserves of more than 5.9 billion bbl of oil and 4.5 trillion ft³ of gas.

Devonian sedimentary rocks unconformably overlie an eroded Cambrian section in the southeast part of the Swan Hills region; in the northwest part of the region, Devonian rocks lap onto the Precambrian granite of the Peace River arch.

Three positive features—the Tathlina uplift, Peace River arch, and the Western Alberta ridge—profoundly influenced Middle Devonian Upper Elk Point and Late Devonian Beaverhill Lake sedimentation. An embayment, shielded on the north by the emergent Peace River arch and on the south and west by the nearly emergent Western Alberta ridge, provided an environment conducive to reef development in the central Swan Hills region. Carbonate-bank deposition flanking the Western Alberta ridge in the south and southwestern part of the study area persisted throughout the time of Beaverhill Lake deposition. These beds merge with the overlying Woodbend reef system.

Recent changes proposed in Beaverhill Lake nomenclature include the elevation of the Beaverhill Lake to group status and the Swan Hills Member to formation status. The term Swan Hills Formation, as used herein, refers to the reef and carbonate-bank facies of the Beaverhill Lake Group, whereas the term Waterways Formation is applied to the offreef shale and limestone facies. The Swan Hills Formation is considered to be equivalent in age to the Calmut and younger members of the Waterways Formation.

The Swan Hills Formation is divided into Light Brown and Dark Brown members. Swan Hills reefs attained a thickness greater than 300 ft, whereas the carbonate-bank facies commonly exceeds 400 ft in thickness. Changing sedimentary and environmental conditions produced a complex reef facies; six major stages are postulated in the development of the undolomitized reef from which the Swan Hills field is producing. Stromatoporoids are the dominant reef-building organisms; abundant *Amphipora* characterize the restricted lagoonal facies.

Although the total impact of Swan Hills production on the provincial economy is difficult to determine, the $184 million paid by the industry to acquire Crown lands in the region during the 10-year period after the initial discovery attests to the economic importance of the Swan Hills producing region.

INTRODUCTION

Before 1957, oil production from the Swan Hills Formation of the Beaverhill Lake Group was unknown. Early in that year, the Virginia Hills field was discovered and focused the attention of industry on a remote and heavily timbered area, approximately 125 mi northwest of Edmonton, Alberta (Fig. 1). Abundant rainfall, muskeg, lack of roads, and rugged topography (1,900–4,200 ft above sea level) made exploration difficult and costly.

The exposed stratigraphic sequence consists of the Late Cretaceous Edmonton Formation and early Tertiary Paskopoo Formation, capped with unconsolidated gravel similar to the Cypress Hills Conglomerate of southern Alberta and Saskatchewan (Russel, 1967).

The Swan Hills Beaverhill Lake reef discovery was made on a large farmout block by the Home Union H. B. Virginia Hills 9–20–65–13–W5M well. On January 31, 1957, a drill-stem test of a 30-ft interval in the Upper Devonian Beaverhill Lake section flowed 40° API oil to the surface. Within 30 days another successful Beaverhill Lake well (the Home *et al.* Regent Swan Hills 8–11–68–10–W5M) was drilled 25 mi northeast of the original discovery. This was the first well of the Swan Hills field. Paradoxically, the confirmation well for the Virginia Hills discovery was a dry hole, even though it was less than 1 mi away; and the 8–11–68–10–W5M well at Swan Hills found oil in what since has proved to be one of the poorest producing areas in the entire complex. However, further drilling around Virginia Hills and Swan Hills, and the impressive number of

[1] Read before the 53rd Annual Meeting of the Association, Oklahoma City, Oklahoma, April 24, 1968. Manuscript received, May 7, 1968; accepted, October 14, 1968.

[2] Home Oil Co. Ltd.

We thank our employer, Home Oil Co. Ltd., for permitting the preparation and publication of this paper, and J. L. Carr, chief geologist, for his encouragement and helpful suggestions. We are particularly appreciative of the long hours spent by W. C. Mackenzie and his drafting department in preparing the illustrations, and also thank W. Hriskevich who compiled the necessary statistical data. R. Sears was most helpful with the parts pertaining to reservoir data, as was A. B. Van Tine with the pressure-depth relations. Finally, we are grateful to Home Oil geologists, particularly H. H. Suter, for criticisms and helpful suggestions after reading the manuscript.

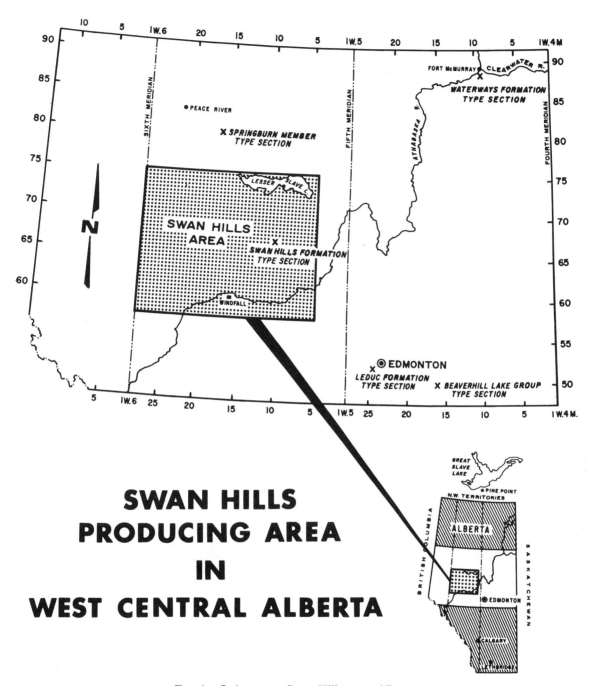

FIG. 1.—Index map, Swan Hills area, Alberta.

new-field discoveries assured continued activity through the region.

Within the map area few more than 12 wells had penetrated the Beaverhill Lake Group before the Home Oil discovery (Fig. 2). By the end of 1967, more than 1,800 wells had been drilled into or through rocks of the Beaverhill Lake Group, and resulted in the discovery of 12 oil fields and two gas fields in the Swan Hills Formation. Total initial in-place hydro-carbons as of January 1, 1968, are estimated to have been in excess of 5.932 billion bbl and 4.512 trillion ft³ of gas (Oil and Gas Conservation Board, 1967).

The Swan Hills reef complex is an excellent example of an undolomitized Devonian carbonate bank-reef development. The availability of numerous cores has facilitated detailed facies studies on several fields. The main purpose of this paper, therefore, is to relate the entire pro-

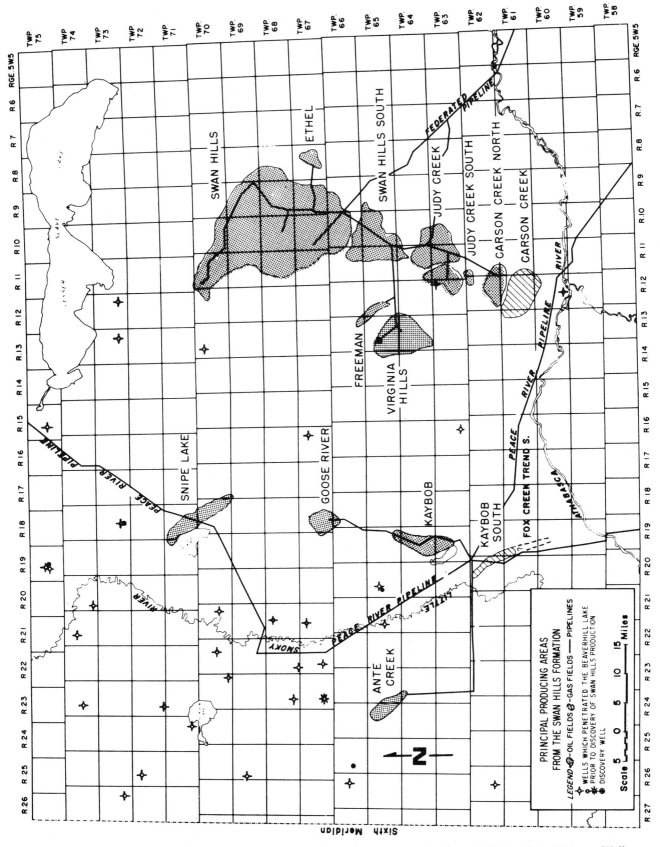

FIG. 2.—Location of Late Devonian Swan Hills Formation fields in Swan Hills region, Alberta. Well symbols are locations where rocks of Beaverhill Lake Group were penetrated before discovery. Gathering system and two major pipeline systems also are shown.

ductive area of the Swan Hills reef complexes to the regional, structural, and stratigraphic setting of the western Canadian sedimentary basin. A secondary purpose is to synthesize the published data on the Swan Hills and related beds and to propose some changes in the nomenclature to reduce confusion.

PREVIOUS WORK

Since Fong's (1959, 1960) original proposal of the type section and discussion of the geology of the Beaverhill Lake Formation, several other excellent studies have been made by Koch (1959), Carozzi (1961), and Thomas and Rhodes (1961). Edie (1961) was the first to make a detailed facies study of the Swan Hills field. Other outstanding studies include those by Fischbuch (1962), Brown (1963), Murray (1964, 1966), Jenik (1965), and Leavitt (1966).

MAP AREA

All present Swan Hills production is from a rectangular area in west-central Alberta, the dimensions of which are approximately 69 by 105 mi, or roughly 7,200 mi². The area is bounded on the east by 115° W long. and on the west by 118° W long. The south boundary is at 54° N lat and the north boundary is at 55° N lat. The area also can be described as lying between T60 and T71 and R8 and R25, W5M.

FIELDS

The Beaverhill Lake oil fields within the map area are, in order of decreasing importance, Swan Hills, Judy Creek, Swan Hills South, Virginia Hills, Kaybob, Carson Creek North, Snipe Lake, Goose River, Freeman, Ante Creek, Judy Creek South, and Ethel (Fig. 2). There are only two gas fields, Carson Creek and Kaybob South. Although most of the present fields are considered to be developed fully, some higher risk, marginal locations could be, and are being, drilled. In the Kaybob South field an important extension toward the southeast recently has been drilled at Fox Creek. Details are not available because of a highly competitive land situation.

DISCOVERY METHODS

Even though well control was extremely sparse for the deeper part of the section, several factors accounted for the gradual increase of exploration.

1. Decline in new discoveries of Leduc reefs (equivalent in age to the Woodbend) in the more accessible country, and the possibility of the presence in this area of other Leduc reef chains with large reserves.

2. Successful exploration along the Mississippian subcrop edges farther south and the possibility of similar conditions in this northern district.

3. An indicated thinning of the underlying Elk Point interval, possibly caused by the presence of a basement high which might have been the locus for reef development.

4. The numerous possibilities for the presence of Mesozoic sandstone bodies which were known to be productive in other parts of the province.

5. The presence of oölites, stromatoporoids, and *Amphipora* in the lower part of the Beaverhill Lake section in some of the older wells. (This indicated shallow-water shoaling and the potential for reef buildup.)

6. The large acreage blocks which could be assembled and the availability of additional offsetting Crown acreage through competitive bidding.

7. The discovery of oil in the "granite wash" just north of the Snipe Lake field.

Seismic studies have been of limited value in exploration for Beaverhill Lake reefs. A geophysical program was conducted on the acreage before the selection of the wellsite and, although the choice of drilling location was influenced by seismic information, the site selection in Virginia Hills was based on a somewhat nebulous feature. Similar seismic data were used to locate the first well in the Swan Hills field.

The value of seismic work in defining areas of Beaverhill Lake reefs has been argued strongly ever since. The arguments may be attributed to uncertainty in identification of the reflecting horizon, lack of velocity contrast between the reef and the enclosing rocks, and the absence of differential compaction or draping over the biohermal buildups. Both seismic structure and isochron maps have been used to select drillable locations, but generally the results have not permitted a good correlation of the well information with the seismic data.

GEOLOGIC HISTORY[3]

The western Canadian sedimentary basin is underlain by the westward continuation of Precambrian rocks of the Canadian shield. The oldest Cambrian basin on the shield rocks was restricted to the Cordilleran trough. It was a long, narrow marine trough in what is now northern British Columbia. The seaway straddled the British Columbia–Alberta boundary in the vicinity of the present Rocky Mountains.

[3] This discussion is mostly a synthesis of work by Van Hees and North (1964), Porter and Fuller (1964), Grayston *et al.* (1964), and Moyer *et al.* (1964).

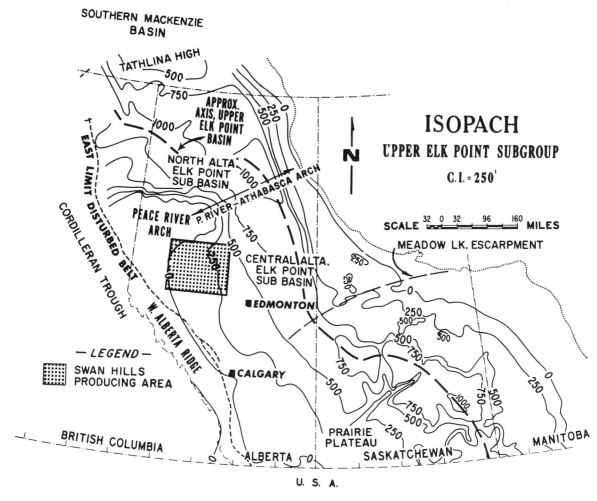

FIG. 3.—Isopach map, Upper Elk Point subgroup. Maximum thickness of Upper Elk Point strata is in northern Alberta and in Williston basin, Saskatchewan. Position of basinal axis is similar to position of Beaverhill Lake basinal axis shown in Fig. 4. Note emergent Western Alberta ridge and Peace River arch. CI = 250 ft. Redrawn from Grayston et al. (1964).

From this trough, Middle and Late Cambrian seas transgressed eastward over the cratonic shelf.

Sub-Devonian erosion removed Upper Cambrian rocks from all but the southeast part of the Swan Hills producing area. No Ordovician or Silurian rocks are present, presumably because of removal by pre-Devonian erosion. Basal Devonian strata unconformably overlie the eroded Cambrian section and in the northwest, lap onto Precambrian granite of the Peace River arch.

Thus, before the Devonian, Caledonian tectonism produced uplift and erosion which led to the widespread destruction of sedimentary beds. Three important positive areas—the Peace River–Athabasca arch, the Tathlina uplift, and the Western Alberta ridge—were present (Fig. 3), and were to exert an impor-

tant influence on sedimentation during Early, Middle, and part of Late Devonian times.

At the beginning of the Early Devonian, the Tathlina uplift and the Western Alberta ridge separated the moderately subsiding basin in Alberta from the MacKenzie basin on the north and the Cordilleran basin on the west. This interior basin, subdivided by the Peace River–Athabasca arch into northern and central Alberta Elk Point subbasins, received 600–1,300 ft of Lower Elk Point clastic and evaporite sediments. Topography controlled the lateral extent of the shallow Lower Elk Point sea.

Collapse of the eastern part of the Peace River–Athabasca arch near the end of Lower Elk Point deposition caused the northern and central Alberta subbasins to merge. The resulting basinal configuration remained essentially unaltered through the rest of the Elk Point de-

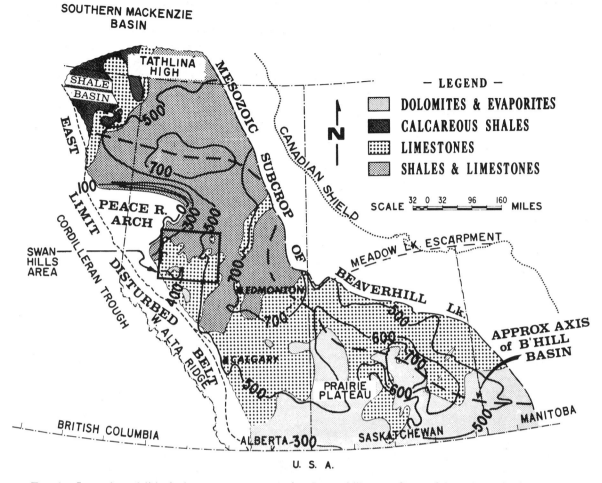

Fig. 4.—Isopach and lithofacies map, Late Devonian Beaverhill Lake Group. More than 700 ft of Beaverhill Lake present in northern and central Alberta and central Saskatchewan. Note progression of facies from outer marine shale in northeastern British Columbia to dolomite and evaporite in Southern Alberta, Saskatchewan, and Manitoba. Swan Hills producing area is southeast flank of Peace River arch. CI = 200 ft. Redrawn from Moyer (1964).

position, and for all of Beaverhill Lake deposition (Figs. 3, 4).

As subsidence and widespread incursion of more normal seawater continued, carbonate material was deposited early in Elk Point deposition. Keg River carbonate banks fringed the basin and patch reefs developed within the basin. For the first time during the Devonian the sea crossed the Meadow Lake escarpment, and sediments were deposited in southern Alberta, southern Saskatchewan, and southwest Manitoba.

The most rapid reef growth was in northeastern British Columbia; the resulting reef complex restricted the circulation of seawater into the Upper Elk Point basin. Reef growth terminated and evaporite accumulation predominated until near the end of Elk Point deposition in central Alberta. In northwestern Al-

berta fluctuating marine conditions prevailed generally near the end of Elk Point deposition. Crickmay (1957), Campbell (1950), and Law (1955) reported a hiatus between the Muskeg and Watt Mountain or Amco Formations in the Fort McMurray, Pine Point, and Steen River areas, respectively. Subaerial erosion and penecontemporaneous shallow-water deposition probably characterized the start of Watt Mountain deposition throughout the Upper Elk Point basin.

The sands of the Gilwood were derived from the Peace River arch and, according to Kramers and Lerbekmo (1967), represent a regressive-deltaic environment during Watt Mountain sedimentation. Other writers, including Guthrie (1956) and Fong (1960), believe the Gilwood Sandstone was deposited during shallow marine transgression. In the writers'

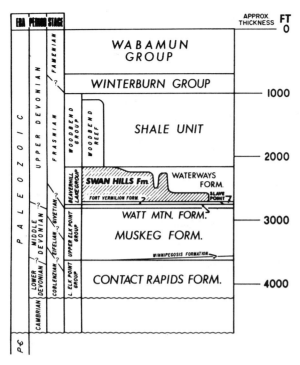

FIG. 5.—Devonian stratigraphic nomenclature, west-central Alberta. Chart shows position of Swan Hills Formation in geologic column. Relation of Swan Hills reef complex to enclosing Waterways Formation and younger Woodbend reefs is shown diagrammatically.

ment and subsequent reef growth. Emergence followed by deepening of water during late Beaverhill Lake deposition terminated Swan Hills reef growth. However, in places, notably the Windfall area, deposition of bank carbonate continued throughout Beaverhill Lake deposition. These carbonate beds grade into the overlying Woodbend reef. During the final deposition of the Beaverhill Lake, a fringing reef developed in the Springburn area on the south flank of the Peace River arch.

The overlying Woodbend consists of carbonate reefs and, in the offreef areas, green shale. The Winterburn consists of silty carbonate, red and green shale, and some anhydrite. The youngest Devonian unit, the Wabamun, is entirely carbonate.

The Devonian rocks within the study area and adjacent regions are subdivided in order of ascending age into Elk Point, Beaverhill Lake, Woodbend, Winterburn, and Wabamun Groups (Fig. 5). The total Devonian section along the eastern margin of the Swan Hills region is approximately 3,600 ft thick and thins markedly northwestward to 300 ft over the Peace River arch (Fig. 6). Within the east-central part of Alberta (i.e., over the central part of the Elk Point and Beaverhill Lake basins) the Devonian is thicker than 4,500 ft.

ELK POINT GROUP

McGehee (1949) first defined the Elk Point Formation. Belyea (1952) elevated it to group rank and Crickmay (1954) described the type section. The writers, following Grayston et al. (1964), subdivide the group into the Lower and Upper Elk Point subgroups, the exact ages of which are not settled. However, we tentatively follow Grayston et al. (1964), Basset (1961), and Hriskevich (1966), and place the Middle-Lower Devonian boundary at the Upper-Lower Elk Point subgroup division.

Opinion is divided concerning the boundary between the Elk Point and overlying Beaverhill Lake Group. The writers prefer to include the Watt Mountain Formation in the Beaverhill Lake Group. However, we recognize the fact that the controversy is far from resolved, hence do not assign the Watt Mountain Formation to either the Elk Point Group or Beaverhill Lake Group. Murray (1964) and Thomas and Rhodes (1961) placed the boundary at the top of the Fort Vermilion anhydrite. Fong (1960), Edie (1961), and Jenik (1965) considered the boundary to be at the top of the Watt Moun-

opinion, an oscillatory environment probably characterized Watt Mountain deposition and we agree with Griffin (1965), who wrote that the Fort Vermilion was deposited during final oscillatory conditions before the main transgressive phase which began during the time of Slave Point deposition.

The Late Devonian Beaverhill Lake transgression began with the deposition of Slave Point carbonate on a broad shelf in northeastern British Columbia, northern Alberta, and the adjacent part of the Northwest Territories. A carbonate reefoid-front facies, similar to the underlying Elk Point reefoid carbonate bank, developed in northeastern British Columbia.

Climatic changes and regional tectonism, accompanied by an influx of argillaceous material, destroyed the Slave Point biota (Griffin, 1965). Alternating limestone and shale characterize the rest of the Beaverhill Lake in northern Alberta and in the Edmonton area. However, in the Swan Hills region a shallow-water embayment, protected on the north by the emergent Peace River arch and flanked on the southwest by the Western Alberta ridge, provided a setting conducive to bank develop-

tain Gilwood Sandstone Member. Grayston *et al.* (1964) believed that the Watt Mountain Formation belongs with the Beaverhill Lake depositional cycle and thus favored its inclusion with the Beaverhill Lake Group.

The Lower Elk Point subgroup consists mainly of terrigenous clastic and evaporite beds and ranges in thickness from 200 to 1,300 ft.

The Upper Elk Point subgroup is subdivided into the Winnipegosis Formation below and the Muskeg Formation above. The Winnipegosis is mainly carbonate, in contrast to the salt and anhydrite which predominate in the Muskeg Formation. Total thickness of this subgroup ranges from 400 ft in the eastern part of the Swan Hills region to more than 1,000 ft in the central part of the basin.

BEAVERHILL LAKE GROUP

The Beaverhill Lake section originally was defined by the Imperial Oil Geological Staff (1950). Common industry practice in the Swan Hills area has been to include basal carbonate beds of the Cooking Lake Formation with the Beaverhill Lake Group. The proper correlation is shown in Figure 7. The writers follow Leavitt (1966) and raise the Beaverhill Lake section to group rank. As used in this paper, the Beaverhill Lake Group includes, from base to top, the Fort Vermilion, Slave Point, Swan Hills, and Waterways Formations (Fig. 5).

The Beaverhill Lake type section in the Edmonton area was defined by Imperial Oil (1950) to include the interval from 4,325 to 5,047 ft (722 ft) in Anglo Canadian Beaverhill Lake (Lsd. 11, Sec. 11, T50, R17, W4M). Fong (1960) defined the Beaverhill Lake section in the Swan Hills producing area as the interval 8,020–8,543 ft (523 ft) in Home Regent "A" Swan Hills (Lsd. 10, Sec. 10, T67, R10, W5M). The top of the interval coincides with the highest beds of dark-brown limestone below bituminous shale. Subsequent drilling between the Swan Hills region and the Beaverhill Lake type section well has indicated that the upper boundary of the Beaverhill Lake as defined in the type section has been miscorrelated with the upper boundary established by usage in the Swan Hills region. This discrepancy is shown in Figure 7.

Beaverhill Lake strata are present through much of the western Canadian sedimentary basin, but are of maximum thickness in central Alberta, where drilling has established the presence of up to 750 ft of section (Fig. 4). Fig-

ure 4 shows that the axis of the Beaverhill Lake basin closely parallels that of the Upper Elk Point basin (Fig. 3). A marked thickness reduction of the Beaverhill Lake in northeastern British Columbia, though due partly to depositional thinning, has been interpreted by Griffin (1965) as primarily the result of post-Beaverhill Lake erosion.

Beaverhill Lake beds may be divided into five main facies: the outer shale facies, carbonate-front facies, inner alternating limestone and shale facies, inner reef facies, and shelf-margin carbonate-evaporate facies.

The carbonate-front facies, extending northeastward through northern British Columbia into the Northwest Territories, formed an effective barrier to the Beaverhill Lake shelf basin, profoundly influencing sedimentation there and on the shelf margin. Seaward, the carbonate-front facies is in abrupt contact with dark, deep-water marine shale, and the central part of the Beaverhill Lake basin contains the alternating limestone and shale facies of the Waterways Formation. Southward, the limestone and shale facies grades into shelf-margin carbonate and evaporite, limestone, and primary dolomite. In southern Alberta, Saskatchewan, and Manitoba, anhydrite is a major constituent of the carbonate-evaporite facies (Fig. 4).

The inner reef facies, designated the Swan Hills Formation, is restricted to an embayment south of the Peace River arch.

McLaren and Mountjoy (1962), on faunal evidence, correlate the Moberly and Mildred —upper members of the Waterways Formation —with the Flume Formation exposed in the front ranges of the Rocky Mountains south of the Athabasca River. Lower beds are correlative with the Flume Formation north of the Athabasca River.

Agreement is lacking on the age of the Beaverhill Lake Group. Mound (1966) established a Late Devonian age from conodont studies. Clark and Ethington (1965), on the basis of conodont data, placed all but the top few feet of the Flume Formation in the Middle Devonian. Norris (1963) assigned the Waterways Formation to the early Late Devonian and the Slave Point Formation in northeastern British Columbia to the Middle Devonian. McGill (1966), on the basis of ostracods, placed the Givetian-Frasnian boundary at the base of the Waterways Formation in the Lesser Slave Lake area in the northeastern corner of the Swan Hills region. Loranger (1965) placed the entire Beaverhill Lake and the lower part

C. R. Hemphill, R. I. Smith, and F. Szabo

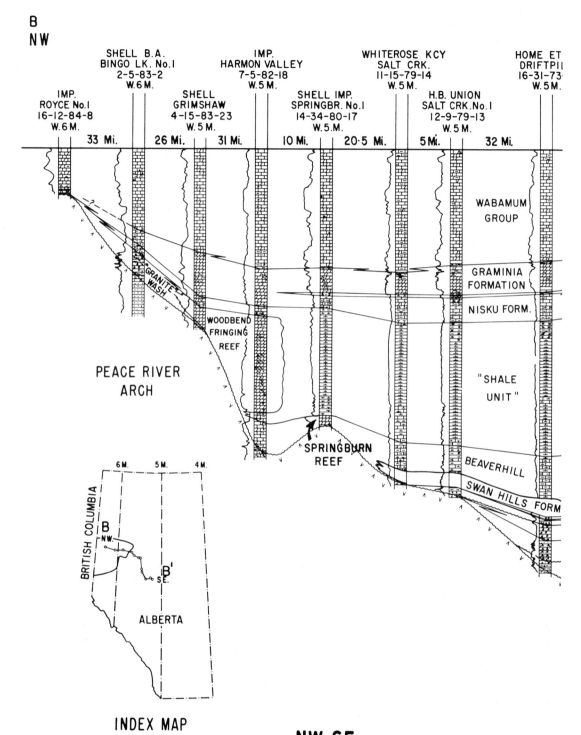

INDEX MAP

**NW-SE
GENERALIZED DEVONIAN-CAMBRIAN
CROSS SECTION B-B'
OFF THE PEACE RIVER ARCH**

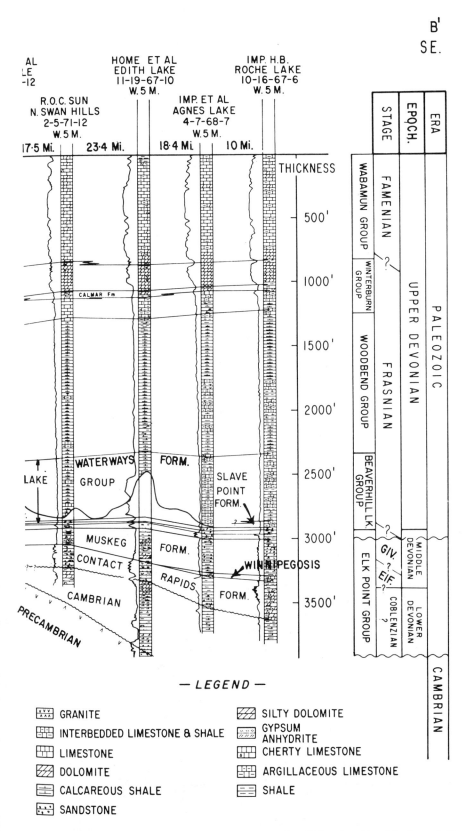

FIG. 6.—Northwest-southeast cross section **B-B'** showing thinning of Devonian over Peace River arch. Crest of arch was emergent until time of Wabamun deposition. Swan Hills Formation, deposited during early Late Devonian transgression, onlaps southeastern flank of Peace River arch. Vertical scale in feet.

C. R. Hemphill, R. I. Smith, and F. Szabo

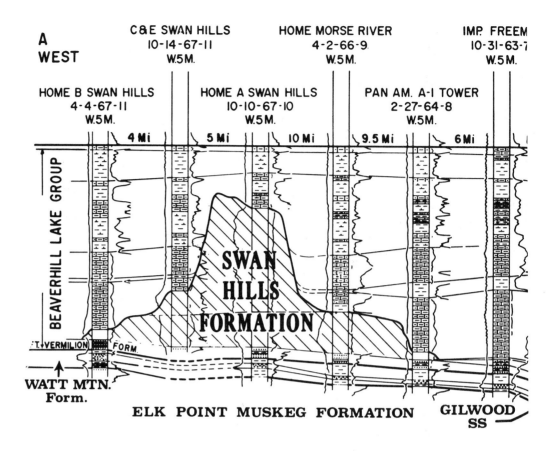

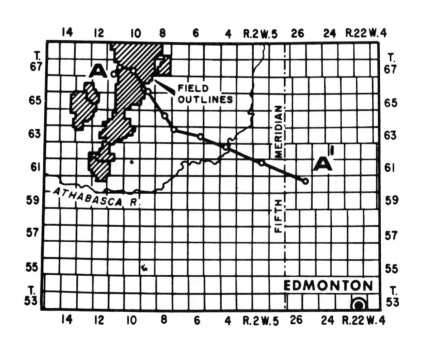

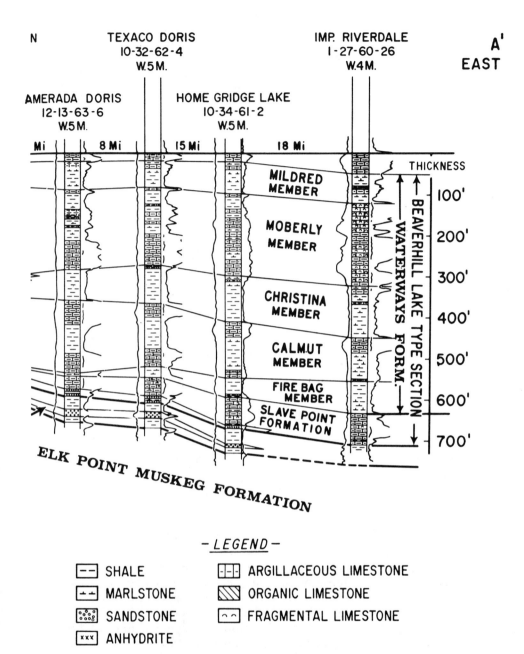

- LEGEND -

`--` SHALE		`---` ARGILLACEOUS LIMESTONE	
`--` MARLSTONE		`\\` ORGANIC LIMESTONE	
`°°` SANDSTONE		`⌒⌒` FRAGMENTAL LIMESTONE	
`xxx` ANHYDRITE			

FIG. 7.—West to east cross section **A-A'** of Beaverhill Lake Group, through Swan Hills field, Alberta, into offreef basin area. Section demonstrates that Swan Hills Formation is time equivalent of Calmut and younger members of Waterways Formation, and not of Slave Point Formation as believed by some authors. Relation of upper Beaverhill Lake limestone in Swan Hills area to Beaverhill Lake type section should be noted. Thicknesses in feet.

of the overlying Woodbend Group in northeastern Alberta in the Middle Devonian.

The writers assign a Late Devonian age to the Beaverhill Lake Group in the Swan Hills region. This does not conflict with the findings of Norris, because the Slave Point Formation is the product of initial deposition of an advancing sea from the north, and thus as a transgressive deposit should become progressively younger southward.

Fort Vermilion Formation.—The name Fort Vermilion Member was introduced by Law (1955) for an evaporite unit at the base of the Slave Point Formation, which overlies the Watt Mountain Formation in the subsurface of northwestern Alberta. This anhydrite unit is only 23 ft thick in the California Standard Steen River type well (Lsd. 2, Sec. 22, T117, R5, W6M) but thickens eastward to 120 ft at the Hudson's Bay No. 1 Fort Vermilion (Sec. 32, T104, R8, W5M). Fort Vermilion anhydrite is widespread; subsequent drilling has established Fong's (1960) "Basal Beaverhill Lake" anhydrite unit as correlative with the Fort Vermilion.

Norris (1963) raised the Fort Vermilion Member to formation status. The writers concur with Norris, because the Fort Vermilion beds are widespread and because we agree with Griffin (1965), that the Fort Vermilion was deposited during final oscillatory conditions before the main Slave Point transgressive phase began.

Slave Point Formation.—Cameron (1918) assigned the name Slave Point Formation to scattered exposures of thin-bedded, medium-grained, dark-gray bituminous limestone, which he concluded were of Middle Devonian age. The limestone crops out on the north and south shores of Great Slave Lake. Cameron originally estimated the total section to be 160 ft thick, but in 1922 revised this estimate to 200 ft.

Campbell (1950) correlated a 310-ft section of limestone, shale, and dolomite penetrated in bore holes at Pine Point on the south shore of Great Slave Lake with Cameron's Slave Point section. Campbell's subsurface section includes 170 ft of fine-grained, stromatoporoidal limestone overlying 11 ft of dark-greenish-gray shale, termed the Amco Formation, which unconformably overlies 129 ft of fossiliferous limestone and dolomitic limestone. Law (1955) noted a hiatus between the Watt Mountain and Muskeg Formations penetrated in the California Standard Steen River (2–22–117–5–W6M),

in northwest Alberta, and he correlated the Watt Mountain with the Amco Formation. Law concluded that only the upper 170 ft of the carbonate rocks in the Pine Point area subsurface correlate with the Slave Point Formation of the plains (informal usage).

The Slave Point Formation of the plains attains a maximum thickness of 500 ft, in northwestern Alberta, north of the Peace River arch, northeastern British Columbia, and the District of MacKenzie. In northeastern British Columbia and the District of MacKenzie, there is an abrupt change in facies between the Slave Point carbonate facies front and the basinal Otter Park Shale. Gray and Kassube (1963) describe the facies front as being rich in stromatoporoids and *Amphipora*. Griffin (1965) defined five rock types in the vicinity of the facies front: dark stromatoporoid-bearing calcilutite, stromatoporoid biosparite, stromatoporoid biomicrite, light-colored micrite (and fossil micrite), and white and gray dolomite. Dolomitization has destroyed many of the original lithic characteristics, but the abundance of stromatoporoids suggests that the front facies is in fact a reef chain.

The Slave Point limestone thins eastward, and also south of the Peace River arch. Crickmay (1957) noted 5.5 ft of magnesium limestone between the Waterways and Muskeg Formations at the Bear Biltmore 7–11–87–17–W4M well near Fort McMurray. On the basis of fossil content, Crickmay considered this unit to be correlative with the Slave Point Formation of the outcrop area. Subsequently, Belyea (1952) correlated the unit with the basal limestone member of the Beaverhill Lake Formation type section (*i.e.,* she considered the Beaverhill Lake section in the Edmonton area to be equivalent to the Waterways Formation plus the 5.5-ft magnesium limestone section at the Waterways type locality). Norris (1963) disputed Crickmay's correlation of these beds with the Slave Point Formation and proposed the term "Livock River Formation" to include this section. Griffin (1965) confirmed Crickmay's interpretation, however, by establishing, on the basis of mechanical-log and lithologic analysis, a convincing correlation between Norris's Livock River Formation and the Slave Point Formation in northeastern British Columbia. Thus, the basal 35-ft limestone unit in the type Beaverhill Lake Formation correlates with the Slave Point Formation.

Swan Hills Formation.—Fong (1960) applied the name Swan Hills Member to the productive unit of the Beaverhill Lake in the Swan Hills area. The type section is the interval at 8,167–8,500 ft (333 ft) in the Home Regent "A" Swan Hills 10–10–67–10–W5M well. Murray (1966) proposed that the Swan Hills Member be raised to formation rank, a view also held by Leavitt (1966). The writers concur. Drilling since 1960 has established the presence of Swan Hills rocks across an area of approximately 7,000 mi². Its widespread occurrence and the economic importance support elevation to formation rank.

Fong defined the Swan Hills as an organic bioclastic limestone unit overlying and gradational with anhydrite and shale now known to be correlative with the Fort Vermilion Formation. The Swan Hills Formation is divisible into a lower Dark Brown member and an upper Light Brown member. In a typical section (Fig. 8) the Dark Brown member is 110 ft thick but may range from 80 to 160 ft, and consists of dark-brown calcarenite and calcilutite with some reef-building organisms. The Dark Brown member grades upward into the Light Brown member, which has a maximum thickness of 320 ft, and consists of calcarenite, calcilutite, and biogenic carbonate.

The Swan Hills Formation consists of organic carbonate deposits. The central part of the Swan Hills region contains organic reefs as thick as, or slightly thicker than, 300 ft. The carbonate-bank deposits which characterize the southwestern part of the region are more than 400 ft thick and grade upward into the overlying Woodbend reef deposits.

The Swan Hills is transgressive, becoming younger northwestward in the direction of the Peace River arch (Fig. 6). The unit wedges out against the lower flank of the Peace River arch. An upper Beaverhill Lake reef termed the Springburn Member is interpreted to be a local fringe reef. The Springburn is a forerunner to the extensive Woodbend fringe reef and is considered by the writers to be separate from the Swan Hills Formation.

Murray (1966) observed a very sharp contact between the reef and offreef facies in the Judy Creek area and suggested that most of the basin facies is younger than the Swan Hills Formation. Leavitt (1966) found in the Carson Creek area, in addition to a sharp reef-offreef boundary, a deeper water fauna in the adjacent offreef facies and concluded that there was considerable relief between the reef complex and the offreef basin. Apparently, this relation is not characteristic of the entire Swan Hills re-

gion, for Fischbuch (1962) reported an interfingering of Swan Hills carbonate rocks with offreef strata near the south end of the Kaybob field.

Waterways Formation.—The name Waterways originally was suggested by Warren (1933) for the limestone and shale which crop out at the confluence of the Clearwater and Athabasca Rivers in northeastern Alberta. Warren's Waterways section, as determined from outcrop and nearby salt wells, totals 405 ft. On the basis of the section penetrated in the Bear Biltmore 7–11 well, Crickmay (1957) also included beds removed by erosion in the Fort McMurray area. The type subsurface section penetrated at Bear Biltmore (Lsd. 7, Sec. 11, T87, R17, W4M), is 740 ft and was subdivided by Crickmay (1957) into five units designated in ascending order the Firebag, Calmut, Christina, Moberly, and Mildred Lake Members.

The Waterways Formation, consisting of alternating brown fragmental limestone and greenish-gray shale, is separated from the underlying Elk Point Group by the 5.5-ft Slave Point magnesian limestone bed. Griffin (1965) traced Crickmay's (1957) five-member subdivision of the Waterways Formation as far north as California Standard Mikkwa (Lsd. 12, Sec. 23, T98, R21, W4M). In addition, Griffin's correlation of the top of the Calmut Member in northeastern British Columbia provided convincing support for his suggestion that thinning of the shale-limestone sequence in northwesternmost Alberta and northeastern British Columbia is due to (1) facies equivalence of the Slave Point Formation (carbonate) with the Waterways Formation (limestone and shale) and (2) progressive northwestward erosional truncation of the Beaverhill Lake Group. These relations are illustrated in Figure 9.

AGE RELATIONS OF SWAN HILLS, SLAVE POINT, AND WATERWAYS FORMATIONS

Warren (1957) observed that ". . . the oil-bearing horizon within the Beaverhill Lake Formation [*i.e.,* the Swan Hills Formation] south of Lesser Slave Lake carries a fauna younger than that of the Firebag Member of the Waterways Formation [Fig. 7] and thus is younger than the Slave Point Formation."

According to Crickmay (1957), *Atrypa* aff. *A. independensis* is a guide fossil everywhere for the Slave Point. Koch (1959) noted the abundance of this species within the basal part of the Swan Hills Formation. Its presence sug-

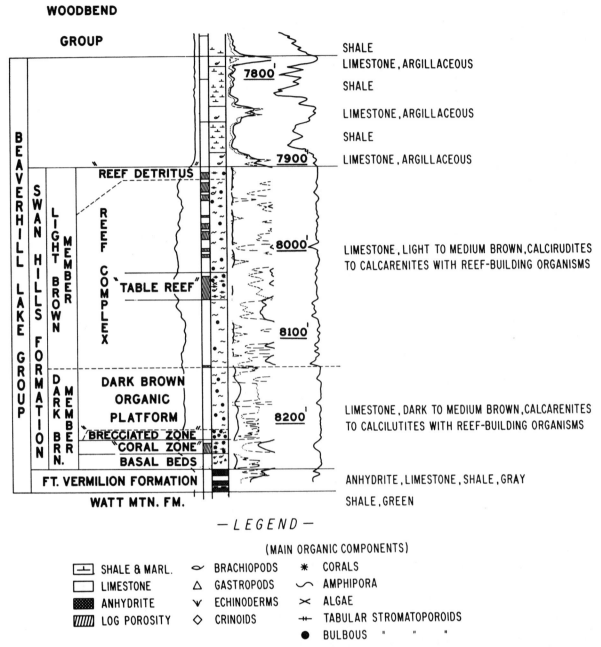

FIG. 8.—Typical Beaverhill Lake section in Swan Hills field, Alberta. Well is Home Regent Swan Hills 4-28-67-10-W5M. KB elev. = 3,322.1 ft. Well was cored almost completely through Swan Hills Formation, permitting excellent control for facies divisions shown. Depths in feet below KB.

gests that the lower beds of this formation correlate with the Slave Point Formation. Koch considered, however, the tendency for *Atrypa* of the *A. independensis* type to migrate with the nearshore carbonate facies (*i.e.*, it is a facies-controlled organism), and accordingly he preferred to correlate the basal Swan Hills with the Firebag Member.

The writers concur with Warren and believe that the basal part of the Swan Hills Formation in the Swan Hills field correlates with the Calmut Member of the Waterways Formation. Moreover, the Slave Point Formation is known to wedge out east of the Swan Hills fields and thus was not deposited within the designated Swan Hills region. These relations are illustrated in Figure 7.

Woodbend Group.—The contact between the Beaverhill Lake and Woodbend Groups is gradational. Bioclastic limestone and shale of

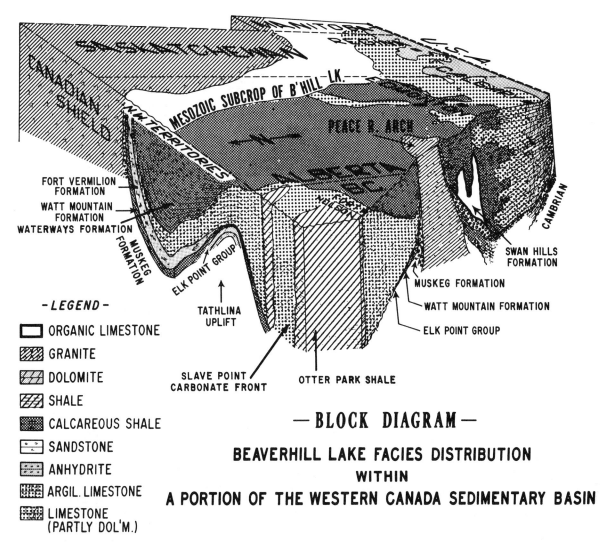

- LEGEND -

☐ ORGANIC LIMESTONE

▨ GRANITE

▨ DOLOMITE

▨ SHALE

▦ CALCAREOUS SHALE

▱ SANDSTONE

▦ ANHYDRITE

▦ ARGIL. LIMESTONE

▦ LIMESTONE
    (PARTLY DOL'M.)

— BLOCK DIAGRAM —

BEAVERHILL LAKE FACIES DISTRIBUTION
WITHIN
A PORTION OF THE WESTERN CANADA SEDIMENTARY BASIN

FIG. 9.—Block diagram showing Beaverhill Lake facies distribution, and position of Swan Hills Formation with respect to Peace River arch and to various Beaverhill Lake facies. View is toward southeast. Basinal Otter Park shale facies is in abrupt contact with Slave Point carbonate-front facies. Calcareous shale and argillaceous limestone of Waterways Formation occupy much of Alberta and grade southeastward into primary dolomite, anhydrite, and limestone facies.

the Cooking Lake Formation compose the basal part of the Woodbend Group. The Cooking Lake Formation formed a platform for subsequent Leduc reef development in central Alberta (Fig. 10) where, in the offreef areas, it is overlain by the Duvernay bituminous shale and interbeds of dark-brown limestone. The Duvernay Formation is overlain by the Ireton Formation, a green shale section flanking and overlying the Leduc reefs.

In the Swan Hills region, the Duvernay-Cooking Lake section is represented principally by greenish-gray calcareous shale; accordingly, the Ireton-Duvernay-Cooking Lake section is referred to herein as the "shale unit." Unlike Fong's (1960) "shale unit," it does not include the Nisku Formation of the Winterburn Group. Thus, the section in this paper is identical with the "shale unit" of Murray (1966).

Younger Devonian.—The Woodbend is overlain by the Winterburn and Wabamun Groups. The Woodbend includes shale, siltstone, anhydrite, dolomite, and silty dolomite. The Wabamun consists mostly of dolomitic limestone and some anhydrite.

STRATIGRAPHY OF SWAN HILLS FORMATION

The Swan Hills Formation is divided into two members on the basis of color and morphologic characteristics—a lower or Dark

185

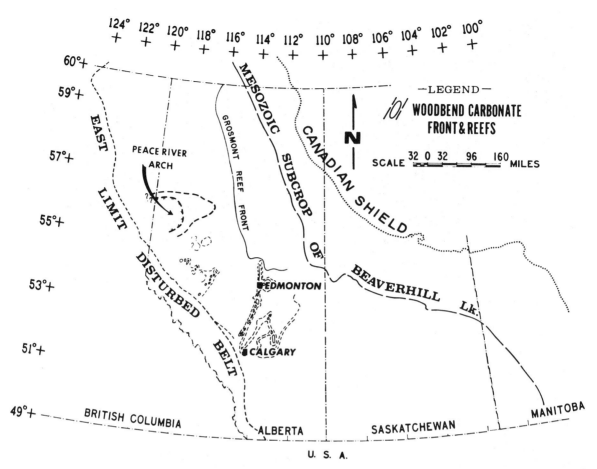

Fig. 10.—Distribution of Late Devonian Woodbend reef systems. Devonian carbonate front and reefing in Alberta reached maximum development during Late Devonian (Woodbend) time. Grosmont reef complex underlies northeastern Alberta and is followed in clockwise direction by Leduc-Meadowbrook reef chain in central Alberta. Simonette-Windfall reef system and Sturgeon Lake atoll cover large part of west-central Alberta including western and southern part of Swan Hills region. Peace River arch is flanked by broad, continuous, fringing reef system. Redrawn from Belyea (1964).

Brown member and an upper or Light Brown member (Fig. 8).

The Dark Brown member is a widespread organic platform or carbonate bank, covering an extensive area of west-central Alberta. It is fringed on the northeast, north, and northwest by stromatoporoidal limestone, which is interpreted as an organic reef (Fig. 11). This member represents the first transgressive phase of the Late Devonian sea.

Generally there is an abrupt facies change toward the east and north between the Dark Brown member and the calcareous shale and argillaceous limestone of the Waterways Formation. On the west the entire Swan Hills Formation is terminated by onlap on the Peace River arch.

The thickness of the Dark Brown member ranges from 80 ft in the Deer Mountain–House Mountain area to approximately 160 ft in the vicinity of Carson Creek (Fig. 12). Thin-section studies show that the dark color is caused mainly by the presence of pyrobitumen and residual oil in the fine matrix rather than by a high argillaceous content.

Frame-building organisms became established at numerous localities on the top of the Dark Brown member and formed a bioherm or "reef complex," as used in this paper. This complex probably grew on topographic highs or positive areas of the underlying carbonate bank, hence its lateral extent is more restricted than that of the platform on which it rests.

The Swan Hills, Judy Creek, Carson Creek North, Kaybob, and Goose River fields are separate reeflike buildups on the platform, separated from each other by surge channels and the associated offreef shale and limestone beds

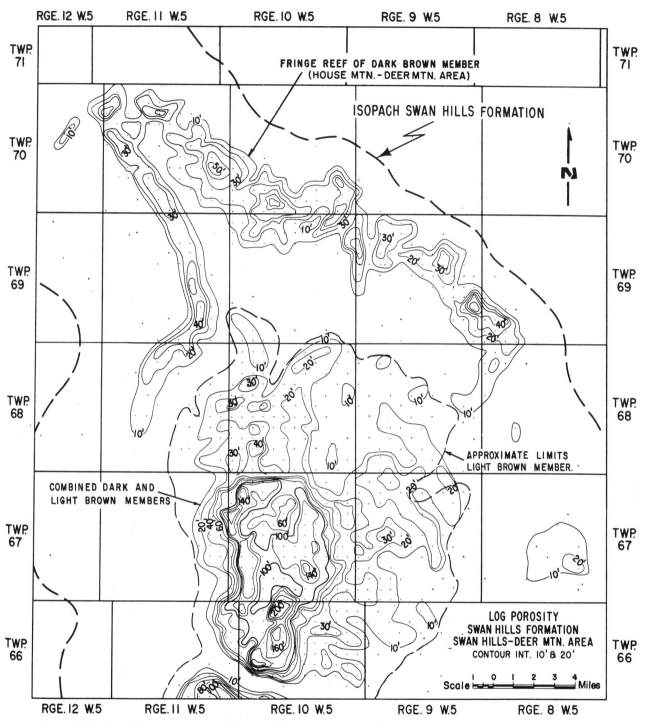

FIG. 11.—Isoporosity map of Swan Hills Formation, based on logs. Map emphasizes productive reef-run and main buildup area. Patch reefs and reef outwash are loci of porosity in intervening area. Porous trend along northern edge originally was called House Mountain–Deer Mountain area but now is included in area designated as Swan Hills field. Note line showing limits of Light Brown member. CI = 10 and 20 ft.

of the Waterways Formation (Figs. 13–15).

South and southwest of the Swan Hills region, the Light Brown member formed a massive carbonate bank with reef-building organisms around the rim of the widespread buildup. The Carson Creek, Virginia Hills, and Ante Creek fields are associated with this reef-rimmed bank (Fig. 14). Farther south and

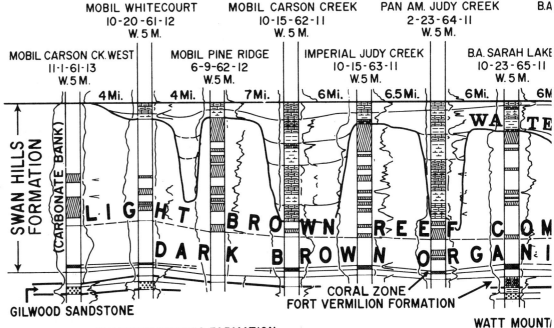

Fig. 12.—South-north cross section **C-C'** of Beaverhill Lake Group, north-central Alberta. Section passes through fields of eastern reef chain. Shale and argillaceous limestone of Waterways Formation separate reef buildups. Gradual southward thickening of Dark Brown and Light Brown members, combined with thinning of gross Beaverhill Lake, results in entire Beaverhill Lake interval consisting of Swan Hills carbonates. Deer Mountain field now is northern part of Swan Hill field. Area through which section passes is shown on Figures 2 and 14; exact location shown on Figure 21.

southwest the bank formed the foundation for the Late Devonian Woodbend reefs, following deposition of the Beaverhill Lake.

Although there are facies variations from field to field in the area, several facies types predominate in all fields (Fig. 8).

Dark Brown Member

"Basal beds."—The lowermost bed of the "Basal beds" overlies the Fort Vermilion anhydrite and shale, and consists of an argillaceous, cryptograined, dark-brown to black limestone

ranging in thickness from 3 to 14 ft. Anhydrite and pyrite are common in the lower part. The "Basal beds" are sparsely fossiliferous, and contain the following fauna in decreasing order of predominance: brachiopods (*Atrypa*), ostracods, echinoderms, and crinoids.

The uppermost units of the "Basal beds" are transitional with the overlying "Coral zone," and have similar organic content. These beds were deposited in aerated seawater of normal salinity. The presence of reef-building organisms in the upper part, the higher percentage

188

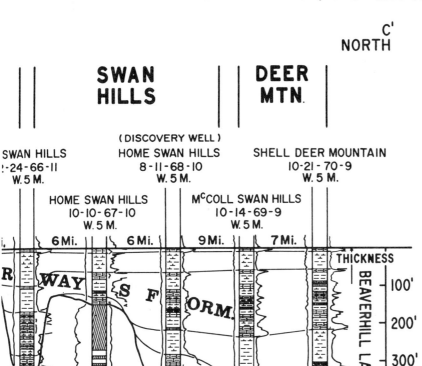

of skeletal material, and the local presence of sparry calcite cement suggest deposition in shallower water than the adjacent offreef basal beds of the Waterways Formation on the east (Fig. 7).

"Coral zone."—Overlying the "Basal beds" is light-brown, fossiliferous limestone of great lateral extent, called the "Coral zone" by Fong (1959). The thickness ranges from 4 to 20 ft. The lower and upper contacts are transitional. This *in situ* reefal biofacies contains the lowest zone of porosity in the productive areas. The

thickness, porosity, and permeability of the "Coral zone" improve where the Light Brown member also is well developed. In the marginal areas of the dark-brown organic platform this facies is not recognizable, but is replaced by stromatoporoid limestone. Organic components of the "Coral zone" in decreasing order of predominance are *Thamnopora* corals, massive and tabular stromatoporoids, *Amphipora,* and *Stachyoides.* Skeletal fragments and grains, and mud-supported carbonate rocks (calcilutite) are the main matrix components.

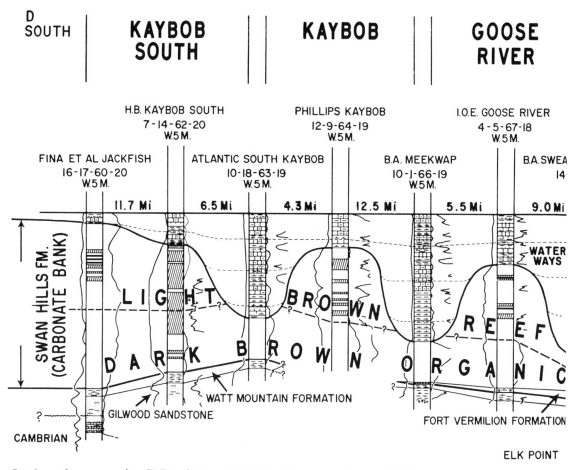

FIG. 13.—South-north cross-section **D-D'** of Beaverhill Lake Group, north-central Alberta. Section passes through western Swan Hills fields and shows relations similar to those in Fig. 12. Gradual pinchout of Fort Vermilion Formation causes basal beds of Beaverhill Lake to overlie directly Watt Mountain Formation. Gross Beaverhill Lake thins markedly between Kaybob and Kaybob South fields. Area through which sections passes is shown on Figures 14 and 21; exact location on Figure 21.

Above the "Coral zone" are dense beds which completely separate it from the Light Brown member. As a result, the "Coral zone" in the Swan Hills field is a separate producing zone of the Swan Hills Formation.

Stromatoporoid reef front.—Around the northern and northeastern margins of the Dark Brown member platform high water energy fostered the growth of a reef body with a rigid framework. The slightly different organic composition in the different reefs probably is related to local energy conditions and food supply. Along the eastern edge of the Swan Hills field, stromatoporoids predominate, but the increasing numbers of *Amphipora* and the greater amount of carbonate-mud matrix suggest a lower energy environment. Toward the central part of the platform, the reef front interfingers with the dark brown *Amphipora*

beds, and the offreef Waterways Formation is discordant. Organic components, in order of decreasing importance, are tabular and massive stromatoporoids, algae, brachiopods, cup corals, *Amphipora*, and crinoids. The matrix ranges from coarse reef detritus to fine carbonate mud and in places consists of skeletal grains, intraclasts, and some sparry calcite.

The porosity common to these beds is associated with both the organic framework and matrix. Hydrocarbon production is from the stromatoporoidal reef front at the edge of the platform north and northeast of the Swan Hills and Virginia Hills fields. Similar beds also are present in the marginal areas of the Light Brown member in the upper part of the reef complex and, where such beds are developed, they also are productive.

"Brecciated zone."—This term applies to a

190

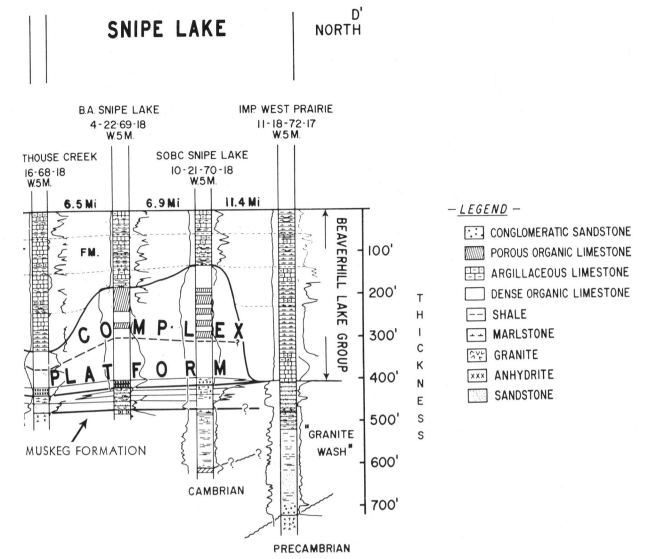

SNIPE LAKE | D' NORTH

THOUSE CREEK
16-68-18
W.5M.

B.A. SNIPE LAKE
4-22-69-18
W.5M.

SOBC SNIPE LAKE
10-21-70-18
W.5M.

IMP. WEST PRAIRIE
11-18-72-17
W.5M.

6.5 Mi   6.9 Mi   11.4 Mi

FM.

COMPLEX
PLATFORM

MUSKEG FORMATION

CAMBRIAN

PRECAMBRIAN

BEAVERHILL LAKE GROUP

"GRANITE WASH"

THICKNESS

100' 200' 300' 400' 500' 600' 700'

- LEGEND -

CONGLOMERATIC SANDSTONE
POROUS ORGANIC LIMESTONE
ARGILLACEOUS LIMESTONE
DENSE ORGANIC LIMESTONE
SHALE
MARLSTONE
GRANITE
ANHYDRITE
SANDSTONE

stromatoporoid bed overlying the "Coral zone" that has been observed in most of the Swan Hills–Ante Creek area. The name refers to the brecciated appearance of the cores, which contain numerous circular and semicircular, partly broken stromatoporoid colonies. However, examination of the cores proved that the appearance is caused by the growth pattern of the stromatoporoids. Only very limited mechanical transport is indicated, and the zone is not truly "brecciated."

The "Brecciated zone" interfingers with the stromatoporoidal reef front on the north and northeast, and with the *Amphipora* beds of the Dark Brown member on the west and southwest. The thickness ranges from 0 to 50 ft. Organic components in order of decreasing abundance include bulbous stromatoporoids, *Amphipora,* and corals. The matrix consists of fine

skeletal grains, intraclasts, and carbonate mud. Pyrobitumen also is common. The organisms of the "Brecciated zone" apparently accumulated in a lower energy environment than those of the stromatoporoidal reef front—probably in a restricted shelf lagoon. Porosity is sparingly present in association with stromatoporoids.

Dark brown Amphipora *beds.*—The dark-brown *Amphipora* beds extend throughout the Swan Hills region, and form the most common unit of the Dark Brown organic platform. The thickness ranges from 40 to 100 ft, generally increasing southward. There also are local variations in the thickness over areas where the Light Brown member is developed. The base of the unit is at the top of the "Coral zone," or at the top of the "Brecciated zone" where the latter is present. The unit interfingers with the stromatoporoidal reef front facies on the east

191

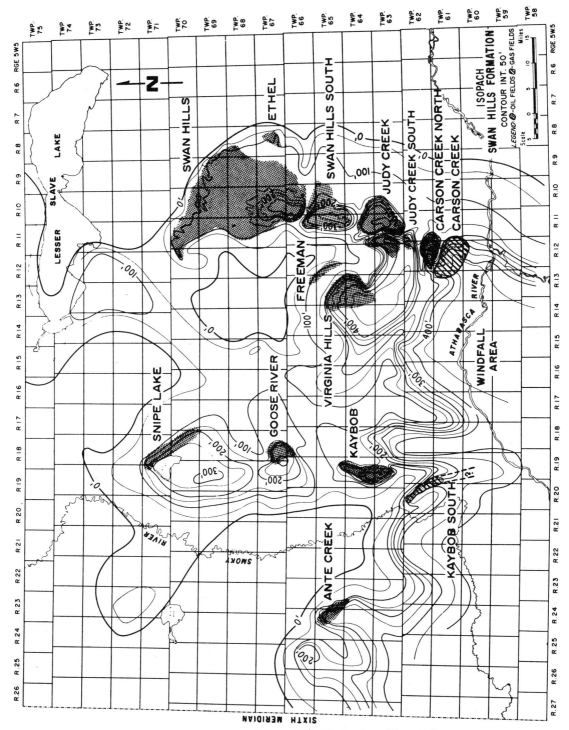

FIG. 14.—Isopach map of Swan Hills Formation. CI = 50 ft.

and northeast. The *Amphipora* content ranges from 0 to 60 percent, and the unit generally is less fossiliferous near the base. A few bulbous stromatoporoids also are present. Organic components, in order of decreasing importance, are *Amphipora*, bulbous stromatoporoids, brachiopods, and gastropods. Laminated limestone beds with sparry calcite also are common and generally devoid of organisms. The matrix ranges from poorly sorted silt- and sand-size grains and pellets to carbonate mud, or calcilutite. Skeletal grains also are recognizable.

Porosity is developed only locally in these beds, and is associated with some *Amphipora*-

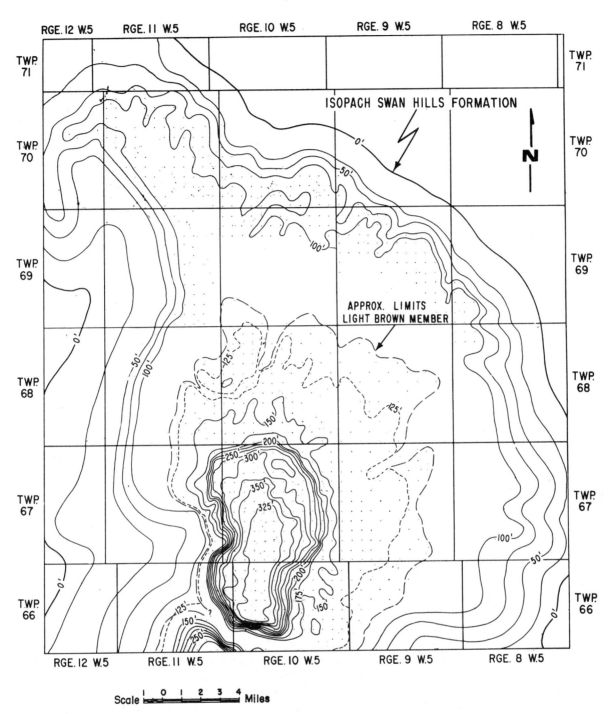

FIG. 15.—Detailed isopach map of Swan Hills Formation in Swan Hills field–Deer Mountain area. Broad platform is present between edge and Swan Hills main buildup. Rigid wave-resistant stromatoporoid reef wall first developed on northeast edge of platform.

rich zones. A little void space is present in a few places in granular or pelleted limestone, but for all practical purposes the dark-brown *Amphipora* beds do not contribute significantly to the pore volume of the Swan Hills Formation. This unit reflects sedimentary and environmental conditions of a restricted shelf lagoon, and the laminated limestone indicates occasional supratidal-flat conditions.

Light Brown Member

The thickness of the Light Brown member changes more abruptly than that of the gently sloping Dark Brown platform on which it lies.

The member ranges in thickness from a few feet to approximately 320 ft south and west of Carson Creek. The alternation of higher energy conditions which produced reef buildups with brief erosional episodes and stable sea-level conditions caused considerable variation in sedimentation.

Light-brown Amphipora *beds.*—These are the most widespread beds of the Light Brown member and cover part of the Dark Brown platform. They range in thickness from 0 to approximately 200 ft, the thickest sections being over the area of the Swan Hills–Carson Creek reef complex.

The light-brown *Amphipora* beds consist of two main limestone types: (1) light-brown to medium-brown limestone, fragmental, with calcarenite matrix and *Amphipora;* and (2) light-olive-brown limestone, with *Amphipora*, calcilutite, and sparry calcite matrix. Limestone of the first type comprises the first sediments deposited on the top of the Dark Brown platform and contains only scattered stromatoporoids. The matrix is fragmental in appearance and consists of skeletal grains and intraclasts, suggesting a shallow, agitated, open-marine environment. In areas where only this fragmental *Amphipora* bed is present over the Dark Brown platform, the contact between the reef and overlying offreef Waterways Formation is sharp. In most places the organisms of the Swan Hills Formation are truncated at the contact, and only a thin zone of pyrite is present; thus very shallow postdepositional erosion is suggested.

Limestone of the second type is found most commonly in the center part of the Light Brown member, *Amphipora* being the predominant organic component. Dendroid, tabular, and bulbous stromatoporoids are present in a few areas, together with ostracods, gastropods, and calcispheres. The matrix consists of sparry calcite, carbonate mud, skeletal grains, and intraclasts. These sediments appear to have been deposited in shallow, slightly agitated water with low turbidity, typical of a restricted shelf-lagoon environment.

Porosity development commonly is associated with the matrix and organisms of the light-brown fragmental *Amphipora* beds, whereas in the light-olive-brown limestone porosity is irregular and generally poorly developed.

"Table reef" (Edie, 1961).—This easily recognizable, widespread zone of the Light Brown member overlies the light-brown fragmental *Amphipora* beds and ranges in thickness from 15 to 25 ft. Its broad lateral extent is characterized by a relatively uniform organic content. In the marginal areas north and northwest of the buildup the organisms are severely broken and reworked, forming a fragmental zone which interfingers with the *in situ* material. The presence of the broken rocks indicates short erosional periods. The principal organisms, in order of decreasing abundance, are dendroid stomatoporoids, Solenoporoid algae, brachiopods, cup corals, and *Amphipora*.

The matrix is considerably varied, but the most common constituents are the skeletal grains and debris of various sizes, with negligible amounts of carbonate mud. Porosity is common in this zone and is associated with both organisms and matrix. Interorganic vuggy porosity present between stromatoporoids suggests secondary, postdepositional leaching. The "Table reef" zone of the Swan Hills Formation probably represents an environment of shallow, agitated water. Local emergence or erosion occurred at the end of deposition, as indicated by the presence of organic debris on the slopes of the reef.

Porous calcarenite beds.—A porous zone, consisting of skeletal grains and reworked and transported organic fragments, overlies the light-brown *Amphipora* beds and, in places, the upper slopes of the Light Brown buildup. In a few areas this zone contains almost no organisms, and consists of well-sorted carbonate sandstone; in the upper slopes, most of the zone consists of well-rounded stromatoporoid and *Amphipora* fragments. The reef rubble on the top and slopes of the buildup suggests that growth terminated as a result of shallowing of the seawater rather than submergence. Porosity generally is excellent, and is intragranular. Intraorganic void space also is common, decreasing slightly toward the central part of the buildup. This unit is one of the most important reservoir rocks of the area. The porous calcarenite beds probably were formed in the most exposed parts of the organic buildup, in shal-

Fig. 16.—South-north sections showing stages I-VI in development of Swan Hills reefing, as interpreted for Swan Hills field. Similar stages can be postulated for other Beaverhill Lake fields.

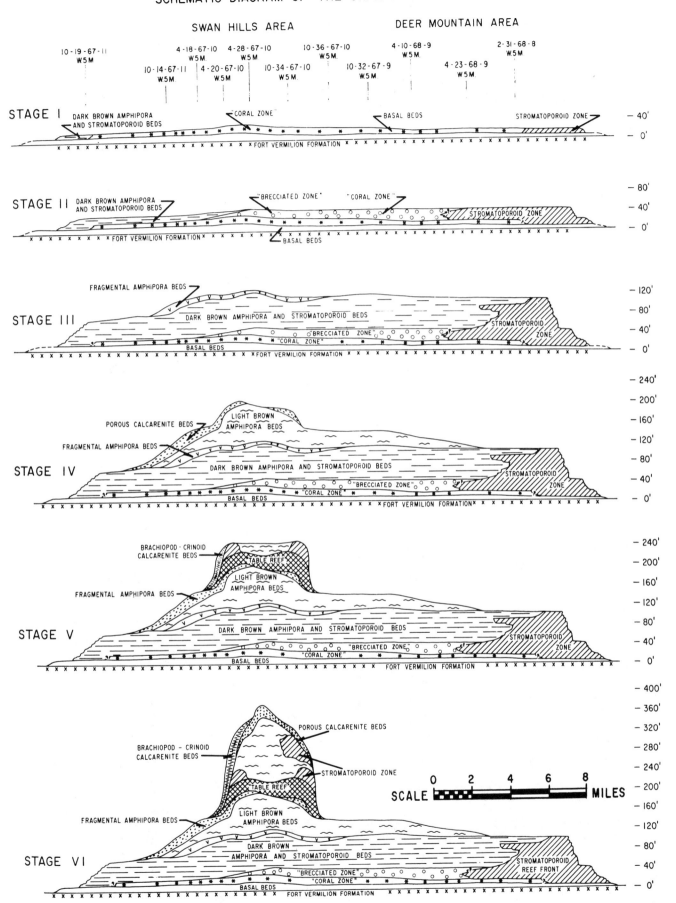

SCHEMATIC DIAGRAM OF THE SWAN HILLS REEF GROWTH

SWAN HILLS AREA   DEER MOUNTAIN AREA

195

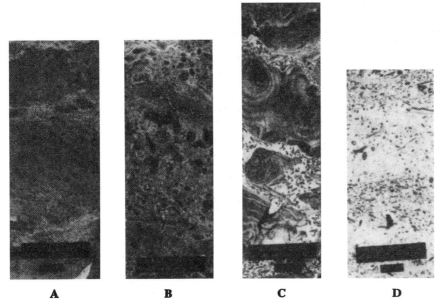

FIG. 17.—**A,** Fine-grained, dark-brown to black argillaceous beds of basal Beaverhill Lake, 3 ft below "Coral zone." Fine clasts of similar material can be seen scattered throughout, suggesting reworking of beds. **B,** *Thamnopora*-type corals in matrix of buff calcarenite, representing "Coral zone" of Swan Hills Formation. Unit forms lowermost productive bed in Swan Hills field. In this sample "Coral zone" has 8.2 percent porosity and 3.6 md permeability. **C,** Broken, bulbous and massive stromatoporoids of "Brecciated zone." Fine *Amphipora* in dark-brown micritic matrix fill interstices between larger stromatoporoid fragments. Some fractured stromatoporoids have calcite infilling. **D,** Dark-brown micritic *Amphipora* beds. Banded appearance results from long axis of *Amphipora* being deposited parallel with normal bedding, and different energy levels which prevailed in carrying *Amphipora* fragments into lagoon.

low turbulent water. The upper boundary with the offreef Waterways Formation is unconformable and appears to be an erosion surface.

Brachiopod-crinoid beds.—The brachiopod-crinoid beds are found most commonly on the west slopes of the Light Brown member, and consist of calcarenites of skeletal origin. In order of decreasing importance the organic components are brachiopods, crinoids, stromatoporoids, and *Amphipora*. The stromatoporoids and *Amphipora* gradually decrease in number, and the calcarenite grades into calcisiltite and fine carbonate mudstone farther away from the main buildup. This brachiopod-crinoid zone with fine calcarenite derived from skeletal material is typical of the reef flanks and is considered to have been deposited in open-marine, aerated water of medium energy. The beds generally are devoid of porosity and therefore have no economic significance.

DEPOSITIONAL HISTORY OF SWAN HILLS FORMATION

In the Swan Hills field area the depositional history of the Swan Hills Formation can be described in six stages, each indicating a major change in sedimentation and environment (Fig. 16). Stages I-III cover the time of Dark Brown member deposition, and Stages IV-VI cover the time of Light Brown member deposition.

Stage I.—The Beaverhill Lake sea gradually transgressed westward from the deepest part of the basin. The area between the Peace River arch and West Alberta ridge was flooded, and the widespread shoal conditions that developed favored organic growth. The depositional environment of associated sediments was similar to that of the offreef Waterways Formation, except for the higher percentage of skeletal material and the first appearance of slightly shallower water organisms (Fig. 17A).

After the appearance of the corals in the upper part of the "Basal beds," the area became densely colonized by frame-building organisms that created a solid base and good foothold for additional organic growth (Fig. 17B). Along the northeast rim, in areas of slightly higher wave energy, a stromatoporoid zone began to develop, whereas farther south along the platform edge in the areas of lower energy, the corals and stromatoporoids were replaced by *Amphipora*-rich beds.

Stage II.—A slight but persistent rise in sea level caused continued stromatoporoid reef growth on the northeast rim. This growth created a broad shelf lagoon behind the

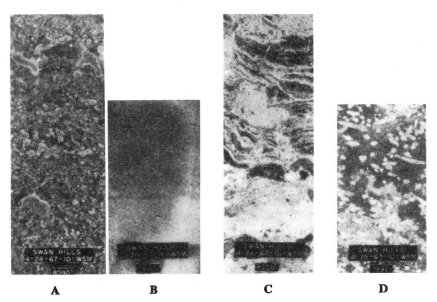

A   B   C   D

FIG. 18.—**A,** Light-brown beds of *Amphipora* bank deposits. In addition to abundant *Amphipora* there are some *Stachyoides* and very few bulbous stromatoporoids in calcarenite matrix. Because of reworking of these beds, *Amphipora* are more randomly oriented and widely scattered. **B,** Fine skeletal reef calcarenite of reworked *Amphipora* bank deposits shows no bedding or recognizable organisms, and is very homogeneous rock. Sample has 9.1 percent porosity and 18 md permeability. **C,** Almost solid framework of massive stromatoporoids bound by algal mats; forms "table reef" (Fig. 5). Calcarenite fills spaces between larger organisms. In sample shown porosity is 12.2 percent with 12 md permeability. **D,** *Amphipora* in light-buff calcarenite. This is typical lagoon deposit just behind reef rim. In sample porosity is 5.5 percent and permeability 10 md.

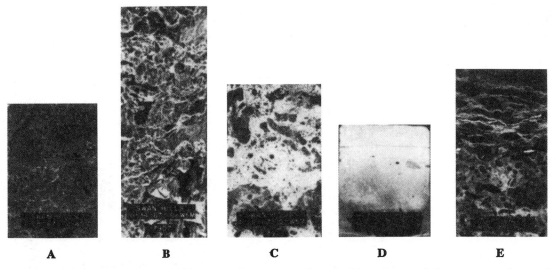

A   B   C   D   E

FIG. 19.—**A,** Pelletoid unfossiliferous carbonate mud deposited in quieter and deeper water of central lagoon. **B,** Stromatoporoid reef-wall material with abundant sparry calcite filling interstices. Organic porosity is 9 percent and permeability is 11 md. **C,** Bioclastic material of "Detrital zone." Reef rubble composed of large and small organic fragments enclosed in dark-brown calcarenite matrix common to top and upper slopes. Porosity is 12.6 percent and permeability 64 md. **D,** Dark-gray, cryptograined, dense, argillaceous, finely bedded calcareous shale of Waterways Formation. **E,** Contact between Swan Hills Formation and overlying Waterways Formation. Close examination of contact shows truncated organisms indicative of erosion before burial. *Boudinage* structure common to some of Waterways beds is present here

C. R. Hemphill, R. I. Smith, and F. Szabo

Table 1. Parameters of Swan Hills Oil Fields

Fields, Pools	Disc. Date	Name and Loc. Disc. Well	Total Wells Drld. to 12/31/67	Av. Well Depth (Ft)	Prod. Area (Acres)	Maximum Reservoir Thickness (Ft)	Av. Net Pay (Ft)	Av. Por. (%)	Av. Perm. (Md)	Water Sat. (%)	Est. Oil in Place (1,000 Bbl)	Cum. Prod. to 12/31/67 (1,000 Bbl)	Remaining Recoverable Crude Oil (1,000 Bbl)
Swan Hills A and B pools	3/ 2/57	Home et al. Regent 8-11-68-10-W5M	520	8,299	102,479	395	52.8	7.8	20	18.6	1,944,000	83,309	688,621
C pool	3/21/58	Texcan Mic Mac Deer Mtn. 10-14-69-9-W5M	353	7,475	55,058	120	29.8	6.2	5	10.2	553,000	17,684	141,616
Judy Creek A pool	2/25/59	Imperial Judy Creek 16-31-63-10-W5M	184	8,665	28,200	358	66.7	9.3	43	16	809,000	42,018	322,032
B pool	9/ 6/59	Imperial Virginia Hills 10-13-63-12-W5M	75	8,842	12,380	453	58.4	9.0	111	17	256,000	13,123	109,757
Swan Hills South	2/27/59	B.A. Pan Am. Sarah Lake 2-13-65-11-W5M	243	8,345	35,680	405	69.8	8.0	26	18.3	897,600	43,639	348,512
Virginia Hills	1/31/57	Home Union H.B. Virginia Hills 9-20-65-13-W5M	134	9,283	24,350	503	48.4	8.1	35	20.4	450,200	25,288	148,992
Kaybob	4/22/57	Phillips Kaybob 7-22-64-19-W5M	107	9,780	18,000	232	59.9	7.4	23.3	22	300,000	22,346	97,654
Carson Creek North	9/ 6/58	Mobil P.R. Carson N 6-1MU 6-1-62-12-W5M	46										
A pool—oil pool —gas cap				8,632 8,580	6,916 3,720		31 11	8.3 8.0	56 15	13	66,000	4,100	24,940
B pool—oil pool —gas cap				8,736	12,372 377		51 7.8	9.1 9.9	167 21	21 21	215,000	8,939	74,911
Snipe Lake	10/24/62	S.O.B.C. Snipe Lake 10-21-70-18-W5M	114	8,534	17,297	272	33.3	7.3	35	27	198,000	10,587	66,633
Goose River (B pool excl.)	8/28/63	B.A. Goose River 10-4-67-18-W5M	28	9,185	7,700	200	49	8.2	103	19	145,000	2,613	20,587
Freeman	10/31/62	H.B. Union Home Freeman 2-1 2-1-66-13-W5M	20	9,184	5,390	160	29	5.9	20	25	40,400	874	3,934
Ante Creek	10/15/62	Atlantic Ante Creek 4-7 4-7-65-23-W5M	24	11,270	7,510	260	25.3	6.3	6	22	34,800	1,838	3,730
Judy Creek South	3/19/60	Mobil Carson Creek 14-31 14-31-62-11-W5M	8	8,925	3,009	136	22.6	6.3	24.4	25	15,000	557	2,443
Ethel	1/28/64	Mobil Atlantic Ethel 10-11 10-11-67-8-W5M	4	7,522	1,289	87	23.6	5.7	10.2	17	8,100	19	62

198

wave-resistant reef front. In the quieter and more restricted waters behind the reef, the less wave-resistant bulbous stromatoporoids of the "Brecciated zone" flourished (Fig. 17C). Farther west, in the quiet, semistagnant waters of the lagoon, *Amphipora* beds were deposited in precipitated carbonate mud.

Stage III.—The rigid stromatoporoid reef wall continued to grow on the northeast side of the carbonate bank, while *Amphipora*-rich beds were deposited in the shelf lagoon behind the reef front (Fig. 17D). The writers assume that carbonate-bank growth ceased when slight eastward tilting deepened the water above the reef-front area, and at the same time part of the carbonate bank became emergent on the west. Brief exposure of the east-central part of the carbonate platform caused reworking of the upper part of the bank in the Swan Hills area and created a thin fragmental zone on this part of the carbonate platform. This marked the termination of deposition for the Dark Brown member and provided a substratum for further organic buildup on the slightly higher, emergent western side.

Stage IV.—After the submergence of the stromatoporoidal reef front, *Amphipora* grew abundantly in the quieter, slightly deeper water of the lagoon bank. The periods of quiescence were interrupted by frequent storms, which transported *Amphipora* fragments into localized carbonate-bank deposits that formed loci for the reef growth of Stage V (Fig. 18A). The presence of porous calcarenite near the top and along the flanks of these bank deposits suggests near emergence at this time, and deposition ceased (Fig. 18B).

Stage V.—Stage V is the "Table reef" which developed on the beds of Stage IV. A temporary stillstand of the sea permitted lateral growth of the organic lattice, which resulted in this widespread, fairly homogeneous buildup (Fig. 18C). Slow subsidence resulted in the formation of a circular stromatoporoidal reef atoll enclosing a central lagoon (Fig. 18D). Fluctuating water level or occasional storms caused the deposition of small amounts of fragmental material in zones within the lagoon and the reef rubble of the outer slopes. Local emergence probably terminated this stage of growth.

Stage VI.—The final stage of Swan Hills reef development was marked by a return to supratidal conditions. This caused renewed growth of *Amphipora* and the deposition of pelletoid and unfossiliferous carbonate-mud beds (Fig. 19A). Carbonate-mud accumula-

Table 2. Parameters of Swan Hills Gas Fields

Fields, Pools	Disc. Date	Name and Loc., Disc. Well	Total Wells Drilled to 12/31/67	Av. Well Depth (Ft)	Prod. Area (Acres)	Max. Reservoir Thickness (Ft)	Av. Net Pay (Ft)	Av. Por. (%)	Av. Perm. (Md)	Water Sat. (%)	Initial GIP (Bcf)	Marketable Gas Prod. (Bcf)	Remaining Marketable Gas to 12/31/67
Kaybob South	9/11/61	H.B. Union Kaybob 11-27 11-27-62-20-W5M	8	10,560	16,440	325	43	10	294.2	15	670	0	370
Carson Creek	2/26/57	Mobil Oil Whitecourt 12-13 12-13-61-12-W5M											
A pool			8	8,550	15,840	89	20	8	21	20	210	8	142
B pool			5	8,610	6,980	83	24	8	75	20	110	-15	95

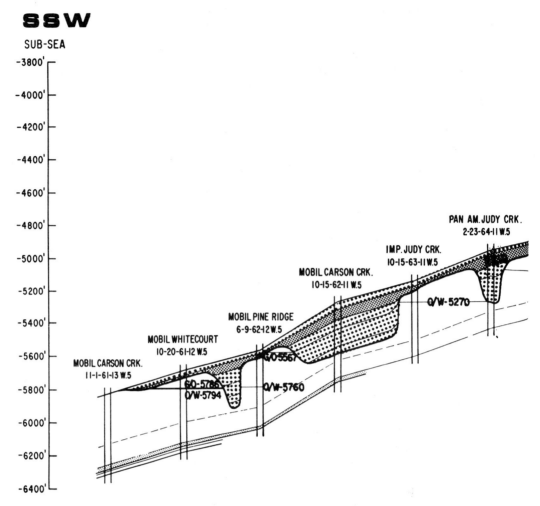

SSW

SUB-SEA

-3800'
-4000'
-4200'
-4400'
-4600'
-4800'
-5000'
-5200'
-5400'
-5600'
-5800'
-6000'
-6200'
-6400'

PAN AM. JUDY CRK.
2-23-64-11 W.5

IMP. JUDY CRK.
10-15-63-11 W.5

MOBIL CARSON CRK.
10-15-62-11 W.5

MOBIL PINE RIDGE
6-9-62-12 W.5

MOBIL WHITECOURT
10-20-61-12 W.5

MOBIL CARSON CRK.
11-1-61-13 W.5

O/W-5270
G/O-5567
O/W-5760
G/O-5786
O/W-5794

FIG. 20.—SSW-NNE structural cross section through eastern fields (Carson Creek to Swan Hills) shows lateral progression of gas-oil and oil-water interfaces. This is prime example illustrating Gussow's (1954) hypothesis of differential entrapment of hydrocarbons. Vertical scale in feet; sea-level datum. Trace is same as C-C′ (Fig. 12), shown on Figure 21.

tion alternated with the deposition of *Amphipora*-rich beds and a few thin terrigenous mud stringers. Incipient reef growth is present locally (Fig. 19B). Upward organic growth was prevented by emergence and strong erosion. The result of this activity was the formation of calcarenite beds and coarse reef rubble on the top and upper slopes of the buildup. These carbonate clastic rocks are the youngest strata of the Swan Hills Formation (Fig. 19C). Penecontemporaneous deposition of carbonate mud and clay (Waterways Formation) in the offreef areas is indicated (Fig. 19D). Sedimentation proceeded at a slightly slower rate relative to the development of the Swan Hills section.

On the east side of the buildup the contact with the stratified Waterways equivalent of the reef is sharp and unconformable (Fig. 19E). On the west side the brachiopod-crinoid beds, which show a gradual decrease in grain size and organic content away from the reef, form a transitional zone between the two formations.

At the end of Stage VI, a sudden increase in water depth drowned the reef complex. Waterways-Woodbend clay and carbonate-mud deposits covered most of the area.

DIAGENESIS OF SWAN HILLS CARBONATES

Dolomitization.—The absence of significant dolomitization in the Swan Hills Formation makes it possible to conduct detailed facies, textural, and environmental studies of the reefs. Most of the formation is in the initial stage of compaction-current dolomitization, which involves decreased porosity and reduced pore volume. Where dolomitization has occurred, it generally plugs void space of primary organic and matrix porosity. Presumably the magne-

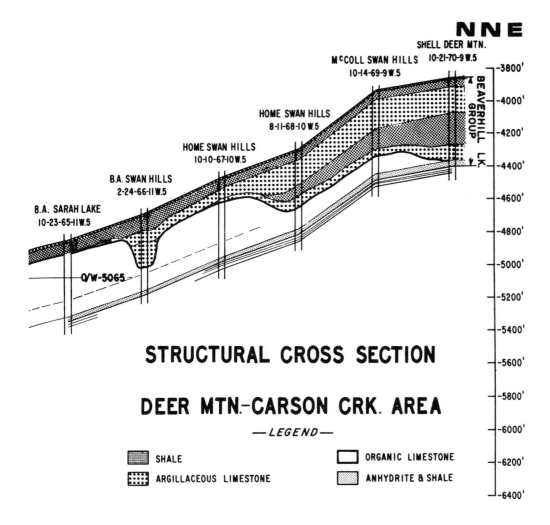

STRUCTURAL CROSS SECTION

DEER MTN.-CARSON CRK. AREA

— LEGEND —

SHALE

ARGILLACEOUS LIMESTONE

ORGANIC LIMESTONE

ANHYDRITE & SHALE

sium ions were brought to the site of precipitation by percolating waters after lithification.

Another form of dolomitization is associated with the organisms; *e.g.*, the axial canals in *Amphipora* commonly contain dolomite infilling in varied amounts. Generally this dolomite infilling is associated with calcite crystals. Presumably the source of magnesium was the organisms, or magnesium ions from areas of higher ion concentration which migrated to and accumulated in the fossils.

A less common form of dolomite is a scattering of perfect dolomite rhombs, about 20–50 μ in diameter, occupying space once filled by much finer calcite grains and microorganisms in a generally dense matrix.

Complete dolomitization of the Swan Hills Formation is observed south and west of the Kaybob area, and south of the Carson Creek gas field, where it is associated mainly with the upper part of the Light Brown member. The reason for this complete dolomitization is not known to the writers, but is assumed to be related to tectonism.

Silicification.—Scarce light-gray to brown-gray dolomitic chert lenses are found in the Swan Hills Formation. Thickness ranges from 1 to 3 in. The material partly replaces skeletal and nonskeletal grains, and possibly was precipitated from chemical solutions of organic origin.

Recrystallization.—Recrystallization is here defined as the formation of new mineral grains in the limestone in the solid state, without the introduction of other elements. Recrystallization is not very important in the Swan Hills area, but it has been observed in a few places in the upper part of the Light Brown member,

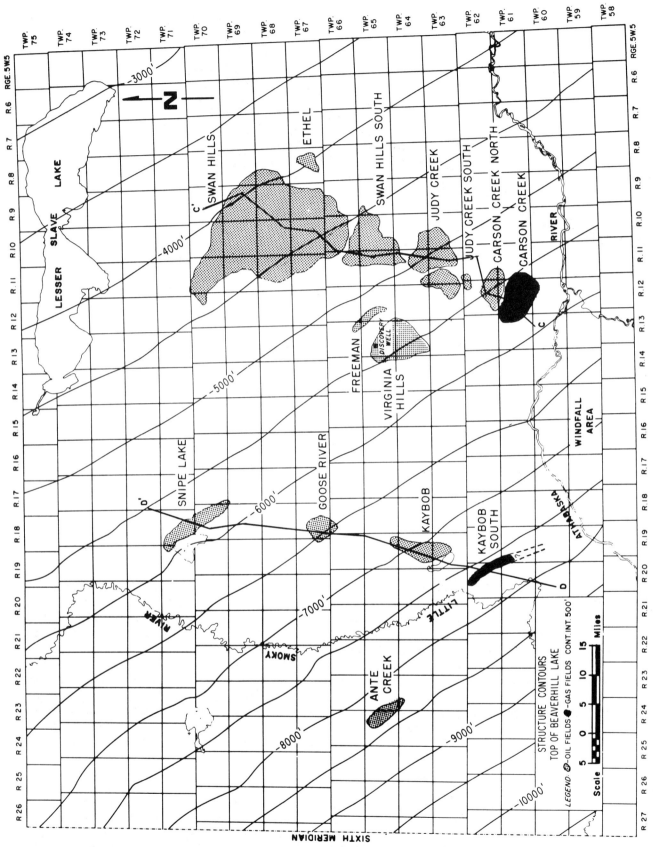

FIG. 21.—Structural contours, datum top of Beaverhill Lake, show southwesterly regional dip. Lack of differential compaction is shown by absence of contour deviation as contours pass through areas of reef buildup. CI = 500 ft. Depth subsea. Shows locations of Figures 12, 13, and 20.

generally in association with the finer grained sediments of the lagoon facies. The matrix is converted into a limestone of saccharoid texture, and the original structure of the *Amphipora* and other organisms is blurred or lost.

Fracturing.—A study was made of numerous cores from the Swan Hills field and the fractured intervals were recorded. Generally, fractures are very scarce in the Dark Brown member, but are more common in the upper part of the Light Brown member.

The matrix of the Dark Brown member consists of very fine-grained calcarenite and calcilutite, which commonly are recemented by secondary calcite. Recementation caused considerable compaction. The lithologic character of the Light Brown member reflects sedimentation in high-energy waters; the matrix ranges in lithologic character from loosely consolidated, fine-grained calcarenite to calcirudite which yielded to stress more easily by fracturing.

Detailed cross-section studies of the Beaverhill Lake Group show that this limestone interval thins slightly over areas of maximum reef buildup. The reduction in thickness probably resulted from the collapse of porous sections accompanied by the development of fractures. Most fractures are vertical or high-angle oblique and many of the fracture planes are lined with secondary calcite crystals. The importance of these fractures is that they may permit communication between otherwise isolated porous zones within the Light Brown member.

Reservoir Characteristics

Reservoir parameters for each field are shown in Tables 1 and 2. Within the map area, approximately 377,000 acres is underlain by productive Swan Hills Formation. In individual wells net pay thickness is as much as 250 ft. However, the best field average is 69.8 ft and the poorest 22.6 ft. Field-weighted average-po-

Table 3. Properties of Natural Gas, Carson Creek and Kaybob South Fields

Field	Sp. Gr. Gas	Recoverable NGL Content (Bbl/MMcf)[1]	Recoverable Sulfur Content (Long Tons/ MMcf Raw Gas)
Carson Ck. A	0.930	131.48	—
B	0.972	142.17	—
Kaybob South	1.0096	107	6.2

[1] Based on raw gas volume with 85 percent recovery of propane, 95 percent recovery of butane, and 100 percent recovery of heavier ends.

Table 4. Beaverhill Lake Fields, Oil Gravity and Saturation Pressures

Field	API Gravity Produced Liquid (°API)	BHT (°F)	Ps @ BHT (psig)	Ps @ 225° F (psig)
Ante Creek	46	233	4,140	4,116
Carson Creek oil leg[1]	43	198	3,767	3,848
Carson Creek gas	61	198	(3,850 DP)[2]	—
Carson Creek N. A	44	187	3.391	3,493
Carson Creek N. B[3]	44	191	3,312	3,414
Freeman	40	221	1,738	1,750
Goose River	40	233	2,052	2,028
Judy Creek A	43	205	2,290	2,350
Judy Creek B	43	206	2,940	2,997
Kaybob S.	60	238	(3,643 DP)	—
Kaybob	43	235	3,019	2,989
Snipe Lake	37	183	1,300	1,426
Swan Hills S.	42	225	2,219	2,219
Swan Hills (Main)	43	225	1,820	1,820
Swan Hills (Inverness)	42	211	1,726	1,768
Swan Hills C	42	189	1,370	1,478
Virginia Hills	38	218	1,812	1,833

[1] There is a thin (8 ft) noncommercial oil leg in this reservoir.
[2] DP denotes dew-point pressure for gas fields.
[3] Reservoir has a small gas cap but oil leg appears undersaturated.

rosity values range from a low of 5.7 percent to a high of 10 percent, whereas weighted average horizontal permeability values for the fields range from 5 to 167 md.

All of the oil pools are undersaturated and there is a general increase in API gravity toward the gas-condensate pools. An odd situation is present in the Carson Creek North field, where both the A and B pools have a gas cap but samples of the reservoir oil indicate undersaturation. Snipe Lake has the lowest gravity of 37° and Ante Creek has the highest of 46°.

Water saturation ranges from 10 to 30 percent, but generally is about 20 percent. The structural relations and the differing water tables along the eastern reef trend are depicted in Figure 20.

Oil is carried to market by the Peace River pipeline and the Federated pipeline. Posted wellhead price is approximately $2.55/bbl (Canadian).

In several fields, the presence of separate pools has been established. Four factors account for this situation:

1. Presence of oil in separate reef facies within the same reef complex (the best example is Swan Hills).
2. Vertical separation by an impermeable green shale barrier, such as is found in the Carson Creek and Carson Creek North fields.
3. Horizontal and vertical permeability barriers caused by completely "tight" reef facies; e.g., the Judy Creek–Judy Creek South relation.

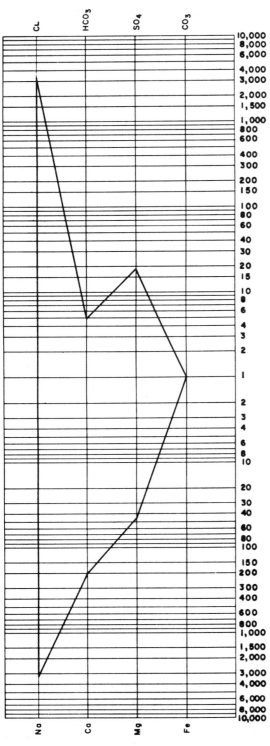

FIG. 22.—Logarithmic plot of typical water analysis, Swan Hills Formation, in milligram equivalents per unit. There is considerable uniformity in chemical composition of formation waters from Swan Hills fields throughout area. This illustration is thought to represent typical logarithmic pattern of a Swan Hills Formation water analysis.

4. Scour or surge channels between reef masses typical of the eastern reef chain, as shown on Figure 14. Dissection of this chain is apparent from the number of separate pools along the trend.

The Light Brown member of the Swan Hills Formation has the best reservoir characteristics. In fact, it is only in the Swan Hills field that the Dark Brown member and the "Coral zone" are of economic importance. There the porous beds of the Dark Brown and Light Brown members are grouped and designated as one pool for proration purposes. The "Coral zone" is separated by the impermeable carbonate of the Dark Brown member and forms the B pool of the field. Probably the principal reason for production in the Dark Brown member is the lack of definite water table, which suggests that the entire basal part of the Swan Hills Formation is above the oil-water interface as calculated from regional pressure data.

Depth of drilling is controlled primarily by the position of the well in the Alberta basin. Regional dip is southwest at approximately 40 ft/mi in the eastern part of the area and steepens to 50 ft/mi in the western part (Fig. 21). Accordingly, drilling depth ranges from 7,475 ft at the northeast end of the Swan Hills field to more than 11,000 ft in the Ante Creek area.

Reservoir Fluids

Gas.—The primary difference in gas composition between the two Beaverhill Lake gas fields (Table 2; Carson Creek and Kaybob South) is in the sulfur content. Gas from Carson Creek is sweet, whereas Kaybob South gas has 16.59 percent H_2S by volume. Both fields are rich in natural gas liquids. Table 3 gives the most important gas properties. Small gas caps are present in the Carson Creek North A and B pools.

Oil.—Oil properties differ somewhat among fields. This is to be expected because of the many reservoirs, each of which has different physical properties. Generally, the Beaverhill Lake reservoirs contain an undersaturated paraffin-base crude, with a low sulfur content ranging from a trace to 0.42 percent. The variations in API gravity are given in Table 4, together with the bottomhole temperature and saturation pressure for each field. The saturation pressures also have been adjusted to a common temperature of 225° F, to permit easier correlation of the reservoir-fluid properties.

Water.—There is an oil-water or gas-water

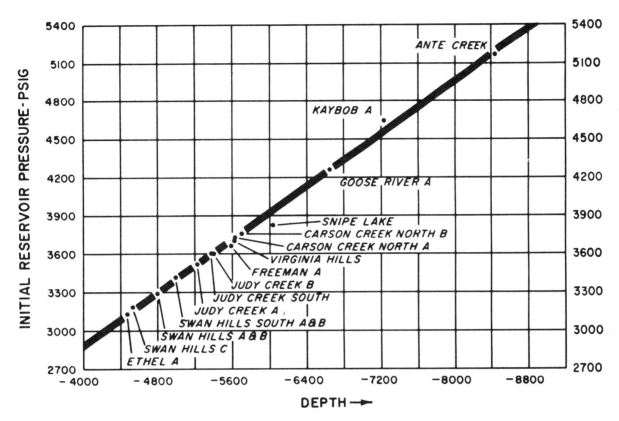

FIG. 23.—Plot of initial pressure *versus* datum (depth subsea), Beaverhill Lake Group. This pressure-depth plot of Swan Hills fields shows close relation of initial pressures, suggesting common pressure system for area. Data from Oil and Gas Conservation Board, Calgary (1967).

interface in most of the developed fields. Notable exceptions are Swan Hills, Ethel, Kaybob, and Freeman. Where a variable water table is reported, it generally can be attributed to a change in lithologic character within the interval in which the water table normally would be. A very few water occurrences which are difficult to explain have been found in the Swan Hills field. Probably rock-geometry is involved (Stout, 1964).

The chemical compositions of different formation waters can be compared by plotting the components as a logarithmic pattern. The figures used are milligram equivalents per unit, obtained by multiplying parts per million (mg/l) by the following factors: Na + K × 0.0435; Ca × 0.0499; Mg × 0.0822; SO_4 × 0.0208; Cl × 0.0282; CO_3 × 0.0333; and HCO_3 × 0.0164.

Water analyses from several Beaverhill Lake fields were plotted in this manner and a typical logarithmic pattern for Swan Hills Formation waters is shown in Figure 22.

PRESSURE RELATIONS

A pressure-depth plot has been made by the Oil and Gas Conservation Board using the original pressures for several of the Swan Hills fields. The close relation of these initial pressures suggests a common pressure system (Fig. 23).

A similar detailed plot was made by selecting the most reliable pressure data from the individual wells in the gas, oil, and water phases of the Swan Hills Formation. To it have been added pressure data from the fluid phases of the widespread Gilwood Sandstone, which is the highest reservoir rock below the Swan Hills Formation. The data indicate a connection between these two reservoirs (Fig. 24).

It is not the writers' intention to speculate on the various stages of fluid migration or to explain the reason for the Gilwood-Swan Hills relation. Many factors are involved, such as rate of compaction, thickness of overburden, regional tilting, fracturing, juxtaposition of porous beds, *etc.* In some areas the Beaverhill

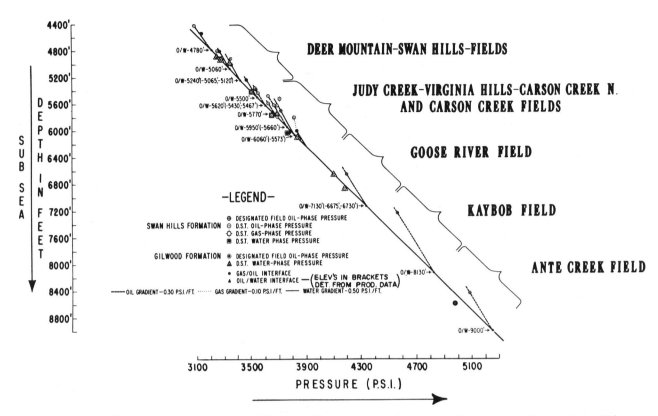

FIG. 24.—Pressure-depth plot, Beaverhill Lake–Gilwood Sandstone system, illustrates relation of Swan Hills oil and gas pools to a common reservoir system. Pressures measured within water phase in both Gilwood Sandstone and Swan Hill Formation reservoirs also relate to single system, suggesting that Swan Hills reservoir continuity is achieved through widespread Gilwood aquifer

Lake directly overlies the Gilwood Sandstone, or the "granite wash" with which the Gilwood Sandstone merges in the vicinity of the Peace River arch. Presumably these areas constitute the principle points for pressure communication.

On Figure 24 two oil-water interfaces are shown for some of the fields. One is the hypothetical interface based on the fluid-phase pressure data relation and represents the level at which the pore space is filled with 100 percent formation water. The other (in brackets) is the operating field oil-water interface, which has been established from production, drill-stem tests, and log interpretations. Field oil-water interfaces generally are established in the transition zone at a point where water-free production is obtained. This probably accounts for the differences between the calculated and reported oil-water contacts.

LOGGING

The most common logging devices used in Swan Hills fields are the induction electric log, microcaliper log, and gamma-sonic log. The microlog is an excellent tool for establishing effective net pay, whereas the sonic log has been used widely for calculating porosity in uncored wells and sections of uncored wells.

SECONDARY RECOVERY

Without exception, early production history of the field and the pools producing oil from the Swan Hills Formation presaged rapid pressure decline. None of the fields has been characterized by a strongly active water drive, and all indicated a low recovery of approximately 16 percent of the oil in place based on primary depletion. The primary-depletion recovery mechanism is rock and fluid expansion down to

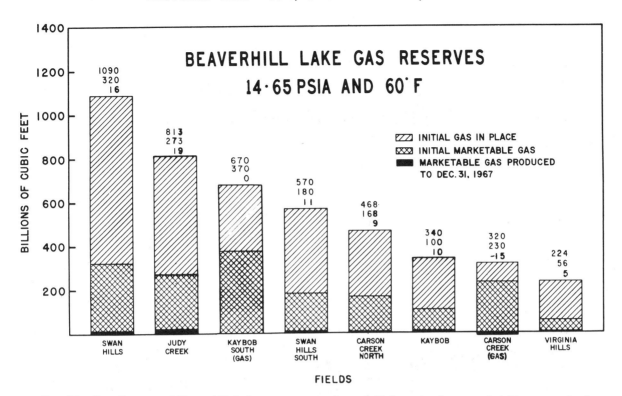

FIG. 25.—Bar diagram of Beaverhill Lake gas reserves shows initial gas in place, marketable gas, and where significant gas is produced for each of principal Swan Hills fields. Gas produced from Carson Creek North field was reinjected into Carson Creek gas field and accounts for negative value.

the bubble point, followed by a relatively inefficient solution gas drive.

Concern of the operating companies and the Alberta Oil and Gas Conservation Board resulted in exhaustive reservoir studies to find the most efficient means of maintaining field pressures above the bubble point, and thus obtain the maximum ultimate recovery of oil. The findings prompted unitization of most of the fields and, with one exception, water injection has been selected as the method of secondary recovery. For the Ante Creek field, a miscible flood is planned because of the higher initial reservoir pressure and solution gas-oil ratio resulting from greater depth of burial.

To the writers' knowledge, for every principal Swan Hills oil accumulation a secondary-recovery scheme is planned or is in effect. Secondary-recovery techniques are expected to result in recoveries of 35–60 percent of the original oil in place.

Water is obtained from surface sources, with approval of the Water Resources Branch of the Department of Agriculture, and is treated chemically for purity and bacteria control before injection. In most cases a line-drive method is adopted using downdip wells and injecting water below the field oil-water interface.

Where no field water table is present and it is necessary to inject into previously oil-producing wells, injection characteristics generally are not as good.

RESERVES

Gas.—Dissolved gas provides most gas reserves from the Swan Hills Formation. This, plus the reserves at Carson Creek and Kaybob South gas fields, accounted for a total of 4.512 trillion ft^3 initial raw gas in place to December 31, 1967. Conservation practices that require gathering most of the casinghead gas for marketing or reinjection into the formation result in an estimated initial marketable gas reserve of 1.097 trillion ft^3 from this source. Addition of the estimated marketable reserves from the two gas fields places the total for the area at 1.697 trillion ft^3.

Figure 25 shows the initial gas in place, marketable gas, and gas produced for each of the Swan Hills fields. Gas produced from the Carson Creek North field has been reinjected into the Carson Creek gas field and accounts for the negative value shown for marketable gas produced for that field.

Oil.—Total recoverable oil for all the fields producing from the Swan Hills Formation has

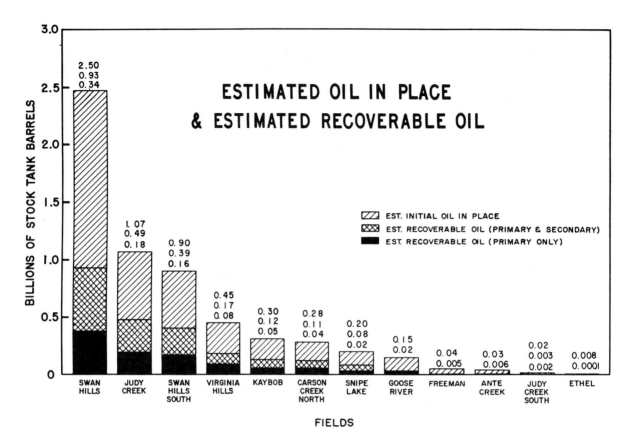

FIG. 26.—Estimated oil in place attributable to each of Swan Hills fields. Estimated recoverable oil by primary recovery and combination of primary- and secondary-recovery methods also is shown.

been estimated at 2.331 billion bbl to December 31, 1967. If only light- and medium-gravity crudes are considered, this figure amounts to 24.78 percent of the total known recoverable oil reserves for the Province of Alberta. The estimated oil in place attributable to each field, together with reserves expected to be produced by primary recovery and a combination of primary- and secondary-recovery methods, is shown in Figure 26.

The history of the petroleum industry in the Province of Alberta has been cyclical, characterized by a series of intermittent major discoveries. The effect of additional reserves from a major strike, such as that of the Swan Hills Formation, is shown in Figure 27.

ECONOMIC IMPACT

An estimate of the total funds spent within the map area after the initial discovery would require compiling figures on seismic work, road building, drilling and completion of wells, townsites, land acquisition, and pipelines. Though such a study is beyond the scope of this paper, it is interesting to examine the factor of land acquisition.

In Alberta, exclusive of early settlement areas, the Crown owns the petroleum and natural gas rights. These rights are disposed of by a system of closed bidding at periodic government sales. There were no freehold lands within the Swan Hills area. In the 2 years before the discovery, approximately $140,291 had been spent for the acquisition of petroleum and natural gas rights within the map area. In the years 1957–1967 this figure rose to $184,034,918.

SELECTED REFERENCES

Andrichuk, J. M., 1958, Stratigraphy and facies analysis of Upper Devonian reefs in Leduc, Stettler, and Redwater areas, Alberta: Am. Assoc. Petroleum Geologists Bull., v. 42, no. 1, p. 1–93.

Bassett, H. G., 1961, Devonian stratigraphy, central Mackenzie River region, Northwest Territories, Canada, in Geology of the Arctic, v. 1: Toronto, Toronto Univ. Press, p. 481–498.

——— and J. G. Stout, 1967, The Devonian stratigraphy of western Canada, in International Symposium on the Devonian System, v. 1: Alberta Soc. Petroleum Geologists, p. 717–752.

Beales, F. W., 1957, Bahamites and their significance in oil exploration: Alberta Soc. Petroleum Geologists Jour., v. 5, no. 10, p. 227–231.

——— 1958, Ancient sediments of Bahaman type:

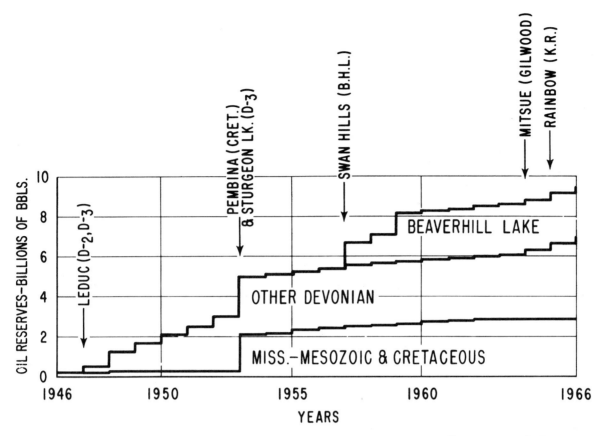

FIG. 27.—Alberta's cumulative ultimate proved oil reserves distributed according to year of discovery to Dec. 13, 1966. Share of Alberta's cumulative ultimate proved oil reserves attributable to Swan Hills fields is shown. By end of 1966 Swan Hills fields accounted for 25.87 percent of Alberta total.

Am. Assoc. Petroleum Geologists Bull., v. 42, no. 8, p. 1845–1880.

——— 1960, Limestone peels: Alberta Soc. Petroleum Geologists Jour., v. 8, no. 4, p. 132–135.

Beard, D. E., 1959, Selective solution in the Devonian Swan Hills Member: Alberta Soc. Petroleum Geologists Jour., v. 7, no. 7, p. 163–164.

Belyea, H. R., 1952, Notes on the Devonian System of the north central plains of Alberta: Canada Geol. Survey Paper 52–27, 66 p.

——— 1955, Correlations in the Devonian of southern Alberta: Alberta Soc. Petroleum Geologists Jour., v. 3, no. 9, p. 151–156.

——— 1964, Upper Devonian, pt. II, in Geological history of Western Canada: Alberta Soc. Petroleum Geologists, p. 66–81.

——— and A. W. Norris, 1962, Middle Devonian and older Paleozoic formations of southern District of Mackenzie and adjacent areas: Canada Geol. Survey Paper 62–15, 82 p.

Bonham-Carter, C. F., 1963, A study of microscopic components of the Swan Hills Devonian reef: Unpub. M.S. thesis, Toronto Univ.

Brown, P. R., 1963, Some algae from the Swan Hills reef: Bull. Canadian Petroleum Geology, v. 11, no. 2, p. 178–182.

Burwash, R. A., H. Baadsgaard, Z. E. Peterman, and G. H. Hunt, 1964, Precambrian, chap. 2, in Geological history of Western Canada: Alberta Soc. Petroleum Geologists, p. 14–19.

Cameron, A. E., 1918, Explorations in the vicinity of

Great Slave Lake: Canada Geol. Survey Summ. Rept., pt. C, 1917, p. 21–28.

——— 1922, Hay and Buffalo Rivers, Great Slave Lakes, and adjacent country: Canada Geol. Survey Summ. Rept., pt. B, 1921, p. 1–44.

Campbell, N., 1950, The Middle Devonian in the Pine Point area, N.W.T.: Geol. Assoc. Canada Proc., v. 33, p. 87–96.

Carozzi, A. V., 1961, Reef petrography in the Beaverhill Lake Formation, Upper Devonian, Swan Hills area, Alberta, Canada: Jour. Sed. Petrology, v. 31, no. 4, p. 497–513.

Century, J. R. (ed.), 1966, Oil fields of Alberta, supplement: Alberta Soc. Petroleum Geologists, 136 p.

Clark, D. L., and R. L. Ethington, 1965, Conodont biostratigraphy of part of the Devonian of the Alberta Rocky Mountains: Bull. Canadian Petroleum Geology, v. 13, no. 3, p. 382–389.

Crickmay, C. H., 1954, Paleontological correlation of Elk Point and equivalents, in Western Canada sedimentary basin: Am. Assoc. Petroleum Geologists, p. 143–158.

——— 1957, Elucidation of some Western Canada Devonian formations: Imperial Oil Ltd., unpub. rept.

De Mille, G., 1958, Pre-Mississippian history of the Peace River arch: Alberta Soc. Petroleum Geologists Jour., v. 6, no. 3, p. 61–68.

Edie, E. W., 1961, Devonian limestone reef reservoir, Swan Hills oil field, Alberta: Canadian Inst. Mining and Metallurgy Trans., v. 64, p. 278–285.

Fischbuch, N. R., 1960, Stromatoporoids of the Kaybob reef, Alberta: Alberta Soc. Petroleum Geologists Jour., v. 8, p. 113–131.

——— 1962, Stromatoporoid zones of the Kaybob reef, Alberta: Alberta Soc. Petroleum Geologists Jour., v. 10, no. 1, p. 62–72.

Folk, R. L., 1959, Practical petrographic classification of limestones: Am. Assoc. Petroleum Geologists Bull., v. 43, no. 1, p. 1–38.

Fong, G., 1959, Type section Swan Hills Member of the Beaverhill Lake Formation: Alberta Soc. Petroleum Geologists Jour., v. 7, no. 5, p. 95–108.

——— 1960, Geology of Devonian Beaverhill Lake Formation, Swan Hills area, Alberta, Canada: Am. Assoc. Petroleum Geologists Bull., v. 44, no. 2, p. 195–209.

Galloway, J. J., 1960, Devonian stromatoporoids from the lower Mackenzie Valley of Canada: Jour. Paleontology, v. 34, no. 4, p. 620–636.

Gray, F. F., and J. R. Kassube, 1963, Geology and stratigraphy of Clarke Lake gas field, northeastern British Columbia: Am. Assoc. Petroleum Geologists Bull., v. 47, no. 3, p. 467–483.

Grayston, L. D., D. F. Sherwin, and J. F. Allan, 1964, Middle Devonian, chap. 5, *in* Geological history of Western Canada: Alberta Soc. Petroleum Geologists, p. 49–59.

Griffin, D. L., 1965, The Devonian Slave Point, Beaverhill Lake, and Muskwa Formations of northeastern British Columbia and adjacent areas: British Columbia Dept. Mines and Petroleum Resources Bull., no. 50, 90 p.

Gussow, W. C., 1954, Differential entrapment of oil and gas, a fundamental principle: Am. Assoc. Petroleum Geologists Bull., v. 38, no. 5, p. 816–853.

Guthrie, D. C., 1956, Gilwood Sandstone in the Giroux Lake area, Alberta: Alberta Soc. Petroleum Geologists Jour., v. 4, no. 10, p. 227–231.

Hriskevich, M. E., 1966, Stratigraphy of Middle Devonian and older rocks of Banff Aquitaine Rainbow West 7–32 discovery well, Alberta: Bull. Canadian Petroleum Geology, v. 14, no. 2, p. 241–265.

Illing, L. V., 1959, Deposition and diagenesis of some upper Palaeozoic carbonate sediments in Western Canada: 5th World Petroleum Cong., Sec. 1, Paper 2, p. 23–52.

Imperial Oil Limited, Geological Staff, 1950, Devonian nomenclature in the Edmonton area, Alberta, Canada: Am. Assoc. Petroleum Geologists Bull., v. 34, no. 9, p. 1807–1825.

Jenik, A. J., 1965, Facies and geometry of the Swan Hills Member, Alberta: Unpub. M.Sc. thesis, Alberta Univ.

Koch, N. G., 1959, Correlation of the Devonian Swan Hills Member, Alberta: Unpub. M.Sc. thesis, Alberta Univ.

Kramers, J. W., and J. E. Lerbekmo, 1967, Petrology and mineralogy of Watt Mountain Formation, Mitsue-Nipisi area, Alberta: Bull. Canadian Petroleum Geology, v. 15, no. 3, p. 346–378.

Law, J., 1955, Rock units of northwestern Alberta: Alberta Soc. Petroleum Geologists Jour., v. 3, no. 6, p. 81–83.

Leavitt, E. M., 1966, The petrology, paleontology and geochemistry of the Carson Creek North reef complex, Alberta: Unpub. Ph.D. thesis, Alberta Univ.

LeBlanc, R. J., and J. G. Breeding (eds.), 1957, Regional aspects of carbonate deposition: Soc. Econ. Paleontologists and Mineralogists Spec. Pub. 5, 178 p.

Loranger, D. M., 1965, Devonian paleoecology of northeastern Alberta: Jour. Sed. Petrology, v. 35,

no. 4, p. 818–838.

McGehee, J. R., 1949, Pre-Waterways Paleozoic stratigraphy of Alberta plains: Am. Assoc. Petroleum Geologists Bull., v. 33, no. 4, p. 603–613.

McGill, P., 1966, Ostracods of probable late Givetian age from Slave Point Formation, Alberta: Bull. Canadian Petroleum Geology, v. 14, no. 1, p. 104–133.

McGrossan, R. G., and R. P. Glaister (eds.), 1964, Geological history of Western Canada: Alberta Soc. Petroleum Geologists, 232 p.

McLaren, D. J., and E. W. Mountjoy 1962, Alexo equivalents in the Jasper region, Alberta, Canada: Canada Geol. Survey Paper 62–63, 36 p.

Mound, M. C., 1966, Late Devonian conodonts from Alberta subsurface (abs.): Am. Assoc. Petroleum Geologists Bull., v. 50, no. 3, p. 628.

Moyer, G. L., 1964, Upper Devonian, pt. I, chap. 6, *in* Geological history of Western Canada: Alberta Soc. Petroleum Geologists, p. 60–66.

Murray, J. W., 1964, Some stratigraphic and paleoenvironmental aspects of the Swan Hills and Waterways Formation, Judy Creek, Alberta, Canada: Unpub. Ph.D. thesis, Princeton Univ.

——— 1966, An oil producing reef-fringed carbonate bank in the Upper Devonian Swan Hills Member, Judy Creek, Alberta: Bull. Canadian Petroleum Geology, v. 14, no. 1, p. 1–103.

Norris, A. W., 1963, Devonian stratigraphy of northeastern Alberta and northwestern Saskatchewan: Canada Geol. Survey Mem. 313, 168 p.

Oil and Gas Conservation Board, 1967, Pressure-depth and temperature-depth relationships, Alberta crude oil pools: OGCB Rep. 67–22.

——— 1968, Reserves of crude oil, gas, natural gas liquids, and sulphur, Province of Alberta: OGCB Rept. 68–18; 175 p.

Porter, J. W., and J. G. C. M. Fuller, 1964, Ordovician-Silurian, pt. 1, chap. 4, *in* Geological History of Western Canada: Alberta Soc. Petroleum Geologists, p. 34–42.

Russel, L. S., 1967, Palaeontology of the Swan Hills area, north-central Alberta: Toronto Univ. Press, Life Sci. Contr. no. 71, 31 p.

Stout, J. L., 1964, Pore geometry as related to carbonate stratigraphic traps: Am. Assoc. Petroleum Geologists Bull., v. 48, no. 3, p. 329–337.

Thomas, G. E., and H. S. Rhodes, 1961, Devonian limestone bank-atoll reservoirs of the Swan Hills area, Alberta: Alberta Soc. Petroleum Geologists Jour., v. 9, no. 2, p. 29–38.

Uyeno, T. T., 1967, Conodont zonation, Waterways Formation (Upper Devonian), northeastern and central Alberta: Canada Geol. Survey Paper 67–30, 20 p.

Van Hees, H., 1958, The Meadow Lake escarpment— its regional significance to lower Paleozoic stratigraphy, *in* 1st Internat. Williston Basin Symposium: North Dakota Geol. Soc. and Saskatchewan Geol. Soc., p. 131–139.

——— and F. K. North, 1964, Cambrian, chap. 3, *in* Geological History of Western Canada: Alberta Soc. Petroleum Geologists, p. 20–33.

Walker, C. T., 1957, Correlations of Middle Devonian rocks in western Saskatchewan: Saskatchewan Dept. Mineral Resources Rept. 25, 59 p.

Warren, P. S., 1933, The age of the Devonian limestone at McMurray, Alberta: Canadian Field-Naturalist, v. 47, no. 8, p. 148–149.

——— 1957, The Slave Point Formation: Edmonton Geol. Soc. Quart., v. 1, no. 1, p. 1–2.

White, R. J. (ed.), 1960, Oil fields of Alberta: Alberta Soc. Petroleum Geologists, 272 p.

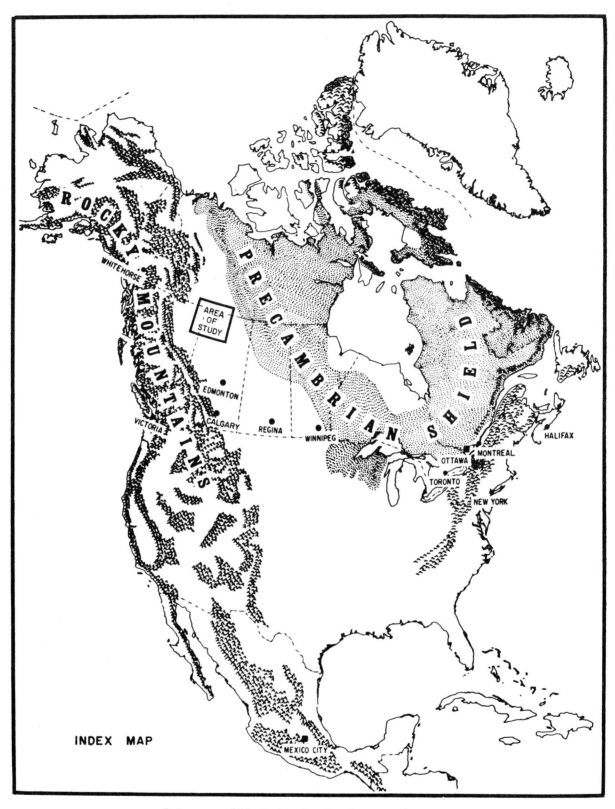

INDEX MAP

Index map of North America showing area of study.

American Association of Petroleum Geologists Memoir 14, *Geology of Giant Petroleum Fields,* copyright 1970, pp. 19-49.

Geology of Middle Devonian Reefs, Rainbow Area, Alberta, Canada[1]

D. L. BARSS,[2] A. B. COPLAND,[2] and W. D. RITCHIE[2]

Calgary, Alberta

Abstract Data obtained from exploration for hydro-carbon-bearing Middle Devonian Rainbow Member reefs in northern Alberta, Canada, provides an excellent opportunity to examine the regional geological history of the Black Creek basin and the evolution of varied reef forms within the Rainbow portion of this basin.

During pre-Middle Devonian and Middle Devonian time, a cyclical sequence of redbeds (clastics and evaporites) of the Lower Elk Point subgroup were deposited in a shallow epicontinental sea. The incipient development of the Black Creek basin occurred at this time. Subsequently, negative epeirogenic movement resulted in widespread deposition of fine-grained, dark carbonate rocks of the Lower Keg River Member. Local faunal changes, as well as local changes in thickness (interpreted to be caused by "lime-mud" mounds), occur in the upper part of this unit. During the time of Upper Keg River deposition crinoidal beds were deposited in the form of a bank about 50 ft thick. The crinoidal bank extended throughout and beyond the Rainbow part of the Black Creek basin. In this bank, reef-constructing organisms flourished in several localities. The loci for concentration of reef organisms which led to rapid Rainbow Member reef growth, are believed to be the mud-mound topographic highs that were present in the underlying Lower Keg River Member. Structural control of reef growth, if it did exist, was subtle.

The reefs that grew in the Rainbow subbasin are characterized by pinnacle and atoll forms having vertical relief of up to 820 ft. Relatively rapid basin subsidence, combined with directional aspect of climate, paleography of the sea floor, and local bathymetry, controlled the external geometry, and to some extent, the internal facies of the reefs.

Detailed lithologic studies reveal 14 facies representing six depositional environments—basin, bank, forereef, organic reef, backreef, and lagoon. Superimposed on the original facies is a variable diagenetic history.

The growth of Shekilie barrier-reef complex across northwestern Alberta and adjacent areas of British Columbia and the Northwest Territories, and regional tectonic movements, altered depositional patterns late in Elk Point sedimentation. The barrier formed by the reef complex and structure prevented the free flow of normal marine waters southeast into the Black Creek basin. In this basin salinity of the water increased and the Black Creek Member salt was deposited, followed by the Muskeg anhydrites. The Muskeg evaporites completely infilled the Black Creek basin and covered the Rainbow Member reefs except for those present in the Shekilie barrier complex. The evaporite cover provided an effective seal for hydrocarbon entrapment. Reserves from the Rainbow Member pools in the Rainbow field and the Rainbow South field are estimated to be in excess of 1.2 billion and 165 million bbl of oil-in-place, respectively.

INTRODUCTION

The importance of Upper Devonian carbonates in providing reservoirs for hydrocarbon accumulation in Western Canada is well known. This paper reports on another Devonian carbonate reservoir—the Rainbow Member—which was discovered early in 1965. The stratigraphic relationship and distribution of significant Devonian units is shown on Figure 1. Reefs of Upper Devonian Leduc Formation occur in a basin covering about 70,000 mi² in central Alberta and contain in the order of 4.6 billion bbl of oil-in-place and 9.75 trillion cu ft of gas. The Swan Hills reefs of Middle-Late Devonian (Beaverhill Lake) age occur in a basin about 27,000 mi² in area and contain about 6.1 billion bbl of oil-in-place and 9 trillion cu ft of gas. The Rainbow Member reefs occur within the Black Creek basin, an area of about 11,000 mi², and contain in excess of 1.5 billion bbl of oil-in-place and 1 trillion cu ft of gas.[3] The significance of the recent discovery at Rainbow is evident from its large hydrocarbon reserves, and from the fact that a discovery of this kind leads to new cycles of exploration and discoveries in other strata. Regional studies indicate that reefs of equivalent age are present in an unexplored area exceeding 156,000 mi² in Western Canada.

The Rainbow Member reefs are of late Middle Devonian (Givetian) age. In the Rainbow fields area in the Black Creek basin they develop both a pinnacle and atoll form with a height up to 820 ft and an areal dimension up to approximately 6 mi². In several wells the entire reef sections were cored throughout and many wells penetrated not only the reef section but also the pre-reef beds. The good quality of the data available has provided an excellent opportunity to study the lithology of the Rainbow

[1] Modified from a paper read by D. L. Barss before the 53d Annual Meeting of the Association, Giants Symposium, Oklahoma City, Oklahoma, April 24, 1968. Manuscript received, February 6, 1969.

[2] Banff Oil Ltd.

The authors thank G. E. Chin, M. E. Hriskevich, J. R. Langton, and S. Machielse for their assistance and constructive criticism. Acknowledgment is also due the Banff Oil Ltd. Drafting Department for their cooperation in preparation of the illustrations.

[3] Reserve figures from Canadian Petroleum Association 1967 Statistical Year Book, April 1968.

D. L. Barss, A. B. Copland, and W. D. Ritchie

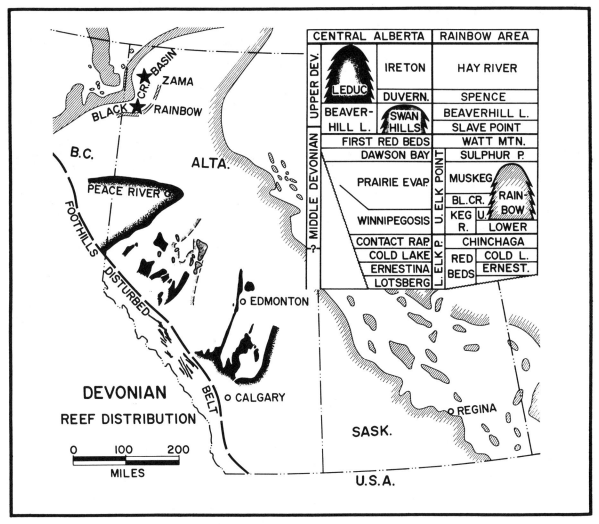

Fig. 1.—Devonian-reef distribution in Western Canada.

Member reefs and the pre-reef geological history.

This paper is divided into three parts—(1) regional geological history prior to, during, and after the time of Keg River deposition; (2) discussion of the inception of reef growth and forms, details of the lithofacies and the effects of post-reef salt solution, normal faults, and compaction; (3) summary of exploration highlights, reservoir factors, and reserves in the Rainbow and Rainbow South fields.

Terminology

The terms "reef," "reef complex," and "organic reef" have been defined by Klovan (1964) and appear to have been used in a consistent manner by recent authors. Klovan (1964), Lowenstam (1959), Cloud (1952), and Nelson *et al.* (1962), all stress the importance of the wave-resistant structure produced by reef organisms. Recently, Stanton (1967) suggested that important independent variables controlling the external form and facies of reefs are directional aspect of climate, paleography of the sea floor, and local bathymetry. He believes that these physiographic factors may exert more control on reef growth than biological factors. A study of the different external forms and facies present in the Leduc, Swan Hills, and Rainbow reef complexes—units that contain very similar organic-reef faunas—indicates that physiographic factors are significant. In addition, rate of basin subsidence may have had an important influence on the external form, and on facies.

Although physiography and subsidence were critical to the external form of the reef complex and, to some extent, the internal facies, we believe that the presence of sediment-binding and carbonate-producing organisms was of prime importance. Also, the study of the distribution of the fauna and detrital components within ancient reefs is a necessary prerequisite to reconstruction of environment of deposition and, as well, to a knowledge of reservoir characteristics. The writers have found the following definitions by Klovan (1964) to be the most satisfactory and have used them in the text.

Reef   Rigid carbonate structure with vertical dimensions significantly larger than the contemporaneous sediments, composed, at least in part, of organisms able to build and maintain the structure as a topographic feature on the sea floor and potentially in the zone of wave action.

Organic reef   That portion of the reef which is or was built directly by organisms and is responsible for the reef's wave resistant character.

Reef complex   The aggregate of reef limestone and related carbonate rocks.

Other definitions concerning reef terminology that are used in this paper are those of Langton and Chin (1968):

Pinnacle reef   Reef or reef complex as defined above, which developed without a backreef or lagoonal facies.

Atoll reef   An organic reef or reef complex which partly or completely surrounds lagoonal and backreef carbonate facies.

Although reef terminology is applied in a reasonably consistent manner by most authors, the conceptual views held by various authors of structures referred to as *banks* and *mounds* are not so clear at the present time. The definition of these terms is important in this paper as the writers examine carbonate masses of diverse shape, organic content, and possible origin.

Grabau (1913) described *banks* as structures formed by growth of shell colonies with little or no relief above the sea floor. He noted that sediments of the bank are interstratified with surrounding sediments. Cummings (1932) used a similar definition to define *shell banks* and recognized their difference from reefs. Lowenstam (1959, p. 433, 434) expanded on Cummings' concept; he differentiated well-defined low-lying structures comprised of unconsolidated banks of sponge, pelecypod, brachiopod, and crinoid material, from true reefs on the basis of the ecologic potential of the biota.

Thus, banks result from accumulation of bioclastic debris where the biota plays a passive role. Lowenstam noted that because of the passive role of the organisms, they do not grow above wave base and, above all, ". . . they lack the regenerative power to deposit carbonate in excess of loss caused by physical forces." Nelson *et al.* (1962, p. 234) followed Lowenstam's definition essentially, describing a ". . .*bank* as a skeletal limestone deposit formed by organisms which do not have the ecologic potential to erect a rigid, wave-resistant structure." Klement (1968) used a definition similar to that of Lowenstam. He indicated that banks are formed by organisms which do not have the ability to build a rigid, three-dimensional frame. He differentiated those banks formed by transport and accumulation of organisms from the in-place accumulations. In the latter group a three-fold subdivision is made into biogenic banks formed—by baffling action of the organisms, by binding action of organisms, and by localized growth of organisms.

Although it may be possible to recognize Klement's subdivisions in modern banks, their identification in ancient beds is not always possible. Moreover, banks may evolve into reefs, or intermittent development of reefs in predominantly bank-type structures may occur. As described by Walther and cited in Lowenstam (1959), an example of growth of sediment-binding organisms on a bank is present on the Pidgeon Bank in the Gulf of Naples. At this locality, calcareous algae are encrusted on Foraminifera-bearing sands. Lowenstam, in discussing this example, mentions that transition of banks to reefs may occur intermittently, and is dependent upon the degree of stability. At Rainbow, crinoidal banks generally underlie the reef, but in some localities stromatoporoid and coral fauna are abundant in the bank. Where this occurs, the concept of evolution from bank to organic-reef as expressed by Lowenstam, appears to be valid. The presence of facies and structures that are transitional from bank to reef does not present problems in terminology at Rainbow. The attributes of "reefs" have been examined previously. The widespread crinoidal unit at the base of the Upper Keg River Member[4] provides a good example of "bank" characteristics. The definition of this term is as follows:

[4] Local usage in stratigraphic nomenclature is followed in this article rather than strict application of rules of the American Commission on Stratigraphic Nomenclature.—Ed.

Bank A carbonate structure formed by organisms which do not build and maintain a structure on the sea floor in the zone of wave action. Sediments of the bank may be interstratified with sediments of either the basin or other carbonate masses and may or may not have topographic expression on the sea floor.

In the present study, several criteria are characteristic of bank sediments—massive encrusting fauna are rare or absent; typical faunas include crinoids, brachiopods, bryozoa, gastropods, branching corals, and *Stromatactis;* the accumulation of carbonate debris, particularly crinoid stems, is determined more by the natural breakdown of the organisms than by the forces of erosion; stratification with laterally equivalent nonbank beds is common and implies that the banks had little or no topographic relief.

Another type of carbonate structure is the *mud mound.* Carbonate structures that fall into this category have been described in the Mississippian of New Mexico by Pray (1958), at the base of the Niagaran reefs in Indiana by Textoris and Carozzi (1964), and in Devonian reefs of the Spanish Sahara by Dumestre and Illing (1967). The Mississippian mounds of Pray range from 25 to 350 ft in thickness and have flank dips up to 35°. The core facies comprises "lime mud"[5] and sparry calcite which passes abruptly into a flank facies of coarse crinoidal material. Fenestrate bryozoans and crinoids are the main faunal components of the core. The bryozoans occur in growth position and as fragmented debris, generally comprising less than 20 percent of the rock. Crinoids are present but in very small amount. Convincing evidence shows that the largest of these mounds had considerable topographic relief at the end of the period of growth. However, they probably did not grow into the zone of vigorous wave action.

The subreef beds of the Niagaran reefs consist of argillaceous calcisiltite which is fossiliferous near the reef structures. According to Textoris and Carozzi (1964, p. 423), the mounds started by "An apparently fortuitous bioclastic accumulation, mostly of crinoids" which "dilutes the calcite mud to form a mound of fossiliferous calcisiltite below wave base." With the addition of fistuliporid bryozoans and *Stromatactis,* the mound continued

[5] The term "lime mud," although slang, has been established in the literature, and is used herein to mean $CaCO_3$ mud.

to grow to wave base. At this stage construction of a reef frame began as a result of an increase in abundance of stromatoporoids.

The "T-bone" Middle Devonian reefs of the Spanish Sahara are thought by Dumestre and Illing (1967) to be mud mounds. These mounds are 60–90 ft thick and show depositional dips on the flanks of up to 34°. They consist essentially of "lime mud"; fauna is comprised of *Stromatactis,* auliporoid corals, finger corals, and crinoid ossicles. The mud mounds are believed to be the loci of later Middle Devonian reef growth.

The processes involved in the growth of mud mounds is not entirely clear. In Pray's (1958) view the fenestrate bryozoans in the New Mexico mounds played a significant role either by forming a current-baffle sediment trap or by forming a mesh-like fragment-constituted sediment-retaining mat. Pray suggested that the "lime mud" is indigenous and favors an algal origin for it even though distinct algal structures have not been recognized. Textoris and Carozzi (1964) favor a physio-chemical origin for the "lime mud" in the Niagaran mounds. The mounds grew simply by accumulating more bioclastic material than the surrounding areas. Binding of the "lime mud" into calcisiltite, they suggest, occurred through simple penecontemporaneous carbonate recrystallization.

A mud mound could therefore be defined as a mud-supported, fossiliferous, carbonate structure. The included organisms themselves do not have the ability to construct a rigid wave-resistant frame and for this reason mounds are believed to have grown mainly below the zone of vigorous wave action. Mud mounds have definite, though in some examples subtle, topographic relief. They are distinguished from banks in that they are mud supported, commonly have topographic relief, and have pronounced flank dips. The muds are generally somewhat argillaceous. Although the faunas present in mud mounds and banks are similar, they comprise a small percentage of mounds as compared with banks.

Local thickening of 20–50 ft in Lower Keg River strata is known beneath several reefs of the Rainbow area. Detailed log correlation has shown that in the 7–32–109–8 W6M well in the "A" pool (Fig. 11) a thickening of about 35 ft occurs in the upper part of the Lower Keg River. In cores of the Lower Keg River in different wells, dips up to 15° have been observed. The faunal components include mainly crinoid material with minor bryozoans, styliolina, and gastropods, and rarely, branching corals and stromatoporoids. These faunas are present in argillaceous calcilutite and calcisiltite and rarely exceed 20 percent of the total rock volume. The evidence at hand favors the view that at least some of these Lower Keg River "thicks" are mud mounds and that they provided favorable sites for the inception of reef growth.

REGIONAL SETTING

Precambrian

Granite and hornblende gneisses, biotite granites, quartz diorites, and metasediments form the Precambrian rocks which constitute the "basement" in this region (Fig. 2). The surface on which sedimentation first began was irregular owing to the topographic expression of major tectonic trends. The three most important tectonic features are—the Tathlina high at the northern edge of the map sheet; the Peace River high in the south; and the Hay River-East Arm fault zone which crosses the area from northeast to southwest. Another feature, the Shekilie fault zone, crosses the western part of the area from northeast to southwest but its subsurface definition is poor. Other major faults are present but there is insufficient subsurface control to establish whether or not they are in trends. Burwash (1957) indicated that there are three Precambrian provinces present in northern Alberta.

Lower Elk Point Subgroup

The Lower Elk Point subgroup comprises a red, evaporitic and clastic sequence ranging from about 500 ft in the deepest part of the basin to zero where these sediments lap onto the Peace River high. Thickness of the Lower Elk Point is less than 50 ft over the Tathlina high (Fig. 3). The Lower Elk Point sediments overlie Precambrian "basement" except in the western part where they overlie Cambrian clastic rocks. The general basin configuration during the time of deposition of the Lower Elk Point persisted into Upper Elk Point deposition.

The Lower Elk Point subgroup comprises the following units in ascending order—"Basal Red Beds," Ernestina, Cold Lake, and Chinchaga (Fig. 4). The age of the Chinchaga Formation has been established by Law (1955) as Middle Devonian. The age of older units is not known, owing to lack of fossils. The Cold Lake and Ernestina Formations probably correlate with beds of the Mirage Point Formation

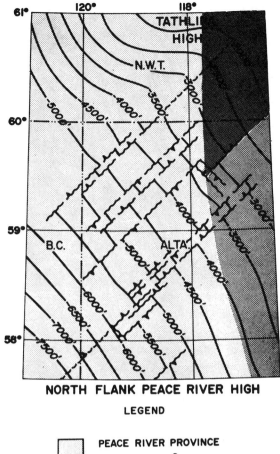

NORTH FLANK PEACE RIVER HIGH

LEGEND

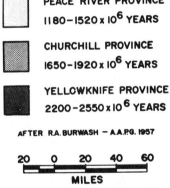

PEACE RIVER PROVINCE
1180–1520 x 10⁶ YEARS

CHURCHILL PROVINCE
1650–1920 x 10⁶ YEARS

YELLOWKNIFE PROVINCE
2200–2550 x 10⁶ YEARS

AFTER R.A. BURWASH — A.A.P.G. 1957

20 0 20 40 60
MILES

FIG. 2.—Precambrian structure map.

which was considered by Norris (1965) to be Middle to Late Ordovician. More recently, Norris (personal commun.) and others have indicated that these beds may be of Middle and possibly Early Devonian age. Near the Tathlina high in the north and the Peace River high in the south, these units grade to coarse clastic deposits. Where definition of the individual units cannot be recognized, the sequence is referred to collectively as the "Elk Point sands."

The Lower Elk Point subgroup comprises several cycles of red clastic and evaporite de-

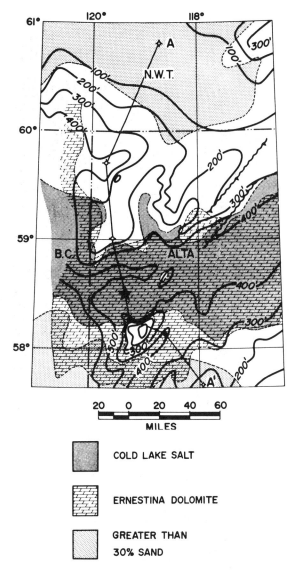

FIG. 3.—Isopach map of Lower Elk Point subgroup. **A-A'** is trace of section in Figure 4. CI = 100 ft.

posits. These sediments were deposited in a widespread, shallow, epicontinental sea during deposition of which periodic restriction or shallowing caused development of evaporitic basins.

The first cycle consists of the "Basal Red Beds"—a red silt and shale unit with, locally, clear quartz "Granite Wash" type sandstone—overlain by red anhydrite and dense, finely crystalline dolomite of the Ernestina Formation. The maximum thickness of the Ernestina is 50 ft. Although relatively thin, it is present across a large area of northern Alberta. The distribution of this unit clearly marks the trend of the east-west Northern Alberta basin (Fig. 3); this trend persisted throughout the deposi-

tion of the rest of Middle Devonian sediments in northern Alberta.

The second cycle of the Lower Elk Point subgroup—the Cold Lake Formation—consists of a thin basal red silt and shale unit overlain by salt which ranges in thickness from 0 to 180 ft. The edge of the Cold Lake is roughly coincident with that of the underlying Ernestina Formation. There are small thickness changes in the Cold Lake Formation; to date, however, there has been no direct evidence of salt solution having taken place.

The third major cycle—the Chinchaga Formation—consists of a basal red shale and silt unit overlain by anhydrite, dolomite, and clastic rock. The anhydrite and dolomite grade into coarse clastic rock toward the Peace River high and toward the Tathlina high (Fig. 4). A clastic cycle can be recognized in the middle part of the Chinchaga and can be correlated regionally; its base is the boundary between the upper and lower members of the Chinchaga Formation as defined by Belyea and Norris (1962).

The regional deposition of the Chinchaga or equivalent strata indicates clearly that a major submergence of this northern region occurred. At the end of the Lower Elk Point deposition, the entire map area was one of very low relief. The general lack of lithologic or thickness changes of the individual units indicates little tectonic activity during deposition of the Lower Elk Point sediments; the only features of prominence were topographic highs that were present when sedimentation in this area began.

Upper Elk Point Subgroup

The Upper Elk Point subgroup contains, in ascending order—Keg River, Muskeg, Sulphur Point and Watt Mountain Formations (Fig. 5). The rocks of the Upper Elk Point subgroup are characterized by numerous facies changes; these can be related primarily to the occurrence of reef growth during the time of Keg River deposition. However, structural movement may have contributed to changes in basin configuration and depositional patterns. The Upper Elk Point subgroup ranges in thickness from about 1,100 ft in the central part of the basin to less than 100 ft over the Tathlina high (Fig. 6). Upper Elk Point strata onlap the Peace River high on the south.

Keg River Formation

The Keg River Formation is subdivided into lower and upper members which are readily recognized throughout the map area. Further

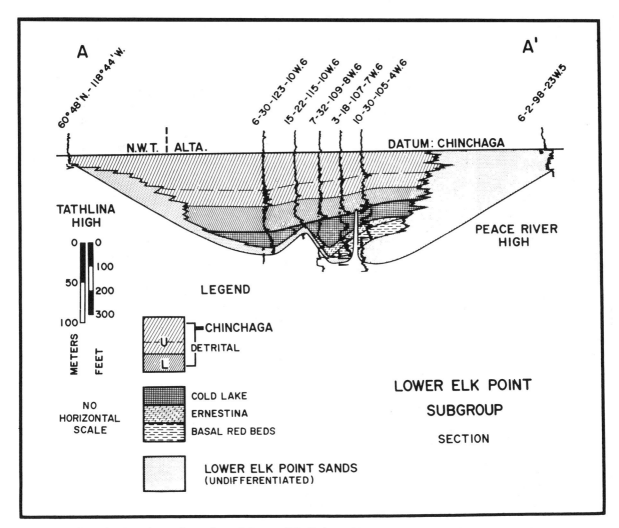

FIG. 4.—North-south section of Lower Elk Point subgroup. Trace is **A-A'** on Figure 3.

subdivision of the lower member into two units, and the upper member into three, is practical within the Rainbow subbasin.

Lower Keg River Member.[4a]—In the Rainbow subbasin the Lower Keg River Member ranges in thickness from 110 to 170 ft (Figs. 7 and 13). It has been divided into a lower and an upper unit by Hriskevich (1966). The lower unit consists of about 70 ft of dark bituminous, argillaceous, very fine-grained limestone and dolomite which contain variable amounts of crinoid debris, sporadic corals, and brachiopods; stromatoporoids are present in minor quantity. Dolomitization appears generally in the lower part of the unit. The basal contact of the lower unit of the Lower Keg River Member is generally recognized by the change from anhydrite to dolomite or limestone. However, in

[4a] *See* footnote 4, p. 21.

some wells a thin primary dolomite is present above the anhydrite of the Chinchaga, and although the age of this dolomite is not clear, it is included with the Chinchaga Formation in this paper. The upper unit comprises beds of dark gray, black, bituminous, argillaceous, very fine-grained limestone with scattered layers of crinoids and brachiopods.

The bituminous, argillaceous limestones of the Lower Keg River or equivalent units are generally less than 150 ft thick but are present in a broad area of Western Canada sedimentary basin. The widespread nature of the carbonate rocks indicates that negative epeirogenic movement followed deposition of the Chinchaga evaporites. The bituminous content in the limestones suggest euxinic conditions—possibly produced by poor circulation on a continent-wide shelf.

Upper Keg River Member.—Within the

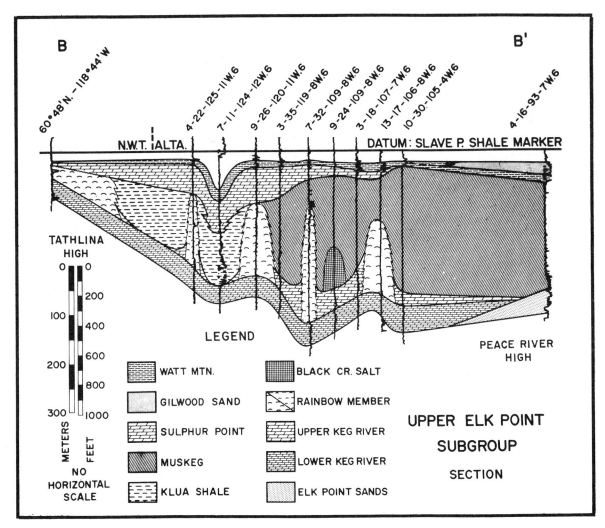

FIG. 5.—North-south section of Upper Elk Point subgroup. Trace is **B-B'** on Figure 6.

Black Creek basin, the Upper Keg River Member ranges from less than 40 ft west of the Zama area to about 150 ft in the Rainbow subbasin. Northwest of the basin, in British Columbia and the Northwest Territories, and east of the basin in the Fort McMurray area, there is evidence that Upper Keg River beds were not deposited.

Hriskevich (1966) proposed a threefold subdivision of the Upper Keg River. Correlations of these units is readily possible throughout the Rainbow subbasin and in some other parts of the Black Creek basin. The lower unit consists of black to light-brown carbonate. It is generally dolomitized and contains a fauna comprising crinoid stems and fragments, corals, brachiopods, *Stromatactis,* and minor stromatoporoids. The lower unit ranges in thickness from zero where organic-reef facies are present, to about 50 ft in offreef sections. The middle unit

is about 50 ft thick and consists of interbedded dark, very bituminous and argillaceous, very fine-grained limestone and very fine-grained, slightly bituminous and argillaceous limestone. Tentaculites is the only significant fossil. The upper unit, also approximately 50 ft thick, is similar lithologically to the middle unit except that the bituminous and argillaceous content is less; laminated bedding is common. Fossils are sparce in the upper unit, with only a few isolated Amphipora and thin-shelled brachiopods present.

At some localities abundant massive stromatoporoids and corals became established in the lowermost unit of the Upper Keg River Member, and a wave-resistant structure was formed. These structures subsequently developed into Rainbow Member atoll and pinnacle-reef complexes. Figure 8 shows the regional distribution of reef complexes. The north-trending reef complex in the western part of the area is termed the "Shekilie barrier." This barrier-reef complex restricted circulation of water in post-reef time with the result that evaporites were deposited southeast of it. Normal marine shales were deposited on the west. The evaporites are included in the Muskeg Formation and the shales in the Klua Formation. East of the Rainbow locality another barrier-reef trend, referred to as the Hay River barrier by McCamis and Griffith (1967), is believed to be present. Definition of this barrier-reef trend is not well documented by subsurface control, but its presence is indicated by the distribution of the Black Creek salt and marked differences in anhydrite content of the lower part of the Muskeg Formation, between wells in the Rainbow area and wells southeast of the postulated barrier. The Black Creek basin, bordered by these regional barrier reefs, is the area in which pinnacle and atoll reefs grew. Figures 8 and 9 illustrate the positions of the barriers enclosing

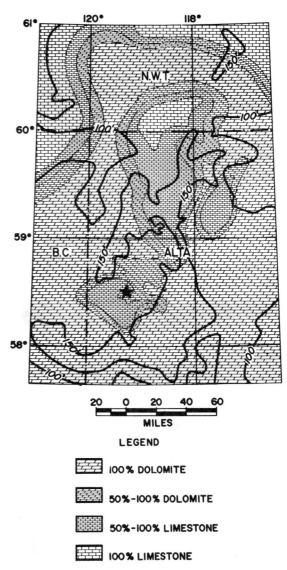

FIG. 7.—Isopach and lithofacies map of Lower Keg River Member. CI = 50 ft.

the Black Creek basin and the stable-shelf carbonates east of the Hay River barrier.

Muskeg Formation

In offreef wells in the Rainbow subbasin, the Muskeg Formation is readily subdivided into the Black Creek Member consisting of salt with minor anhydrites; and an unnamed upper member consisting of interbedded anhydrites, primary dolomites and secondary, locally porous dolomite of organic and detrital origin.

The Black Creek salt ranges locally in thickness from a maximum of 271 ft to 0, owing to solution and removal. At the base of the Black Creek salt, two thin anhydrite beds, 10–15 ft thick, are generally present and directly overlie the limestone of the Upper Keg River Member.

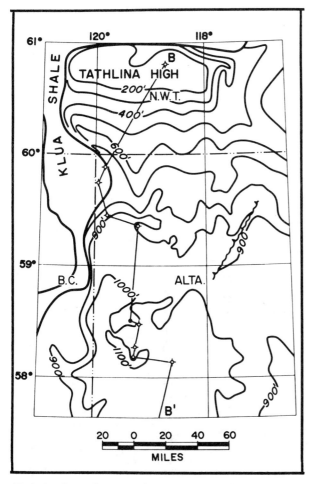

FIG. 6.—Isopach map of Upper Elk Point subgroup. **B-B'** is trace of section in Figure 5. CI = 100 ft.

D. L. Barss, A. B. Copland, and W. D. Ritchie

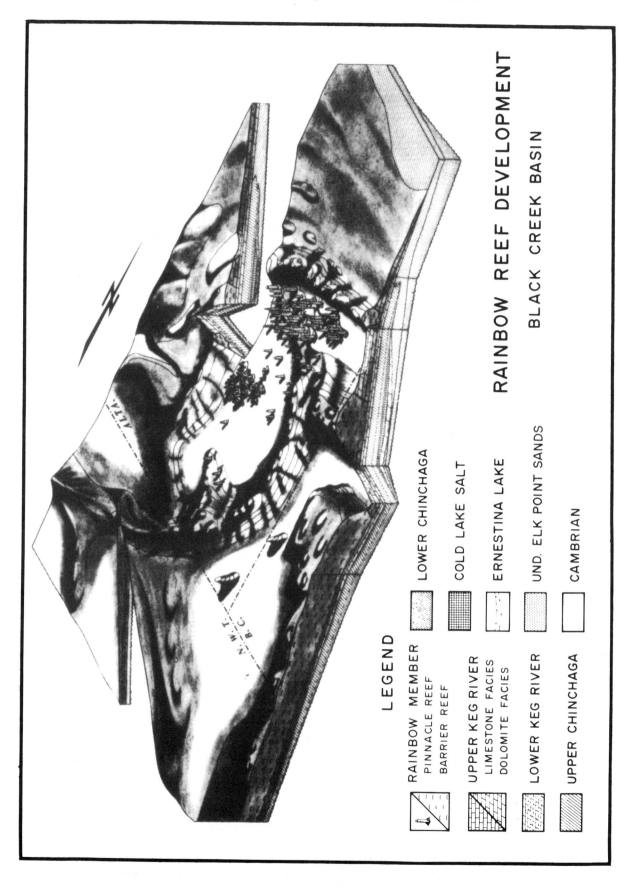

LEGEND

RAINBOW MEMBER
PINNACLE REEF
BARRIER REEF

UPPER KEG RIVER
LIMESTONE FACIES
DOLOMITE FACIES

LOWER KEG RIVER

UPPER CHINCHAGA

LOWER CHINCHAGA

COLD LAKE SALT

ERNESTINA LAKE

UND. ELK POINT SANDS

CAMBRIAN

RAINBOW REEF DEVELOPMENT

BLACK CREEK BASIN

Fig. 8.—Rainbow reef development in Black Creek basin.

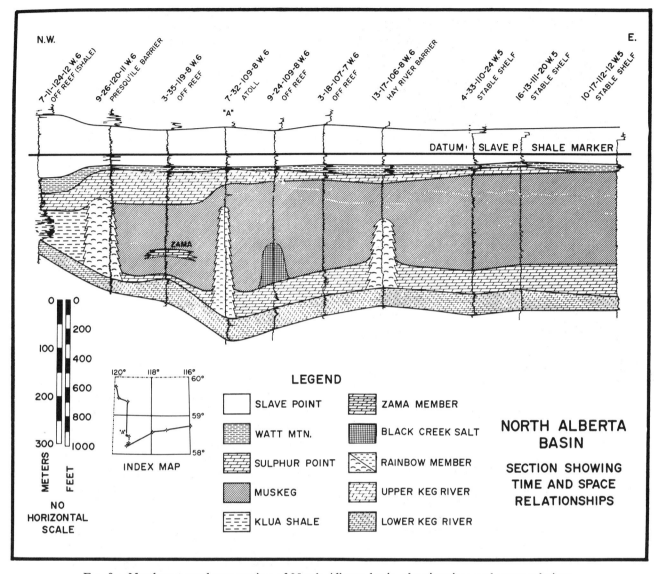

FIG. 9.—Northwest-southeast section of North Alberta basin, showing time and space relations. "A" well is in "A" pool, Figure 13.

The contact between the Black Creek Member and the overlying anhydrite is sharp.

In the Zama part of the Black Creek basin, the Muskeg sequence has been subdivided, in ascending order, into the Black Creek, Lower Anhydrite, Zama, Upper Anhydrite and Bistcho Members by McCamis and Griffith (1967). Black Creek salt, as much as 124 ft thick, is established by well control in the Zama area, and salt ranging in thickness from 0 to 170 ft is also present further to the northeast in the Steen River area. The Lower and Upper Anhydrite Members of the Muskeg Formation are separated by porous, brown, laminated, fine- to medium-crystalline saccharoidal dolomites of the Zama Member which range up to 110 ft thick. The Zama Member lies directly on the high part of reef buildups and overlies the Lower Anhydrite Member in the off-reef areas as illustrated by McCamis and Griffith

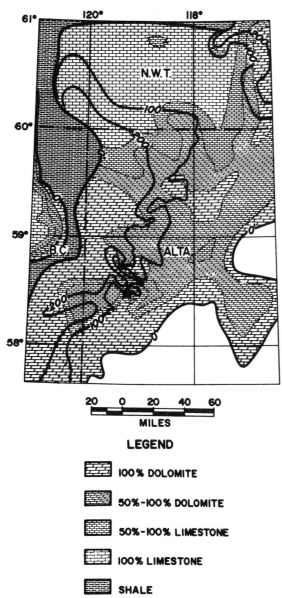

20   0   20   40   60

MILES

LEGEND

100% DOLOMITE

50%-100% DOLOMITE

50%-100% LIMESTONE

100% LIMESTONE

SHALE

Fig. 10.—Isopach and lithofacies map of Sulphur Point Formation. CI = 100 ft.

(1967, Fig. 20). Fossils are sparse and consist of amphiporoids, stromatoporoids, crinoids, and brachiopods. The Zama Member is not recognized in the Rainbow subbasin.

Sulphur Point Formation

McCamis and Griffith (1967) have introduced the name Bistcho Member to include the beds which previously were included in the Sulphur Point and Presqu'ile Formations. The Bistcho Member is considered by them to be a facies variant of part of the Muskeg evaporites.

The writers favor retention of the name Sulphur Point for the following reasons—the name and type section of the Sulphur Point

Formation was established by Norris (1965); the name has priority and has gained wide acceptance within the industry; and the upper and lower contacts are easily identified on lithologic and mechanical logs in a wide area of northern Alberta, northeast British Columbia, and the Northwest Territories. As there is evidence of a disconformity between the Sulphur Point Formation and underlying Muskeg Formation, we conclude that the Sulphur Point carbonate rocks were deposited subsequent to the Muskeg evaporites and are not a facies variant of the upper part of the Muskeg Formation.

The name Presqu'ile is not used consistantly by different authors in the area, particularly in subsurface correlations. Generally, the name refers to the dolomitized part of the Sulphur Point Formation. When used in this sense, the upper and lower contacts of the dolomites are commonly poorly defined. For these reasons, we recommend discontinuing use of the name Presqu'ile.

The Sulphur Point Formation ranges in thickness from zero along its eastern edge, to more than 250 ft in the western part of the map area (Fig. 10). The lower contact with the anhydrite of the Muskeg is sharp. The presence of a disconformity is indicated by the relations of the lower contact of the Sulphur Point with the Muskeg as shown in Figure 23. At the top of the Muskeg Formation the presence of an extra unit in a low reef-flank position (4–9 well, Fig. 23) may indicate compaction in the underlying Muskeg beds or possibly erosion before deposition of the Sulphur Point. Abrupt local changes in thickness of the Sulphur Point, as seen in several wells in the Rainbow locality, also indicate the presence of a disconformity at the base of the Sulphur Point. It is possible, however, that these local changes in thickness reflect normal fault movements following Muskeg deposition or during deposition of the Sulphur Point. The upper contact with the Watt Mountain is placed at the first prominent occurrence of green shale. The presence of irregular and rounded limestone fragments at this contact suggests a disconformity between the Watt Mountain and the Sulphur Point.

The Sulphur Point Formation consists of fine- to medium-grained limestones with biota consisting of scattered Amphipora, calcispheres, gastropods, and algae. Although regional dolomitization trends can be mapped (Fig. 10), dolomitization is variable in local

224

detail. Intergranular and vuggy porosity ranges from poor to very good. The Sulphur Point unit is a secondary, but important reservoir in northern Alberta.

Watt Mountain Formation

The Watt Mountain Formation is 10–20 ft thick locally, and ranges up to a maximum of 65 ft. It consists of apple-green and gray shale and very fine-grained, dense limestone. There is a marked increase in the relative amount of clear-quartz sandstone and siltstone in the formation as the Peace River high is approached.

RAINBOW AREA

Inception of Reef Growth

In Western Canada detailed geologic information on the incipient stage of reef growth

generally has been very meager. As a result there have been different opinions expressed on whether structure or sedimentary features, or both, have been responsible for initiation of reef growth. The entire reef section has been cored in many wells in the Rainbow field and the pre-Rainbow Member-reef beds have been drilled in many other wells. The subsurface data that have been obtained are pertinent to questions concerned with initial stages of reef growth.

In the Rainbow area there was an abundance of organisms having the ecologic potential to construct rigid wave-resistant structures that grew on the Lower Keg River surface. In the 7–32–109–8 W6 well, for example, massive stromatoporoids and corals are well developed in beds of the basal crinoidal bank of the

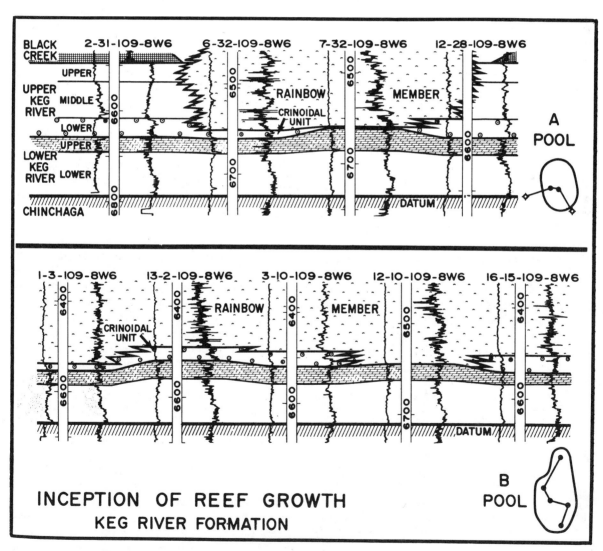

FIG. 11.—Inception of reef growth in Keg River Formation, Rainbow area.
Location of "A" and "B" pools is on Fig. 13.

Upper Keg River Member (Fig. 11). Once the massive stromatoporoids and corals became established they grew upward and outward into the present form of the reef structure. The crinoidal bank contains, in addition to abundant crinoid stems and fragments, numerous branching and massive tabulate corals, even in wells not overlain by reefs. This is evident at the 2–28–109–8 W6 well about 1 mi south of the "A" pool. The presence of a fauna with the ecologic potential to construct a wave-resistant structure at this nonreef well indicates that factor(s) other than presence of reef-building organisms are critical in triggering reef growth at specific locations.

Another possible explanation for localizing reef growth is structural highs. Structural mapping of the top of the Chinchaga indicates that normal faults with displacement in the order of 20 to 80 ft are common. However, there is no coincidence between the normal fault highs and earliest reef growth. Nor is there correspondence of Lower Keg River "thicks" with residual lows on the Chinchaga surface (Fig. 12).[6] Correlation of markers within the Lower and Upper Keg River Members in off-reef wells indicates that normal faults occurred after the inception of reef growth. For example, at the "A" pool (Fig. 11), markers directly overlying the Lower Keg River–Chinchaga contact can be readily correlated. The conformity of these markers with the contact and absence of onlap of beds indicates that the basal Keg River beds were deposited on a regular surface. The evidence indicates that local changes in thickness in the Lower Keg River were not caused by faults.

As local Lower Keg River "thicks" commonly underlie the Rainbow Member reefs (Fig. 13), the possibility that they controlled inception of reef growth has been examined. In the 7–32–109–8 W6 well in the "A" Pool,

[6] Figure 12 is a modified first-order residual map on the top of the Chinchaga surface. The modification refers to adjustments that were made to the present day elevations of the Chinchaga surface. These changes were necessary in order to remove the effects of post-Devonian normal-fault movements and to isolate fault movements of post-Chinchaga to pre-Watt Mountain age. The top of the Hay River Shale is used because it is unaffected by post-Chinchaga, pre-Watt Mountain normal-fault movement and because it is not affected significantly by drape over the Rainbow Member reefs nor by solution of salt from the Black Creek Member. First order residual values obtained from the top of Hay River Shale structure-contour map were used to adjust the present-day elevations of the Chinchaga surface from which a modified first-order residual Chinchaga map was prepared.

and in other wells in the "B" Pool, thickening occurs in the upper part of the Lower Keg River Member. The top of the Lower Keg River has dips of up to 15°. Faunal evidence indicates that these features, *i.e.* the local "thicks," may be mud mounds and that they provided the locus for reef growth during Upper Keg River deposition.

Figure 11 illustrates that the earliest concentration of reef-constructing organisms occurred at the 7–32 location overlying the mud mound-like topographic high. Thus, at the "A" pool, it appears that a mud-mound high was important in controlling the inception of reef growth. As shown in Figure 13, the areal size of the mud mound and the overlying concentration of reef fauna is small compared with the reef complex that subsequently developed. In the other pools, where well control is poor compared with the "A" pool, the details of reef inception are less obvious. However, the presence of local Lower Keg River "thicks" in many of the other pools (Fig. 13) suggests a history of reef inception similar to that of the "A" pool.

Although a mud-mound control of inception of reef growth appears valid, it should be emphasized that reef growth did not start when a certain specific thickness of the Lower Keg River was present. There may have been gentle depressions and highs present on the sea floor that combined with local Lower Keg River "thicks" to initiate reef growth.

Reef Form Evolution

Once the massive stromatoporoids and corals with rigid wave-resistant characteristics became established, lateral and vertical growth proceeded rapidly. Inception of reef growth is considered to have been similar for the different reefs at Rainbow, but the size and geometric form that these reefs reached at maturity differs markedly. In the Rainbow area, two basic geometric forms are present. These are oval, characterized by the pinnacle reef, and elliptical, characterized by the atoll reef. The atoll reef may be further subdivided, on the basis of size and internal geometry, into the crescent-atoll and large-atoll forms.

The independent variables that controlled the final form and dimension of the reefs are —size of the mound or groups of mounds on which the reef growth started, rate of subsidence of the basin, and the physical environment in which the reefs grew. The presence of a massive stromatoporoid and coral fauna is of

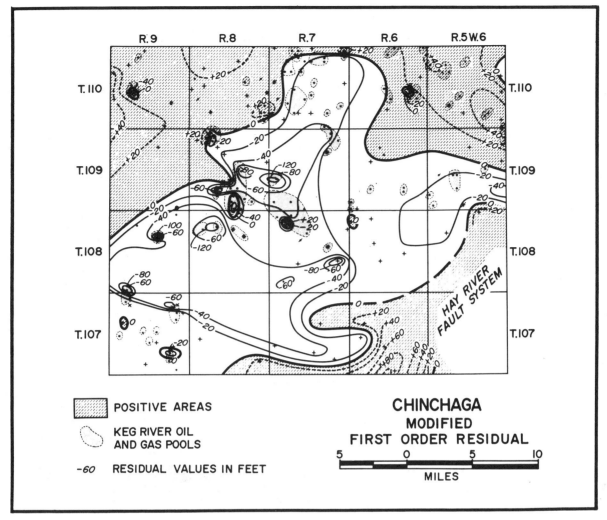

FIG. 12.—Chinchaga modified first order residual.

course important, but as similar faunas are common in other units of the Devonian, the concern here is with factors that produced reefs of the size and geometric form at Rainbow. A comparison of Rainbow Member reefs with part of the great Barrier Reef, Australia described by Fairbridge (1950) is thought to be useful. Figure 14 shows this comparison. Fairbridge illustrated the evolution of a reef from a small initial patch to a crescentic reef. The size of the patch reefs is reflected in the final shape and size of the complete reef masses. The larger ones developed into large horseshoe or crescent atolls and large atoll reefs, whereas the smaller patches developed into small horseshoe atolls or oval reefs.

At Rainbow, the size of the mound(s) on which reef growth began appears to have played an integral part in determining the size and shape of the final reef. Relative subsidence of the basin was rapid at Rainbow, so that from a small mound the reef grew vertically to produce a small oval-shaped pinnacle reef, less than ½ mi in diameter. On larger mounds the crescent-atoll reef developed. Crescent-atoll reefs are generally more than 2 mi long and more than 1½ mi wide, and are characterized by an organic rim which is generally located on the north and east sides. Sediments typical of quiet-water deposition were deposited in a lagoonal area behind the rim. The "O" pool (Fig. 14) is a good example of the crescent atoll. Large atolls developed on a larger mound or on several small, closely spaced mounds. A more or less continuous outer organic rim developed along which growth of massive stromatoporoids and corals was prolific, owing to the presence of abundant food and oxygenated wa-

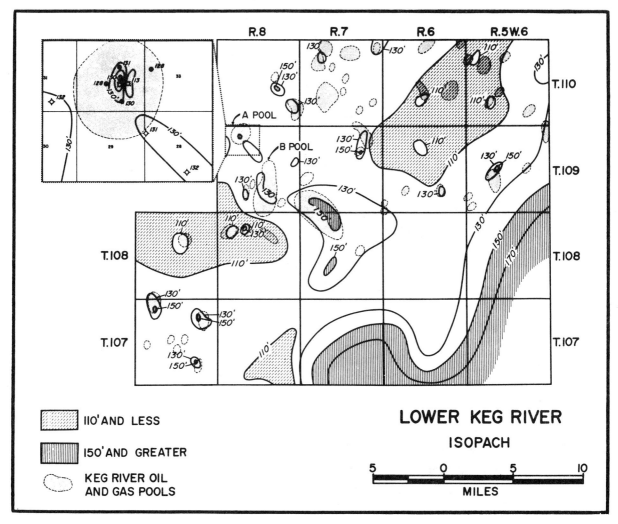

Fig. 13.—Isopach map of Lower Keg River Member, Rainbow area. Sections of "A" and "B" pools are shown in Figure 11. CI = 20 ft.

ters. A lagoonal quiet-water environment was present in the central low area. The "B" pool (Fig. 14) is typical of this form.

In addition to the influence of size of the mounds and rapid subsidence of the basin, the physical environment is believed to have exerted a strong influence on the reef forms that developed at Rainbow. Stanton (1967) indicated that three aspects of the physical environment are of major importance—directional climatic effects, particularly the wind; the paleogeographic setting (position of major shoals or land barriers, extent of deep open ocean); and local bathymetry.

The effect of wind in shaping the reef forms is illustrated in the comparison of some of the modern reefs of the Great Barrier with Rainbow Member reefs. Figure 14 illustrates the development of horseshoe-shaped reefs with con-

vexity in a windward direction. Several authors have shown this to be due to the relatively high organic growth and sediment production on the windward side. Moreover, movement of sediment by currents and waves results in a windswept crescent shape. These are convex toward the north-northeast and so indicate the prevailing wind direction at the time of reef growth.

The paleogeography of an area, that is, the position of the basin or deep water relative to the shelf and shore, determines the effectiveness of directional climatic factors and is therefore of critical significance with regard to reef forms. Because water depth is a significant factor in the consideration of paleogeography, it is inevitable that questions pertaining to the depths in which the Rainbow Member reefs grew, and relative age of the reef and offreef beds be examined also. Wave energy available for the

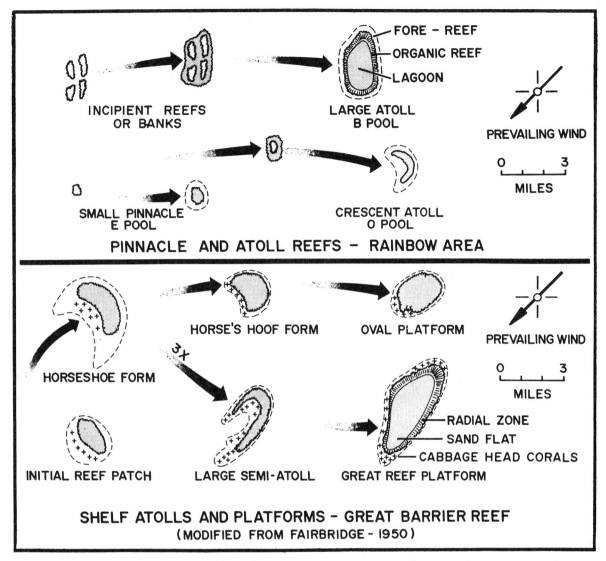

FIG. 14.—Pinnacle and atoll reefs, Rainbow area, compared with Great Barrier Reef, Australia (modified after Fairbridge, 1950).

erosion and transport of carbonate sediments is shown to be related to the length of fetch, that is, the distance between windward barriers and the reef(s) in question. Another factor which limits wave energy is water depth. In shallow seas, the energy imparted by winds to waves is severely restricted. It is possible to estimate the maximum height of waves as a measure of wave energy. The interrelations of fetch length, water depth, and wave height is illustrated by Bretschneider (1954). Stanton (1967) discussed an example of a low-energy situation in the shallow water of the Bahama platform. Water depths of 2–3 fm over the platform effectively limit the wind energy that can be transmitted to the waves. As a result, patch reefs protected by the shallow water alone are

roughly circular and are without the downwind "tails" of detritus.

The abundance of coarse clastic material and the elliptical or crescent shape of the reefs in the Rainbow area indicate that growth took place in an environment characterized by high wave energy. It is improbable that land barriers of a shallow shoaling sea were present in the windward direction. Evidence indicates that deep water surrounded the buildups. Carbonate rudites and arenites were transported to the edge of the organic reef and were deposited on the forereef slope. The lack of transport of reef-derived detritus away from the reef complex indicates low-energy deep-water conditions in the offreef area. Dips in the order of 25° in the forereef beds provide further evidence that

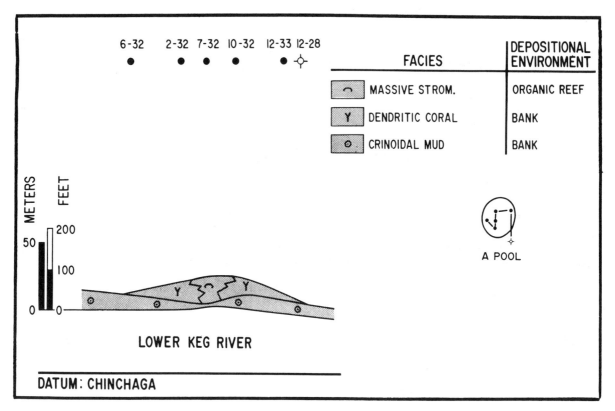

FIG. 15.—Pinnacle reef, deep-bank phase. Location of "A" pool shown on Figure 13.

strong topographic relief existed between the zone of organic production and destruction and bottom edge of the buildups.

In summary, all of the physiographic factors —directional aspect of climate, the paleogeography of the sea floor, and local bathymetry, indicate that the Rainbow Member reefs, for the most part of their growth, exhibited strong topographic relief. From this it follows that deposition of Muskeg evaporites occurred later than the laterally equivalent reef beds. However, deposition of the Black Creek salt and Muskeg anhydrites could have occurred in the offreef deeper water contemporaneously with reef growth near the surface.

Lithofacies

In the following section the lithologic and faunal constituents of Rainbow Member reefs are discussed briefly, largely on the basis of the work of Langton and Chin (1968) who examined in detail approximately 15,000 ft of core. These authors were able to recognize 14 lithofacies which they assigned to six depositional environments.

Following is a discussion of the lithofacies and depositional environments during the significant phases of reef growth at Rainbow.

Early bank phase (pinnacle).—Facies present in this phase are massive stromatoporoids, dendritic corals, and crinoidal muds (Fig. 15). The massive stromatoporoid facies containing massive corals, branching stromatoporoids, brachiopods, and gastropods, is of specific interest because of its reef-building potential. However, during the early bank phase the crinoidal mud facies was dominant and there was only minor topographic relief on the sea floor. The writers consider that the term "deep bank" most accurately describes the structure and environment at this time.

Bank to organic-reef transition phase (pinnacle).—Figure 16 illustrates the facies which are present and their distribution. The massive stromatoporoids continued to flourish and were joined by the reef-constructing massive coral facies. Coral rudite facies is present and consists of fragments derived from the massive stromatoporoids and corals. The rudites have an arenite matrix. This facies is present in the forereef position of the pinnacles. Skeletal arenites are widely distributed throughout the different environments of the reef complex, however, they are commonly abundant in this phase. In addition to the arenites and rudites, other facies present include the massive syrin-

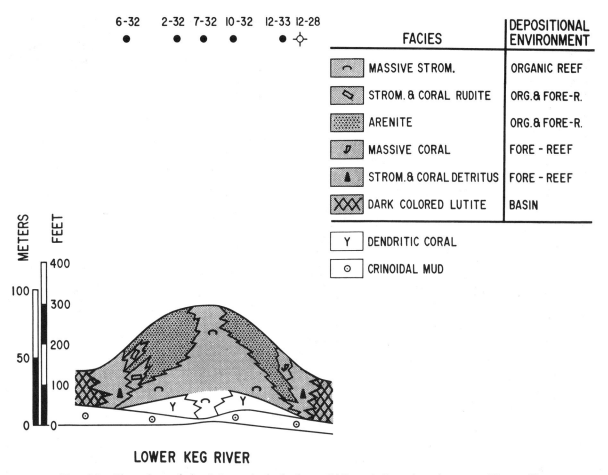

FIG. 16.—Pinnacle reef, bank/organic-deef phase, "A" pool. Location shown on Figure 13.

goporoid corals and organic-reef detritus. The organic-reef-detritus facies consists of fine fragments of stromatoporoids and branching corals in a brown, very finely crystalline dololutite matrix. The dark color, presence of fine matrix, and occurrence in a low flank position indicate deposition in relatively deep, quiet water. This facies is transitional with the dark-colored lutites. The latter consist of finely laminated bituminous calcilutites and are found in the offreef basin environment.

There was a delicate balance between reef-building ability of the organic-reef organisms and the destructive forces of the physical environment. The reef was subjected to strong turbulent water conditions, as indicated by the presence of abundant arenites and rudites.

Organic-reef phase (pinnacle).—During this phase (Fig. 17) the massive stromatoporoids and corals of the organic-reef environment became dominant and the major part of the reef

complex was constructed. The presence of rudite and arenite zones attests to the turbulent water conditions to which the organic reef was exposed. However, movement of derived detritus was restricted to the forereef slope, where deposition occurred below wave base.

Two additional facies, the light-colored lutites and the skeletal rudites, were deposited during this phase. The light-colored lutites consist of micro- to very fine-grained limestone and constitute a minor part of the organic reef; they were deposited in local, protected areas on the reef. The skeletal rudite consists of fragmented gastropods, brachiopods, colonial septate corals, and Stachyodes in the calcilutite to calcisiltite matrix. Massive stromatoporoid fragments are present sporadically.

Langton and Chin (personal commun.) have recorded the presence of massive stromatoporoid facies at the top of small pinnacle reefs and to within 50 ft of the top of some of

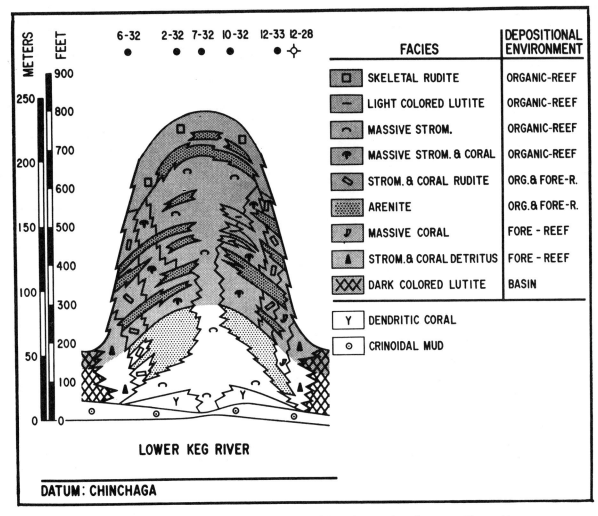

FIG. 17.—Pinnacle reef, organic-reef phase, "A" pool. Location shown on Figure 13.

the larger pinnacle and atoll reefs. However, in the uppermost part of the organic-reef phase, the areal extent of this facies became much smaller and there is an increase in the amount of skeletal rudite facies. The decrease in massive stromatoporoid facies indicates a less favorable environment—possibly an increase in salinity of the seas.

Shallow-bank phase (pinnacle).—The terminal stage of reef growth is illustrated on Figure 18. The Amphipora facies, though not extensive, is generally present in this phase. Massive stromatoporoid facies is present also, but the areal extent of this facies was reduced further from that present in the organic-reef phase. In addition to the Amphipora facies, the laminite facies is generally present in this phase. The laminites are thought to be, in part, of algal origin. There is some difficulty in separating the skeletal rudites of the organic-reef phase from

those present in the shallow-bank phase. The distinction is made mainly on the basis of contained fossils and associated lithology. The skeletal rudite of the organic-reef environment contains minor massive stromatoporoids and rare amphiporoids, whereas the skeletal rudite of the shallow shoal environment contains numerous amphiporoids and rare stromatoporoids.

During this late phase, the population and variety of the fauna decreased. These changes and the presence of laminites and primary anhydrites in the upper 20–50 ft of some reefs, indicate an environmental change to shallow, highly saline waters. Minor occurrences of massive stromatoporoids probably indicate isolated topographic features where favorable salinities were maintained by introduction of fresh meteoric waters.

Atoll Reefs.—The growth of atoll reefs was

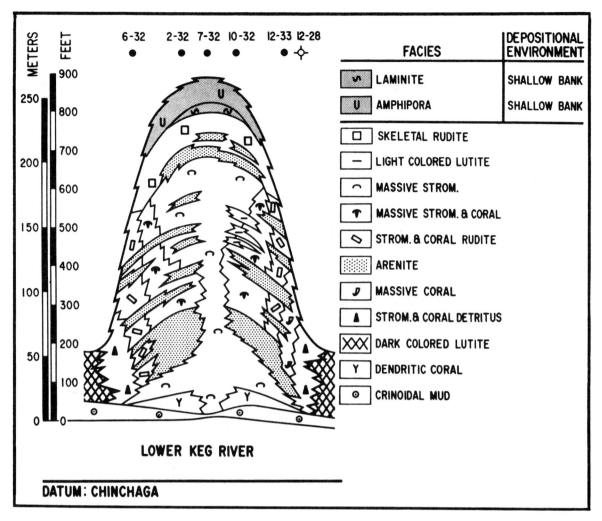

FIG. 18.—Pinnacle reef, shallow-bank phase, "A" pool. Location shown on Figure 13.

very similar to that of the pinnacles (Fig. 19). The main difference occurred during the early bank phase when the organic-reef fauna became established over a much broader area. As reef growth proceeded, the organic-reef facies grew toward the outside edge of the reef complex where nutrients were more plentiful. The upward growth of the organic-reef facies was more rapid than the other parts of the reef complex. As a result, a rim developed around the edge of the reef and resulted in the development of a relatively quiet-water interior lagoon. In the atoll reefs, in addition to the deep-bank, organic-reef, forereef, and shallow-bank environments of the pinnacle reefs, a lagoon environment is present. The facies which are present in the lagoon comprise stromatoporoid and coral rudites, arenites, nonskeletal rudites, laminites, and light-colored lutite. The arenites and stromatoporoid and coral rudites

are very similar to those found on the forereef position of the pinnacle, and represent material moved by wave action behind the rim of the reef. Nonskeletal rudites consist of fragments of lutite torn from the lagoon floor during storms and then redeposited in a quiet-water environment. The laminites and light-colored lutites are important constituents in the lagoonal environment.

Diagenesis

In addition to recrystallization of $CaCo_3$ and deposition of calcite cement, other diagenetic processes which have affected the reef are solution, dolomitization, anhydritization, and deposition of carbon. The effect of the different diagenetic processes is beyond the scope of this paper and only generalized comments on certain aspects of anhydritization, dolomitization, and solution are presented.

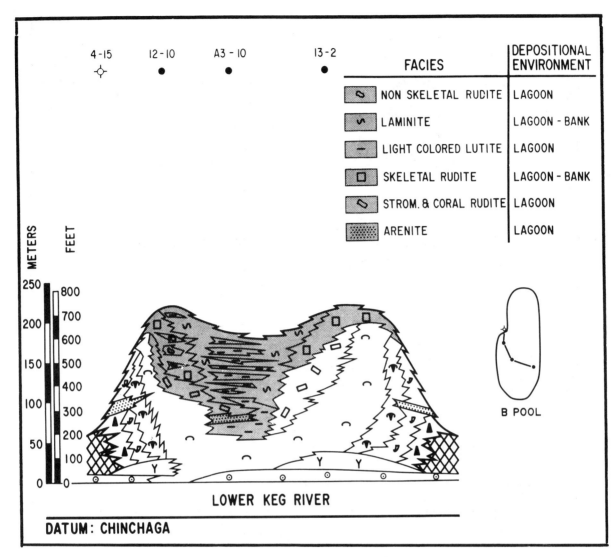

FIG. 19.—Atoll reef, lagoon-bank phase, "B" pool. Location shown on Figure 13.

Anhydritization, although present, is not appreciable except in the upper 20–50 ft of the shallow-bank phase. In addition to the presence of anhydrite in the uppermost beds of the reef, some thin beds of anhydrite have been found more than 300 ft from the top of full reef buildup. The anhydrite occurs disseminated through the carbonate, as nodules in vugs, and as a replacement mineral. It ranges from clear white to gray and microcrystalline to medium crystalline. Machielse (personal commun.) has noted that the anhydrite is associated with brown, cryptocrystalline to microcrystalline dolomite which appears to be of primary origin. Thus, although the possibility of primary origin for some of the anhydrite exists, the anhydrite which is most common is the clear-white, finely crystalline, secondary variety. Also, as this vari-

ety of anhydrite infills pores and vugs, and as it cannot be correlated from well to well, it is believed to be of secondary origin.

Three factors are important with regard to dolomitization. Langton and Chin (1968) pointed out that the degree and type of dolomitization appear to be related to reef size and its internal makeup. Also, they recognized early and late stages of dolomitization.

The relation between dolomitization and reef size is illustrated by the fact that pinnacle reefs are not dolomitized, crescent atolls are moderately to strongly dolomitized, and all large atolls are strongly dolomitized. An exception is the small pinnacle "D" pool (Fig. 24) reef which is completely dolomitized. In this case, the dolomitization may be the result of its proximity to the "B" pool atoll reef and its

234

connate water environment. Dolomitization is characteristic of atolls and not pinnacles because, as Langton and Chin believe, the larger reefs created their own evaporitic environment. Evaporation resulted in an increase in density of the water. Movement of this relatively dense saline water through the reef resulted in dolomitization by the seepage refluxion process.

Two stages of dolomitization are clearly evident. The early stage resulted in selective replacement of the fine-grained matrix material and did not obliterate sedimentary and faunal textures to any significant extent. In its final form it resulted in a fine-grained dolomite. The color of the host rock was retained and is generally light to dark brown. The late phase of dolomitization is characterized by coarse crystallinity and light color. The white, coarsely crystalline dolomite replaces and occurs as veins in the early, brown, finely crystalline dolomite. The presence of the two kinds of dolomitization gives the rock a "marbled" appearance. In other types of late-stage dolomitization, terms such as *banded, nebulous,* and *pseudo-brecciated,* describe the texture very well. The presence of the late-stage of dolomitization is the result of late, postdepositional introduction of magnesium-rich waters, along fracture zones or zones of porosity. The magnesium-rich waters may have been displaced from the Muskeg evaporite beds, by compaction.

There has been solution, but the degree of solution was highly variable in all reefs. Large fragments of gastropods and brachiopods locally have been completely leached. Also, there was selective leaching of stromatoporoids, corals, and laminites. In addition, abundant stylolitization and the pronounced degree of compaction in atoll reefs (*e.g.* "F"-pool reef [Fig. 24]) indicate that considerable solution occurred in certain beds. The relation and relative timing of solution processes and late-stage dolomitization has not been fully established. It is evident that initial porosity was important in controlling both solution and infill processes. The porous laminates show the effects of the two processes (Langton and Chin, 1968, Pl. 3, Fig. 10). In this illustration solution occurred along the porous and permeable layers and was followed by infilling of the late-stage dolomites. Late-stage dolomitization continued inward to the less permeable parts of the rock. In some places only large remnants of the host rock are left in a matrix of coarse-grained, white dolomite. It appears that late-stage dolomitization

occurred simultaneously or immediately following solution.

Rainbow Member—Time and Space Relations

The major part of the Rainbow Member reefs are considered to be the time equivalent of the Upper Keg River Member. The facies containing the massive stromatoporoid and coral fauna and the erosional products occur from the base of each reef to the top of full reef buildups in some reefs and to within 50 ft in most other reefs. Strong topographic relief is indicated by dips on forereef beds in the order of 25°. Also, there is no evidence of intertonguing of the reef detritus with the laterally equivalent Muskeg evaporite beds; the arenites, rudites and organic-reef detritus were carried only to the reef slope where they were deposited in deep water in which the energy was too low to carry away the detritus from the reef structure. All of the factors of the physical environment indicate that deep water surrounded the Rainbow Member reefs. It follows that deposition of Muskeg evaporites occurred either subsequently to reef growth or in deep water at the same time as reef growth was taking place near the surface.

Although the age relation for the major part of Rainbow Member reef growth and Muskeg evaporite deposition seems clear, there appears to be direct evidence of contemporaneity of reef growth and evaporite deposition in the terminal, shallow-bank phase at the top of the reef. The presence of laminites and Amphipora indicate relatively quiet water. The reduction in both number and variety of fossils, and the presence of dark-brown, very finely crystalline, primary-type anhydrite, indicates a highly saline water environment. Deposition of Muskeg evaporites was probably taking place at the same time as the Rainbow Member reef growth in the terminal shallow-bank phase.

In the Zama subbasin (Figs. 9 and 20) the geologic history was different from that at Rainbow. The Upper Keg River reefs (McCamis and Griffith, 1967) are probably the age equivalent of the Rainbow Member reefs. However, the correlation between Rainbow and Zama is not clear. A distinctive stratigraphic unit—the Zama Member—overlies either reef where there is full buildup, or Muskeg evaporites in the offreef area (Fig. 20). The Zama Member consists of fine-grained dolomite with low fossil content, commonly bedded or laminated. Because it is underlain and overlain by Muskeg evaporites in offreef areas, it is

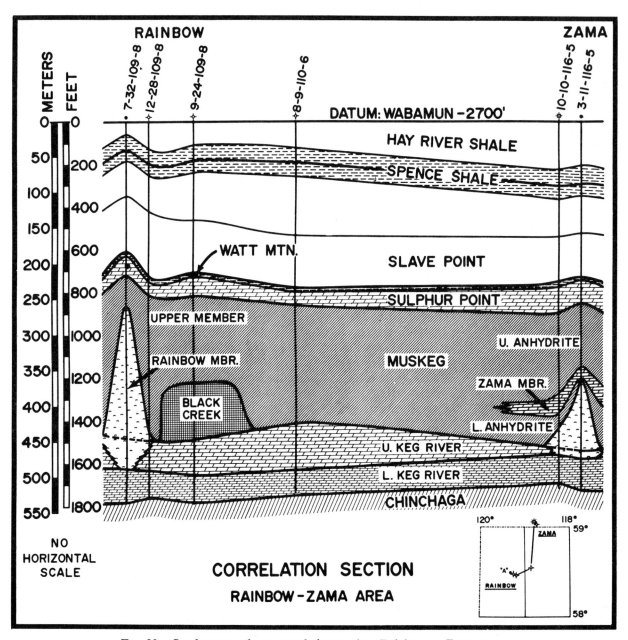

FIG. 20.—Southwest-northeast correlation section, Rainbow to Zama area.

known to be of Muskeg age. It is not recognized in the Rainbow subbasin. McCamis and Griffith (1967), however, consider it to be equivalent to the upper 150–250 ft of the Rainbow Member. We consider this correlation to be untenable. At Rainbow, growth of the organic reef occurs either to the top or to within 50 ft of the top of most of the reefs. The strong regional shoal environment which led to deposition of the Zama Member in the Zama subbasin was not present in the Rainbow subbasin.

Post-Rainbow Member Structure

The present structural attitude of the beds is due to four factors—normal fault movement, solution of the Black Creek salt, compaction of Muskeg evaporite beds, and post-Laramide regional tilting. Regional tilting is not significant in the context of this paper and is not discussed.

Normal fault movement.—In the Rainbow map area, two systems of faulting are present (Fig. 12). One system, the Hay River-East Arm, is about 10 mi south of the Rainbow

236

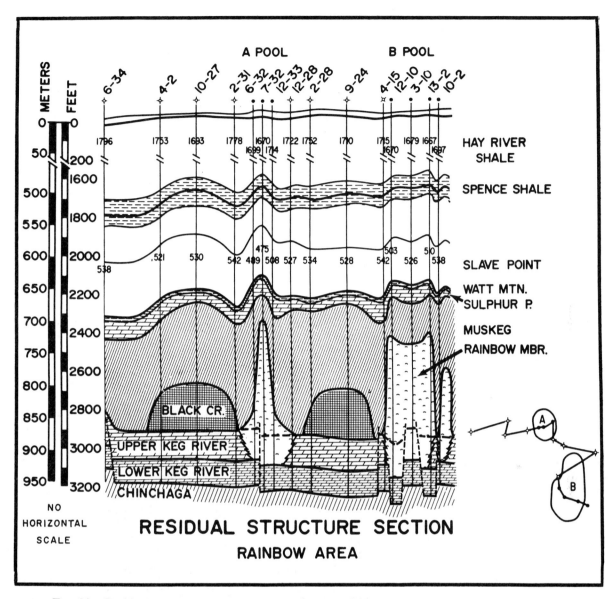

Fig. 21.—Residual structure section, showing effects of salt solution, Rainbow area. "A" and "B" pools shown in Figure 13.

field. It is composed of normal faults with throws of 100 ft or greater. The strike of the individual faults is variable, but the overall trend of the fault system is southwest-northeast. The second system of faulting is present in the general Rainbow field area. Faults in this system have throws in the order of 20–80 ft and they have no apparent trend (Figs. 12, 21).

The Hay River–East Arm fault system represents an early Precambrian movement that probably was modified later by erosion. When deposition started in this area, only prominent topographic features were present. Reactivation of normal and transcurrent faults occurred dur-

ing different periods as late as the end of Mississippian time. Other normal faults and fault systems of this nature are present in northern Alberta.

Normal faults in the second system, with present-day throws of 20–80 ft are of more direct interest as they commonly underlie the edges of the reefs in the Rainbow field (Fig. 12). Because of this relation there has been a tendency to view these faults as the loci for reef growth. As stated, the writers do not subscribe to this view. The loci for reef growth appears to have been Lower Keg River mud mounds. Detailed correlation of markers within

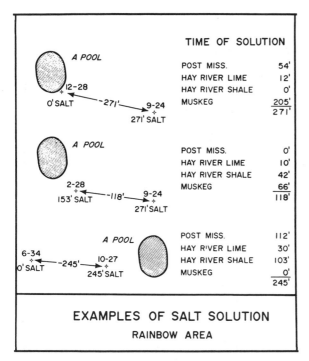

FIG. 22.—Examples of salt solution. Rainbow area.

the Lower and Upper Keg River Members (Fig. 11) indicates that normal-fault movement occurred subsequent to deposition of the lower part of the Upper Keg River Member, that is, sometime during Rainbow Member reef growth or later. As normal faults are not evident at the Watt Mountain level, the time of movement can be dated between that of the deposition of the upper Rainbow Member and the Watt Mountain Formation. Regional structure analysis supports the interpretation of normal fault movements at this time. In the area of the Tathlina high in the Northwest Territories it is clear that the Lower Keg River and probably the Upper Keg River were deposited as a normal sequence. The section between the Watt Mountain and Lower Keg River, however, is considerably thinner than that in the rest of the basin on the south. Uplift of the Tathlina high before Watt Mountain deposition—possibly during Muskeg deposition—resulted in non-deposition and possibly erosion. Along the She-kilie trend on the western edge of the Black Creek basin, the interval from the top of the Muskeg Formation to the top of Lower Keg River Formation is thin. The overlying Sulphur Point Formation, however, is regionally thick. It appears that uplift occurred toward the end of Muskeg deposition, which resulted in a thin Muskeg section, or possibly some erosion of the Muskeg beds. During Sulphur Point deposition

down-warping occurred which resulted in a relatively thick Sulphur Point section along the western part of the basin (Figs. 5, 9, and 10).

Within the Rainbow subbasin the normal faults with movement of 20–80 ft may be rebound features related to the emplacement of individual reef masses. The aggregate weight of the individual reefs and the regional barrier-reef complexes could have caused both local and broad crustal readjustments. The combination of the barrier-reef complexes and crustal readjustments along the Shekilie and Tathlina features formed a regional barrier that prevented the flow of normal marine waters southeast to the Black Creek basin.

Solution of Black Creek Member salt.—The distribution of the Black Creek Member salt is best known in the Rainbow subbasin. The salt beds have been found in nine wells, with thickness up to 271 ft. Seismic data are also useful in defining the presence of salt. Presence of the Black Creek Member in the Zama subbasin has been confirmed at the Hudson's Bay Zama No. 10–21–116–5 W6 well which penetrated 124 ft of salt. Salt is also present on the northeast in the Steen River area where six wells have penetrated up to 180 ft of the Black Creek Member. The original extent of the Black Creek salt was probably much greater than it is today.

In the Rainbow subbasin, solution of salt occurred in early Late Devonian (upper Muskeg, Hay River) and post-Mississippian times. A summary of the amounts and periods of salt solution of specific examples is shown on Figure 22 and illustrated on Figure 21. Solution during upper Muskeg deposition can be seen in many wells. For example, from the 9–24 well to the 12–28 well, 271 ft of salt was removed; 205 ft can be accounted for in the thicker Muskeg section. Salt solution in the amount of 103 ft occurred during Hay River deposition between wells 6–34 and 10–27. Between wells 2–28 and 9–24, 52 ft of thickening of the Hay River is attributed to salt solution. There was no apparent post-Mississippian salt solution between these two wells. However, 54 ft of post-Mississippian solution is indicated between the 12–28 and 9–24 wells, and 112 ft between wells 10–27 and 6–34.

Compaction.—In Figure 23, compaction of Muskeg beds is well illustrated in the 4–9, 5–9, and 15–9 wells. Detailed correlation of Muskeg units 2 to 5, inclusive, shows aggregate thickening from 260 ft over nearly full reef at well 15–9 to 330 ft over high reef flank at well

5–9. Because of the high flank position of these two wells, it is apparent that thickening of Muskeg units 2 to 5 was not influenced by salt solution. The combined thickness of units 2 to 5 on the reef flank in well 4–9 is 400 ft. Although thickening due to salt solution is possible at 4–9 because it is in a low flank position, uniformity and consistency of correlation with the other wells suggests that changes in thickness are due to differential compaction. From 15–9 to 4–9 thickening due to compaction amounts to 35 percent. In addition to the thickening described above, Figure 23 also shows 115 ft of onlap of basal Muskeg beds onto the reef between 4–9 and 5–9 and an additional 68 ft in unit 6 between wells 5–9 and 15–9, for a total of 183 ft. This fact and absence of intertonguing of reef and Muskeg beds indicate the topographic relief that existed on the reef late in Muskeg deposition.

Figure 21 illustrates compaction and thinning of beds overlying the "A" and "B" reef pools of the Rainbow field. In the interval Spence Shale "A" marker to Watt Mountain, thickening in the order of 35–55 ft occurs from a full reef buildup to completely off-reef, full Black Creek Member salt section. Also shown in Figure 21 is thickening of 15–20 ft in this same interval in the 3–10 well of the "B" atoll—a well which is located in the lagoonal area. This indicates that solution and compaction of lagoonal reef carbonates occurred during the time represented by this interval.

In the interval, top Hay River Shale to Spence "A" marker, thickening from reef to offreef is in the order of 30–50 ft. As this amount of thickening occurs where there has been no solution of the Black Creek salt, it is evident that compaction of the offreef Muskeg evaporites continued until the end of Hay River Shale deposition. However, there was only insignificant compaction following Hay River deposition. Rainbow Member reefs are not readily discernible on structural contour maps of the top of Hay River Shale.

EXPLORATION AND RESERVES—RAINBOW LAKE

Prior to the discovery of the Rainbow field by the Banff *et al.* Rainbow No. 7–32–109–8 W6 well, there had been no drilling within the outline of the current field limits (Fig. 24). Subsequent exploration and development drilling has resulted in 55 Keg River oil pools, 7 Keg River gas pools, 6 Muskeg oil pools, and 4 Sulphur Point oil pools. About 72 of the 126

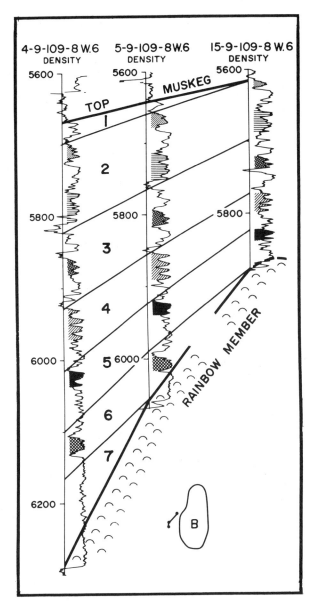

FIG. 23.—Correlation of Muskeg markers, "D" pool (wells 4–9, 5–9) and "EEE" pool (15–9), Rainbow area (Fig. 24). No horizontal scale.

exploratory wells, or approximately one well in two, have been discoveries. The exceptionally high success rate of 56 percent for exploratory wells resulted from use of modern geophysical techniques. Draping over reef structures and salt solution near reefs are identifiable seismically, particularly when "stacking" or common-depth point shooting is used. The use of seismic data has met with similar success in other areas within the Black Creek basin. In the Virgo and Zama Lake fields north of Rainbow, 110 Keg River oil pools have been found. The number of oil and gas completions in the Keg River Formation within the Black Creek

D. L. Barss, A. B. Copland, and W. D. Ritchie

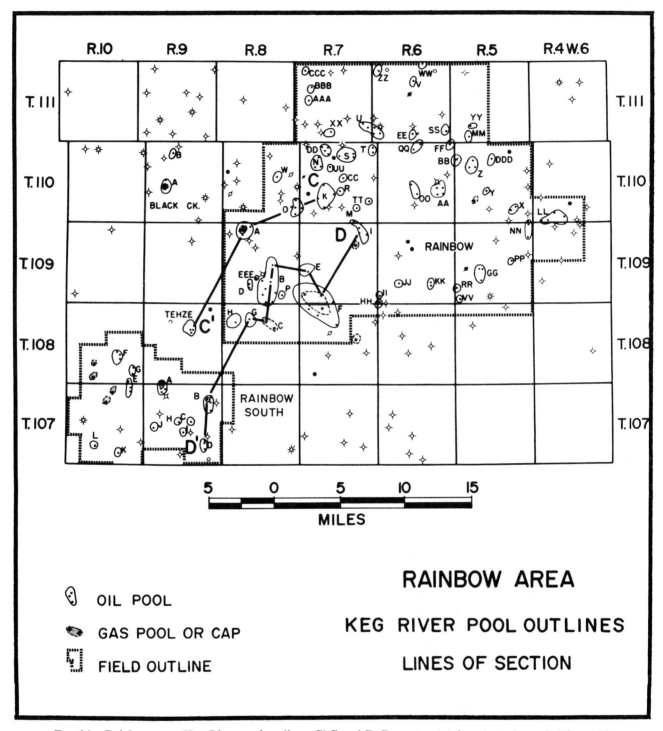

FIG. 24.—Rainbow area, Keg River pool outlines. C'-C and D'-D are traces of sections shown in Figure 25.

basin probably will increase, inasmuch as the area is still under active exploration and development.

Figure 25 illustrates two schematic sections across Rainbow and Rainbow South fields, showing reservoir parameters and recoverable reserves for 10 of the 55 Keg River oil pools.

The "C" pool on D'-D section is a gas pool and the Tehze pool is outside the current field boundaries. At the present ime, these 10 pools account for approximately 60 percent of the estimated primary recoverable reserves of the two fields.

Figure 25 also illustrates the wide range of

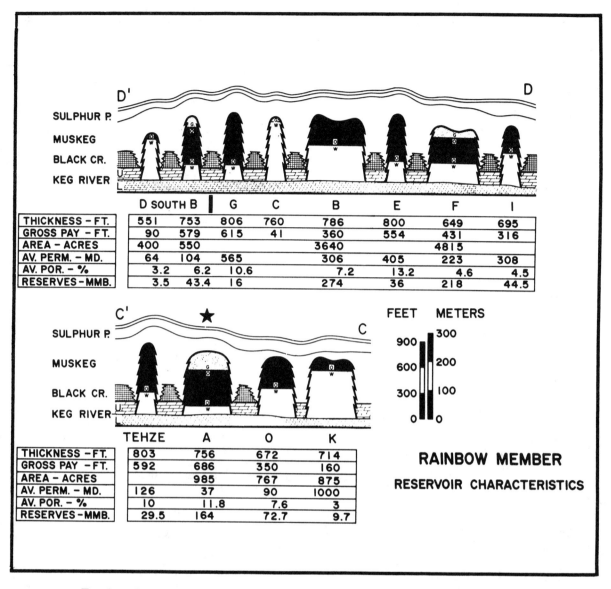

	D SOUTH	B	G	C	B	E	F	I
THICKNESS – FT.	551	753	806	760	786	800	649	695
GROSS PAY – FT.	90	579	615	41	360	554	431	316
AREA – ACRES	400	550			3640		4815	
AV. PERM. – MD.	64	104	565		306	405	223	308
AV. POR. – %	3.2	6.2	10.6		7.2	13.2	4.6	4.5
RESERVES – MMB.	3.5	43.4	16		274	36	218	44.5

	TEHZE	A	O	K
THICKNESS – FT.	803	756	672	714
GROSS PAY – FT.	592	686	350	160
AREA – ACRES		985	767	875
AV. PERM. – MD.	126	37	90	1000
AV. POR. – %	10	11.8	7.6	3
RESERVES – MMB.	29.5	164	72.7	9.7

RAINBOW MEMBER

RESERVOIR CHARACTERISTICS

FIG. 25.—Reservoir characteristics and reserves, Rainbow Member, in Rainbow area. Traces of sections **D′-D** and **C′-C** are shown in Figure 24.

net pays and reservoir factors found in Rainbow. Generally, the net pays are better in the western part of the Rainbow field, generally in the 200–500 ft range. However, there are exceptions such as the "C" pool which has only 41 ft of Keg River gas pay. Porosity and permeability values vary greatly from pool to pool, as well as within individual pools and are related to the different lithofacies and to diagenetic effects. An example of the broad relation of reservoir facies to lithofacies is illustrated in Rainbow South "A" pool and in the Rainbow "B" pool; in these, porosity is lower in the center of the atolls, averaging 6 percent, and increases significantly to about 11 percent along the flanks of the reefs. This increase is attrib-

uted to the higher percentages of arenites and rudites in the forereef position. Another example is the South "G" pool in Rainbow South field where the 4–12–108–10 W6 central well has an average of 7.4 percent porosity and the 12–1–108–10 W6 flank well has an average of 10.4 percent porosity. In addition to varying reservoir properties within a reef as a result of different lithofacies, there is variation in the same lithofacies between reefs owing to the variation in content of "lime mud" in the matrix and owing to the degree to which solution and dolomitization processes were active. Dolomitization has had a variable effect on reservoir properties. In some pools, the "F" pool, for example, dolomitization is essentially complete.

The average porosity in this pool is 4.5 percent which is lower than many of the less dolomitized reefs. The effect of solution on porosity and permeability is evident in nearly all reefs in varying degrees. An example of extreme solution is found in the small pinnacle reef at well 15–9–109–8 W6. In this well, 400 ft of dolomite and limestone was penetrated in which the porosity averaged 23 percent. Approximately 100 ft of this section had porosity values in the range of 40 to 50 percent. This type of cavernous porosity is uncommon in the Rainbow area; in the Rainbow Member reefs, porosity ranges more commonly from 5 to 11 percent. Further details on geologic and reservoir properties and their correlation were presented by Langton and Chin (1968).

With the Rainbow and South Rainbow fields, there are estimated to be 1.2 billion and 165 million bbl of oil-in-place, respectively. The reef reservoirs have good vertical relief and lend themselves to enhanced recovery schemes. Primary recovery factors range from 20 to 52 percent whereas secondary recovery methods, such as miscible flood and water or gas injection, are realistically estimated to boost the ultimate recoveries to 80–95 percent.

CONCLUSIONS

1. Rainbow Member reef growth began in the lower unit of the Upper Keg River Member. This unit contains, in addition to crinoids and brachiopods, a reef-prone fauna consisting of massive stromatoporoids, branching septate, and tabulate corals. The inception of reef growth was sensitive to "highs" on the underlying Lower Keg River sea floor. These "highs," with relief of about 20–40 ft, are thought to be mud mounds which formed in the upper part of the Lower Keg River Member. No direct evidence of structural control of reef growth exists, although there may have been gentle warping of the Lower Keg River surface which combined with mud mounds to control reef growth.

2. The major part of Rainbow Member reefs are of late Middle Devonian (Upper Keg River) age. There is the possibility that the uppermost part of the reefs, that is, the shallow-bank deposits, may be of "Muskeg age."

3. Within the Rainbow part of the Black Creek basin are Rainbow Member reefs of both pinnacle and atoll forms with vertical dimensions up to 820 ft. The size and geometry of the reefs are related to the initial size of the mound(s) on which they grew, rate of basin subsidence, directional influence of climate, paleogeography of the sea floor, and local bathymetry.

4. Numerous normal faults with 20–80 ft of throw are present. Movement occurred after Rainbow Member reef growth began and prior to deposition of the Watt Mountain sediments.

5. Salt solution and compaction features are recognized in the Black Creek basin. Solution of Black Creek salt occurred mainly in the Rainbow area late in Muskeg deposition, and in early Late Devonian, and post-Mississippian times. Compaction of the Muskeg evaporites occurred from the time of Muskeg deposition until at least the end of Hay River Shale deposition. Detailed correlation of markers in the Rainbow area indicates compaction of approximately 35 percent within the Muskeg Formation.

6. Modern seismic techniques have been used in exploration for Rainbow Member reefs. These techniques have proved to be remarkably successful as shown by a discovery of one Keg River oil or gas accumulation for every two wells drilled.

SELECTED REFERENCES

Adams, J. E., and Rhodes, M. L., 1960, Dolomitization by seepage refluxion: Am. Assoc. Petroleum Geologists Bull., v. 44, no. 12, p. 1912–1930.

Andrichuk, J. M., 1961, Stratigraphic evidence for tectonic and current control of Upper Devonian reef sedimentation, Duhamel area, Alberta, Canada: Am. Assoc. Petroleum Geologists Bull., v. 45, no. 5, p. 612–632.

Belyea, H. R., 1959, Devonian Elk Point Group, central and southern Alberta, parts of 72, 73, 82, 83: Canada Geol. Survey Paper 59–2, 14 p.

——— and D. J. McLaren, 1962, Upper Devonian formations, southern part of Northwest Territories, northeastern British Columbia, and northwestern Alberta: Canada Geol. Survey Paper 61–29, 74 p.

——— and A. W. Norris, 1962, Middle Devonian and older Palaeozoic formations of southern district of MacKenzie and adjacent areas: Canada Geol. Survey Paper 62–15, 82 p.

Bretschneider, C. L., 1954, Generation of wind waves over shallow bottom: U.S. Army Corps Engineers, Beach Erosion Board Tech. Memo. no. 51, p. 1–24.

Burwash, R. A., 1957, Reconnaissance of subsurface Precambrian of Alberta: Am. Assoc. Petroleum Geologists Bull., v. 41, no. 1, p. 70–103.

Brown, P. R., 1963, Some algae from the Swan Hills reef: Bull. Canadian Petroleum Geology, v. 11, no. 2, p. 178–182.

Cloud, P. E., 1952, Facies relationships of organic reefs: Am. Assoc. Petroleum Geologists Bull., v. 36, no. 11, p. 2125–2149.

Cummings, E. R., 1932, Reefs or bioherms?: Geol. Soc. America Bull., v. 43, no. 1, p. 331–352.

Douglas, R. J. W., 1959, Great Slave and Trout River map-areas, Northwest Territories, parts of North halves of 85 and 95: Canada Geol. Survey Paper 58–11, 57 p.

——— and D. K. Norris, 1959, Fort Liard and La Biche map-areas, Northwest Territories and Yukon, 95B and 95C: Canada Geol. Survey Paper 59–6, 23 p.

——— and ——— 1961, Camsell Bend and Root River map-areas, District of MacKenzie, Northwest Territories, 95J and K: Canada Geol. Survey Paper 61–13, 36 p.

——— and ——— 1963, Dahadinni and Wrigley map-areas, District of MacKenzie, Northwest Territories, 95N and O: Canada Geol. Survey Paper 62–33, 34 p.

Dumestre, A., and L. V. Illing, 1967, Middle Devonian Reefs in Spanish Sahara, in Internat. Symposium on the Devonian System: Calgary, Alberta Soc. Petroleum Geologists, v. 2, p. 333–350.

Fairbridge, R. W., 1950, Recent and Pleistocene coral reefs of Australia: Jour. Geology, v. 58, no. 4, p. 330–401.

Fischbuch, N. R., 1960, Stromatoporoids of the Kaybob reef, Alberta: Alberta Soc. Petroleum Geologists Jour., v. 8, no. 4, p. 113–131.

Fong, G. 1960, Geology of the Devonian Beaverhill Lake Formation, Swan Hills area, Alberta, Canada: Am. Assoc. Petroleum Geologists Bull., v. 44, no. 2, p. 195–209.

Grabau, A. W., 1913, Principles of stratigraphy: New York, A. G. Seiler, 1185 p. 2d ed., 1924, 1185 p.; Dover reprint, 1960, 2 v.

Gray, F. F., and J. R. Kassube, 1963, Geology and stratigraphy of Clarke Lake gas field, northeastern British Columbia: Am. Assoc. Petroleum Geologists, Bull., v. 47, no. 3, p. 467–483.

Grayston, L. D., D. F. Sherwin, and J. F. Alan, 1964, Middle Devonian, Chap. 5, in Geological history of Western Canada: Alberta Soc. Petroleum Geologists, p. 49–59.

Hriskevich, M. E., 1966, Stratigraphy of Middle Devonian and older rocks of Banff–Aquitaine Rainbow West 7–32 discovery well, Alberta: Bull. Canadian Petroleum Geology, v. 14, no. 2, p. 241–265.

Ingels, J. J. C., 1963, Geometry, paleontology, and petrography of Thornton reef complex, Silurian of northeastern Illinois: Am. Assoc. Petroleum Geologists Bull., v. 47, no. 3, p. 405–440.

Jenik, A. J., and J. R. Lerbekmo, 1968, Facies and geometry of Swan Hills Reef Member of Beaverhill Lake Formation (Upper Devonian), Goose River field, Alberta, Canada: Am. Assoc. Petroleum Geologists Bull., v. 52, no. 1, p. 21–56.

Klement, K., 1968, Reefs and banks; bioherms and biostromes (abs.): Oilweek, Feb. 19, p. 14, 20.

Klovan, J. E., 1964, Facies analysis of the Redwater reef complex, Alberta, Canada: Bull. Canadian Petroleum Geology, v. 12, no. 1,, p. 1–100.

Kornicker, L. S., and D. W. Boyd, 1962, Shallow-water geology and environments of Alacran reef complex, Campeche Bank, Mexico: Am. Assoc. Petroleum Geologists Bull., v. 46, no. 5, p. 640–673.

Kuenen, Ph. H., 1950, Marine geology: New York, John Wiley & Sons, 568 p.

Langton, J. R., and G. E. Chin, 1968, Rainbow Member facies and related reservoir properties, Rainbow Lake, Alberta: Bull. Canadian Petroleum Geology, v. 16, no. 1, p. 104–143; modified version, Am. Assoc. Petroleum Geologists Bull., v. 52, no. 10, p. 1925–1955.

Law, J., 1955, Geology of northwestern Alberta and adjacent areas: Am. Assoc. Petroleum Geologists Bull., v. 39, no. 10, p. 1927–1978.

Lowenstam, H. A., 1959, Niagaran reefs of the Great Lakes area: Jour. Geology, v. 58, no. 4, p. 430–487.

McCamis, J. G., and L. S. Griffith, 1967, Middle Devonian facies relationships, Zama area, Alberta: Bull. Canadian Petroleum Geology, v. 15, no. 4, p. 434–467.

Murray, J. W., 1965, Stratigraphy and carbonate petrology of the Waterways Formation, Judy Creek, Alberta, Canada: Bull. Canadian Petroleum Geology, v. 13, no. 2, p. 303–326.

Nelson, H. F., C. W. Brown, and J. H. Brineman, 1962, Skeletal limestone classification, in W. E. Ham, ed., Classification of carbonate rocks—a symposium: Am. Assoc. Petroleum Geologists Mem. 1, p. 224–252.

Norris, A. W., 1965, Stratigraphy of Middle Devonian and older Palaeozoic rocks of the Great Slave Lake region, Northwest Territories: Canada Geol. Survey Mem. 322, 180 p.

Oliver, T. A., and N. W. Cowper, 1963, Depositional environments of the Ireton Formation, central Alberta: Bull. Canadian Petroleum Geology, v. 11, no. 2, p. 183–202.

Pray, L. C., 1958, Fenestrate bryozoan core facies, Mississippian bioherms, southwestern United States: Jour. Sed. Petrology, v. 28, no. 3, p. 261–273.

Sherwin, D. F., 1962, Lower Elk Point section in east-central Alberta: Alberta Soc. Petroleum Geologists Jour., v. 10, no. 4, p. 185–191.

Sikabonyi, L. A., 1959, Paleozoic tectonics and sedimentation in the northern half of the West Canadian basin: Alberta Soc. Petroleum Geologists Jour., v. 7, no. 9, p. 193–216.

Soderman, J. W., and A. V. Carozzi, 1963, Petrography of algal bioherms in Burnt Bluff Group (Silurian), Wisconsin: Am. Assoc. Petroleum Geologists Bull., v. 47, no. 9, p. 1682–1708.

Stanton, R. J., Jr., 1967, Factors controlling shape and internal facies distribution of organic carbonate buildups: Am. Assoc. Petroleum Geologists Bull., v. 51, no. 12, p. 2462–2467.

Stout, J. L., 1964, Pore geometry as related to carbonate stratigraphic traps: Am. Assoc. Petroleum Geologists Bull., v. 48, no. 3, pt. 1, p. 329–337.

Textoris, D. A., and A. V. Carozzi, 1964, Petrography and evolution of Niagaran (Silurian) reefs, Indiana: Am. Assoc. Petroleum Geolgists Bull., v. 48, no. 4, p. 397–426.

——— and ——— 1966, Petrography of a Cayugan (Silurian) stromatolite mound and associated facies, Ohio: Am. Assoc. Petroleum Geologists Bull., v. 50, no. 7, p. 1375–1388.

Thomas, G. E., and H. S. Rhodes, 1961, Devonian limestone bank–atoll reservoirs of the Swan Hills area, Alberta: Alberta Soc. Petroleum Geologists Jour., v. 9, no. 2, p. 29–38.

Van Hees, H., 1956, Elk Point Group: Alberta Soc. Petroleum Geologists Jour., v. 4, no. 2, p. 29–37.

BULLETIN OF THE AMERICAN ASSOCIATION OF PETROLEUM GEOLOGISTS
VOL. 48, NO. 3 (MARCH, 1964), PP. 329-337, 11 FIGS.

PORE GEOMETRY AS RELATED TO CARBONATE STRATIGRAPHIC TRAPS[1]

JOHN L. STOUT[2]
Denver, Colorado

ABSTRACT

Stratigraphic entrapment of oil in carbonate is a function of petrophysics of the reservoir and trap rock. These petrophysical characteristics can be observed from sample examination without extensive laboratory measurements.

Petrophysics is an essential addition to the physical measurements of total porosity and permeability routinely collected from reservoir rock samples. Total porosity is a ratio of the rock's void space to its bulk volume. Under subsurface reservoir conditions, this porosity is occupied by fluid of two phases. Commonly the non-wetting oil phase occupies this porosity according to the size and distribution of the rock's pore system. The displacement of interstitial water by oil depends on the size of pore throats. That part not effectively displaced by oil remains as irreducible water saturation within the reservoir. These reservoir properties can be determined from capillary pressure measurements conducted in the laboratory. The capillary pressure curves may be investigated by the same statistical methods used on cumulative curves from sieve analysis of unconsolidated sands.

Seven distinctive petrophysical characteristics were evident from 200 samples of Williston basin carbonate rocks studied. These characteristics may be classified by effective porosity, displacement pressure, and pore distribution. Representative examples from this study show good and intermediate reservoir rock as well as reservoir-trap rock. The concept of low effective porosity can explain high water-cut production from carbonate reservoirs.

INTRODUCTION

With the increasing interest in searching for oil in carbonate reservoirs, it is becoming more necessary to understand completely the stratigraphic entrapment of oil. Understanding stratigraphic entrapment of oil in carbonate rocks involves differentiating the reservoir and trapping properties of rocks such as the two samples in Figure 1. These two photomicrographs are to the same scale and taken under crossed nicols. The rock of Figure 1-A is a clastic limestone with numerous disconnected vugs. Figure 1-B is a tight appearing cryptocrystalline dolomite with few visible pores. However, sample B is reservoir rock from Clear Creek field on the Nesson anticline, capable of flowing 400 barrels of oil and only 20–30 barrels of water per day. This field is a typical undersaturated reservoir under fluid expansion control (Craft and Hawkins, 1959, p. 107–109). Sample A of Figure 1 is the cap rock from the same field. Thus, we see a need to understand more about two rocks such as these; not just that they

[1] Manuscript received, March 14, 1963. Paper presented on April 24, 1963, at the Rocky Mountain Section meeting of the American Association of Petroleum Geologists in Casper, Wyoming.

[2] Geophysicist, Potential Methods Section, California Oil Company, Western Division. The writer thanks The California Oil Company for permission to publish this paper. He is indebted to various members of The Standard Oil Company of California for their contributions to some of the basic research.

are different, but all about their interstices. What makes one a reservoir rock, whereas the other is not? This study has been left to the reservoir engineer as his tool for producing oil from known reservoirs (Amyx et al., 1960, p. 133–174, and Pirson, 1958, p. 97–133). The petrology learned from such a study (Archie, 1952, p. 285–289) would greatly benefit the exploration geologist in his search for new oil in carbonate provinces.

PORE GEOMETRY STUDY

Most studies of pore geometry have been restricted to sandstone-reservoir rock. These results have been applied to the Mission Canyon age carbonate-reservoir rocks of the Williston basin. The schematic enlargement of Figure 2 shows the internal pores and pore throats of a reservoir rock. The large, hachured areas of the illustration are rock fragments. The areas of fine hachuring are isolated patches of secondary mineralization, or recrystallized cement binding the grains together. The sum of the areas between the rock fragments, including the intercrystalline pore space of the cryptocrystalline binder, is porosity in this example. The ratio of this area between the grains to the total area of the illustration is the per cent porosity. This ratio is unduly large in this schematic diagram because of the emphasis on pores and pore throats. It is this total porosity that is measured by core analysis (Pirson, 1958, p. 42–44).

Under subsurface reservoir conditions, this

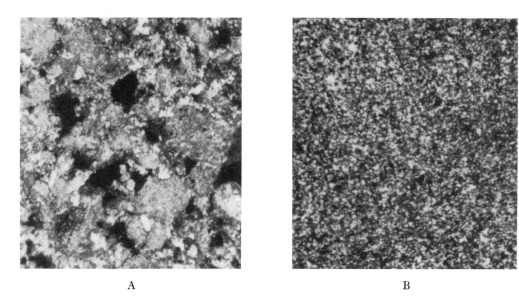

A   B

FIG. 1-A.—Photomicrograph of clastic limestone, cap rock in Clear Creek field, ×15, crossed nicols. Skelly's Larson well No. 1, depth 9,216 feet. Location: NW. ¼, SE. ¼, sec. 34, T. 152 N., R. 96 W., McKenzie County, North Dakota.

B.—Photomicrograph of cryptocrystalline dolomite, reservoir rock in Clear Creek field, ×15, crossed nicols. Skelly's Larson well No. 1, depth 9,233 feet.

porosity would be occupied by fluid of two phases—interstitial water (as defined by Levorsen, 1954, p. 298), and oil, shown by the shaded areas in Figure 2. Since this is a water-wet reservoir, there is a thin film of water around all the rock grains. Although the oil is the non-wetting phase, there may be a few patches where oil will adhere to the rock surfaces (Mattax and Kyte, 1961, p. 119) and will remain in pendular sus-

pension even when fluid is produced from this rock. There are also insular globules of oil that may be produced only if the pressure drop is sufficient to overcome the Jamin effect (Pirson, 1958, p. 72), forcing the globule through the pore throats.

In the framework of this pore system, there are pores and interconnecting throats. These throats are present between the rock fragments and also

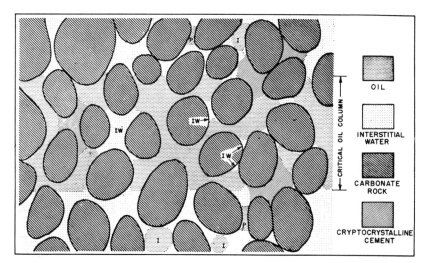

FIG. 2.—Schematic diagram of reservoir rock. Some oil remains isolated from reservoir accumulation in pendular suspension (P) and insular globules (I). Irreducible water (IW) remains within reservoir oil accumulation.

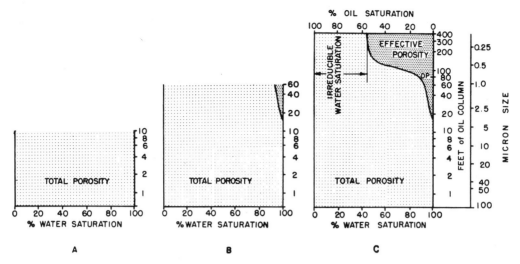

FIG. 3.—Capillary pressure curves showing progressive pore invasion of non-wetting fluid under increasing pressure. Little invasion occurs before displacement pressure (DP) is exceeded.

between the crystals of the cement (Aschenbrenner and Achauer, 1960, p. 236–239). The insular globules of oil can not migrate through these throats because of the high interfacial tension between oil and water under reservoir conditions. If enough globules coalesced by the accretion of smaller globules, the combined effect of the buoyancy of this oil column and dynamics of the reservoir would cause migration of the oil. The critical oil column of Figure 2 is that oil column required to cause oil to displace the interstitial water, or migrate through the smaller pore throats of this example's upper area. The reservoir exhibits a finite permeability to oil at this displacement pressure.

As oil invades the rock's pore space, it forces the wetting fluid out by funicular flow (Mattax and Kyte, 1961, p. 119–122). There will be areas where the water becomes disconnected from the rest of the interstitial water and will remain within the oil reservoir as irreducible water saturation (Amyx et al., 1960, p. 163). This is illustrated by areas of interstitial water within the reservoir oil of Figure 2. This pore space is not available to the oil in this reservoir-rock example. Thus, the total porosity is reduced by this irreducible water saturation to some effective porosity (Amyx et al., 1960, p. 39) that is a much smaller part of the rock's bulk volume. This is an important aspect of reservoir-rock petrophysics.

All reservoir properties depicted by the schematic diagram of Figure 2 can be represented by

the series of capillary pressure graphs shown in Figure 3. The area of the graph in part A represents total porosity of the rock sample. This porosity is 100 per cent saturated with interstitial water, although nearby there is an impending continuous column of oil 10 feet high outside this rock. This oil column may be extensive laterally, but its vertical buoyancy is the effective pressure that must overcome the displacement pressure of this rock sample.

At the time of the graph in Figure 3-B additional oil has accumulated to create a 60-foot oil column. Oil has invaded 5–10 per cent of the total porosity through some of the vugs of the rock, but, unless these vugs are connected, there is no appreciable flow of oil into this rock until a critical height of the oil column exceeds the rock's displacement pressure (point DP, Fig. 3-C). At this time, oil imbibes into the effective pore space of the rock. Additional height of the invading oil column does not increase the effectiveness of the rock's porosity because of the isolated patches of irreducible water. The capillary pressure curve rises asymptotically to an irreducible water saturation of 44 per cent of the rock's total porosity. The area to the right of the curve in Figure 3-C is now the effective part of this sample's total porosity.

Muskat (1937, p. 61) points out the necessity to know the pore geometry of a reservoir rock in order to study the movement of fluid through it. Mercury-capillary-pressure measurements are experimental determinations of this pore geometry. The laboratory procedure for capillary pressure measurement by mercury injection is briefly described by Scheidegger (1957, p. 51–52).

The mercury injection method best reproduces reservoir conditions. Displacement methods where water is the wetting phase are affected by the techniques of extracting mud-filtrate and drying core samples. The mercury injection method has been found to be reliable and to have good reproducibility (Burdine et al., 1950, p. 198–199 and 201). Samples of Williston basin carbonate reservoir rock were investigated by this method. As the surface tension of mercury and the applied pressure are known, the size of pore throats invaded can be calculated (Levorsen, 1954, p. 531–532). These sizes are shown on the graph of Figure 3-C in micron measure. The physical relation of laboratory injected mercury to crude oil under reservoir conditions depends on the individual reservoir. An average relation of specific gravity, dissolved gas, and bottom-hole pressure in Mission Canyon formation reservoirs of North Dakota may be considered as $3\frac{2}{3}$ pounds per square inch for each foot of vertical oil column. Feet of oil column is given on the graphs of Figure 3.

The capillary pressure curve is somewhat similar to a cumulative curve from the sieve analysis of sands. The median size, sorting, and other statistical parameters can be calculated and compared with data from other rock samples. This similarity is not completely true since the indurated carbonate rock is examined rather than a disaggregated sample. Not only is the influence of the deposited carbonate fragments measured, but the effects of all diagenesis are evident in the resulting capillary pressure curve.

Pores deep within the indurated rock sample are invaded by the mercury only through the pore throats. Mercury-capillary-pressure curves differ when measured through pressure decline rather than pressure buildup. Mercury is left in the deeply imbedded pores in much the same manner as residual oil in a depleted oil reservoir. The nature of this residual mercury, or hysteresis, resulting from pressure buildup and pressure decline has not been extensively studied. Limitations of the measuring apparatus used could be alleviated through use of newly developed pressure sensing devices similar to the one described by Studier (1962, p. 94–95). These pressure sensors exhibit high volume impedance and instantaneous response to pressure change. Use of an apparatus equipped with such a pressure sensor would be beneficial in both the study of internal pore-space distribution and residual oil in reservoirs.

Carbonate pore geometry in this paper does not include the effects of various fracture systems that may be present in reservoir rock. Naturally occurring fractures, as well as fractures resulting from stimulation techniques used on completing a productive well, would add to the effective pore system discussed here.

PETROPHYSICAL CHARACTERISTICS

Several distinctive characteristics were developed from this study of more than 200 Williston basin carbonate reservoir rocks. Seven photomicrographs and corresponding mercury-capillary-pressure curves are described in this paper to give the most diagnostic features of these rocks. These photomicrographs are all 15-power enlargements and were taken under crossed nicols. The one-millimeter scale is divided into tenths, or 100 microns, to give a visual correlation of the photomicrograph to the capillary curve.

This study was principally concerned with the productivity variations in carbonate reservoir rock. Sampling with this purpose in mind necessarily limited the suite of rocks studied to the zones of high energy deposit. The samples used, therefore, are limestones deposited in moderately to highly agitated waters (Plumley et al., 1962, p. 89–91). Recrystallization and dolomitization of rock deposited in an agitated environment modifies the distinguishing characteristics of this rock.

The most efficient classification of these samples would be one based on the rock's reservoir properties such as Archie's carbonate reservoir rock classification (1952, p. 280–289). The following examples are listed according to the effectiveness of the rock's total porosity, the relative displacement pressure, and the pore distribution.

LOW EFFECTIVE POROSITY

Well sorted pore distribution—low displacement pressure.—Figure 4 is a photomicrograph of a recrystallized pelletoidal limestone in a very fine matrix. The few pores have generally been plugged by clear anhydrite crystals. Dolomitization along a few of the channels has not improved the rock's permeability.

The mercury-capillary-pressure curve shows a well sorted pore distribution and a low displace-

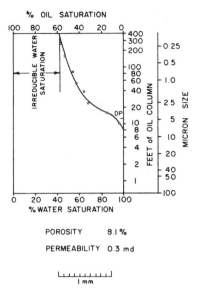

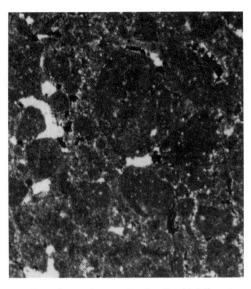

Fig. 4.—Mercury-capillary-pressure curve and corresponding photomicrograph of pelletoidal limestone, ×15, crossed nicols. Cardinal's Brace well No. 1, depth 4,505 feet. Location: NE. $\frac{1}{4}$, SE. $\frac{1}{4}$, sec. 33, T. 159 N., R. 81 W., Bottineau County, North Dakota.

ment pressure. The irreducible water saturation is greater than 40 per cent in this sample. The water saturation is considered high if it remains above the quantity of 22 per cent within the applied pressure range in the laboratory (1,015 pounds per square inch). This rock sample would be the compact-crystalline Type I of Archie's carbonate reservoir rock classification.

Poorly sorted pore distribution—high displace-

ment pressure.—Dolomitization may occur to the detriment of a rock's reservoir capacity. The rock example of Figure 5 shows an improvement of permeability over the previous example, but the tightly interlocking rhombs, seen in the photomicrograph, have created a high displacement pressure. Displacement pressure is considered high if the size of the first pore throats that must be invaded is less than $2\frac{1}{2}$ microns in diameter.

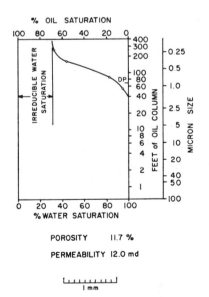

Fig. 5.—Mercury-capillary-pressure curve and corresponding photomicrograph of calcareous dolomite with considerable anhydrite replacement, ×15, crossed nicols. Johnson's Nelson-Durnin well No. 3, depth 4,121 feet. Location: SW. $\frac{1}{4}$, NW. $\frac{1}{4}$, sec. 30, T. 161 N., R. 81 W., Bottineau County, North Dakota.

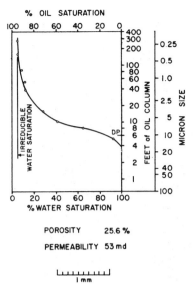

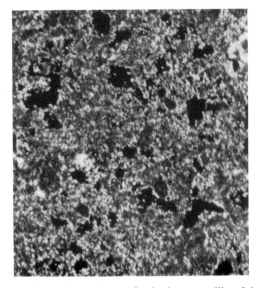

FIG. 6.—Mercury-capillary-pressure curve and corresponding photomicrograph of microcrystalline dolomite, ×15, crossed nicols. California Oil's Hofland No. 1, depth 4,088 feet. Location: SW. ¼, NE. ¼, sec. 24, T. 161 N., R. 82 W., Bottineau County, North Dakota.

High displacement pressure reduces the possibility for this rock to contain oil, the non-wetting phase of a two-phase fluid. Replacement by anhydrite has also reduced the reservoir capacity of this rock.

LOW DISPLACEMENT PRESSURE

Well sorted pore distribution—high effective porosity.—When dolomitization has occurred to the extent it has in Figure 6, the reservoir quality can be optimum. Figure 6 is of a microcrystalline dolomite. This is an interrhombohedral pore space, Type IV, of the classification referred to by Aschenbrenner and Achauer (1960, p. 237), or the saccharoidal, Type III of Archie's classification. In this rock the 25.6 per cent porosity is 95 per cent available to oil saturation. The displacement pressure is low, and the pore distribution is very well sorted. This means the effective porosity is saturated to more than 50 per cent with oil under a small increase in pressure or slightly increased invading oil column. This is a very good reservoir rock.

Poorly sorted pore distribution—low effective porosity.—In Figure 7 the total porosity is only 35 per cent effective, reducing the 15.1 per cent total porosity to about 5 per cent. Murray (1960, p. 72) suggests a rock sample should exhibit 50 per cent effective porosity at a capillary pressure equal to 60 feet of oil column to be considered a commer-

cial reservoir rock. The effective porosity of the sample in Figure 7 is less than this requirement for a reservoir rock. The photomicrograph shows some insight into the high water saturation of this example. The fine, cryptocrystalline material around the more dense pellets holds the interstitial water.

If an oil reservoir contained much of this type of rock, it would be natural to expect a high water-cut with the produced oil. This highly water-saturated rock is subjected to a pressure drop at the well-bore annulus as is the oil-saturated rock of Figure 6. Both types of rock occur commonly in the same carbonate reservoir. This is one reason for selectively perforating this type of reservoir.

Poorly sorted pore distribution—high effective porosity.—Figure 8 shows a rock similar to that of Figure 7, but dolomitization has improved the effective porosity. There are all sizes of throats and pores through what was cement material. The cumulative curve is poorly sorted, but the porosity becomes increasingly more effective with pressure. This rock would probably contain commercial oil in a reservoir, although an oil column of 10–12 feet would saturate this rock to only about 5 per cent of the total rock volume.

HIGH DISPLACEMENT PRESSURE

Well sorted pore distribution—low effective porosity.—Figure 9 is an example of the reservoir

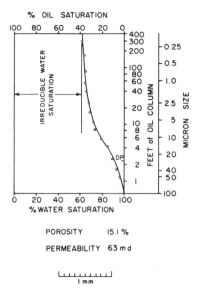

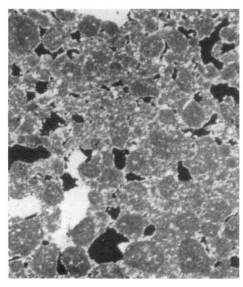

POROSITY 15.1%

PERMEABILITY 63 md

Fig. 7.—Mercury-capillary-pressure curve and corresponding photomicrograph of pelletoidal limestone, ×15, crossed nicols. Johnson's Nelson-Durnin well No. 6, depth 4,123 feet. Location: SW. ¼, SW. ¼, sec. 30, T. 161 N., R. 81 W., Bottineau County, North Dakota.

capacity of a cryptocrystalline dolomite. The porosity of 17.7 per cent is not effective. The low permeability and extremely high displacement pressure are concluding evidence that this would not be reservoir rock.

Well sorted pore distribution—high effective porosity.—In Figure 10, the "pin-point porosity," is more effective although the total porosity is

slightly less than the porosity in the example of Figure 9. The permeability is much improved, but, still, the displacement pressure is so high that under similar reservoir conditions this rock would be essentially a reservoir cap to any of the previous samples of this paper. It is well to point out the porosity of this rock is good, 12.3 per cent and 8.5 millidarcys permeability, but with such a high

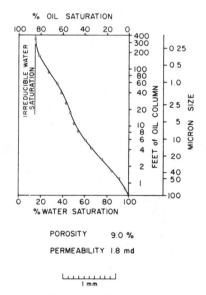

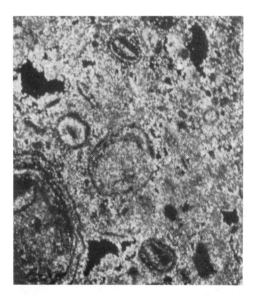

POROSITY 9.0%

PERMEABILITY 1.8 md

Fig. 8.—Mercury-capillary-pressure curve and corresponding photomicrograph of dolomitic limestone, algal oolites, ×15, crossed nicols. Johnson's Nelson-Durnin No. 1, depth 4,120 feet. Location: NE. ¼, SE. ¼, sec. 25, T. 161 N., R. 82 W., Bottineau County, North Dakota.

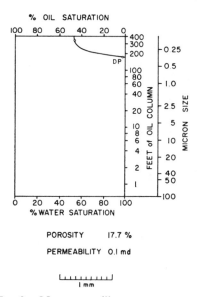

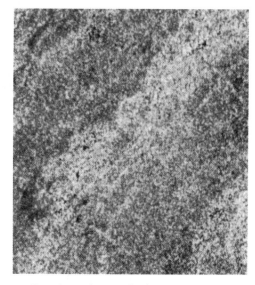

% OIL SATURATION

PROSITY   17.7 %

PERMEABILITY 0.1 md

FIG. 9.—Mercury-capillary-pressure curve and corresponding photomicrograph of cryptocrystalline dolomite, massive algae, ×15, crossed nicols. California Oil's Hofland No. 1, depth 4,086 feet. Location: SW. ¼, NE. ¼, sec. 24, T. 161 N., R. 82 W., Bottineau County, North Dakota.

displacement pressure an oil column of 100 feet would be required to force any appreciable oil into this rock.

CONCLUSIONS

The cap rock-reservoir rock relation is portrayed in the two photomicrographs of Figure 1. In Figure 11, the pore distribution curves of the two rocks in Figure 1 are superimposed to show the capacity of the reservoir. The reservoir rock has a low displacement pressure equivalent to sustaining a 14-foot column of oil, with a 4-foot transition beneath it. Thirty per cent effective saturation by oil is considered here as the economic limit; the height of the transition zone is dependent on this definition of the economic limit. There is 12 per cent interstitial water in the reservoir, and the water-wet cap rock will sustain up

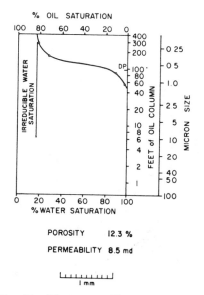

% OIL SATURATION

PROSITY   12.3 %

PERMEABILITY 8.5 md

FIG. 10.—Mercury-capillary-pressure curve and corresponding photomicrograph of cryptocrystalline dolomite. ×15, crossed nicols. Johnson's Nelson-Durnin well No. 2, depth 4,100 feet. Location: NE. ¼, NW. ¼, sec. 30, T. 161 N., R. 81 W., Bottineau County, North Dakota.

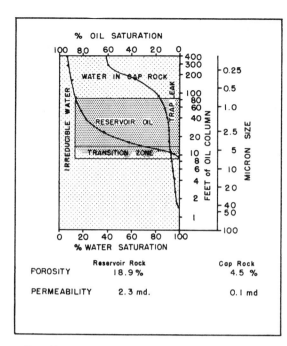

FIG. 11.—Pore geometry of stratigraphic trap. Upper capillary curve represents pore distribution of cap rock shown in Figure 1-A. Lower capillary curve represents pore distribution of reservoir rock shown in Figure 1-B.

to 66 feet of oil in the reservoir. Note there may be 5–10 per cent leakage through the cap if a favorable hydrodynamic gradient were not known to be present. Such a leak is not uncommon in carbonate reservoirs (Hill, 1961, p. 52–56). The extended coarse portion of the cap rock's capillary curve is related to the vugs seen in Figure 1-A.

The seven petrophysical classes suggested in this paper are distinct and can easily be recognized in thin-section and drill-cutting examination. Pore geometry is independent of porosity and permeability measurements, but it is a necessary addition to these data routinely collected. Only a few mercury-capillary-pressure

measurements need to be run in a carbonate province to correlate the petrophysics and petrology of the area. The time and expense of extensive laboratory measurements are unwarranted. It is necessary, however, to understand the petrophysics of stratigraphic traps if they are to be explored for with confidence.

REFERENCES

Amyx, J. W., Bass, D. M., Jr., and Whiting, R. L., 1960, Petroleum reservoir engineering; physical properties: New York, McGraw-Hill Book Co., Inc.
Archie, G. E., 1952, Classification of carbonate reservoir rocks and petrophysical considerations: Am. Assoc. Petroleum Geologists Bull., v. 36, no. 2, p. 278–298.
Aschenbrenner, B. C., and Achauer, C. W., 1960, Minimum conditions for migration of oil in water-wet carbonate rocks: Am. Assoc. Petroleum Geologists Bull., v. 44, no. 2, p. 235–243.
Burdine, N. T., Gournay, L. S., and Reichertz, P. P., 1950, Pore size distribution of petroleum reservoir rocks: Trans. Am. Inst. Min. Metall. Engineers, Petroleum Br., v. 189, T.P. 2893, p. 195–204.
Craft, B. C., and Hawkins, M. F., 1959, Applied petroleum reservoir engineering: Englewood Cliffs, N. J., Prentice-Hall, Inc.
Hill, G. A., 1961, Reducing oil-finding costs by use of hydrodynamic evaluations: Economics of petroleum exploration, development and property evaluation: p. 38–69, Dallas, Tex., Internatl. Oil and Gas Educ. Center, SW. Legal Found.
Levorsen, A. I., 1954, Geology of petroleum: San Francisco, Calif., W. H. Freeman and Co.
Mattax, C. C., and Kyte, J. R., 1961, Ever see a water flood?: Oil and Gas Jour., v. 59, no. 42, p. 115–128.
Murray, R. C., 1960, Origin of porosity in carbonate rocks: Jour. Sed. Petrology, v. 30, no. 1, p. 59–84.
Muskat, M., 1937, The flow of homogeneous fluids through porous media: Ann Arbor, Mich., J. W. Edwards, Inc.
Pirson, S. J., 1958, Oil reservoir engineering: 2d ed., New York, McGraw-Hill Book Co., Inc.
Plumley, W. J., Risley, G. A., Graves, R. W., Jr., and Kaley, M. E., 1962, Energy index for limestone interpretation and classification, in Classification of carbonate rocks, W. E. Ham, editor: Am. Assoc. Petroleum Geologists Mem. 1, p. 85–107.
Scheidegger, A. E., 1957, The physics of flow through porous media: New York, The MacMillan Co.
Studier, Walter, 1962, Quartz pressure sensors: Instruments and Control Systems, v. 35, no. 12, p. 94–95.

American Association of Petroleum Geologists Memoir 24,
North American Oil and Gas Fields, copyright 1976,
pp. 121-135.

Altamont-Bluebell—A Major, Naturally Fractured Stratigraphic Trap, Uinta Basin, Utah[1]

PETER T. LUCAS and **JAMES M. DREXLER**[2]

Abstract The Altamont-Bluebell trend is composed of a highly overpressured series of oil accumulations in naturally fractured, low-porosity, Tertiary lacustrine sandstones. It now covers more than 350 sq mi (907 km²) located across the deeper part of the Uinta basin of northeastern Utah. Postdepositional shift of the structural axis of the basin in late Tertiary time produced a regional updip pinchout of northerly derived sandstones into a lacustrine "oil-shale" sequence. Facies shifts during the deposition of more than 15,000 ft (4,570 m) of lacustrine sediments have resulted in a changing pattern of reservoir distribution and hydrocarbon charge at various stratigraphic levels. About 8,000 ft (2,440 m) of stratigraphic section is oil bearing, and up to 2,500 ft (760 m) of section contains overpressured producing zones in the fairway wells.

Reservoir performance is significantly enhanced by vertical fractures and initial fluid-pressure gradients, some of which exceed 0.8 psi/ft. The crude has a high paraffin content resulting in pour points above 100°F (37.78°C), gravities of 30–50° API, and an average GOR of 1,000 cu ft/bbl. This unique combination of geologic and hydrocarbon conditions makes it difficult to evaluate the ultimate recovery of the field, which could be more than 250 million bbl.

INTRODUCTION

The Altamont-Bluebell field, in the Uinta basin of northeast Utah, produces crude oil with high pour points from highly overpressured accumulations. Production is from multiple thin Tertiary sandstones, mainly at depths between 8,000 and 17,000 ft (2,440–5,200 m), within an area 45 by 15 mi (72 by 24 km). The field is being developed on 1-mi (1.6 km) spacing. As of June 1974, 210 wells had been completed, and production had increased to an average of 60,000 BOPD. Further significant increases in production will depend on the solution of transportation problems related to the movement of large volumes of extremely waxy crude oil with high pour points from this relatively remote location.

SIGNIFICANCE

The Altamont-Bluebell trend contains a major accumulation which is unique because of the character of reservoir and fluid properties and the highly disseminated occurrence of producible oil. These factors have resulted in unusual engineering and evaluation problems. The most important of these factors are: (1) very low matrix porosities enhanced by postlithification fractures; (2) multi-

ple thin productive zones with abnormally high fluid pressures; (3) undersaturated waxy crude with pour points of over 100°F (37.78°C); (4) production derived from intervals up to 2,500 ft (760 m) thick in the central part of the field; and (5) difficulty in defining field limits laterally and vertically because the trap is purely stratigraphic and there are no simple downdip water levels or sharp facies boundaries to the producing intervals. The presence of sandstone matrix porosity, fractures, high fluid pressures, and multiple producing zones is the key to commercial production. The factors limiting production are (1) facies changes from sandstone to nonporous redbeds, lacustrine shale, and dense carbonate rocks, or (2) the occurrence of capillary water in the sandstones with lowest porosity. A thick series of organic-rich shales and dense carbonate mudstones provides the cap seal.

The field is significant not only as a recent major hydrocarbon discovery, but also as an example of oil accumulation near the center of a deep basin. The pertinence of the latter observation is that Altamont-Bluebell may be an example of a group of deep-basin, organic-shale-related, overpressured accumulations where hydrocarbons are (at least initially) the dominant movable fluid within a large volume of rock. Evidence of significant hydrocarbon potential in this type of setting exists in many deep-basin wildcats; their exploitation awaits only the recognition of sufficient reservoir volume.

EXPLORATION HISTORY

The Uinta basin long has been recognized as one of the most petroliferous basins in the United States, because of the exposures around the basin

[1]Manuscript received, January 17, 1975. Published with permission of Shell Oil Company.

[2]Shell Oil Company, Houston, Texas 77001.

We recognize with appreciation the numerous members of the Shell Oil Company staff, in both the Exploration and Production Departments, who contributed significantly to our knowledge during the formative stages of the exploration effort and the field development. We also thank the Shell management for permission to publish.

255

Peter T. Lucas and James M. Drexler

FIG. 1—Index map, Uinta basin, Utah, showing structural elements and location of major hydrocarbon occurrences.

margins of thick "oil shales," gilsonite dikes, and "tar sands" with billions of barrels of oil in place (Fig. 1). The first truly commercial oil discovery in Utah and the Uinta basin was in 1948 at Ashley Valley near Vernal. This field produces from a structural trap in Paleozoic reservoirs on the northeast margin of the basin. Several later discoveries were made in the Tertiary section, such as Roosevelt in 1948 and Red Wash and Duchesne in 1951. Exploratory activity waned and then increased again in the early 1960s, but no significant reserves were found except for the western extension of Red Wash to the Wonsits area in 1964. During 1967, shallow productive zones at Bluebell were discovered on a low-relief structural nose. Development in this field could not be supported at the original well spacing, and spacing was subsequently increased to 320 acres.

Prior to the discovery of Altamont in 1970, Red Wash was the only major field in the basin. Red Wash field also produces waxy crude from a stratigraphic trap in Tertiary rocks; however, the productive sandstones occur at much shallower depths than those at Altamont and have normal fluid and reservoir conditions (Koesoemadinata, 1970). The generally poor development-success ratios in the basin were due to rapid productivity declines from thin producing zones and to severely reduced porosities at the greater depths.

Abundant noncommercial shows were found. A few wells in the central part of the basin indicated elevated formation-fluid pressures but did not prove to have further producing potential. Most published views of the Tertiary stratigraphy based on the earlier drilling activity suggested that a rather limited Tertiary stratigraphic potential remained to be tested. No doubt the reason for these views was the layer-cake stratigraphic concept whereby basin-margin redbed clastic units were correlated across the basin. From such an understanding of the stratigraphy, the top of the Wasatch facies became the general base of drilling for several years.

In October 1969, Mountain Fuel Supply initially completed their Cedar Rim No. 2, Sec. 20, T3S, R6W, which discovered what now appears to be the western edge of the Altamont trend. Shell Oil Company, in December 1969, spudded the Miles No. 1, Sec. 35, T1S, R4W, which was completed in May 1970 for 1,150 BOPD from 79 ft (24 m) of perforations between 12,341 and 12,942 ft (3,761–3,944 m) with an estimated 26 ft (7.9 m) of net "pay."

The Miles No. 1 established production at abnormally high pressure (over 0.7 psi/ft) from multiple zones, and also had an extensive interval of shows. Of greater importance is the fact that the well indicated the presence of only thin Wasatch redbeds with oil-bearing lacustrine strata below. The fact that oil shows and pressures were still increasing at total depth led to subsequent deeper drilling, which firmly established the presence of continuous lacustrine facies in the Green River Formation (Eocene) and the underlying Flagstaff (Eocene-Paleocene) in the central basin area (Fig. 2). This expanded section with oil-bearing potential disproved the belief that there was no opportunity for Tertiary production below the Wasatch, and it became clear that the Colton, Wasatch, and North Horn Formations were basin-margin facies and that the limited area of Flagstaff lacustrine facies seen in outcrops at the western margin of the basin expanded in the subsurface to the east.

In April 1971, Gas Producing Enterprises deepened their Powell No. 3 in the Bluebell field from shallow producing zones in the Green River Formation to a new total depth of 12,530 ft (3,819 m). This deepening extended the Altamont producing zones eastward 13 mi (21 km), into the Bluebell field. Current wildcat and development drilling continues along the field margins in an effort to determine economic limits of these producing zones. Greatest potential for extension is now toward the east.

UINTA BASIN GEOLOGIC HISTORY

Concurrent with the Laramide orogeny along the Wasatch Mountains on the west, the generally north-south-trending Late Cretaceous shoreline receded eastward (Young, 1966). Within the evolving Late Cretaceous coastal plain, increasing subsidence in northeastern Utah, southern Wyoming, and adjacent Colorado created a major area of internal drainage. In the Uinta basin, the lacustrine environment initially attracted inflow of clastic material primarily from the south. A thick wedge of northward-thinning coarse clastic sediment (North Horn, Colton, Wasatch facies) was deposited, which interfingered with organic-rich lacustrine clays and carbonate muds (Green River, Flagstaff facies). Later, clastic inflow from the Uinta Mountains to the north became dominant (Fig. 2). The total thickness of the sequence deposited from initiation of internal drainage to the first major Eocene expansion of lacustrine facies (Tgr 3 marker; Fig. 2) exceeds 8,000 ft (2,440 m; Fig. 3). Widespread carbonate and shale beds which resulted from this lacustrine expansion provide a good series of markers which aid in reconstruction of the more complicated relations of underlying facies.

During late Paleocene and early Eocene time, northerly derived clastics were deposited as a wedge of southward-thinning redbeds (Fig. 4). Because of the general textural similarity, color, and associated marginal-lacustrine sediments, it has been the practice to assign these rocks in the subsurface to the Wasatch Formation. They are only partly equivalent in age to the smaller Colton tongue of clastic redbeds on the south flank

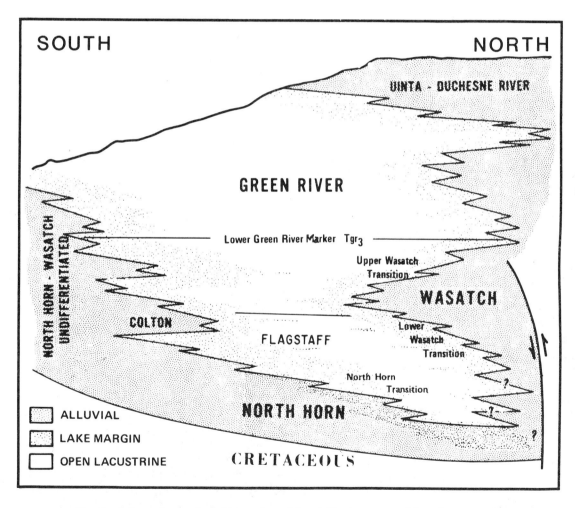

FIG. 2—Schematic stratigraphic section of lower Tertiary across Uinta basin, Utah.

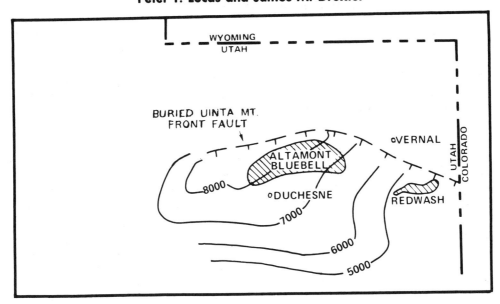

FIG. 3—Isopach map of lower Tertiary from base of Tertiary to lower Green River marker, northern Uinta basin, Utah.

of the basin and, in contrast, are underlain by an open-lacustrine, organic shale facies of the Flagstaff. The Altamont-Bluebell trap exists where the northerly derived Wasatch sandstones pinch out within the lacustrine depocenter. Where north- and south-derived sediments have crossed the lake and intermingled, as at the eastern and western ends of the basin, oil entrapment is erratic and volumes are small; the accumulations depend on geometry of individual sandstone units rather than on a more regional facies relationship.

During middle Eocene time, subsidence continued but clastic influx waned and organic-rich, "oil-shale" carbonate sediment spread more widely over the basin. This intermontane continental sedimentation continued into the Oligocene and then virtually ceased until Quaternary uplift of the Colorado Plateau. Additional differential uplift of the basin margins resulted in truncation of the entire Tertiary section across the south flank of the basin and deposition of Quaternary gravels across the north flank. This latest uplift produced a shift of the structural axis of the basin from the more central east-west trend which existed during most of the depositional phase to an axial deep directly adjacent to the presently buried Uinta Mountains frontal fault (Fig. 3).

The Uinta basin area has had a long and varied structural history. Early tectonism preserved a thick belt of Precambrian rocks across parts of the Uinta Mountains. To the south during middle Paleozoic time, the Uncompahgre became a major basement uplift. However, most important in framing the Uinta basin and influencing fault and fracture trends was the Laramide orogeny and later rejuvenation of the Uinta Mountains. Compression in the west caused extensive thrusting which provided a source of sediment; later, relaxation resulted in extension faulting, fracturing, and localized igneous intrusions. Recurrent uplift in the Uinta Mountains area from early Tertiary to Pleistocene time was expressed by pulses of clastic sedimentation and by burial of major mountain-front faults.

Within the central part of the basin, no significant faulting is documented at the surface, although orthogonal fracture sets are found throughout. However, more interesting for evaluation of subsurface conditions are the solid-hydrocarbon-filled fractures which generally parallel the trends of the major structural elements (Crawford and Pruitt, 1963). These veins or dikes are nearly vertical and are filled with ozocerite, gilsonite, and wurtzilite originating from organic-rich layers generally shallower (and less mature chemically) than those related to the Altamont trend. The dikes suggest that significant hydraulic pressures were generated in the rich source-rock layers so that extensive vertical fractures up to 10 ft (3 m) wide were opened. However, analogy with subsurface fractures of the Altamont-Bluebell field must be tempered with the observation that most of the Altamont-Bluebell fractures have

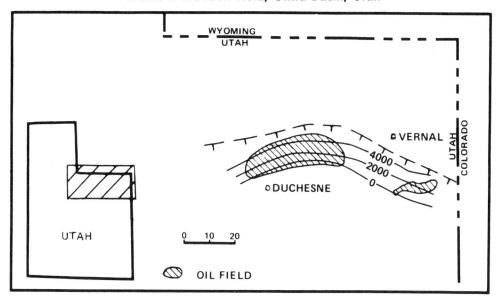

FIG. 4—Isopach map of north-derived Wasatch redbeds below lower Green River marker.

significant mineralization lining the fracture walls. This is not the general case in the dikes filled with solid hydrocarbons.

FIELD CHARACTERISTICS

Structure

Structural closure plays no part in entrapment of hydrocarbons at Altamont-Bluebell; however, the regional dip provides the setting for updip porosity pinchouts. The productive limits of the field extend more than 5,000 ft (1,525 m) downdip to essentially the deepest part of the basin (Fig. 5). To the north, a seismically mapped, buried fault is present adjacent to the basin deep and parallel with the Uinta Mountains front. Seismic records show over 6,000 ft (1,830 m) of displacement at the base of the Tertiary section, but little, if any, displacement is recognizable in the stratigraphic column above the level of the top of abnormal formation pressures. The fault appears to be a steep reverse fault along which major displacement was more or less synchronous with initiation of deposition of the Wasatch redbed wedge. Some growth continued during deposition of the lower Green River beds. Both the fault displacement and the impermeable redbeds probably contribute to retention of the abnormal fluid pressures of the field in the mountainward direction.

South of the area of major mountain-front faulting, neither seismic nor well data indicate faulting which might control the field outline. A few minor faults indicated by seismic data may influence local reservoir continuity, but this has not yet been established. In addition, faulting is not sufficiently common or of such magnitude as to have a meaningful causal relationship to the fracturing which occurs throughout the field.

Fractures

Fractures in the reservoirs of Altamont-Bluebell are essential for commercial flow rates. Core studies indicate a significant frequency of nearly vertical, mineral-lined open fractures throughout the field. Most nonmineralized breaks in recovered cores have been interpreted as having been induced by the coring process. In a collection of cores from 3 to 4.5 in. (7.6–10 cm) in diameter, more than 90 percent of the fractures are within 10° of vertical and 5 percent fall within 40° of horizontal. The pattern of occurrence is about 50 percent single fractures within any core diameter (Fig. 6) and 13 percent as crossing sets (Fig. 7). The remainder occur as multiple, subparallel fractures (Fig. 8). No significant results were obtained from limited attempts to determine fracture directions. The dominant directions of reservoir communication probably will not be determined until more data are available on asymmetry in reservoir drainage and pressure communication between wells. In the older, shallow Bluebell accumulation (northeast corner of the Altamont trend), the communication proved

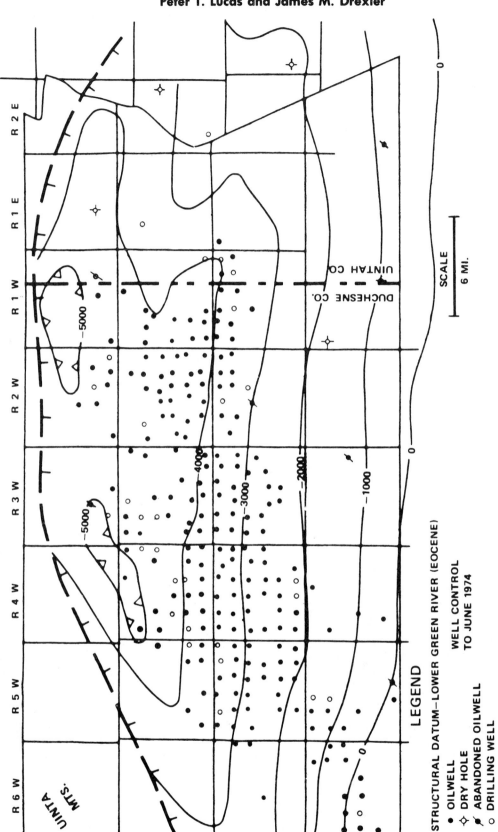

LEGEND

STRUCTURAL DATUM—LOWER GREEN RIVER (EOCENE)

- ● OILWELL
- ✧ DRY HOLE
- ✦ ABANDONED OILWELL
- ○ DRILLING WELL

WELL CONTROL
TO JUNE 1974

FIG. 5—Structure map of top of lower Green River marker, Altamont-Bluebell field, Utah. Structural nose across east end of field is site of original shallow Bluebell production.

to be east-west, approximately parallel with the axis of the structural nose.

Fracture frequency is controlled significantly by lithology. The dense carbonate mudstones are most fractured, sandstones less so, and shales are least fractured. Because vertical fractures are dominant, nearly 90 percent of fractures examined terminate within the recovered cores, either abruptly at stylolites or depositional lithologic boundaries (Figs. 9, 10), or gradually at more subtle lithologic transitions from sandstone or carbonate rock to shale. From 25 to 30 percent of the original fractures observed are open and fluid filled. These open fractures have average wall separations of about 0.5 mm and are approximately 50 percent infilled with coarse crystalline minerals. The mineral fill is predominantly calcite but, where fractures cut through sandstones, significant amounts of quartz were deposited prior to calcite deposition.

The maximum frequency of fractures appears to be in the zone of interbedded thin sandstones, carbonate rocks, and lacustrine shales. The redbeds on the north and the more massive shale and carbonate mudstone sections on the south are relatively unfractured. However, the optimum producing trend is offset northward from the area of thin sandstone interbeds toward the redbed wedge, where thicker bedded and coarser sandstones provide essential matrix porosity for economically adequate reservoir storage capacity. Production from the fractured mudstones in the central lacustrine facies declines rapidly in spite of high initial reservoir pressures.

As is common in all attempts to analyze subsurface fracture distributions, the interpretation of results can be significantly biased and limited by the nature of material available for study. Because small-diameter cores rarely recover rock with large fractures preserved in subsurface position, determination of actual spacing and orientation of open fractures in the horizontal plane becomes an imprecise statistical exercise. It is concluded from the best data available that *open* fractures in the central Uinta basin are probably oriented in a more or less orthogonal network, spaced a few feet apart. A more subtle observation is that porous sandstones appear to have

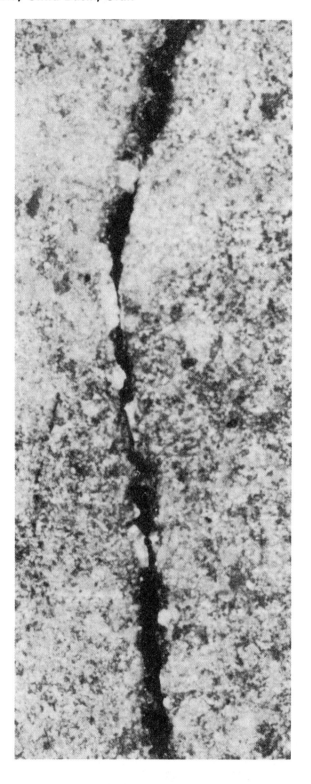

FIG. 6—Photograph of plug (1 in. or 2.5 cm in diameter) from core taken at 10,742 ft (3,275 m), showing vertical fracture in sandstone. Oil-stained fracture is lined with euhedral quartz overgrowths overlain by patches of calcite druse. Fracture permeability is 600 md through estimated average opening of 0.3 mm. Sandstone matrix porosity is 5.5 percent and permeability is less than 0.1 md. Core from Shell 1-13B4 Myrin Ranch, Sec. 13, T2S, R4W, Duchesne County, Utah.

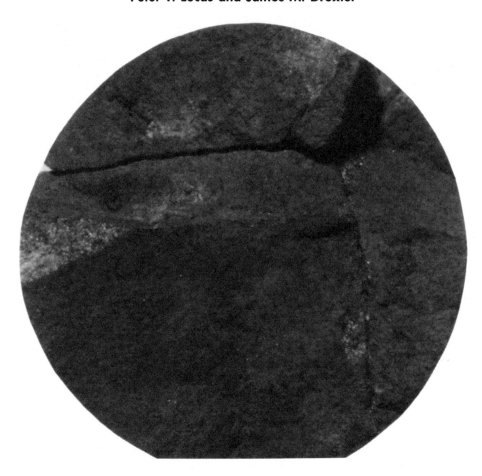

FIG. 7—Vertical view of core (2.5 in. or 6.3 cm diameter), showing nearly vertical intersecting fractures in nonreservoir sandstone with matrix porosity of 2.8 percent. Walls of partially open fractures are lined with quartz and calcite druse. Core from Shell 1-36A3 Ute, Sec. 36, T1S, R3W, Duchesne County, Utah.

higher frequencies of both total fractures and open fractures than nonporous beds of equal thickness. If this observation is true, it might suggest that fracture generation is related to pore-fluid pressure as well as to rock strength.

No unique mode of fracture generation has been proved. Fractures clearly postdate rock burial, compaction, and lithification, but they predate hydrocarbon emplacement. This timing is determined on the basis of the similarities of fracture-filling mineralization and reservoir-rock cementation, and on the basis of textural fabrics, which indicate no oil inclusions or interference phenomena. The unusually systematic change in oil composition and gravity in producing zones at increasing burial depths suggests "primary" migration into the fractured reservoirs from the adjacent sequence of organic lacustrine shales. No additional reservoirs or migration phenomena are

known which could have provided opportunity for secondary migration of oil into the presently producing zones. As a result, fracture propagation, fracture mineralization, and entry of the oil charge probably occurred during a relatively short time span during the later phases of basin subsidence and sediment loading. The preferred interpretation for creation of the deep subsurface fractures, in order to satisfy both the timing and the observed physical character and distribution, is that they formed during subsidence through a process combining preferential fracturing of less ductile, porous, generally thin beds in an area of post-thrusting crustal relaxation and high pore-fluid pressures.

Rock-mechanics theory indicates that high pore pressures contribute to yield. Pore pressure would be provided most readily by hydrocarbon generation from the load-bearing kerogen layers in the highly organic-rich rock sequence. However, this mechanism alone does not satisfy the interpreted sequence of (1) rock compaction and lithification, (2) fracturing, (3) fracture mineralization, and (4) oil migration. As a consequence, the concept that pore-fluid pressures are an essential stimulus to rock yield within the constraints of the Altamont-Bluebell setting requires the initiation of abnormal fluid pressures during expulsion of connate waters in the late phases of compaction but before significant oil generation has taken place. By this time, permeability must have been low enough to cause fluid pressure buildup. The mechanism implies that the stratigraphically deepest fractures were formed first as sediment overburden increased and the overpressured envelope expanded upward with increasing maturation of the source rocks. Continued burial maintained the oil-generation process and increased the magnitude of abnormal pressures.

Reservoir Porosity

Sufficient open fracture space has not been observed to account for the volumes produced from the better wells in the field (19 wells had produced more than 500,000 bbl of oil each as of January 1, 1975). Core and mechanical-log data indicate that reservoir storage capacity is provided primarily by multiple low-porosity sandstones averaging 5 percent porosity (range, 3–10 percent). The highest matrix porosities are in sandstones with low clay content and little or no calcite cement. These are either fine-grained, well-sorted (Fig. 11), or poorly sorted sandstones with pebbles. In contrast, the uniformly lower porosities are present in the more lacustrine calcite-cemented sandstones (Fig. 12). All sands were highly compacted prior to quartz cementation.

Figure 13 shows the electric-log characteristics of a typical productive interval in the field. Producing zones are dispersed over long intervals in

→

FIG. 8—Sandstone core with partially open, oil-stained fracture. Bifurcating, irregular fracture occurs within a 21-ft (64 m), very fine- to medium-grained sandstone interval. Sandstone matrix porosity is 4 percent, permeability is less than 0.1 md, and residual oil saturation is 20 percent. Fracture recovery was 5 ft (1.5 m) before it reached core margin at top and base. Original fracture opening is about 30 percent infilled and healed with quartz and calcite druse. Note slight vertical displacement across fracture. From 11,526 ft (3,513 m), Shell 1-21B4 Hunt, Sec. 21, T2S, R4W, Duchesne County, Utah.

FIG. 9—Polished core slab of lower Green River dolomitic limestone with fracture partially filled with calcite. Subtle downward compositional and textural changes from nonorganic to organic carbonate rock control fracture terminations at lower stylolite. Multiple hairline fractures are completely healed. From 8,376 ft (2,553 m), Shell 1-23B4 Brotherson, Sec. 23, T2S, R3W, Duchesne County, Utah.

the central part of the field or are limited to a single zone of several thin productive intervals at the field margins. Logs of potential producing zones are characterized by a subdued SP curve and by the high resistivity normally associated with hydrocarbon saturation. Commercial production appears to depend on the presence of fractures, which are not reflected on the logs. Because normal logging techniques have not made it possible to recognize fractures, indication of total producing potential requires completion and production testing. Full-bore spinner surveys taken several times over an extended period in the same borehole commonly indicate varying flow rates and different contributing zones from within the gross productive interval.

FIG. 10—Core slab of interbedded black, organic-rich shale and carbonate mudstone showing discontinuous, open vertical fracture with only hairline indication of fracture interconnection. These short open-fracture segments are calcite-druse lined, but are open horizontally through entire core width (4 in. or 10 cm). From 10,643 ft (3,244 m), Shell No. 1 Murdock, Sec. 26, T2S, R5W, Duchesne County, Utah.

Reservoir Pressures

Fluid pressures are essentially hydrostatic from a depth below near-surface topographic effects to below the lower Green River marker (about 8,000 –10,000 ft [2,400–3,000 m]). Below this, pressure gradients increase to a maximum of over 0.8 psi/ft (equivalent to 16 lb/gal mud weight) in the central part of the field. In the transitional interval, the rate of increase of pressure is as high as 3 psi/ft. As a result, even closely spaced reservoirs in this interval have considerable pressure differential. The areal distribution of maximum fluid pressures encountered throughout the field is shown in Figure 14. The indicated values encompass all stratigraphic levels and are based on a combination of borehole pressure measurements, drill-stem-test pressures, and drilling-mud weights. The base of the Tertiary stratigraphic column is generally nonpermeable; therefore, few

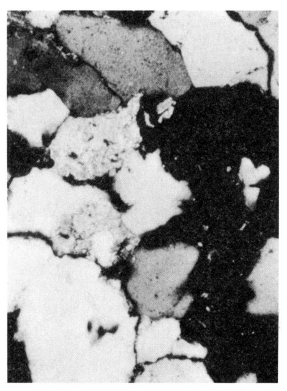

FIG. 12—Photomicrograph of highly compacted, fine-grained quartz sandstone with quartz overgrowths and pore-filling calcite cement. Nonreservoir rock has porosity of about 3 percent. From 11,571 ft (3,527 m), Shell No. 1 Miles, Sec. 35, T1S, R4W, Duchesne County, Utah. Magnification 20×.

reliable data are available as to the distribution of pressures below the producing zones. Indications are that reduced fluid pressures again approaching hydrostatic occur in the lowermost Tertiary. Additional information on the problems of drilling and completion at Altamont-Bluebell caused by the combination of multiple-fractured producing zones with differential pressures is given by Baker and Lucas (1972), Findley (1972), and Bleakley (1973).

Production

Initial well productivities are at flow rates up to 5,000 bbl/day with gas/oil ratios ranging from 1,500 cu ft/bbl (4,250 m³/bbl) in the updip part of the field to 500 cu ft/bbl (1,415 m³/bbl) downdip. Wells are currently flowing at an average of about 600 bbl of fluid per day. Reservoir drive mechanism is liquid expansion–solution gas. Gas-saturation pressures range from about 5,000 psi in the oldest reservoirs to 2,600 psi in the shallow,

FIG. 11—Photomicrograph of highly compacted, fine- to medium-grained quartz sandstone with relict oolitic dolomite rock fragment. Secondary quartz overgrowths have reduced porosity to 8 percent. Sandstone was oil stained. From 12,545 ft (3,824 m), Shell No. 1 Miles, Sec. 35, T1S, R4W, Duchesne County, Utah. Magnification 8×.

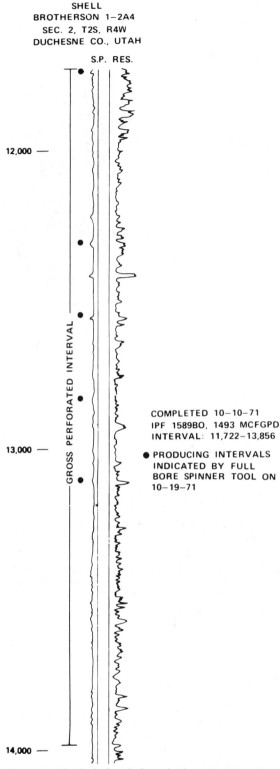

SHELL
BROTHERSON 1–2A4
SEC. 2, T2S, R4W
DUCHESNE CO., UTAH

S.P. RES.

12,000 —

13,000 —

14,000 —

GROSS PERFORATED INTERVAL

COMPLETED 10–10–71
IPF 1589BO, 1493 MCFGPD
INTERVAL: 11,722–13,856

● PRODUCING INTERVALS
INDICATED BY FULL
BORE SPINNER TOOL ON
10–19–71

FIG. 13—Typical electric log of Altamont-Bluebell producing well.

more normally pressured reservoirs. With minor exceptions, the initial fluids produced under elevated reservoir pressures are hydrocarbons essentially free of water. As the pressure is drawn down, water production commences and gradually increases. The lower limit of matrix porosity for production of oil is about 3 percent. However, residual oil saturation is not evident in all sandstones with porosity above this value, although fractures passing through these sandstones produce oil. Because identifiable free-water levels are absent in the field, water production is assumed to result from drainage of the sediment blocks between fractures as the pressure is reduced. Liquid expansion provides the drive mechanism. Water is probably contributed from both the unstained sandstones and from the partial water saturation of oil-bearing sandstones. At present, water is about 30 percent of the total fluid produced, and the percentage is increasing.

Oil gravities range from 30 to 50° API and pour points range from 95 to 115°F (35–46°C). In general, oil gravity, pour-point temperature, and gas-saturation pressure increase with depth. The increase in gravity is coincident with a change from a naphthenic "immature" crude oil to a highly paraffinic crude oil. The more naphthenic crude oils are black, whereas the paraffinic oils become lighter in color grading to olive brown and then to honey yellow as the wax content increases. The thermal gradient for the field is approximately 1.5°F/100 ft (0.87°C/30 m).

FIELD MODEL

The trap and reservoir concepts are shown diagrammatically in Figures 15 and 16. Lacustrine shales and carbonate mudstones provide the source rock, updip seal, and cap seal for the accumulation. Transitional facies between continental redbeds and the sealing rocks are characterized by thin beds (generally 10 ft [3 m] or less) of very fine- to medium-grained, white sandstones, ostracodal limestones, and light gray to pale green waxy shales. The pressure gradient of the field reaches a maximum of slightly over 0.8 psi/ft in the central area. Away from this area, maximum pressures decline to hydrostatic updip where the facies change to nonreservoir lacustrine shales and "tight" carbonate rocks and, laterally, where the clastic beds from northerly and southerly sources intertongue.

The unusual reservoir conditions (vertically separated, fractured, low-porosity producing zones

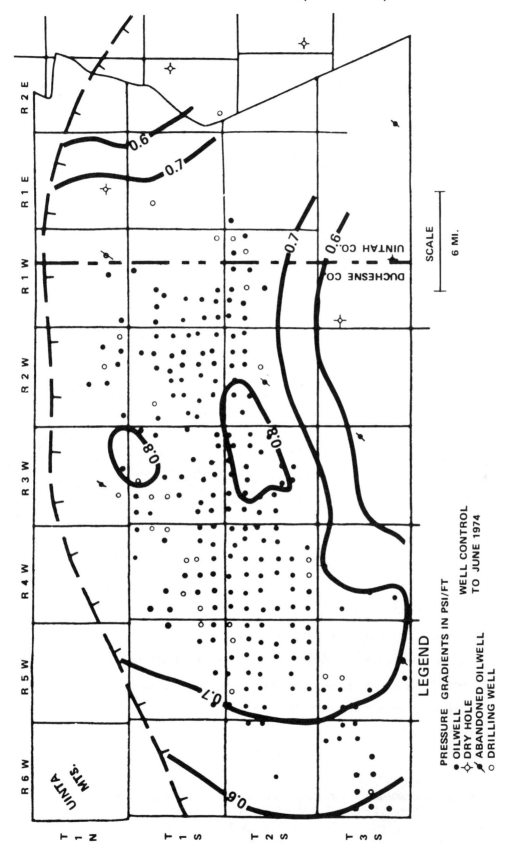

FIG. 14—Maximum observed pressure gradients, Altamont-Bluebell field, Utah. Data based on mud weights, hydrocarbon shows, lost circulation, and measured pressures. Maximum values occur at different stratigraphic levels within gross measured interval.

Peter T. Lucas and James M. Drexler

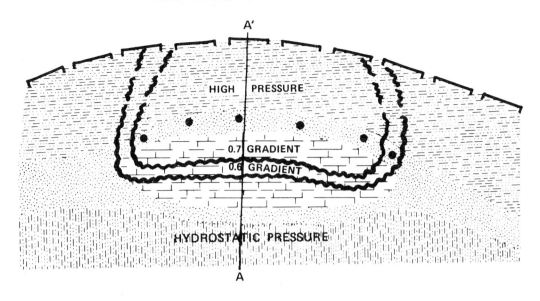

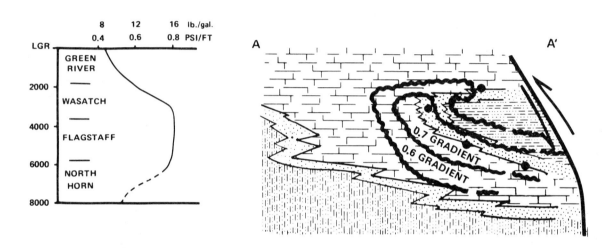

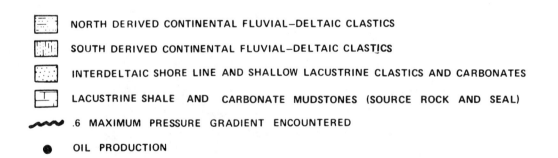

NORTH DERIVED CONTINENTAL FLUVIAL–DELTAIC CLASTICS

SOUTH DERIVED CONTINENTAL FLUVIAL–DELTAIC CLASTICS

INTERDELTAIC SHORE LINE AND SHALLOW LACUSTRINE CLASTICS AND CARBONATES

LACUSTRINE SHALE AND CARBONATE MUDSTONES (SOURCE ROCK AND SEAL)

.6 MAXIMUM PRESSURE GRADIENT ENCOUNTERED

OIL PRODUCTION

FIG. 15—Altamont-Bluebell trap and reservoir model, Uinta basin, Utah,
showing pressure distribution relative to lithology.

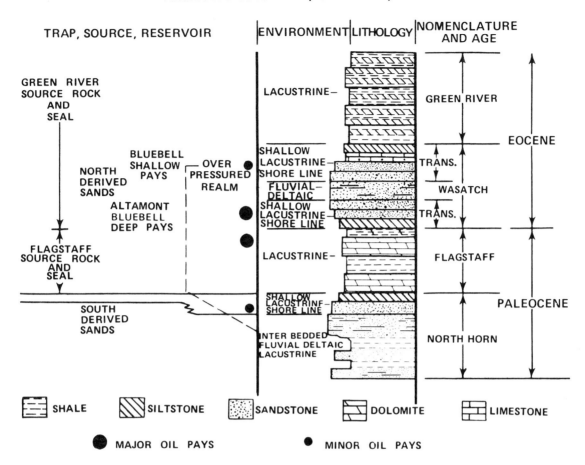

FIG. 16—Stratigraphy, nomenclature, and environmental relations of northern Uinta basin.

of variable lithology) and fluid properties (undersaturated, high-pour-point oil) do not lend themselves to normal reservoir-engineering techniques. Ultimate recoveries for individual wells are predictable only after a considerable production history. Even then, the ultimate oil/water ratio is uncertain. Production history to date indicates that the larger well recoveries (over 500,000 bbl) will occur mostly within the 0.7-psi/ft pressure-gradient envelope and updip from the maximum thickness of the wedge of Wasatch redbeds.

The field encompasses over 350 sq mi (907 km²) and may have an ultimate production of more than 250 million bbl (Lucas, 1973). Because many facets of reservoir character and performance are not fully understood, the final economic value and ultimate drilling density are still uncertain. It is clear, however, that the Altamont-Bluebell field represents one of the most widely dispersed major oil accumulations to be exploited.

REFERENCES CITED

Baker, D. A., and P. T. Lucas, 1972, Major discovery in Utah: Strat trap production may cover over 280 square miles: World Oil, v. 174, no. 5, p. 65-68.

Bleakley, W. B., 1973, How Shell solves Uinta basin problems: Oil and Gas Jour., v. 71, no. 6, p. 45-50.

Crawford, A. L., and R. G. Pruitt, 1963, Gilsonite and other bituminous resources of central Uintah County, Utah: Utah Geol. and Mineralog. Survey Bull. 54, p. 215-224.

Findley, L. C., 1972, Why Uinta basin drilling is costly, difficult: World Oil, v. 174, no. 5, p. 77-81.

Koesoemadinata, R. P., 1970, Stratigraphy and petroleum occurrence, Green River Formation, Red Wash field, Part A: Colorado School Mines Quart., v. 65, no. 1, 77 p.

Lucas, P. T., 1973, Altamont—a major fractured and overpressured stratigraphic trap (abs.): AAPG Bull., v. 47, no. 4, p. 791.

Young, R. G., 1966, Stratigraphy of coal-bearing rocks of Book Cliffs, Utah-Colorado: Utah Geol. and Mineralog. Survey Bull. 80, p. 7-21.

American Association of Petroleum Geologists Memoir 16,
Stratigraphic Oil and Gas Fields, copyright 1972, pp. 47-81.

Primary Stratigraphic Traps in Sandstones[1]

DAVID B. MacKENZIE

Marathon Oil Company, Denver Research Center, Littleton, Colorado 80120

Abstract Primary stratigraphic traps in sandstone involve lateral termination of the reservoir as a direct or indirect result of factors related to the depositional environment. Red Wash, Coalinga East, Pembina, Mitsue, Bell Creek, Cut Bank, Burbank, and Bradford are among the very few giant oil accumulations found in such traps. As these traps rarely can be detected by surface measurements, other discovery methods are essential. The understanding of depositional process and environment is a promising approach.

Primary stratigraphic traps in sandstone are present in many facies, including fluvial, deltaic, shallow marine, and deeper marine. The largest sizes and greatest number occur in shallow-marine and shoreline environments. Knowledge of sandstone models of all kinds may provide valuable clues in interpreting fragmentary well data in terms of size, shape, trend, and characteristics of the reservoirs being sought.

The distribution of many sandstone bodies may be controlled in part by underlying, commonly inconspicuous, erosional surfaces. Reconstruction of the paleotopography of the unconformity thus may commonly delineate prospective trends. The distribution of trap barriers may be controlled by environment. For example, discrete shoreline sandstone bodies replaced updip by lagoonal shales are better prospects than those replaced updip by sandy ("leaky") deltaic deposits. Such sandstones are more likely to be related to interdeltaic rather than deltaic areas.

Most progress will come from further development and refinement of depositional models. A greater understanding of shallow-marine sandstone bodies is especially needed. Moreover, as exploration emphasis shifts offshore, there will be a growing premium on ability to recognize depositional models in the absence of cores and outcrops.

INTRODUCTION

Oil-filled stratigraphic traps in sandstone are hard to find. Most known ones were found either unintentionally in the course of exploration for structural accumulations or through intensive drilling programs based on scarce subsurface leads. Yet, the presence of large traps like those at the Burbank and Pembina fields and the decreasing number of economically attractive onshore structural prospects have spurred great research efforts over the past 15 years to develop techniques for finding stratigraphic traps with only a minimum of subsurface control. Although some of this research has been aimed at empirical correlations and some at development of geophysical and geochemical approaches sensitive to lateral variations of thin stratigraphic units, the emphasis has been on environmental types of sandstone bodies. Each environmental type has a characteristic size, geometry, paleogeographic orientation, and relation to enclosing facies; it also has characteristic internal porosity and permeability distributions. As we learn to distinguish each type in cores, cuttings, and logs, we can explore more effectively for oil and gas. Better knowledge of the nature of the target allows more effective design of the appropriate set of exploration approaches and techniques, thus enhancing exploration success.

A knowledge of depositional models has been applied to exploration at the prospect level and, particularly, to efficient field development and extension.

Equally important is the regional delineation of favorable areas. On this scale of stratigraphic-trap exploration, depositional environment of sandstones is only one of many relevant geologic factors in fairway delineation: others include structural history, postdepositional alteration, adequacy of source rocks, *etc.*

PRINCIPAL FINDINGS

Primary stratigraphic traps in sandstone-shale sequences are commonly present in alluvial, deltaic, and shallow-marine deposits and are less common in deep-water sandstones. They are especially common in the Pennsylvanian of the Mid-Continent and the Cretaceous of the Rocky Mountains, but are scarce in the Tertiary of California and the Gulf Coast. Major oil accumulations in primary stratigraphic

---

[1] Manuscript received, January 14, 1971.

Marathon's work in the interpretation of sandstones has been concentrated heavily in the Cretaceous of the Rocky Mountains, partly because of outstanding surface exposures and available subsurface cores, and partly because of Marathon's historic exploration interest there. Consequently, many of the examples in this paper are drawn from the Rocky Mountain Cretaceous. However, the sandstone depositional models and exploration approaches discussed should be broadly applicable. Only for deep-water sandstones does one have to seek elsewhere for appropriate models.

In preparing this paper, I have drawn on the work of my Marathon colleagues both in exploration and in research, particularly that of J. C. Harms and D. G. McCubbin. I acknowledge with thanks their contributions and comments.

Critical reviews by H. R. Gould, G. Rittenhouse, and W. K. Stenzel led to significant improvements in the paper.

traps in sandstone are scarce compared to those in structural, reef, and unconformity-sealed traps.

Individual primary stratigraphic-trap prospects cannot be defined by any combination of present surface geophysical or geochemical methods, except that large traps in thick sandstone bodies less than 3,000 ft (915 m) deep possibly are detectable by gravity or seismic methods. The principal reason is that primary stratigraphic traps occur at the lateral edges of sandstone bodies usually less than 100 ft (30 m) thick. Even if lateral variations in porosity or saturation in such thin genetic units could be detected from the surface, they would be obscured by other unrelated changes in shallower and deeper rocks.

On the basis of subsurface information, lateral proximity to stratigraphic traps may be indicated by lateral changes in clay content, formation-water salinity, or hydrodynamic gradient. Highs related to inferred differential-compaction effects in beds overlying a discontinuous sandstone body may encourage deeper drilling to reach it. However, since such proximity indicators are rare—and successful use is even more rare—a more general approach is needed. A review of data from wildcat failures in light of knowledge of many different kinds of sandstone depositional models is still a very useful guide to finding traps.

For delineation of the more prospective fairways, reconstruction of the paleogeography has been one successful technique. It probably should be augmented more commonly by considerations of postdepositional processes such as solution, cementation, clay-mineral dehydration, and fluid-potential gradients.

Many primary traps are present in sandstones whose distribution is partly or wholly controlled by an underlying erosional surface. Examples include alluvial-valley and marine strike-valley fills (Fig. 6). In view of this relation, paleotopographic and paleogeologic maps of the erosional surfaces are important aids to exploration for many primary stratigraphic traps. Even where an erosional surface is not a factor or cannot be identified, shoreline sandstones deposited during an overall transgression may be more prospective than those deposited during an overall regression (Fig. 5).

In future exploration, our growing knowledge of stratigraphy, including both depositional and diagenetic aspects of rock interpretation, will have important application in delineating prospective fairways in which to concentrate effort on finding structural and pre-unconformity traps, as well as primary stratigraphic traps.

PRIMARY TRAPS

In the stratigraphic traps in sandstone discussed in this paper, lateral variation in the lithology of the reservoir rock, or a break in its continuity, has been a major factor in entrapment. In most of them, the lateral variation of permeability is a direct result of the depositional environment, rather than of postdepositional selective solution or cementation. In those unusual and still poorly understood examples in which selective solution of fossil debris or less stable detrital grains has played a significant role in creating the reservoir, the selective solution itself probably is governed by factors related to the depositional environment. For these reasons, I shall refer to all of them as primary stratigraphic traps.

Following this definition, I exclude from consideration all those pools trapped beneath an angular unconformity, because the main trapping mechanism is unrelated to deposition of the reservoir sands. The East Texas field, with 6 billion bbl of ultimately recoverable oil in the Upper Cretaceous, is the best example in this category. A related group which is not considered includes those pools in which a tar seal at a surface unconformity is an important trapping element. Among the largest examples are the Bolívar Coastal field (Venezuela), with 30 billion bbl in the mid-Tertiary, and the Quirequire field (Venezuela), with 1 billion bbl in the Pliocene-Pleistocene; the Kern River field of California, with 700 million bbl in the Pliocene-Pleistocene; and the Athabasca tar sands of northeastern Alberta. Fields such as these may be relatively easy to find, either because of surface seeps or because of convergence along unconformities detectable by the reflection seismograph.

Not only are primary stratigraphic traps harder to find, but giant fields in such traps are much more scarce. A few giant fields in this category are shown in Table 1.

TRAP REQUIREMENTS

Reservoir rocks capable of containing significant amounts of oil must be juxtaposed with barrier beds capable of acting as effective seals in a three-dimensional configuration that encloses a volume of rock of low energy potential. This trap must form at the right time and place to intercept migrating oil. In a primary

Primary Stratigraphic Traps in Sandstones

Table 1. Selected Giant Primary Sandstone Stratigraphic Traps

Name	Location	Depth (ft)	Age	Year of Discovery	Ultimate Recoverable Reserves (million bbl)	Depositional Type	References
Greater Red Wash	Utah	5500	Eocene	1951	135[1]	Lacustrine delta	Koesoemadinata (1970)
Coalinga East	Calif.	8000	Eocene	1938	520[1]	Shallow marine (?)	Chambers (1943)
Pembina	Alta.	5000	L. Cret.	1953	1800[2]	Shallow marine	Michaelis (1957) and Nielson (1957)
Bell Creek	Mont.	4500	E. Cret.	1967	114[1]	Barrier bar and Delta front Barrier bar	McGregor and Biggs (1963) Berg and Davies (1968)
Cut Bank	Mont.	3000	E. Cret.	1926	200[1]	Alluvial valley	Blixt (1941), Shelton (1967)
Burbank	Okla.	3000	Penn.	1920	500[1]	Shoreline Alluvial valley	Bass et al. (1942) Marathon (unpub.) and D. R. Baker (personal commun., 1969)
Bradford (partly structural)	Pa.	1500	L. Dev.	1871	660[1]	Shallow marine Turbidite	Fettke (1938) (by comparison with nearby N. Y. Devonian outcrop)
Mitsue	Alta.	5700	M. Dev.	1964	300[3]	Deltaic	Kramers and Lerbekmo (1967)
Nipisi	Alta.	5500	M. Dev.	1965	200[3]	Deltaic (?)	Kramers and Lerbekmo (1967)

[1] Oil and Gas Journal, January 26, 1970.
[2] Oil and Gas Journal, January 13, 1969.
[3] H. W. Nelson (personal commun., 1970).

stratigraphic trap, the boundary between the reservoir and the enclosing rocks may be slightly older than, contemporaneous with, or slightly younger than the reservoir itself.

Except for isolated lenses of sandstone enveloped in shale, most primary stratigraphic traps in sandstone have some structural elements; the configuration of the trap boundary is commonly governed by tilting or gentle arching as well as by stratigraphic variation. All gradations and combinations of structural and stratigraphic conditions have been observed.

Although the updip lateral change commonly is from reservoir sandstone to impermeable shale, the lithologic contrast need not be so great. In an area of western Nebraska, for example, the updip barrier to an alluvial valley-fill sandstone reservoir is itself a fine-grained sandstone with some permeability (Harms, 1966). The key is not contrast in permeability but rather in capillary-pressure characteristics.

One of the most difficult problems is how to predict the capacity of an updip stratigraphic-trap barrier. The problem arises because barriers commonly consist of complexly interlaminated sandstones, siltstones, and shales. In Figure 1, the block in the upper diagram represents a sandstone bed (overlain and underlain by shale beds not shown) dipping to the left. Oil is trapped downdip from the barrier. The lower diagram is a schematic representation of the same block, in which the innumerable tortuous channelways through the barrier are represented by three tubes.

The capacity of the barrier is determined by the "critical throat," the tightest part of the particular channelway that requires the least pressure for passage of oil through it. We have no way of sampling, measuring, or even estimating reliably the capacity of the throat; however, a minimum value possibly might be established by observation of the length of oil column in a nearby similar trap (Harms, 1966).

If hydrodynamic conditions prevail, the situation may be modified somewhat. If the hydrodynamic gradient is downdip, the capacity of the barrier to hold oil or gas is enhanced; if the gradient is updip, the barrier capacity is diminished correspondingly. In this way, local hydrodynamic conditions may influence the occurrence of stratigraphic traps or the size of the stratigraphic pools.

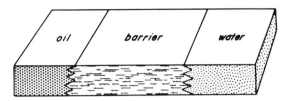

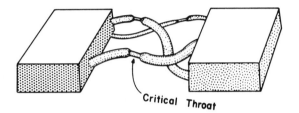

Fig. 1—Schematic diagram of stratigraphic-trap barrier.

Examples of fields with unusually effective stratigraphic-trap barriers—as indicated by the height of continuous oil column in the trap—are Horseshoe Canyon, northwest New Mexico (oil column 2,750+ ft, capillary pressure ~500 psi; McCubbin, 1969), and Coalinga East, California (1,800 ft, ~180 psi).

LOCATING PROSPECTS

Potter (1967) discussed prediction problems related to sandstone bodies. He considered two aspects: (1) "the location problem"—where to find the sandstone bodies—and (2) "the extension problem"—how to outline efficiently the areal extent with a minimum of drilling. Location of sandstone bodies at the prospect level and, more partciularly, location of traps are the primary concern in this section.

In relatively mature areas of known stratigraphic traps in sandstone, a common, and sometimes effective, exploration approach has been used for many years. This approach involves drilling on a structural nose between a downdip well with porous and permeable, but water-saturated, sandstone and an updip well with no reservoir sandstone. It was applied successfully in the 1950s in the Denver basin. Of course, where the previously drilled wells are far apart, there is considerable uncertainty in predicting where and how the reservoir sandstone terminates updip.

Another empirical approach that may be successful is based simply on trend. Where early exploration in an area indicates that the sandstone bodies are elongate and aligned, an obvious step is to drill on trend. A good example is the Cretaceous "Gallup" or Tocito reservoirs of northwest New Mexico.

The spotty success of these simple methods forced consideration of more scientifically sophisticated techniques. Geologists reasoned that, if the origins of the sandstone bodies being sought were better understood, exploration would be more successful. A central idea was that recognition of the depositional environment, coupled with knowledge of paleoslope and distribution of enclosing facies, would be an effective approach to the location problem. Hence, much industry research on depositional models has been done in the past decade.

Depositional Models

In Table 2, the exploration characteristics of environmental types of sandstone bodies encountered most in exploration are summarized. Because it is difficult to capture the essence of the conceptual model in a brief table of this kind, two or three useful references are cited for each environmental type.

How is the origin of a particular sandstone body determined? No one characteristic, or even any two in combination, is sufficient to permit identification or in some cases to narrow the possibilities significantly. Furthermore, for sandstones, in contrast to carbonate rocks, the kind of information available from cuttings—color, grain composition, and texture—has been of limited value when used alone. The characteristics best determined in cores or outcrops—especially sedimentary structures, nature of contacts, sequence, and relation to enclosing facies—are much more diagnostic.

Because the amount of information on depositional models that is decipherable from cuttings and logs is limited, effective use of them depends on how closely their characteristics can be related to more basic control from cores or outcrops. For example, a marked base and an upward decrease in grain size, characteristic of many valley fills, may be reflected on the SP curve by an abrupt excursion at the base and a gradual decrease upward. If this relation can be established in one or more wells in an area by core-log comparison, the character of the SP can be used as a secondary control in identifying and mapping the sandstone body.

As a second example, two different types of sandstones within a generally sandy interval may be distinguishable by a grain-size difference. In northern Colorado, the contact between the Horsetooth and Fort Collins Members of the Muddy Sandstone (transgressive deltaic sandstone versus regressive shoreline sandstone) was determined by binocular examination of grain size in cuttings after the relations were established by detailed outcrop study (MacKenzie, 1965). No consistent log difference was observable. It should be emphasized that this is purely a local characteristic.

If the paleogeographic setting and paleoslope, as well as the depositional model, are known, sandstone trends may be predictable. The clearest example is that of barrier-island sandstone bodies, which are usually parallel with the shoreline. Alluvial valley deposits trend generally parallel with the dip of the paleoslope, but major departures are common, and marine bars, even if elongate, tend to have diverse orientations.

Some implications of different sandstone depositional models to exploration and field development are illustrated in Figure 2. Although stratigraphic traps in most sandstone depositional types require local structure or updip bends related to regional dip, traps in isolated offshore-bar sandstone bodies require no special structural situation. Traps in porous marine sandstones developed by winnowing of the fines might be sought on local synchronous highs, but stratigraphic traps in sandstone beds deposited by turbidity currents would be sought on the edges of depositional lows.

Halos

In the carbonate realm, the detection of debris beds in an exploratory well may indicate the presence of a nearby reef. In the sandstone realm, comparable proximity indicators are less common. However, some types of offshore bars are surrounded by facies halos in which clay content increases gradually outward. In the marine bars of the Willson Ranch field, Nebraska, for example (Exum and Harms, 1968), the gradient in clay content can be detected for a distance of 1–2 mi (1.6–3.2 km) beyond the field boundaries. These proximity indicators may be useful in locating low-clay (high-sand) areas (see "offshore bar," Fig. 2).

A similar example is related to the seaward margins of barrier islands. These sandstone bodies disappear seaward by becoming progressively thinner and finer grained until they merge imperceptibly into the enveloping shales. Recognition of this kind of seaward edge in two or more wells might provide a basis for determining the direction in which the porous and permeable part of the sandstone body lies.

In discontinuous sandstone units of the Rocky Mountains that have been partly flushed by meteoric waters, the salinity of formation waters—as indicated by normal or induction devices for example—may be a proximity indicator to oil accumulations. The hypothesis is that oil will be trapped only in the places protected from flushing. These relatively unswept parts should be characterized by higher formation-water salinities. Consequently, encountering waters with normal salinity in a formation otherwise known to be flushed fairly pervasively may be a clue to the proximity of oil. Although this technique has been widely discussed, I know of no demonstrable case where it has been applied successfully to oil finding.

Differential Compaction

Wherever a sandstone body is replaced laterally by a shale section, some degree of differential compaction almost certainly will result (Mueller and Wanless, 1957). An example is shown in Figure 3. The effect of differential compaction normally persists upward in the section for a few hundred or more feet. This drape is reflected as structure in the younger beds and, in places, can be used to infer the presence of discontinuous sandstone bodies in a section deeper than that penetrated.

Rittenhouse (1961) illustrated the problem of restoring the cross section of an isolated sandstone body to its original shape. It is clear that, where differential compaction has been a significant factor, cross-sectional shape is an unreliable index of the depositional model. Truncation of older markers and onlap of younger markers are more useful criteria.

Seismic Responses

There have been many attempts to prospect for stratigraphic traps in sandstone with the reflection seismograph, and the results have been disappointing. The principal problem has been that the wavelength of seismic energy returned to the surface is greater than the thickness of the sandstone bodies being sought. Most individual sandstone bodies are less than 100 ft (30 m) thick. Where such sandstone bodies are deeper than a few thousand feet, their lateral variations or terminations are seismically invisible.

A more general but still valuable goal is the detection of other stratigraphic information—such as sandstone-shale ratios and the thickness, continuity, and spacing of individual sandstone beds. By use of the concept of sedi-

Table 2. Summary of Characteristics of

	GENERAL LITHOLOGY	CHARACTERISTICS OF ENTIRE SEDIMENT BODY				CHARACTERISTICS OF		
		THICK-NESS (FEET)	SHAPE, HORIZONTAL DIMENSIONS	DISTRI-BUTION, TREND	RELATIONSHIP TO ADJACENT OR ENCLOSING FACIES	LITHOLOGY, COMPOSITION, TEXTURE, FAUNA	BOUNDING CONTACTS	OVERALL VERTICAL GRAIN-SIZE CHANGE
SUBAERIAL MIGRATED DUNE SANDS	sands, no muds; homogeneous	10 to > 1000	elongate, or sheets up to 1000's of sq. miles in area	downwind from source of sand	commonly the end stage of a regressive sequence	well sorted sands; pebbles & clasts rare	variable	not systematic
ALLUVIAL SANDS (DELTAIC DEPOSITS)	sands, muds, some gravels	usually 30-80; sometimes 200-300	continuous bodies; usually ½ to 5 mi. wide, 10's to 100's of miles long	make large angles with shoreline trends	lower contacts erosional; lateral contacts erosional or indeterminate	pebbles and clasts common; proportion of mud variable	base erosional; top usually transitional	upward decrease
DISTRIBUTARY CHANNEL FILLS (DELTAIC DEPOSITS)	sands, muds	up to 200	continuous sinuous bodies, usually < 1 mi. wide		commonly enclosed in nonmarine or blackish muds			
DELTA-FRONT SHEET SANDS (DELTAIC DEPOSITS)	sands	20-80	sheets		underlain by marine pro-delta muds; overlain by marsh muds			
REWORKED TRANSGRESSIVE SANDS (DELTAIC DEPOSITS)	sands	1-40	sheets		underlain by or adjacent to deltaic deposits	well sorted; may contain coarse sand lag	both sharp	not systematic
REGRESSIVE SHORELINE SANDS (BARRIER-ISLAND SANDS)	sands; rare muds	20-60	elongate or sheets; up to miles wide, 10's of miles long	parallel to shoreline where elongate	transitional downward and seaward into muds, landward into lagoonal or deltaic deposits	well sorted; pebbles & clasts rare; marine fauna, if any	base transitional; top sharp	upward increase; (but middle may have coarsest beds)
OFFSHORE BARS (SHALLOW MARINE SANDS)	sands with mud partings	several to 10's	elliptical lenses, less than a few sq. miles in size	scattered; orientation variable	enclosed in and intertongues laterally with marine muds and silts		sharp, or narrowly transitional	
STRIKE-VALLEY SANDS (SHALLOW MARINE SANDS)	fine to coarse sands and muds; heterogeneous	10-50	elongate; up to several miles wide, 10's of miles long	parallel to pre-unconformity paleo strike	fills erosional strike valleys; intertongues with marine muds seaward; onlaps landward	pebbles, clasts, glauconite, phosphate, marine fauna		not systematic
PROXIMAL TURBIDITES (DEEP-WATER SANDS)	interbedded sands, silts and muds	100's to 1000's	fans or sheets up to 1000's of sq. miles in area	high flanks of deep basins near sand source	may be middle part of regressive sequence from deep to shallow-water deposits	graded bdg; displaced shallow-water fauna; proximal turbidites often with interbedded debris beds	variable	
DISTAL TURBIDITES (DEEP-WATER SANDS)				sumps of deep basins*	sands interbedded with deep-water muds			

Primary Stratigraphic Traps in Sandstones

Some Environmental Types of Sand Bodies

INDIVIDUAL VERTICAL SECTIONS						MISCELLANEOUS REMARKS	REFERENCES
PRIMARY SEDIMENTARY STRUCTURES					DEFORMATIONAL AND ORGANIC SEDIMENTARY STRUCTURES		
STRATIFICATION	CROSS-STRATIFICATION			RIPPLES			
	CONTACTS; SET THICKNESSES	NATURE OF LAMINAE	SHAPE OF SETS				
conspicuous high-angle cross bedding	erosional, horiz. or sloping; sets up to 10's of feet thick	lee dips 25°-34°; commonly tangential to lower boundary	tabular; sometimes enormous troughs	high indices; crests often parallel to dip of lee beds	slumps not uncommon; vertebrate tracks		McKee (1966)
many beds lenticular; abundant cross bedding	erosional, planar or concave up; usually ¼-2 feet thick	maximum dips usually 20°-25°, inclined or tangential to lower boundary	trough	short-crested; linguoid; microtrough in cross-section; abundant	slumps common; burrows uncommon		Harms (1966) Hewitt & Morgan (1965) Fisk (1944) Potter (1967)
					slumps, burrows not uncommon		Frazier (1967) Brown (1969)
similar to barrier sand bodies ——————————————————————>							
high-angle cross bedding; orientation diverse	erosional, planar; ¼-2 feet thick	maximum dips 20°-25°, tangential to lower boundary	wedge or tabular	not conspicuous; microtroughs present	slumps, burrows uncommon	lateral facies changes may provide proximity indicators	Frazier (1967) MacKenzie (1965)
upper & lower: subhorizontal stratification with low-angle truncations, esp. near base; sets < 1 foot thick				most abundant near base; symm.; long-crested	load structures & burrows common at base		Bernard et al (1962) Weimer (1966) McCubbin & Brady (1969)
middle: high-angle cross bedding, ¼-2 feet thick, tangential laminae; cross-laminae dip obliquely shoreward; local scours			wedge or trough	uncommon	uncommon		
low-angle cross bedding	erosional, planar; ~1 ft. thick	most dips < 10°; laminae parallel to lower set boundary	wedge?	common; some symm., long-crested	burrows abundant only in marginal facies	gradual outward decrease in sand/clay may provide proximity indicator	Exum & Harms (1968)
tabular units with high-angle cross bedding dipping parallel to sand body elongation	erosional, planar; ¼-5 feet thick	max. dips 25°-30°; tangential to lower boundary	tabular; sets straight & continuous for 100's of feet	common locally, esp. at toes of x-sets; some are long-crested wave ripples	burrowing common	paleogeologic and paleotopographic maps effective in exploration	McCubbin (1969)
parallel stratified or structureless; may have large mud-lined scours	trough-shaped sets found rarely			asymmetric ripples, both short and long-crested, found at tops of individual beds	burrows uncommon; bedding plane tracks and trails often present	cf. distal beds, proximal beds are thicker, coarser grained, less well graded, less regular, more deformed, and more porous and permeable	Walker (1966, 1967)
1-3' continuous beds; parallel or ripple stratified							

POSSIBLE IMPLICATIONS OF SANDSTONE MODELS

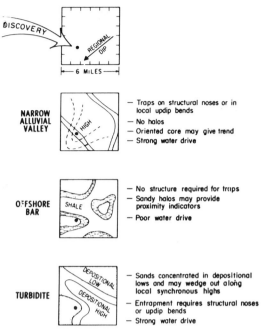

FIG. 2—Possible implications of selected sandstone depositional types to exploration and field development.

mentation models through sequences of hundreds of feet of strata, it may be possible to make the step from seismic field records to stratigraphic interpretations. Sedimentation models are not random stacks of various lithologies; rather, they commonly are organized systematically. The use of reflection seismic data for recognizing and deciphering these systematic stratigraphic relations in areas of sparse well control holds much promise (Harms, 1968).

FIELD DEVELOPMENT

Once oil has been discovered in a stratigraphic trap, efficient field development depends in part on the geologist's ability to predict size, shape, trend, and distribution of internal reservoir characteristics from the early well data. This is "the extension problem" (Potter, 1967).

Even where control is available in wells on a 40-acre or smaller spacing, depositional models have been found to have a major application in reservoir geology. Although well control may be very close, one well commonly cannot be correlated satisfactorily with the next, or the reservoir adequately characterized, in the absence of understanding of the depositional model. Even the spacing of the contours on an isopach map of a sandstone body depends on

the model, and good contouring can contribute substantially to locating the margins of a reservoir beyond the points of dense control.

Reservoir geology is particularly important in successful application of enhanced oil-recovery techniques such as *in situ* combustion and miscible waterflooding. These technically sophisticated operations require large dollar investments for either expensive injections or fluids for injection. Unless the geology of the reservoir is exceptionally well established before the recovery program is formulated, total economic failure may be the result.

One common and important problem in field development is how to decipher the trend of a sandstone body from core observations. The approach assumes a known relation between the orientation of primary sedimentary structures and the trend. The relation probably is best established for alluvial sandstones. However, because of potentially large variability in current direction locally, it is usually important to measure the orientation of several tens of sets of cross-strata. In many other types of elongate sandstone bodies, the relation between trend and sedimentary structures is not reliably predictable. In barrier islands, for example, trough-shaped crossbedding reflects currents flowing obliquely landward (Reineck, 1963) at widely varying angles to the trend of the sandstone body.

Deciphering trend also assumes that a core in the subsurface can be oriented geographically. The best method is by inhole magnetic orientation of the core barrel when the core is cut. Another possibility[2] is to relate the remanent magnetization of the core to the paleomagnetic field prevailing when the sands were deposited. The paleomagnetic method is useful only for sites relatively near the paleomagnetic equator, and successful application is rare.

Hewitt and Morgan (1965) described the reservoir characteristics of a Pennsylvanian alluvial sandstone which forms the reservoir at the Fry *in situ*-combustion site in the Illinois

[2] Much of the logging company literature notwithstanding, our experience in comparing cores with dipmeter surveys suggests that cross-stratification is usually not detectable by dipmeter measurements. The reason is that few cross-strata within sandstones are marked by clay partings or other mineralogic (or resistivity) contrasts (see also Jizba *et al.,* 1964).

Cores with dipping beds can, of course, be oriented geographically by reference either to known regional dip or to a dipmeter survey.

[For a more detailed discussion of use of dipmeter data, see paper by Jageler and Matuszak (this volume). *Editor*]

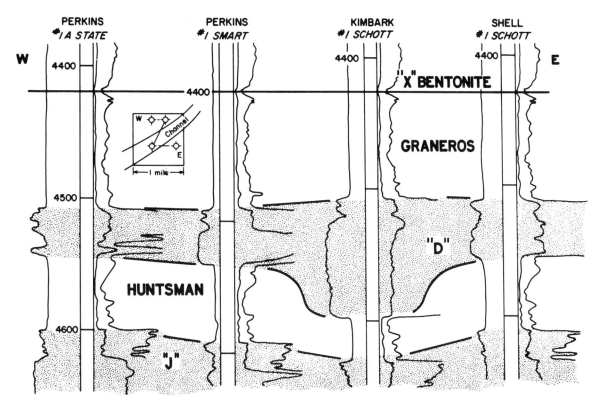

FIG. 3—Differential compaction around Cretaceous "D" alluvial channel sandstone in Sec. 21, T6N, R53W, Logan County, Colorado.

basin. Among the interesting characteristics of the reservoir is the relation of directional horizontal permeability to the current direction of the depositing river. In sandstones with both small-scale and large-scale trough cross-stratification, the horizontal permeability at right angles to the current direction is between 85 and 95 percent of that parallel with the current direction. Montadert (1963) applied knowledge of sedimentary structures gained on outcrop to the establishment of preferred permeability directions within the Hassi-Messaoud field of Algeria.

FAIRWAY DELINEATION

The easiest and most common approach to the question of where exploration should be concentrated is to relate belts of stratigraphic-trap oil occurrence to one or more easily measured stratigraphic parameters. In extension of a fairway within a developing basin, the parameters can be derived empirically from within the area already productive. In less explored areas, they can be sought by comparing oil occurrence with stratigraphic factors in mature basins believed to be similar.

Among the most common of these parameters are ratios of sandstone to shale. In many areas the occurrence of oil tends to be within a belt in which the ratio of sandstone to shale is in the value range known to be optimum for the area (Dickey and Rohn, 1955). On the Gulf Coast, this phenomenon is referred to the "sand-breakup" in a transition from massive sandstone landward to shale seaward.

The next step is to try to reconstruct the paleogeography. This step commonly involves no more than delineation of the broad areas of marine and nonmarine deposition. The more prospective areas generally lie in the belt where marine and nonmarine deposits intertongue. From these general paleogeographic reconstructions, much more detailed ones can be made if appropriate rock materials are available for examination and interpretation. Fisher and McGowen (1969) related oil and gas productive trends in the lower Wilcox Group (Eocene) of the Texas Gulf Coast to many specific depositional environments.

In fairway delineation, undue emphasis has been placed on the depositional environment as the controlling factor. Other important factors

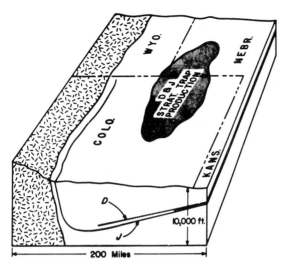

FIG. 4—Fairway of Cretaceous stratigraphic production in Denver basin of eastern Colorado and western Nebraska.

are the postdepositional history of the sediments, adequacy and proximity of source rocks, and the effects of formation-water movement and structural tilting and deformation.

The importance of formation-water fluid potentials (hydrodynamics) in fairway delineation was stressed by Hill *et al.* (1961). They argued that the most favorable sites for belts of productive stratigraphic traps were gentle flanks of sedimentary basins characterized by downdip flow (or downdip potential gradients) through discontinuous sandstone bodies. In such situations, the capacity of updip shale barriers forming the traps would be augmented by the potential gradient.*

An example of a productive stratigraphic-trap trend probably determined by both facies change and favorable potential gradients is the Middle Jurassic Shaunavon trend of southwest Saskatchewan (Christopher, 1964). In this area, 400 million bbl of ultimately recoverable reserves are contained in only 20 fields where oil is trapped stratigraphically in sandstones (Carlson, 1968). Two of these fields, Dollard and Bone Creek-Instow, have reserves of more than 70 million bbl each. All the fields are in shelf deposits. The narrow shelf trends north-northeast and dips eastward; it is bounded on the west by lagoonal deposits and on the east by basinal shales and carbonate rocks. The sandstone reservoirs are marine bars and tidal-channel fills partly enclosed in argillaceous facies. Continuity of the sandstones updip beyond

* Dickey and Hunt (this volume) discuss the theory of formation-water fluid potentials; they conclude that downdip and updip flows are theoretically possible but unlikely to occur commonly in nature. *Editor*

the pools (westward) and the information on formation pressures suggest that eastward decrease in fluid potential has been important in entrapment.

Some caution on the overemphasis of any one technique is certainly warranted. For example, although there was much study of hydrodynamics in the late 1950s and early 1960s, I know of no well-documented case in which its application has led to a significant discovery.

Denver Basin Fairway

The problems of fairway delineation can be illustrated by the production from stratigraphic traps in the Cretaceous on the east flank of the Denver basin (Fig. 4). The initial discovery was made by Marathon Oil Company in western Nebraska in 1949. From that time until the mid-1960s, about 17,000 wells were drilled. This exploration resulted in several hundred fields with a total of about 800 million bbl of ultimately recoverable reserves. Exploration was reactivated in 1970, particularly in the area just east of Denver.

The oil accumulations are in discontinuous sandstone bodies, localized on subtle noses, on the gently westward-dipping east flank of the Denver basin. Structural entrapment is significant only in the northeastern part of the fairway.

Production is from two zones—the "D" and "J" sandstones. Each has an average thickness of several tens of feet and consists of a wide variety of shallow-water marine to nonmarine units, generally lenticular, and complexly arranged in both plan and cross-sectional views. The "D" sandstones were derived entirely from the Canadian shield on the east; they pinch out westward at a depth of about 6,500 ft (1,980 m) in the basin. The "J" sandstones were derived from the Cordilleran region on the west, as well as the Canadian shield; they persist over the entire basin. Alluvial, offshore-bar, and other environmental types of sandstones form the reservoirs.

Even though the "D" and "J" sandstones are everywhere separated by several tens of feet of unfractured shale, the fairway of "D" sandstone production nearly coincides with the fairway of "J" sandstone production. Why the fairways overlap and why they are located where they are is not known. Because a similar assemblage of depositional units persists far east of the fairway in both the "D" and "J," and west of it in the "J," the answer does not lie wholly in considerations of depositional environment. Other important factors must be involved. Hydrodynamics may be one of them. Gradients are low in the fairway compared to those in the "J" sandstone farther west; however, significant downdip gradients cannot be demonstrated in the fairway.

As depth of burial increases west of the fairway in the Denver basin, sandstones tend to become more tightly cemented—thus less porous. The interbedded shales become more fractured. This deterioration in quality of both reservoir and sealing beds may be significant to the western productive limits of the "J" sandstone.

On the east side of the fairway, burial of the source shales may not have been sufficient to allow primary migration. Alternatively, because the sandstones in both the "D" and "J" become thicker, more permeable, and more continuous eastward—with fewer updip shale barriers—oil once generated may have migrated eastward, escaped to the surface, and been dissipated.

The foregoing is an illustration of the complexity of the problem of fairway delineation, even in a mature and geologically well-known area. Further understanding will come probably from research on postdepositional factors rather than from better definition of the sedimentary environments.

Some aspects of paleogeographic reconstruction, particularly in relation to transgression and regression, may provide more help in fairway delineation than has been available in the past.

Overall Transgression Versus Overall Regression

Thick sequences of broadly intertonguing marine and nonmarine beds contain many potential stratigraphic traps, particularly where sandstone bodies are associated with the shoreline. An appropriate model of this type of deposition is the Upper Cretaceous of Wyoming, which is a thick sequence of nonmarine beds and shoreline marine sandstones intertonguing eastward with marine shales. Major source areas on the west supplied sediments to the broad coastal plains along the west side of the seaway. The shoreline migrated back and forth throughout most of Late Cretaceous time, resulting in the intertonguing relations. Exploration is a matter of dividing the sequence into time-parallel zones as thin as possible, and attempting to map the paleogeography, especially the strandline, for each zone.

Periods of overall regression are characterized by a seaward retreat of the various depositional environments. This seaward retreat usually is marked by abundant sediment supply and active seaward-building deltas with associated delta-front sand bodies. In contrast, during periods of overall transgression, the depositional environments shift landward, and because of rising base level the rate of sediment supply is relatively low. Sands are supplied by marine reworking of older deltaic deposits or by longshore transport from local river mouths along the shoreline. Barrier islands with associated landward lagoons are a common development. (The word "overall" is emphasized because, clearly, the barrier islands themselves formed by seaward progradation during episodes of stillstand or minor regression during the overall transgression.)

In parts of the Rocky Mountain area where regional dip of the Cretaceous is eastward,[3] the impact of these considerations on stratigraphic-trap exploration is as follows (D. G. McCubbin, personal commun.). During periods of overall regression, shoreline sand bodies, if present, may be replaced updip by deltaic deposits (Fig. 5). These deltaic deposits, because of their many associated types of sands—particularly distributary-channel sands—probably would be relatively poor barriers to updip migration.

In contrast, during periods of overall transgression, the shoreline sand bodies would be replaced updip by sand-poor lagoonal muds which, when compacted, would be relatively good barriers to updip migration of oil. Furthermore, the sands would be overlain by marine shales, which should be effective barriers.

In applying this concept to stratigraphic-trap fairway delineation in exploration, emphasis would be placed on finding ancient shorelines in interdeltaic areas, especially those of periods of overall transgression. It is not a coincidence that the only significant stratigraphic-trap oil discovery in the many Upper Cretaceous shoreline sandstones of Wyoming (Patrick Draw) is in an overall transgressive rock sequence (Weimer, 1966; McCubbin and Brady, 1969).

Other examples of stratigraphic traps in shoreline sandstones resulting from overall transgressive sedimentation are the basal Pennsylvanian[4] Morrowan sandstones of northwest Oklahoma (Busch, 1959).

---

[3] In areas of west dip, because of gradual eastward (offshore) decrease in thickness and permeability of shoreline sandstones, and ultimate gradation into marine shales, stratigraphic traps would not be expected because reservoir and barrier beds would not be close enough together.

[4] D. C. Swanson, in a preceding paper in this volume, presents evidence that the Morrowan clastic units are not transgressive, but are facies equivalents of the Chesteran carbonate shelf deposits. *Editor*

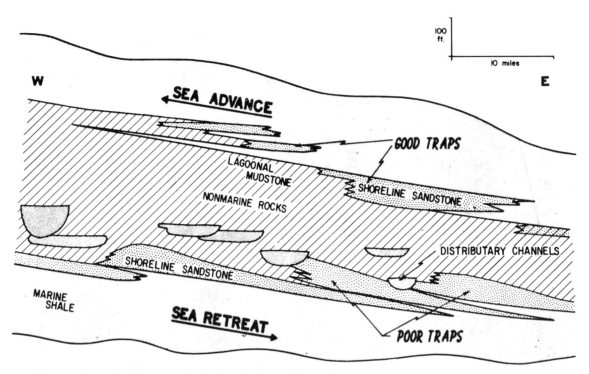

Fig. 5—One schematic cycle in Rocky Mountain Upper Cretaceous showing deltaic distributary channels landward and updip from shoreline sandstones deposited during overall regression (below), and lagoonal mudstones landward and updip from shoreline sandstones deposited during overall transgression (above).

Why Are Few Stratigraphic Traps Found in the Gulf Coast?

In light of the foregoing discussion, and in view of the abundant transgressions and regressions and known heterogeneity of the onshore Gulf Coast Tertiary section, the scarcity of recognized stratigraphic traps there is puzzling. (Although common in the Wilcox and Vicksburg Groups, stratigraphic traps are scarce in the post-Eocene part of the section.) The most probable explanation is that early-formed structures were ubiquitous in the zone of oil generation and migration. In this particular zone of intertonguing sandstones and shales, the following conditions prevailed: (1) The shales were buried deeply enough for oil to have been generated. (2) The sandstones interbedded with the shales provided both the avenues of migration and the reservoirs. (3) Contemporaneous structures formed by gravitational sliding and/or piercement tended to be concentrated there and trapped most of the migrating oil. Even though the distribution of oil might be governed partly by lateral changes from sandstone to shale, the traps would be called "structural."

In contrast, in shallow areas where there are updip pinchouts, as in the updip Frio of coastal Texas, the shales interbedded with the sandstones may not have acted as sources because they were not buried deeply enough. Oil generated farther downdip was trapped mainly in downdip traps.

An alternate explanation assumes there are relatively few updip changes from reservoir to barrier lithology because potential barriers are leaky owing to widespread, deltaic, distributary-channel sandstones located shoreward. This explanation seems unlikely, however, because the distribution of oil on many structures is definitely stratigraphically controlled; thus, the presence of locally effective barriers is implied.

Sandstones Above Unconformities

Although the importance of angular unconformities in the geologic record long has been recognized (Levorsen, 1954), the abundance and importance of stratigraphic breaks with no angular discordance have not always been appreciated. However, increasing knowledge of stratigraphy indicates that disconformities are a common part of most stratigraphic sequences. Some disconformities are related to local epeirogenic movements; some appear to be of continental or intercontinental extent, suggesting eustatic changes in sea level. The waxing and waning of midocean rises (Menard, 1964) provides one explanation for greater frequency of eustatic movements in the geologic record than previously had been supposed.

As late as the early 1950s, for example, the Cretaceous of the western interior seaway was widely regarded as a continuous sequence with relatively few breaks in the stratigraphic record. That concept is no longer valid. Many unconformities and/or disconformities are present.

Waagé (1955) recognized a widespread Early Cretaceous transgressive disconformity in the lower part of the Dakota Group in the northern Front Range foothills of Colorado. What is probably the same surface (MacKenzie, 1965) subsequently was recognized in the Black Hills (Waagé, 1959) and in western Colorado (McCubbin, 1961). In northwest New Mexico, Dane (1960) recognized a widespread erosional surface between Carlile rock (Turonian) and lower Niobrara (Coniacian) rocks. On the basis of faunal evidence by Eicher (1960) and Ellis (1963), as well as physical evidence by Baker (1962), Harms (1966), and MacKenzie (1965), at least one and possibly several erosional surfaces of intra-Albian age are recognized within the Muddy sandstones. The reservoir of the Bell Creek field is directly above the faunal break discussed by Eicher and Ellis. On the basis of detailed ammonite zonation, Gill and Cobban (1966) have found a previously unrecognized unconformity of late Campanian age in western Wyoming. Apparently, several thousand feet of sediment has been eroded locally.

Where angular discordances are involved, the implication to petroleum exploration is clear enough: either pre-unconformity stratigraphic traps or pre-unconformity anticlines not reflected in the overlying beds are possible exploration targets. However, I believe emphasis should be placed on the importance of erosional surfaces—whether angular unconformities or disconformities—in determining the stratigraphic-trap possibilities in the directly overlying transgressive sandstones.

The distribution of sandstones above the surface of unconformity is determined in part by the paleotopography of the surface. In some situations there is a strong tendency for the sandstones to be concentrated in valleys or other lows on the surface. Spooner (1964) cited good examples from the basal Upper Cretaceous Tuscaloosa sandstones of east-central Louisiana. In other places, sandstones may be concentrated over the locally steeper parts of an erosional surface. Where the slopes are slightly steeper, the rate of transgression may have slowed, thus providing adequate time for the accumulation and winnowing of discrete reservoir sandstone bodies.

One approach to predicting the probable distribution of transgressive sandstone bodies overlying an unconformity is the application of quantitative aspects of geomorphology (Horton, 1945; Martin, 1966). These relate to average stream length, average stream spacing, and ratio of length between first- and second-order streams. The application of these quantitative concepts to paleosurfaces allows a general determination of the number of tributaries, length of streams, and channel slope.

An excellent example of oil trapped in sandstones whose distribution was controlled by the underlying erosional surface is the Cretaceous sandstone reservoirs of the San Juan basin in northwest New Mexico (McCubbin, 1969). These sandstones produce from stratigraphically controlled oil accumulations in marine strike-valley sandstones (a term coined by Busch, 1959) deposited during a transgression over the pre-Niobrara erosion surface (Fig. 6). Individual sandstone bodies are localized on the seaward side of cuesta faces formed by the outcrop of relatively resistant beds in the folded and truncated pre-Niobrara sequence. Successively younger sandstones in the overstepping sequence extend farther in the direction of transgression. The sands, transported parallel with the shoreline, were deposited in significant thicknesses where the advance of the sea was slowed by the increase in slope associated with the ridges.

The paleotopography of the erosion surface consisted of northwest-trending cuesta-like ridges and intervening valleys; the steeper slopes faced northeast. Local relief was more than 100 ft (30 m). Individual sandstone bodies are elongate parallel with the ridges and valleys on the erosion surface, and with the direction of sand transport. They thin abruptly to the southwest by onlap against the erosion surface, and thin more gradually in the opposite direction, largely by facies change to shale. Pre-unconformity shales and sandstones provide part of the barrier to updip migration of oil and gas.

The paleotopography of the erosion surface is obviously important in evaluating potential stratigraphic traps and in predicting the geometry of the reservoirs. The paleotopography is obtained from several kinds of maps. One is an isopach map of the interval between the uncon-

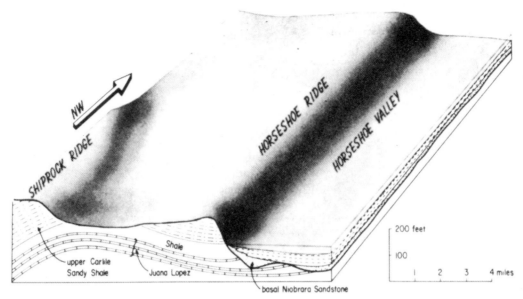

Fig. 6—Setting of basal Niobrara (Upper Cretaceous) strike-valley sandstone, northwest New Mexico (after McCubbin, 1969).

formity and some widespread, originally horizontal, marker above it. Corrections for differential compaction may be needed. However, since interpolation of patterns between control points is difficult, the isopach map should be supplemented by a paleostructural map. Structure at the approximate time the unconformity developed is reflected by an isopach map of an interval bounded by markers directly above and below the unconformity. The two isopach maps considered together make possible the construction of accurate paleogeologic and paleotopographic maps which aid in prediction of the trend and distribution of ridges and valleys, and hence strike-valley sandstones.

Strike-valley sandstones may be more common than presently recognized. The possibility of their existence should be considered wherever marine deposits directly overlie a tilted and truncated sequence of alternating resistant and nonresistant strata.

Although the preceding discussion has dealt only with sandstones overlying erosional surfaces cut subaerially, Yeats (1965) has documented an example in which the topography of a submarine unconformity controlled the distribution of overlying oil-bearing turbidites.

EFFECTIVE USE OF STRATIGRAPHIC CONTROLS IN EXPLORATION

From this review, it should be apparent that any one type of stratigraphic control is inadequate for an effective exploration program for

stratigraphic traps in detrital rocks. Although, for any given subsurface section, a core is commonly the most useful source of data, core control nearly always is spaced too widely to be used alone. However, logs and cuttings, although generally providing closer control, are less useful for many kinds of interpretations. The solution to this apparent dilemma is to integrate, so far as possible, the various types of control. The key to successful interpretation is to establish, within each area of relatively homogeneous stratigraphy, correlations between characteristics as determined by the different types of control available.

With detailed core or outcrop study as a starting point, it is commonly possible to subdivide an interval into correlatable stratigraphic units. The next step is to determine how each of the units can be recognized and correlated, using the more densely distributed but less definitive types of logs in common use in the area. The logs then become satisfactory substitutes for cores. (At this stage, a comprehensive well-data system, in which digitized logs are an integral part, may be meshed with appropriate programs to yield maps useful in exploration.)

In the same way, a particular lithologic unit, with detailed characteristics established by core examination, may be recognizable in cuttings by a diagnostic color, texture, or mineralogy. The examination of cuttings should be tailored to the specific job. By recording only those particular characteristics that serve to distinguish members recognized in the area and zone being studied, sample examination time is greatly shortened and the value of the resulting sample log is greatly increased.

If this approach is used, the reader may ask how the significant data can be recognized. There are, of course, no pat answers; but adherence to one general rule is essential: *interpretation should proceed apace with description*. The tendency is strong to develop a routine that stresses uncritical description and defers interpretation until all the cores or sets of cuttings have been described. To a great extent, one sees only what he is looking for and, unless an attempt is made to interpret the rocks as they are being described, much significant information will be overlooked. The adoption of multiple working hypotheses is even better. Although it is commonly necessary or expedient to follow the lines of inquiry suggested by one hypothesis, evaluation of the observations according to several hypotheses tends to bring into focus the differentiating criteria.

RESEARCH NEEDS

More and Better Models

Although our knowledge of depositional models has grown dramatically in the past decade, there is still a need for a wider spectrum and more detailed information for use in exploration. A particularly large gap is that of the distinguishing characteristics of sandstone bodies deposited in shallow-marine environments offshore. In the range of environmental types from subaerial dunes to deep-water sands, the shallow-marine environments are the least well known.

Offshore

As exploration emphasis shifts offshore, there is a growing need for stratigraphic predictions in the absence of nearby outcrops and continuous cores. Ways must be found to make more effective use of cuttings, logs, and surface geophysical information to predict the distribution of reservoir sandstones, stratigraphic traps, and favorable source-reservoir relations. The application of this exploration technique is most advanced in the Gulf Coast offshore, where the base of the reflection curtain visible on common-depth-point seismic-record sections is being used to map the base of potential reservoir sandstones.

Seismic Energy of Higher Frequency

A third research need is the successful return of higher frequency seismic energy from greater depths and the interpretation of the resulting data. Since primary stratigraphic traps are related to lateral changes in relatively thin intervals (usually less than 100 ft or 30 m), the higher the frequency of seismic energy returned, the more opportunity there will be to detect lateral seismic anomalies relating to time or character that can be interpreted in terms of stratigraphic traps.

Giants Versus Dwarfs

A fourth need is to determine whether the few large primary stratigraphic traps have unique and predictable characteristics as compared with small traps. Are the large ones large because of size of the sandstone body, greater porosity, a more prolific source, a larger trap, a more effective updip seal, or some combination of these factors? The answers might assist in diverting exploratory effort from the small, scarcely economic traps to the more profitable ones.

Many writers have called attention to the common proximity of oil occurrence to unconformities. Of the giant primary traps listed in Table 1, Bell Creek is underlain by a disconformity, and Cut Bank and probably Burbank (D. R. Baker, personal commun., 1969) are underlain by erosional surfaces. Giant stratigraphic traps beneath unconformities are well documented.

Consequently, in a quest for giant fields, more attention to finding subtle unconformities and disconformities and to mapping their paleotopography and paleogeography would be a useful exploration approach.

FUTURE EXPLORATION FOR PRIMARY STRATIGRAPHIC TRAPS IN SANDSTONE

In areas where leasing and drilling costs are low, primary stratigraphic traps in sandstone-shale sequences are still worthwhile exploratory objectives, particularly for independents. In addition, a growing knowledge and appreciation of the habitats of such traps hopefully will increase the ratio of wildcat successes to failures.

As this book testifies, however, cases are rare in which primary stratigraphic traps in sandstone have been found where predicted. Moreover, major accumulations of oil in such traps are few—worldwide. Consequently, a large-scale, major-company search for them specifically would not seem justified. Such an exploratory effort would be warranted only with a vastly improved technology.

Whether or not that degree of improvement takes place, there is a much better alternative which can be applied now. Our increasingly sophisticated understanding of stratigraphy, depositional models, and diagenesis should be applied to finding not only stratigraphic traps, but all kinds of traps. Rather than focusing narrowly on primary traps, we should apply our knowledge to the problem of predicting fairways of favorable source, reservoir, and trap relations. In these fairways, we can concentrate exploration for anticlinal, fault, and pre-unconformity traps, as well as for the primary ones discussed in this paper.

SELECTED REFERENCES

Baker, D. R., 1962, The Newcastle formation in Weston County, Wyoming—a nonmarine (alluvial plain) deposit, in Symposium on Early Cretaceous rocks of Wyoming and adjacent areas: Wyoming Geol. Assoc. 17th Ann. Field Conf. Guidebook, p. 148–162.

Bass, N. W., Goodrich, H. B., and Dillard, W. R., 1942, Subsurface geology and oil and gas resources of Osage County, Oklahoma: U.S. Geol. Survey Bull. 900-J, p. 321–342.

Berg, D. R., and Davies, D. K., 1968, Origin of Lower Cretaceous Muddy Sandstone at Bell Creek field, Montana: Am. Assoc. Petroleum Geologists Bull., v 52, p. 1888–1898.

Bernard, H. A., Leblanc, R. J., and Major, C. F., 1962, Recent and Pleistocene geology of southeast Texas, Field excursion no. 3, in Geology of the Gulf Coast and central Texas and Guidebook of excursions: Houston Geol. Soc., p. 175–224.

Blixt, J. E., 1941, Cut Bank oil and gas field, Glacier County, Montana, in A. I. Levorsen, ed., Stratigraphic type oil fields: Am. Assoc. Petroleum Geologists, p. 327–381.

Brown, L. F., 1969, Late Pennsylvanian paralic sediments, in Guidebook to the Late Pennsylvanian shelf sediments, north-central Texas: Dallas Geol. Soc., p. 21–33.

Busch, D. A., 1959, Prospecting for stratigraphic traps: Am. Assoc. Petroleum Geologists Bull., v. 43, no. 12, p. 2829–2843.

Carlson, C. E., 1968, Triassic-Jurassic of Alberta, Saskatchewan, Manitoba, Montana, and North Dakota: Am. Assoc. Petroleum Geologists Bull., v. 52, no. 10, p. 1969–1983.

Chambers, L. S., 1943, Coalinga East extension area of the Coalinga oil field: California Div. Mines and Geology Bull. 118, p. 486–490.

Christopher, J. E., 1964, The Middle Jurassic Shaunavon Formation of south-western Saskatchewan: Saskatchewan Dept. Mineral Resources Rept. 95, 96 p.

Dane, C. H., 1960, The boundary between rocks of Carlile and Niobrara age in San Juan basin, New Mexico and Colorado: Am. Jour. Sci., v. 258-A, p. 46-56.

Dickey, P. A., and Rohn, R. E., 1955, Facies control of oil occurrence: Am. Assoc. Petroleum Geologists Bull., v. 39, no. 11, p. 2306–2320.

Eicher, D. L., 1960, Stratigraphy and micropaleontology of the Thermopolis shale: Yale Univ. Peabody Mus. Nat. History Bull. 15, 126 p.

Ellis, C. H., 1963, Micropaleontology of the Mowry Shale, Newcastle Formation, and equivalent stratigraphic units: Rocky Mtn. Assoc. Geologists 14th Field Conf. Guidebook, p. 149–155.

Exum, F. A., and Harms, J. C., 1968, Comparison of marine-bar with valley-fill stratigraphic traps, western Nebraska: Am. Assoc. Petroleum Geologists Bull., v. 52, no. 10, p. 1851–1868.

Fettke, C. R., 1938, The Bradford oil field, Pennsylvania and New York: Pennsylvania Geol. Survey Bull. M21, 4th ser., 454 p.

Fisher, W. L., and McGowan, J. H., 1969, Depositional systems in Wilcox Group (Eocene) of Texas and their relation to occurrence of oil and gas: Am. Assoc. Petroleum Geologists Bull., v. 53, no. 1, p. 30–54.

Fisk, H. N., 1944, Geological investigation of the alluvial valley of the lower Mississippi River: Vicksburg, Mississippi, U.S. Army Engineers, Mississippi River Commission, 78 p.

Frazier, D. E., 1967, Recent deltaic deposits of the Mississippi River: their development and chronology: Gulf Coast Assoc. Geol. Socs. Trans., v. 17, p. 287–315.

Gill, J. R., and Cobban, W. A., 1966, Regional unconformity in Late Cretaceous, Wyoming: U.S. Geol. Survey Prof. Paper 550B, p. B20–B27.

Harms, J. C., 1966, Stratigraphic traps in a valley fill, western Nebraska: Am. Assoc. Petroleum Geologists Bull., v. 50, no. 10, p. 2119–2149.

——— 1968, Advances in stratigraphy: detrital sedimentation models: Soc. Exploration Geophysicists 38th Ann. Mtg. Prog., p. 77.

Hewitt, C. H., and Morgan, J. T., 1965, The Fry in situ combustion test—reservoir characteristics: Jour. Petroleum Technology, v. 17, p. 337–353.

Hill, G. A., Colburn, W. A., and Knight, J. W., 1961, Reducing oil-finding costs by use of hydrodynamic evaluations, in Petroleum exploration, gambling game or business venture—Inst. Econ. Petroleum Explor., Devel., and Property Evaluation: Englewood, N. J., Prentice-Hall, Inc., p. 38–69.

Horton, R. E., 1945, Erosional development of streams and their drainage basins, hydrophysical approach to quantitative morphology: Geol. Soc. America Bull., v. 56, no. 3, p. 275–370.

Jizba, A. V., Campbell, W. C., and Todd, T. W., 1964, Core resistivity profiles and their bearing on dipmeter survey interpretation: Am. Assoc. Petroleum Geologists Bull., v. 48, p. 1804–1809.

Koesoemadinata, R. P., 1970, Stratigraphy and petroleum occurrence, Green River Formation, Red Wash field, Utah: Colorado School Mines Quart., v. 65, no. 1, 77 p.

Kramers, J. W., and Lerbekmo, J. E., 1967, Petrology and mineralogy of Watt Mountain Formation, Mitsue-Nipisi area, Alberta: Bull. Canadian Petroleum Geology, v. 15, p. 346–378.

Levorsen, A. I., 1954, Geology of petroleum: W. H. Freeman & Co., San Francisco, 703 p.

MacKenzie, D. B., 1965, Depositional environments of Muddy Sandstone, western Denver basin, Colorado: Am. Assoc. Petroleum Geologists Bull., v. 49, p. 186–206.

Martin, Rudolph, 1966, Paleogeomorphology and its application to exploration for oil and gas (with examples from Western Canada): Am. Assoc. Petroleum Geologists Bull., v. 50, p. 2277–2311.

McCubbin, D. G., 1961, Basal Cretaceous deposits of southwestern Colorado and southeastern Utah: Ph.D. dissert., Harvard Univ.

——— 1969, Cretaceous strike-valley sandstone reservoirs, northwestern New Mexico: Am. Assoc. Petroleum Geologists Bull., v. 53, p. 2114–2140.

——— and Brady, M. J., Depositional environment of the Almond reservoirs, Patrick Draw field, Wyoming: Mtn. Geologist, v. 6, p. 3–26.

McGregor, A. A., and Biggs, C. A., 1968, Bell Creek field, Montana: a rich stratigraphic trap: Am. Assoc Petroleum Geologists Bull., v. 52, p. 1869–1887.

McKee, E. D., 1966, Structures of dunes at White Sands National Monument, New Mexico (and a comparison with structures of dunes from other selected areas): Sedimentology, v. 7, no. 1, p. 3–69.

Menard, H. W., 1964, Marine geology of the Pacific: McGraw-Hill, New York, 271 p.

Michaelis, E. R., 1957, Cardium sedimentation in the Pembina River area: Alberta Soc. Petroleum Geologists Jour., v. 5, no. 1, p. 73–77.

Montadert, L., 1963, La sédimentologie et l'étude détaillée des hétérogénéités d'un réservoir: application au gisement d'Hassi-Messaoud: Inst. Français Pétrole Rev., v. 18, p. 241–257.

Mueller, J. C., and Wanless, H. R., 1957, Differential compaction of Pennsylvanian sediments in relation to sand-shale ratios, Jefferson County, Illinois: Jour. Sed. Petrology, v. 27, p. 80–88.

Nielson, A. R., 1957, Cardium stratigraphy of the Pembina field: Alberta Soc. Petroleum Geologists Jour., v. 5, no. 4, p. 64–72.

Potter, P. E., 1967, Sand bodies and sedimentary environments: A review: Am. Assoc. Petroleum Geologists Bull., v. 51, p. 337–365.

Reineck, Hans-Erick, 1963, Sedimentgefüge im Bereich der südlichen Nordsee: Senckenberg. Naturf. Gesell., Abh. 505, 138 p.

Rittenhouse, Gordon, 1961, Problems and principles of sandstone-body classification, in Geometry of sandstone bodies: Am. Assoc. Petroleum Geologists, p. 3–12.

Shelton, J. W., 1967, Stratigraphic models and general criteria for recognition of alluvial, barrier-bar, and turbidity current sand deposits: Am. Assoc. Petroleum Geologists Bull., v. 51, p. 2441–2461.

Spooner, H. V., 1964, Basal Tuscaloosa sediments, east-central Louisiana: Am. Assoc. Petroleum Geologists Bull., v. 48, p. 1–21.

Waagé, K. M., 1955, Dakota Group in northern Front Range foothills, Colorado: U.S. Geol. Survey Prof. Paper 274-B, p. 15–51.

——— 1959, Stratigraphy of the Inyan Kara Group in the Black Hills: U.S. Geol. Survey Bull. 1081–B, p. 11–90.

Walker, R. G., 1966, Shale Grit and Grindslow shales: transition from turbidite to shallow-water sediments in the upper Carboniferous of northern England: Jour. Sed. Petrology, v. 36, p. 90–114.

——— 1967, Turbidite sedimentary structures and their relationship to proximal and distal depositional environments: Jour. Sed. Petrology, v. 37, p. 25–43.

Weimer, R. J., 1966, Time-stratigraphic analysis and petroleum accumulations, Patrick Draw field, Sweetwater County, Wyoming: Am. Assoc. Petroleum Geologists Bull., v. 50, p. 2150–2175.

Yeats, R. S., 1965, Pliocene seaknoll at South Mountain, Ventura basin, California: Am. Assoc. Petroleum Geologists Bull., v. 49, p. 526–546.

The American Association of Petroleum Geologists Bulletin
V. 58, No. 3 (March 1974), P. 447-463, 13 Figs.

Stratigraphic Trap Accumulation in Southwestern Kansas and Northwestern Oklahoma[1]

GREGORY W. MANNHARD[2] and DANIEL A. BUSCH[3]

Albuquerque, New Mexico 87106, and Tulsa, Oklahoma 74103

Abstract An oil- and gas-bearing inlier of Morrowan sandstones presents a paradoxical situation in that water, oil, and gas are present in an inverted structural relation, some of the gas being structurally lower than oil and water.

A map of the pre-Pennsylvanian topography reveals that a modified trellis drainage system was developed on subaerially exposed Mississippian strata. Southwestward-tilted resistant limestones of the Chesterian Series stood out as subparallel cuestas; intervening erosional valleys developed on interbedded nonresistant shales.

Morrowan strata were deposited under conditions of cyclic marine southwest-to-northeast transgression, which was interrupted by several minor regressions. Sands were deposited during the regressive phases, and shales during transgressive phases.

Small-scale structural noses and closures on the Inola limestone are largely the result of differential compaction of shales deposited over buried pre-Pennsylvanian topography.

Detailed mapping affords a logical explanation for each producing well and all but a few dry holes in the Second Morrow sand of the Harper Ranch field of Clark County, Kansas. The anomalous distribution of some gas accumulation is explained by Gussow's principle, modified to apply to a stratigraphic trap.

INTRODUCTION

In contiguous parts of Harper and Woods Counties, Oklahoma, and Clark and Comanche Counties, Kansas, an inlier of oil- and gas-bearing productive sandstones of Morrowan (Early Pennsylvanian) age occupies a restricted part of the northern shelf of the Anadarko basin (Fig. 1).

Although many geologists have studied this area, there is little unanimity as to the complexities of the structural and stratigraphic factors controlling hydrocarbon accumulation. Some of the Morrow sandstones contain gas in structurally lower positions than oil. Likewise, some of the oil is produced from Morrow sandstones structurally lower than water-bearing Morrow sandstones. A geologic study of this inverted-gravity situation was made by attempting to determine (1) the position of the Mississippian-Pennsylvanian systemic boundary; (2) stratigraphic correlations of units above and below the systemic boundary; (3) relative ages of the strata under consideration; (4) possible relations between pre-Pennsylva-

nian structure and stratigraphy and the pre-Pennsylvanian topographic surface; (5) influence of paleotopography on the distribution of Morrow sandstones; (6) possible relations between the present structural configuration of the Inola limestone and the buried pre-Pennsylvanian landscape; and (7) relation of structure and stratigraphy to accumulation of oil and gas in the Morrow sandstones.

From the outset, it was apparent that a critical problem would be to determine the position of the Mississippian-Pennsylvanian unconformity because of difficulty in distinguishing on mechanical well logs between the fine-grained calcareous sandstones of the Morrowan Series and the sandy and silty limestones common in the basal part of the Chesterian Series. To resolve this problem many mechanical-log cross sections of the Mississippian strata were made to relate them to the stratigraphy of the overlying Pennsylvanian rocks. Two northwest-southeast cross sections and four northeast-southwest cross sections are included in this report. To provide positive identification of the lithologic units, two of these sections were extended beyond the limits of the study area. One cross section extends from the southeast corner of the study area for a distance of over 100 mi to connect with a cross section published by Lukert (1949). Cherokee limestone marker beds, identified on Lukert's type section, were projected into the study area along this extended cross section. Another cross section extends southwestward from the study area into western Harper and Beaver Counties,

[1] Manuscript received, June 26, 1973; accepted, October 3, 1973.

[2] Graduate student, University of New Mexico.

[3] Consulting Geologist and Visiting Professor, University of Oklahoma.

Mechanical well logs, sample logs, and samples generously were provided by Gulf Oil Corporation, Mobil Oil Company, Atlantic-Richfield Company, University of Oklahoma Research Institute, and the University of Oklahoma Geology and Geophysics Library.

Gregory W. Mannhard and Daniel A. Busch

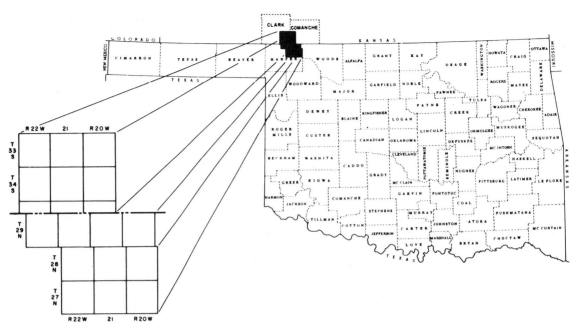

Fig. 1—Location map, Clark and Comanche Counties, Kansas, and Harper and Woods Counties, Oklahoma.

Oklahoma, where stratigraphic units of Atokan, Morrowan, and Chesterian ages are well established. Binocular microscopic studies of rotary cuttings were made of 27 key wells on the cross sections as a means of recognizing significant lithologic changes related to the Mississippian-Pennsylvanian unconformity.

In addition, data were obtained from 335 mechanical well logs and 86 commercially prepared sample logs. Where available, core descriptions provided by operators also were utilized.

A limited area in Kansas, which includes contiguous parts of T33, 34S, R21W, was selected for detailed study with emphasis on producing sandstones of the Harper Ranch field.

Previous Investigations

Regional studies of the lithology and stratigraphy of Mississippian and Pennsylvanian rocks of the Kansas part of the study area have been completed by several investigators. Describing the lithology of Chesterian strata of Meade and southeastern Seward Counties, Kansas, which adjoin the study area on the west, Clair (1948) indicated that the uppermost limestone beds of the Chester are characterized by the presence of fine oolites and abundant fossils, and that limestones lower in the section are typically finely sandy. He also stated that the lower part of the Chester contains beds of fine-grained, white, angular sandstone.

On an index map of southwestern Kansas, Moore (1949) illustrated post-Mississippian–pre-Desmoinesian strata north and west of the study area in Ford and Meade Counties. However, possibly because of lack of adequate well control at the time of his study, he did not indicate the presence of Morrowan or Atokan strata within the study area.

On a northeast-southwest cross section in southwestern Kansas, Lee (1953) confirmed the presence of Morrowan strata in adjacent Meade County, but depicted the Morrowan wedging out southwest of the area. Lee described the basal Morrow beds in Meade County as consisting of glauconitic sandy limestone and calcareous sandstone.

Bowles (1962) completed a generalized study of the subsurface geology of Woods County, Oklahoma. On his structure map of the base of the Pennsylvanian System, he indicated a well-defined nose plunging southwestward across T28N, R20W, and T27N, R21W. He stated that this feature was reflected by all units above the Viola Limestone. In the present investigation this structure is reflected to a limited degree on a structure map of the Inola limestone.

Pate (1959) published a regional study of the stratigraphic traps along the north shelf of the Anadarko basin. His isopach map of the Morrow "formation" shows these strata wedging out along a line across the southwestern part of the area. The map shows general south-westerly thickening of the Morrow.

In a report on the petroleum geology of Harper County, Jordan *et al.* (1959) showed a similar position for the updip limit of Morrowan strata.

A part of the dissertation of Khaiwka (1968), on the geometry and depositional environments of Pennsylvanian reservoir sandstones in northwestern Oklahoma, overlaps the present study. The position of the updip limit of Morrowan strata, as shown by Khaiwka on an isopach map of the Morrow, is in general agreement with the position as shown by Pate (1959) and Jordan *et al.* (1959). The recognition of Morrow sandstones northeast of this previously defined updip limit is primarily the result of an increase in density of well control since these earlier investigations were made.

Oil and gas field studies within the study area have been published by Waite (1956), Kornfeld (1959), and Price (1963). Waite, describing the Harper Ranch field of Clark County, Kansas, wrote (1956), "Production seems to be the result of stratigraphic traps in an embayment where offshore sand bars have been built up." Price (1963) wrote, "The North Buffalo Pool is principally an anticlinal structural reservoir with stratigraphic modification of the producing area." He also stated, "The anticlinal structure was probably formed in post-Hunton–pre-Mississippian time and tilted and rejuvenated several times during geologic history." Price indicated that only three wells within the North Buffalo pool produce from Morrow sandstones and that accumulation within these sandstones is primarily stratigraphic. Kornfeld (1959, p. 67) stated, "Principal feature of the pre-Pennsylvanian structure at North Buffalo is the existence of a 50 foot normal tension fault with a strike of N. 64 degrees W., intersecting the major axis of the anticline which strikes N. 42 degrees E."

STRATIGRAPHY

The stratigraphic interval of primary interest in this study extends from the Inola limestone, at the top, to the Mississippian-Pennsylvanian unconformity, at the base. To establish the position of this unconformity it was necessary to correlate and describe in some detail the upper part of the Mississippian section. Figure 2 is a composite log of the named units of the interval studied.

Named units have been identified in accordance with Jordan's (1957) work. Inasmuch as

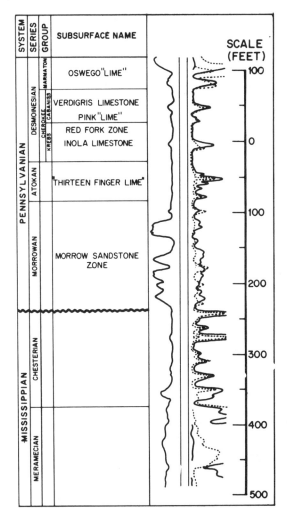

FIG. 2—Composite electric log of study area.

most writers consider the "Thirteen Finger lime" to represent beds of Atokan age, this usage is retained in this report. The Cherokee Group, as originally defined by Haworth and Kirk in 1894, included strata from the base of the Oswego limestone (Fort Scott Limestone) to the top of the Mississippian. It later was redefined to extend from the base of the Fort Scott to the base of the Desmoinesian (Jordan, 1957). Surface Cherokee units were divided into Krebs and Cabaniss Groups by Oakes in 1953 (Jordan, 1957). Although the term "Cherokee" is not used on the outcrop in Oklahoma, it is a widely used term in subsurface geology.

Pennsylvanian System

Desmoinesian Series—Cherokee Group

Within the study area the Cherokee Group

291

consists of a sequence of thin, persistent beds of alternate limestone and shale. No significant sandstone development has been recognized within this interval. The section thins northward, indicating that the major basin of Cherokee deposition was south of the area.

Verdigris Limestone Member (of Cabaniss Formation)—The Verdigris limestone was named by Smith in 1928 for exposures along the Verdigris River in Rogers County, Oklahoma (Jordan, 1957). In the study area the Verdigris is almost directly below the Oswego "lime" and consists of one or two persistent limestone beds. The average thickness of the Verdigris in the area is 10 ft.

Pink "lime"—The Pink "lime" (Jordan, 1957) is the subsurface equivalent of the Tiawah Limestone Member of the Cabaniss Formation, the type locality being near the town of Tiawah, Rogers County, Oklahoma. In the area, the Pink "lime" is approximately 15 ft below the Verdigris limestone and consists of one or more thin limestone beds. It is identified readily on mechanical logs in most of the area.

Red Fork zone—The term "Red Fork" was applied by Hutchison in 1911 to the producing sand of the Red Fork field, Creek and Tulsa Counties, Oklahoma (Jordan, 1957). It is the equivalent of the Taft Sandstone Member of the Boggy Shale in the Krebs Group at the surface. No significant sandstone development was noted within the Red Fork zone in the area of study; the zone is composed mainly of shale interbedded with thin limestone beds and some streaks of calcareous siltstone.

Inola Limestone Member (of Boggy Shale of Krebs Group)—The Inola limestone was named by Lowman in 1932 for exposures near the town of Inola, Rogers County, Oklahoma (Jordan, 1957). Identification of the Inola limestone was established by constructing a regional stratigraphic cross section (not included in this report) eastward from this area to intersect the published cross section which extends from Marion County, Kansas, to Osage County, Oklahoma (Lukert, 1949).

The Inola limestone is well developed throughout the area. It is an excellent lithologic time-marker bed because of its lateral persistence and nearly constant thickness. This unit is a light brownish-gray, compact to very fine-grained, sparsely fossiliferous limestone with an average thickness of about 8 ft.

Atokan Series

"Thirteen Finger lime"—The "Thirteen Fin-ger lime" has a maximum of 13 "fingers" on the resistivity curve of electric logs (Jordan, 1957). In the study area it makes up the entire Atokan Series. Stratigraphically, it is 10–30 ft below the Inola limestone. In the eastern part of the area the "Thirteen Finger lime" is poorly developed, being represented by only one or more thin limestone beds. It is progressively better developed toward the west and southwest where it reaches a thickness of approximately 50 ft and consists of several alternate thin limestone and shale beds.

Lithologically, the "Thirteen Finger lime" is light brownish-gray to gray, mottled, fossiliferous, and finely crystalline. It is in places finely sandy or silty. The interbedded shales are gray to dark gray, pyritic, micaceous, and fossiliferous.

Morrowan Series

Morrow sandstone zone—The name "Morrow formation" was applied by Adams and Ulrich in 1904 to a sequence of shale, thin-bedded limestone, and sandstone exposed near the village of Morrow, Washington County, Arkansas (Jordan, 1957). It subsequently was raised to group rank by Purdue in 1907, and later designated as a series (Jordan, 1957). In the study area the upper limit of the Morrow is defined by the base of the "Thirteen Finger lime" and the lower limit by the Mississippian-Pennsylvanian unconformity. In this area the interval ranges in thickness from 0 to 170 ft, with general thickening southwestward toward the major basin of deposition.

Sandstone and siltstone in this zone are lenticular with individual sandstones as thick as 50 ft. Lithologically the sandstone is light olive gray to brownish gray, commonly glauconitic, medium to very fine grained, calcareous, and silty. It displays a relatively low degree of sorting and roundness. Calcium carbonate as a cement locally constitutes up to 50 percent of the total rock volume. Siltstone is common.

Mississippian System

Chesterian Series

Rocks of the Morrowan Series unconformably overlie Chesterian strata in the study area except along the eastern margin where the Chester has been removed by erosion and the underlying Meramec subcrops at the unconformity surface.

The Chester consists of a sequence of relatively thin limestone strata interbedded with shale. Thin beds of siltstone and very fine-grained calcareous sandstone are common near the base. The limestone is typically light brownish gray, finely oolitic, medium to very fine grained, silty, finely sandy, and locally fossiliferous. Many of the limestone beds have a granular appearance caused by the high content of silt and fine-grained sand.

The abundance of fossils and the presence of fine- to very fine-grained oolites, particularly in the upper part of the section, are distinctive features of the Chester. Consequently, the position of the Mississippian-Pennsylvanian unconformity generally can be identified by the first appearance of oolites and abundant fossils in well cuttings.

Meramecian Series

No attempt is made in this report to subdivide the Meramecian Series into its respective formations. However, distinctive electric-log marker beds have been correlated on the cross sections.

On mechanical logs the Meramec generally can be distinguished from the overlying Chester on the basis of the more massive appearance of its limestones and the decreasing abundance of thick and widespread shale beds common within the Chester.

Upper beds of the Meramec typically are composed of light-gray to brownish-gray, cherty, oolitic, fine-grained, finely sandy limestone. The oolites generally differ from those of the Chester in being darker in color (in places almost black), giving the samples a speckled appearance under the microscope.

Regional Cross Sections

Six regional cross sections (Figs. 3–8) are included in this report. Their locations are shown on Figure 9. All cross sections were constructed using the top of the Inola limestone as the stratigraphic datum. The Inola was selected because of the ease of recognition on mechanical logs and its position just above the investigated stratigraphic interval. Use of the Inola as a reference effectively restores the structural and stratigraphic relations that existed within the Morrowan section during the time of deposition of the Inola, modified only by later slight differential compaction within the Morrowan section.

Figures 3 and 4 extend northwest-southeast, roughly parallel with the regional strikes of Lower Pennsylvanian and Upper Mississippian strata. Figures 5–8 extend northeast-southwest, perpendicular to regional strike. The pre-Pennsylvanian southwest dip of Mississippian strata is shown clearly on the northeast-southwest cross sections. The amount of dip generally is less than 1°. An exception to the southwest dip is apparent in wells 35 and 36 of Figure 6, on which the dip of the Mississippian strata is negligible. This condition reflects the influence of uplift associated with the structure which forms the North Buffalo oil and gas field in T28N, R22W. Also apparent from the northeast-southwest cross section is that Mississippian rocks thin northeast by truncation. On Figure 6 the Chester thins by truncation from a maximum thickness of approximately 250 ft, in well 35, to 30 ft in well 40, and is absent in well 41 (0.75 mi farther east). A similar truncation of the Chester takes place east of well 46 (Fig. 7). Northeast of well 55 (Fig. 8), the Chester also is absent by truncation.

The Morrow zone contains as many as six individual sandstone bodies (well 14, Fig. 4). The more prominent of these have been identified as the First, Second, and Third Morrow sandstones (Fig. 3).

The distribution of Morrow sandstone is controlled to varying degrees by the configuration of the Mississippian erosion surface. The lowermost sandstone beds wedge out rather abruptly against topographic highs. This is illustrated on Figure 7, wells 47–50, where sand filled a large northwest-southeast-trending valley eroded into the underlying Mississippian strata. This situation also is apparent in wells 58 and 59 (Fig. 8), where the Third Morrow sandstone exhibits the characteristics of a typical strike-valley sandstone as described by Busch (1959). It is teardrop in profile with an abrupt seaward and gradual landward pinchout. The sand was deposited in a stream valley on the landward side of a cuesta; the caprock is preserved in well 9. The escarpment controlling this strike valley was formed by erosion of southwestward-tilted, resistant Mississippian limestone underlain by nonresistant shale. The influence of another strike valley in controlling sandstone distribution is shown in wells 56, 57, and 9 of the same cross section. The escarpment for this valley is preserved in well 55. Sandstone beds higher in the section wedge out less abruptly; their geometric characteristics are not so rigidly controlled by erosional depressions.

On Figure 5 Morrow sandstone has been spread out over a gently sloping Mississippian erosion surface which contains only slight undulations.

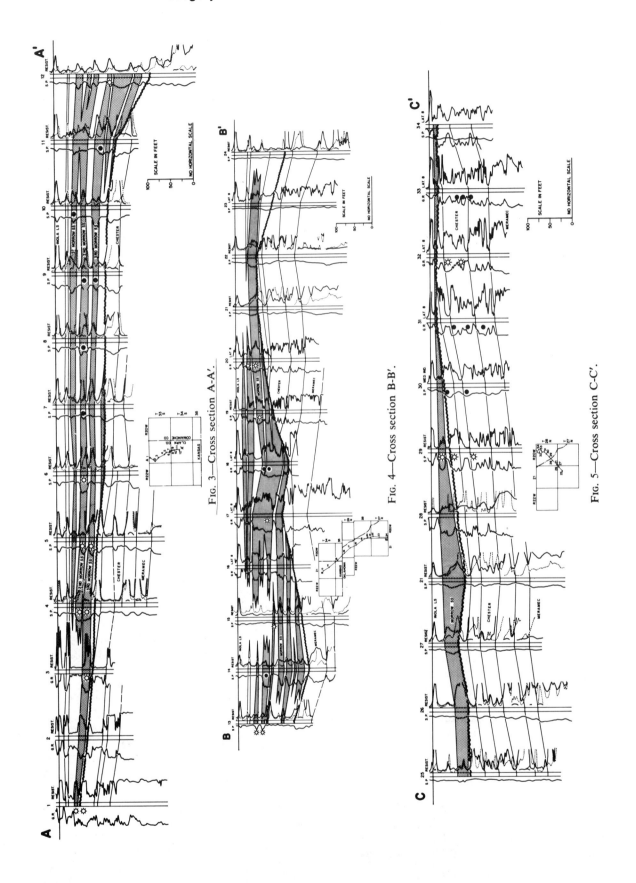

Fig. 3—Cross section A-A'.

Fig. 4—Cross section B-B'.

Fig. 5—Cross section C-C'.

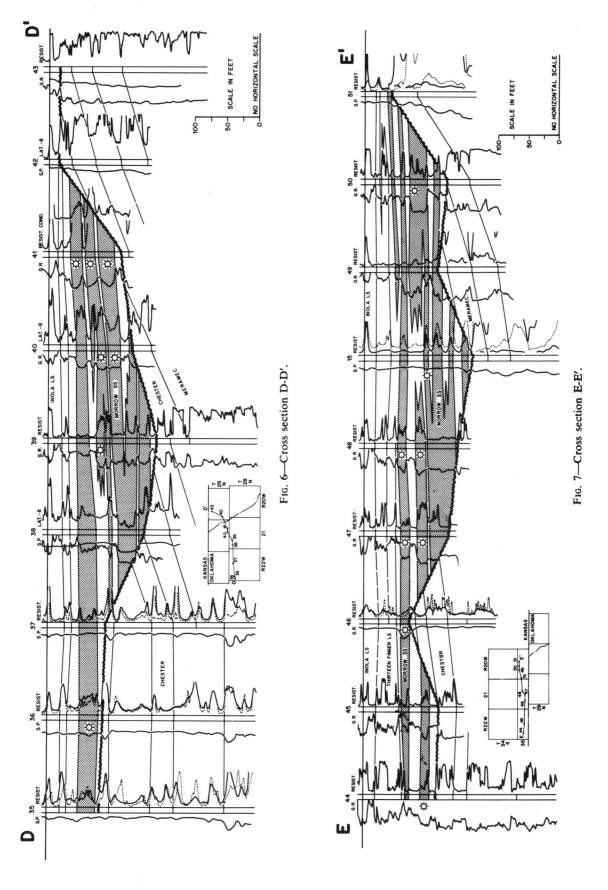

FIG. 6—Cross section D-D'.

FIG. 7—Cross section E-E'.

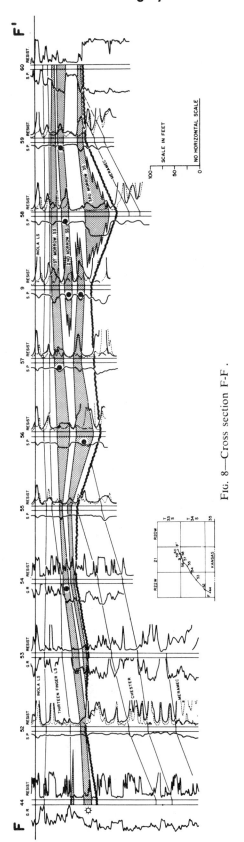

Fig. 8—Cross section F-F'.

The "Thirteen Finger lime" onlapped the area from southwest to northeast and is reflected mostly by northeast thinning and progressive disappearance of the basal members of the limestone sequence. This is illustrated on Figure 8 where the "Thirteen Finger lime" is developed characteristically in wells 44, 52–54, but only the uppermost limestone member is preserved in wells 55–60. Onlap also is apparent on Figure 7; the "Thirteen Finger lime" is developed typically in wells 44 and 45 and thins by progressive disappearance of its basal members in wells 46–51.

Figures 7 and 8 show that most sandstones within the area are Morrowan in age; they occupy the interval between the base of the "Thirteen Finger lime" and the Mississippian-Pennsylvanian unconformity.

The Inola limestone is identifiable on all cross sections. Its near parallelism with the top of the underlying "Thirteen Finger lime" indicates that no significant hiatus is present between the Atokan and Desmoinesian Series.

STRUCTURE

A regional structural contour map was constructed with the Inola limestone as datum (Fig. 9). In Oklahoma, the regional dip on the Inola is southwest and averages 45 ft per mile. In Kansas, the regional dip is south-southwest and averages 30 ft per mile.

Figure 9 illustrates two principal types of structures: (1) closures (such as North Buffalo) caused by tectonism; and (2) small-scale noses and closures caused by differential compaction of the shale within the Morrowan section.

On a structure map of the base of the Pennsylvanian, Bowles (1962) showed a well-defined structural nose plunging southwestward across T28N, R20W, and T27N, R21W. The structural expression of this feature on the Inola limestone may be that of three or four closures aligned in a southwest direction across the two townships. Such a southwestward structural nosing is general in this area.

Kornfeld (1959) indicated that the North Buffalo field (northwest part of T28N, R22W) is on a pre-Pennsylvanian northeast-trending anticline cut by a northwest-trending normal fault. Neither the northeast trend of the structure nor the fault could be confirmed from this study.

Most of the structures in the area are of the

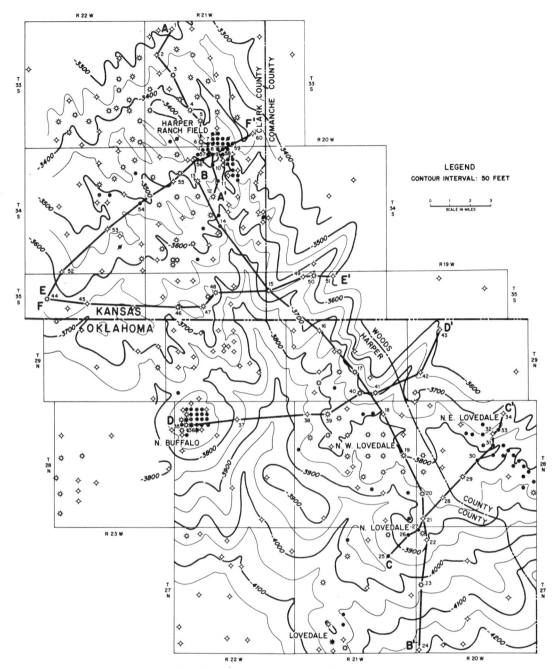

FIG. 9—Structure map of Inola limestone.

differential-compaction type. Many of these are subparallel southeast-plunging noses and northwest-southeast-trending closures. To analyze this structural relief, the Inola structure map (Fig. 9) was compared with the pre-Pennsylvanian topographic map (Fig. 10). The axes of structural lows on the Inola generally coincide with the axes of maximum Inola-to-Mississippian thickness (topographic lows) on the pre-Pennsylvanian paleotopographic map. Structural highs on the Inola, however, are above the axes of minimum thickness (buried hills). The Inola is downwarped along the axes of maximum (Morrow) sediment thickness as a result of the greater effect of compaction in these areas; consequently the Inola is draped over buried hills

Gregory W. Mannhard and Daniel A. Busch

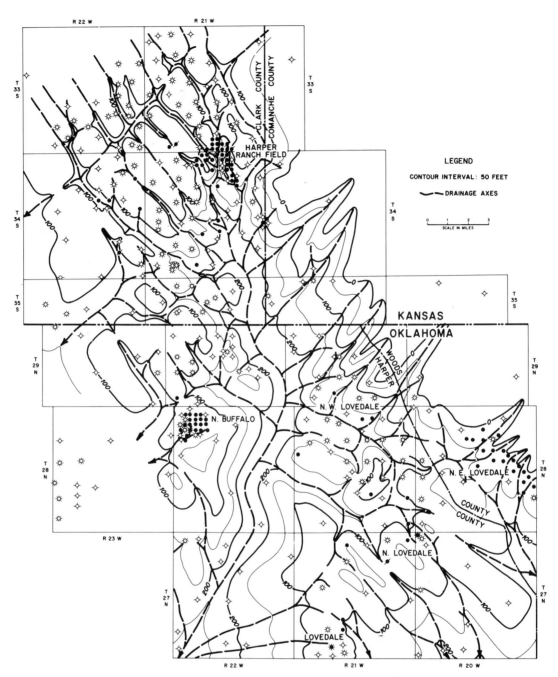

FIG. 10—Pre-Pennsylvanian topography (isopach map, top of Inola limestone to Mississippian unconformity).

of the pre-Pennsylvanian erosional surface.

In considering differential compaction it is assumed that the Mississippian sediments were lithified and compacted prior to Morrowan deposition. Cross sections show that the thicker sandstone section in the valley fills is accompanied by thicker shale sections, and by an increase in the number of shale beds deposited within the valley areas. In contrast, less shale was deposited above erosional remnants. Thus, there was more shale to compact in the valley areas than on erosional remnants. In addition, the individual sandstone beds in the valley-fill sections generally have a higher shale content than "cleaner" sandstones which were spread across paleotopographic highs. As a result, the valley sandstones were slightly more "compactable" than the sheetlike sandstones.

GEOLOGIC HISTORY

At the close of Mississippian time, northwestern Oklahoma and adjacent Kansas were subjected to epeirogenic upwarp. As a result, Upper Mississippian strata were elevated above sea level, homoclinally tilted toward the southwest, and exposed to subaerial erosion. A well-developed drainage system was established on the truncated Mississippian strata.

Figure 10 is a "cast" of the pre-Pennsylvanian paleotopographic surface. Lines connecting places of maximum interval thickness represent the axes of drainage courses, whereas the areas of minimum interval thickness represent the upland (divide) areas separating them.

The general direction of drainage was southwest with the paleotopographically lowest area in T27N, R22W. Khaiwka (1968) indicated the low area by means of an isopach map of the Morrow. As a result of the present study, it is apparent that the valley continues northeast through T28N, R22W. The principal tributaries of the southwestward-flowing stream consisted of several strike-valley streams which flowed northwest and southeast, respectively. A modified trellis drainage system which was well adjusted to the stratigraphy and structure of the underlying Mississippian section was developed within the area. Tilted resistant limestone beds of the Chesterian Series stood out as subparallel cuestas with dip slopes toward the southwest. Intervening strike valleys were eroded into the nonresistant shale beds.

The dominant paleoslope within the area was southwestward, the direction of flow for the major consequent streams. It can be assumed, therefore, that during Early Pennsylvanian time the area of the most extensive subsidence of the Anadarko basin was on the southwest, as demonstrated by Khaiwka (1968).

Two prominent topographic features in areas that probably were rising structurally during Early Pennsylvanian time provided important controls on the development of drainage courses. One of these is associated with the North Buffalo structure in the northwestern part of T28N, R22W. It was an obstacle which diverted the position of the main (consequent) drainage course southeastward. The other was a dissected plateaulike feature associated with the anticlinal structure extending diagonally across T28N, R20W, and T27N, R21W (Figs. 9, 10). It formed a topographic divide across

the southeastern part of the area and caused some streams to flow northwestward.

Subsidence of the Anadarko basin in Early Pennsylvanian time was accompanied by a marine transgression toward the northeast. The strike valleys were drowned progressively as the sea transgressed farther northeast. Transgression was cyclic and interrupted by several minor regressions. During these regressive phases with their associated shallow-water conditions sands were deposited. The interbedded muds were deposited when deeper water conditions returned during transgressive phases. The lowermost sandstone bodies in the section were concentrated largely within the confines of the submerged strike valleys. As the valleys were filled with sediment, the thickness and configuration of succeeding individual sandstone bodies were controlled to a progressively lesser extent by the underlying topography.

The marine transgression in Morrowan time continued into Atokan and Desmoinesian times. Complete submergence of the investigated area was effected with the deposition of the Inola limestone.

PETROLEUM GEOLOGY

The names and general locations of major oil and gas fields within the area are shown on Figures 9 and 10. Most of these fields produce either partly or wholly from Morrow sandstones. At North Buffalo other productive units include the Arbuckle, Viola, Meramec, Chester, Oswego, Tonkawa, Lansing, and Kansas City. The most productive oil field in the area is the Harper Ranch field of Clark County, Kansas, which has produced 2,974,255 bbl of oil (Internat. Oil Scouts Assoc., 1970). The several Lovedale fields in Oklahoma have provided most of the gas production. As of December 1970, they have produced 59,878,684 MCF of gas (Internat. Oil Scouts Assoc., 1970).

Morrow and Chester production in the Lovedale, North Lovedale, Northeast Lovedale, and Northwest Lovedale fields is largely from stratigraphic traps. Oil accumulation in the Morrow of the Harper Ranch field is entirely stratigraphic. Some degree of structural control is suggested for deep production in the Lovedale and North Lovedale fields. Accumulation of hydrocarbons in the Arbuckle, Viola, Oswego, Lansing, and Kansas City of the North Buffalo field primarily is structure controlled.

Harper Ranch Field

The Harper Ranch field was selected for detailed study because it affords the greatest amount of well control, and no adequate explanation for the presence of oil and gas has been provided previously in the literature.

The Harper Ranch field straddles the contiguous boundaries of T33, 34S, R21W, Clark County, Kansas. The discovery well, the United 1 Harper, in Sec. 9, T34S, R21W, was completed in a Morrow sandstone for 6 MCF of gas in June 1953 (Waite, 1956).

Morrow sandstones of the Harper Ranch field have been zoned vertically into the First, Second, and Third Morrow sandstones, as shown on cross sections AA' (Fig. 3) and FF' (Fig. 8). The Second Morrow sandstone has the most favorable reservoir characteristics and the most widespread distribution. For this reason the Second sandstone accounts for more than 90 percent of the production within the area. Porosity calculations (from mechanical logs) range up to a maximum of 27 percent. On the basis of core analysis the porosity ranges from 10 to 26 percent and the permeability ranges from 25 to 350 md (Waite, 1956).

By detailed mapping an attempt was made to arrive at a logical explanation for each dry hole and producing well in the Second sand. Similar procedures should be applicable to the other Morrow sandstones within the study area.

Figure 11 is an isopach map of the interval from the top of the Inola limestone to the base of the Second sandstone. Such a map provides a simulated reconstruction of the depositional surface on which the Second sandstone was deposited. The map shows an undulatory surface which is essentially a subdued replica of the pre-Pennsylvanian topography. There is a series of elongate, subparallel, southeast-trending ridges and intervening valleys. The surface probably was not exposed to subaerial erosion for any significant period of time, for the cross sections show no evidence of channeling into the shale section separating the Second and Third sands. The axes of maximum thickness on this map generally coincide with the axes of drainage courses developed on the pre-Pennsylvanian erosion surface. This is attributed to the fact that many pre-Pennsylvanian valleys were not filled entirely by compensatory deposition when the Second sand was deposited, and consequently the positions of the original valleys continued to be the sites of maximum sediment fill.

Figure 12 is an isopach map of the Second Morrow sandstone. The reliability of the Sec-

ond sandstone isopach map was enhanced greatly by making a paleodrainage map first (Fig. 11). Maximum Second sandstone development is expected along the axes of paleo-drainage.

An important aspect of the isopach map (Fig. 12), insofar as oil and gas accumulation is concerned, is the location of lines of zero thickness. Such lines represent the lateral limits of effective sandstone permeability.

Figure 13 is a structure map of the Second Morrow sandstone on which appropriate symbols have been superimposed to show both the known and probable distribution of oil, gas, and water. The zero sandstone thickness line was traced from the isopach map of the Second sandstone (Fig. 12).

Structural contours drawn on top of the Second Morrow sandstone bend in an upstream direction along the axes of maximum thickness (Inola to base of the Second sandstone interval) rather than in a downstream direction. This suggests that the Second sandstone was downwarped in response to compaction within the underlying section. Maximum compaction, and hence maximum downwarping, occurred over buried pre-Pennsylvanian valleys which received relatively greater thicknesses of compactable shale than did surrounding paleotopographically high areas.

When production data for the Second sandstone are added to the structure map, it is apparent that the distribution of oil and gas is essentially independent of the structure. By superimposing the lines of zero-sandstone thickness from the Second Morrow sandstone isopach map (Fig. 12) onto the structure map, the necessary trapping mechanism for the oil and gas is provided. The updip wedge-outs and shale-outs of the Second sandstone served as effective barriers to the lateral migration of hydrocarbons. Vertical escape of hydrocarbons was prevented by the impermeable shale deposited on the Second sandstone.

In two areas (Fig. 13) gas is being produced from the Second Morrow sandstone at structurally lower elevations than oil, and the gas in these areas is in direct contact with water at the downdip limit of gas accumulation. One such area is in the southwestern part of the mapped area and includes parts of Secs. 7–9, 16–18, T34S, R21W. A much smaller area (occupied by a single abandoned gas well) is in the southeastern part of the mapped area and includes parts of Secs. 12, 13, T34S,

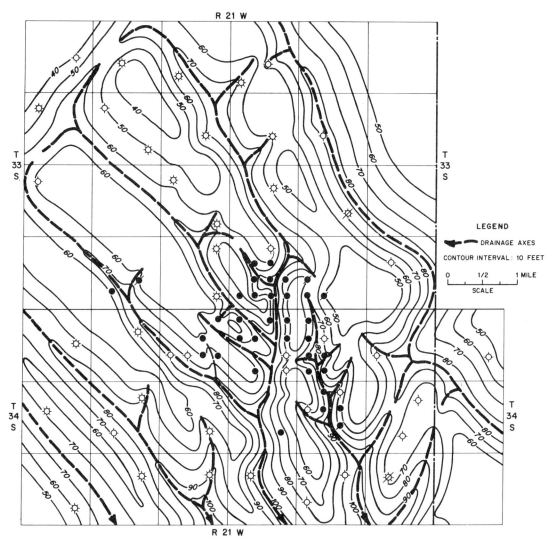

FIG. 11—Isopach map, top of Inola limestone to base of Second Morrow sandstone.

R21W. The anomalous areas of gas production are separated by a larger updip trap with normal reservoir conditions in which gas is present at structurally higher elevations than oil, and oil at structurally higher elevations than water. The anomalous presence of gas in the downdip areas can be explained by Gussow's principle (Gussow, 1953), modified to apply to a stratigraphic trap. It is assumed that the downdip traps which now are producing only gas once were filled partly or entirely with oil. As gas migrated into these areas the oil was forced out at the bottom (spill points) of the traps and moved into the central updip trap. The positions where the oil escaped from the downdip traps into the updip trap are at the most downdip edges of the respective permeability barriers which separate the individual downdip traps from the central updip trap. The locations of the two spill points are shown on the structure map (Fig. 13). As gas entered the downdip traps all of the oil eventually was forced updip at the two spill points. The gas then began to spill out of the downdip traps and partly displaced the oil which had accumulated in the central updip trap, thus establishing the present reservoir conditions.

Another seemingly anomalous situation exists along the eastern margin of the area. Wells in Secs. 15, 23, and 36, T33S, R21W, have tested salt water in the Second Morrow sandstone updip from oil and gas producing areas in the Second sandstone. The single abandoned gas well in Sec. 26 is anomalous and could result

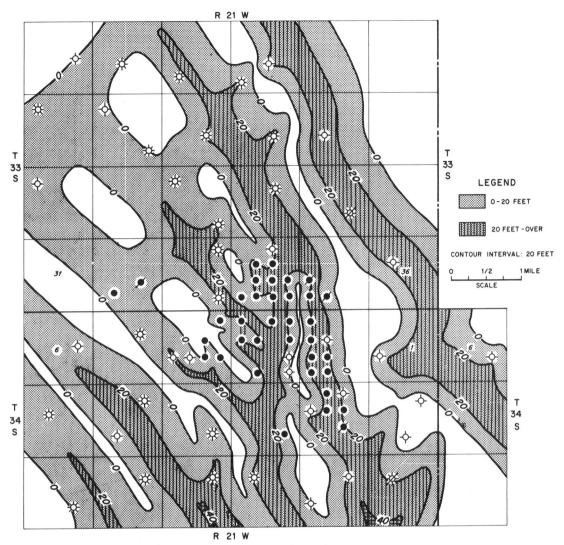

FIG. 12—Isopach map, Second Morrow sandstone.

from a local lateral extension of the gas-producing area on the southwest. This condition suggests the presence of a permeability barrier restricting the migration of hydrocarbons into the area of salt-water accumulation. The existence of such a permeability barrier is confirmed by a gradual shaling out of the Second sandstone in wells at the northeast margin of the oil- and gas-producing area. The Second sandstone is completely absent in the two wells of Sec. 12, T34S, R21W. These are in the zone separating the oil- and gas-producing area from the area of water accumulation.

North Buffalo Field

The North Buffalo field, in Secs. 5–8, T28N, R22W, Harper County, Oklahoma, has pro-

duced 1,823,960 bbl of oil and 1,962,017 MCF of gas (Internat. Oil Scouts Assoc., 1970). The discovery well was the Sinclair 1 Holcomb in Sec. 7, completed in July 1958 in the Arbuckle Formation.

Morrow sandstone production in this field is limited to three gas wells. The sandstone is shaly and erratic in occurrence. Accumulation of gas in the Morrow is controlled primarily by stratigraphic factors.

Lovedale Fields

The name Lovedale has been applied to an oil- and gas-producing area extending from T26N, R20, 21W, into T29N, R21W. The Lovedale producing area includes the Lovedale, North Lovedale, Northeast Lovedale, and

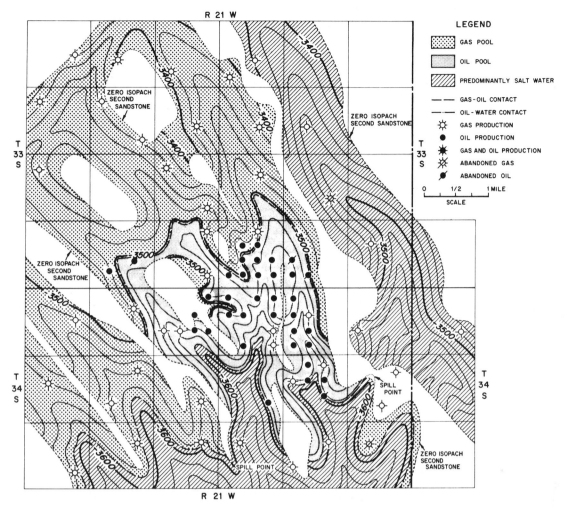

FIG. 13—Structure map, Second Morrow sandstone, showing oil and gas distribution. Contour interval 20 ft.

Northwest Lovedale fields (Fig. 9). Entrapment of oil and gas in the Morrowan Series is mainly the result of updip wedge-outs and shale-outs of sandstones. Chester production is primarily from unconformity traps in which hydrocarbons have accumulated in the tilted and truncated strata below the unconformity.

Only the northernmost part (T27N, R21W) of the Lovedale field is in the study area. Production is from the Arbuckle, Viola, Hunton, Chester, Morrow, Kansas City, and Tonkawa. The Lovedale field has produced 427,056 bbl of oil and 33,528,672 MCF of gas (Internat. Oil Scouts Assoc., 1970).

The North Lovedale field was discovered in 1958, with the completion of the Woods 1 Seevers (Sec. 6, T27N, R20W) as an oil-producing well in the Marmaton (Jordan *et al.*, 1959). This field has yielded 140,595 bbl of oil and 6,116,516 MCF of gas (Internat. Oil Scouts Assoc., 1970).

The Northeast Lovedale field is almost exclusively oil producing. Entrapment occurred in the tilted and truncated edges of the sandy limestones of the basal Chesterian Series.

Production in the Northwest Lovedale field is largely from stratigraphic traps in the Morrow and Chester. This field has yielded 245,959 bbl of oil and 19,836,479 MCF of gas (Internat. Oil Scouts Assoc., 1970).

SUMMARY AND CONCLUSIONS

1. Determination of the position of the Mississippian-Pennsylvanian unconformity was accomplished from mechanical-log cross sections of the Mississippian and Pennsylvanian sections, and by binocular microscopic studies of rotary cuttings of key wells on the cross sections

to recognize significant lithologic changes related to the unconformity.

2. Most of the sandstones considered are of Morrowan age; two regional cross sections indicate that these sandstones are present in the interval between the base of the "Thirteen Finger lime" and the Mississippian-Pennsylvanian unconformity.

3. By correlating Cherokee limestone marker beds (identified on a cross section published by Lukert, 1949) into the study area along an extended cross section it was possible to identify the Inola limestone and other marker beds in the area.

4. Drainage on the pre-Pennsylvanian erosional surface was controlled by a combination of structure and lithology of the underlying Mississippian sediments.

5. A modified trellis drainage system was developed on subaerially exposed Mississippian strata. Tilted, resistant limestones of the Chesterian Series stood out as subparallel cuestas with southwestward dip slopes; intervening erosional valleys were developed on nonresistant shale beds.

6. Small-scale structural noses and closures on the Inola limestone are generally the result of differential compaction of shales within the underlying Morrow section. Maximum downwarping of the Inola limestones occurred over buried pre-Pennsylvanian valleys which were filled with greater thicknesses of shale than surrounding paleotopographically high areas.

7. Detailed subsurface mapping affords a logical explanation for each producing well and all but a few dry holes in the Second Morrow sandstone of the Harper Ranch field of Clark County, Kansas.

8. The anomalous distribution of some gas accumulation in Second Morrow sandstone of the Harper Ranch field can be explained by the use of Gussow's principle, modified to apply to a stratigraphic trap.

9. The principles utilized in the study of the Harper Ranch field should be applicable in future exploration throughout the area for oil and gas in Morrow sandstones.

10. The methods employed in this study have potential application in many areas of the world and all parts of the stratigraphic column.

SELECTED REFERENCES

Abels, A. A., 1962, A subsurface lithofacies study of the Morrowan Series in the northern Anadarko basin: Oklahoma City Geol. Soc., Shale Shaker Digest, v. 3, p. 93–108.

Allen, A. E., Jr., 1955, The subsurface geology of Woods and Alfalfa Counties, northwestern Oklahoma: Oklahoma City Geol. Soc., Shale Shaker Digest, v. 1, p. 261–281; originally published, 1954, Shale Shaker, v. 4, no. 8, p. 5–21, 23–26, 32.

Andresen, M. J., 1962, Paleodrainage patterns—their mapping from subsurface data, and their paleogeographic value: Am. Assoc. Petroleum Geologists Bull., v. 46, p. 398–405.

Barby, B. G., 1958, Subsurface geology of the Pennsylvanian and Upper Mississippian of Beaver County, Oklahoma: Oklahoma City Geol. Soc., Shale Shaker Digest, v. 2, p. 133–154; originally published, 1956, Shale Shaker, v. 6, no. 10, p. 9–12, 15–32.

Barrett, L. W., 1965, Subsurface study of Morrowan rocks in central and southern Beaver County, Oklahoma: Oklahoma City Geol. Soc., Shale Shaker Digest, v. 4, p. 425–443; originally published, 1964, Shale Shaker, v. 14, no. 9, p. 2–12, 13–20.

Bowles, J. P. F., 1962, Subsurface geology of Woods County, Oklahoma: Oklahoma City Geol. Soc., Shale Shaker Digest, v. 3, p. 197–215; originally published, 1959, Shale Shaker, v. 10, no. 4, p. 2–23.

Busch, D. A., 1959, Prospecting for stratigraphic traps: Am. Assoc. Petroleum Geologists Bull., v. 43, p. 2829–2843.

Clair, J. R., 1948, Preliminary notes on lithologic criteria for identification and subdivision of the Mississippian rocks in western Kansas: Kansas Geol. Soc., 14 p.

Gibbons, K. E., 1965, Pennsylvanian of the north flank of the Anadarko basin: Oklahoma City Geol. Soc., Shale Shaker Digest, v. 4, p. 71–87; originally published, 1962, Shale Shaker, v. 12, no. 5, p. 2–10, 12–19.

Gussow, W. C., 1953, Differential trapping of hydrocarbons: Alberta Soc. Petroleum Geologists News Bull., v. 1, no. 6, p. 4–5.

Haworth, E., and M. Z. Kirk, 1894, A geologic section along the Neosho River from the Mississippian formation of the Indian Territory to White City, Kansas, and along the Cottonwood River from Wyckoff to Peabody: Kansas Univ. Quart., v. 2, p. 104–115.

Huffman, G. G., 1959, Pre-Desmoinesian isopachous and paleogeologic studies in central Mid-Continent region: Am. Assoc. Petroleum Geologists Bull., v. 43, p. 2541–2574.

Hutchison, L. L., 1911, Preliminary report on the rock asphalt, asphaltite, petroleum, and natural gas in Oklahoma: Oklahoma Geol. Survey Bull. 2, 256 p.

International Oil Scouts Association, 1970, International oil and gas development, v. 40, pt. 2, p. 331.

Jordan, L., 1957, Subsurface stratigraphic names of Oklahoma: Oklahoma Geol. Survey Guide Book 6, 220 p.

———— J. D. Pate, and S. R. Williamson, 1959, Petroleum geology of Harper County, in Geology of Harper County, Oklahoma: Oklahoma Geol. Survey Bull. 80, p. 69–92.

Khaiwka, M. H., 1968, Geometry and depositional environments of Pennsylvanian reservoir sandstones, northwestern Oklahoma: Ph.D. dissert., Univ. Oklahoma, 126 p.

Kornfeld, J. A., 1959, Geology of North Buffalo oil and gas field, Harper County, Oklahoma: Tulsa Geol. Soc. Digest, v. 27, p. 67–68.

Krumbein, W. C., and L. L. Sloss, 1963, Stratigraphy and sedimentation, 2d ed.: San Francisco, W. H. Freeman, 660 p.

Lee, W., 1953, Subsurface geologic cross section from Meade County to Smith County, Kansas: Kansas Geol. Survey Oil and Gas Inv. 9, 23 p.

Levorsen, A. I., and F. A. F. Berry, 1967, Geology of petroleum, 2d ed.: San Francisco, W. H. Freeman, 724 p.

Lowman, S. W., 1932, Lower and Middle Pennsylvanian stratigraphy of Oklahoma east of the meridian and north of the Arbuckle Mountains (abs.), in Tulsa Geology Society summaries and abstracts, 1932: Tulsa, Oklahoma, Tulsa Daily World, December 19, 1932.

Lukert, L. H., 1949, Subsurface cross sections from Marion County, Kansas, to Osage County, Oklahoma: Am. Assoc. Petroleum Geologists Bull., v. 33, p. 131–152.

Martin, R., 1966, Paleogeomorphology and its application to exploration for oil and gas (with applications from western Canada): Am. Assoc. Petroleum Geologists Bull., v. 50, p. 2277–2311.

Moore, R. C., 1949, Divisions of the Pennsylvanian System in Kansas: Kansas Geol. Survey Bull. 83, 203 p.

Oakes, M. C., 1953, Krebs and Cabaniss groups, of Pennsylvanian age, in Oklahoma: Am. Assoc. Petroleum Geologists Bull., v. 37, no. 6, p. 1523–1526.

Pate, J. D., 1959, Stratigraphic traps along north shelf of Anadarko basin, Oklahoma: Am. Assoc. Petroleum Geologists Bull., v. 43, p. 39–59.

Peterson, J. A., and J. C. Osmond, eds., 1961, Geometry of sandstone bodies: Am. Assoc. Petroleum Geologists Spec. Pub., 240 p.

Powell, B. D. H., Jr., 1955, The subsurface geology of Woodward County, Oklahoma: Oklahoma City Geol. Soc., Shale Shaker Digest, v. 1, p. 195–214; originally published, 1953, Shale Shaker, v. 4, no. 1, p. 4–30.

Price, B. D., 1963, North Buffalo, in Oil and gas fields of Oklahoma: Oklahoma City Geol. Soc. Reference Rept., v. 1, p. 4–A.

Purdue, A. H., 1907, Description of the Winslow quadrangle: U.S. Geol. Survey, Geol. Atlas Winslow folio (no. 154), 6 p.

Rascoe, B., Jr., 1962, Regional stratigraphic analysis of Pennsylvanian and Permian rocks in western Mid-Continent (Colorado, Kansas, Oklahoma, Texas): Am. Assoc. Petroleum Geologists Bull., v. 46, p. 1345–1370.

Swanson, D. C., 1967, Some major factors controlling the accumulation of hydrocarbons in the Anadarko basin: Shale Shaker, v. 17, no. 6, p. 106–114.

Thornbury, W. D., 1965, Regional geomorphology of the United States: New York, John Wiley and Sons, 609 p.

Thornton, W. D., 1965, Mississippian rocks in the subsurface of Alfalfa and parts of Woods and Grant Counties, northwestern Oklahoma: Oklahoma City Geol. Soc., Shale Shaker Digest, v. 4, p. 117–128; originally published, 1962, Shale Shaker, v. 12, no. 8, p. 2–13.

Vance-Rowe Reports, 1969, Production reports of Panhandle and northwestern Oklahoma: Tulsa, Oklahoma, p. 120–123.

Van Siclen, D. C., 1958, Depositional topography—examples and theory: Am. Assoc. Petroleum Geologists Bull., v. 42, p. 1897–1913.

Waite, G., 1956, The Harper Ranch pool, in Kansas oil and gas pools, v. 1: Kansas Geol. Soc., p. 50.

Woodruff, E. G., and C. L. Cooper, 1928, Oil and gas in Oklahoma; geology of Rogers County: Okla. Geol. Survey Bull. 40–U, 24 p.

American Association of Petroleum Geologists Bulletin
V. 62, No. 5 (May 1978) pp. 848-853.

Permeability Traps in Gatchell (Eocene) Sand of California[1]

ROBERT SCHNEEFLOCK[2]

Abstract Early in the development of Coalinga East Extension, Pleasant Valley, and Guijarral Hills fields of California, geologists recognized that numerous permeability variations and traps exist in the Eocene Gatchell sand. These anomalies developed as the sand became plugged with interstitial kaolinite. Kaolinite can be detected easily in cores but is extremely difficult to recognize from electric logs. Kaolinite forms in the Gatchell as a postdepositional alteration product of feldspars, micas, and formation water. Pore spaces filled with hydrocarbons, however, are not subject to this reaction; thus porosity and permeability are preserved. Subsequent structural adjustments could leave these "clay-sealed" oil traps in seemingly unfavorable structural positions.

Permeability traps in the Gatchell are too numerous for *all of the oil* in the Coalinga area to have been found. A thorough understanding of existing production as well as structural, depositional, and diagenetic history is required to find the remaining reserves.

INTRODUCTION

The Gatchell sand is of middle Eocene age and probably is best known in the Coalinga area of Fresno County, California. Cumulative production from the Gatchell is approximately 500 million bbl of oil and 340 Bcf of gas. This production has come mainly from three fields: Coalinga East Extension, Pleasant Valley, and Guijarral Hills.

Three other fields—Kettleman North Dome, Five Points, and Cantua Creek—also have Gatchell production, but in smaller quantities.

The approximate extent of the Gatchell is shown by the stippled area in Figure 1. Its stratigraphic position is shown on the electric log from the Great Basins 1-35 Northeast Kettleman in Sec. 35, T20S, R17E. This well is located in the approximate center of the stippled area.

[1]Manuscript received, June 24, 1977; accepted, November 8, 1977.

[2]Tenneco Oil Co., Bakersfield, California 93309.

The writer thanks Tenneco Oil Co. for permission to publish this work, also Jim Dorman for encouraging the study, and the many colleagues who have reviewed the manuscript.

Article Identification Number
0149-1423/78/B005-0009$03.00/0

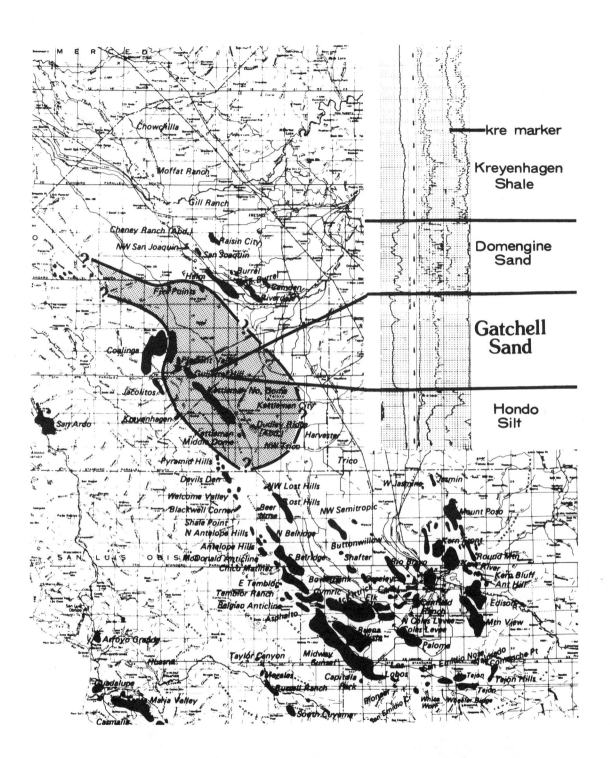

FIG. 1—Gatchell distribution map and type log. Base map is from Munger (1977).

EXISTING PRODUCTION

All known Gatchell fields are stratigraphic traps and are most unusual. For example, Pleasant Valley and Guijarral Hills fields are unusual in that the oil commonly appears to be trapped in the bottom of the sand (Fig. 2). Electric logs show no continuous shale break between the lower productive interval and the upper wet section, and sonic logs show the sand to be almost uniform in porosity throughout. Until one looks very closely at core data, there would seem to be no plausible explanation for the trapping mechanism.

Through detailed study of conventional cores, early geologists were able to recognize the nature of the stratigraphic traps at Pleasant Valley and Guijarral Hills fields (Loken, 1955; Simonson, 1958; Sullivan, 1962). They recognized that the wet sand above the pay zone was plugged with kaolinite. Modern electric-logging techniques generally are unable to distinguish this plugging because the ion exchange potential of kaolinite produces nearly the same SP deflection and resistivity as a clean, wet sand. Sonic or acoustic logs record the presence of kaolinite as "shale porosity." What then appears from electric logs to be a thick section of clean, wet sand with oil at the bottom is actually oil trapped against sand that has its pore spaces completely plugged with kaolinite.

AUTHIGENIC ORIGIN

Merino (1975) has determined from thin section, X-ray diffraction, and electron-microprobe analysis that the kaolinite in the lower McAdams (Gatchell equivalent) is authigenic in origin. He demonstrated this by showing examples in which kaolinite fills postdepositional compaction fractures and other examples in which feldspar and mica grains are only partially altered to kaolinite (Fig. 3). Thin section work by the writer supports Merino's findings. Photomicrographs of authigenic kaolinite, such as the one in Figure 4, show that kaolinite crystals grow in delicate booklike arrangements. Webb (1974) stated that "delicate books of crystals probably could not survive detrital transport, additional evidence that they crystallized in the pore space after deposition of the sand."

ANALOGIES

Although the type of kaolinite trap in the Coalinga area is unusual, it is not unique. Webb (1974), Wilson and Pittman (1977), and Wilson

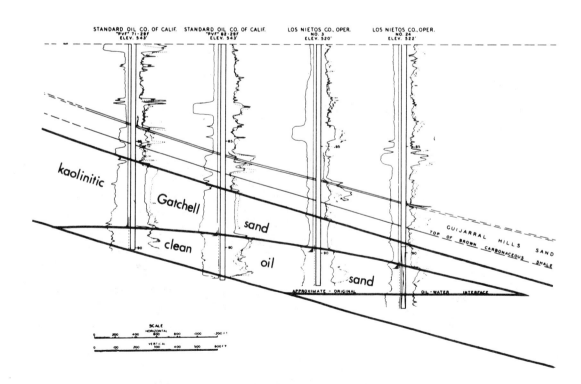

FIG. 2—Cross section of Pleasant Valley oil field showing oil trapped at bottom of Gatchell sand. Modified from Loken (1955).

FIG. 3—Thin section of Gatchell sand showing large plagioclase crystals. Wavy margins of grains are alteration contacts where kaolin has replaced plagioclase. Fractures in grains also are filled with kaolinite. Dark areas are both microcrystalline kaolinite matrix and voids caused by plucking of grains during thin sectioning. ×10.

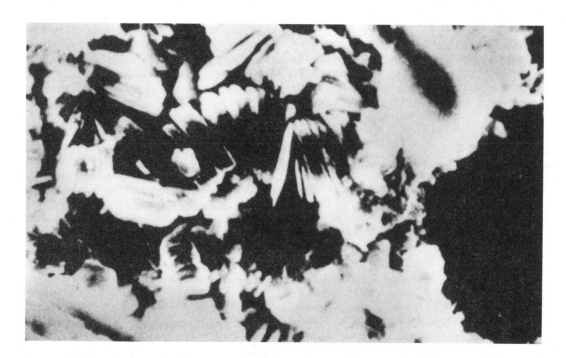

FIG. 4—Scanning electon photomicrograph of pore-filling authigenic kaolin. ×2,300.

(1977) have shown the nature and presence of kaolinite traps in many areas. Shelton (1964) reported authigenic kaolinite in numerous sands ranging in age from Ordovician to Miocene.

KAOLINIZATION

Kaolinite in the arkosic Gatchell sand was an alteration product of formation water, feldspars, and micas. Merino (1975) estimated a volume increase of approximately one and a half times owing to this alteration. Plugging of the pore spaces occurred both from the volume increase and from the solution and reprecipitation of kaolinite. This process appears to have occurred almost ubiquitously in the Gatchell with one notable exception—the oil zone.

Formation waters (a necessary ingredient for this diagenesis) were displaced by the accumulation of oil. This means that kaolinite plugging of the sand did not take place in the reservoir because oil accumulated early enough. Empirically, this is borne out by four existing Gatchell fields: Coalinga East Extension, Pleasant Valley, and the Gatchell production in Guijarral Hills and Kettleman North Dome fields. Only in Coalinga East Extension does the Gatchell have extremely good permeability (average 459 md), which is a result of early hydrocarbon migration and accumulation. Reservoirs in the other three fields have poorer permeabilities (averaging 50 to 75 md) because of late hydrocarbon accumulation.

Early accumulation at Coalinga East Extension is evidenced by several things. (1) Structurally, the reservoir is truncated either by an angular unconformity or a pinchout which can be interpreted to demonstrate structural growth prior to deposition of the overlying beds. The other three fields show a parallel relation between the reservoir and much younger beds (i.e., uplift did not take place in them until the time of Miocene Temblor deposition and later). (2) The original structural high of Coalinga East Extension field was 5 mi (8 km) north of the present high. The present structure is on the axis of the Coalinga nose–Kettleman dome structure which grew much later. As the new structure grew, a saddle developed between it and the old structure. The saddle trapped the original gas cap on the old structure, but oil below the saddle shifted easily to the new structure. The result is that there are two highs in this field, but only the lower one has a gas cap. (3) Other evidence for early accumulation is demonstrated by the three zones of permeability in the oil column at Coalinga East Extension. These zones are roughly parallel with subsea horizons rather than bedding planes and show successive increases in permeability higher in the oil column. This zonation is evidence of at least three stages of oil accumulation in this reservoir. Kaolinization continued until the formation water was replaced with oil. Permeabilities in the zone nearest the water level at Coalinga East Extension field are of the same order of magnitude as permeabilities in the other three downdip fields. This would seem to be logical, as none of the other three fields demonstrates any evidence of early accumulation.

CONCLUSIONS

Evidence presented here indicates that the kaolinite plugging of the Gatchell occurred after the sand was deposited. Kaolinite formed in the pore spaces as a result of chemical reaction between formation water and the feldspars and micas of the arkosic Gatchell. Little, if any, kaolinite is present in hydrocarbon-saturated Gatchell sand, whereas Gatchell sand not filled with hydrocarbons contains large amounts of kaolinite. Therefore, in hydrocarbon-saturated Gatchell, hydrocarbons displaced formation water before kaolinization began. In Gatchell sand not saturated with hydrocarbons, kaolinization continued until pore spaces were plugged, thus destroying permeability. Oil-saturated Gatchell reservoirs with lower permeabilities commonly resulted from hydrocarbon accumulations after some kaolinization already had taken place.

Most nonproductive wells in the Coalinga-Kettleman area which penetrated the Gatchell show this sand to be plugged with kaolinite. Failure to recognize that kaolinization is suppressed in the presence of hydrocarbons may cause this potentially good reservoir to be overlooked. Laterally, the Gatchell may be an excellent reservoir, provided hydrocarbons accumulated prior to kaolinization. Once accumulation has taken place and kaolinization has sealed the oil into a "clay trap," subsequent structural adjustments could place the trap in a seemingly unfavorable position. This would seem quite possible in the Coalinga-Kettleman area which had an active structural history and is in close proximity to the San Andreas fault. It seems unlikely that the Gatchell sand, with its large regional extent and numerous permeability variations, has yielded all of its oil. Successful exploration will require a thorough understanding of its structural, depositional, and diagenetic history.

REFERENCES CITED

Loken, K. P., 1955, Pleasant Valley oil field: California Oil Fields, v. 41, p. 54-59.
Merino, E., 1975, Diagenesis in Tertiary sandstones from Kettleman North Dome, California; 1, diagen-

etic mineralogy: Jour. Sed. Petrology, v. 45, p. 320-336.

Munger, A. H., ed., 1977, Munger map book, 21st edition: Los Angeles, Munger Oil Information Service, Inc.

Shelton, J. W., 1964, Authigenic kaolinite in sandstone: Jour. Sed. Petrology, v. 34, p. 102-111.

Simonson, R. R., 1958, Oil in the San Joaquin Valley, California, in L. G. Weeks, ed., Habitat of oil: AAPG, p. 99-112.

Sullivan, J. C., 1962, Guijarral Hills oil field: California Oil Fields, v. 48, p. 37-51.

Webb, J. E., 1974, Relation of oil migration to secondary clay cementation, Cretaceous sandstones, Wyoming: AAPG Bull., v. 58, p. 2245-2249.

Wilson, H. H., 1977, "Frozen-in" hydrocarbon accumulations or diagenetic traps—exploration targets: AAPG Bull., v. 61, p. 483-491.

Wilson, M. D., and E. D. Pittman, 1977, Authigenic clays in sandstones—recognition and influence on reservoir properties and paleoenvironmental analysis: Jour. Sed. Petrology, v. 47, p. 3-31.

THE AMERICAN ASSOCIATION OF PETROLEUM GEOLOGISTS BULLETIN

APRIL 1977   VOLUME 61, NUMBER 4

"Frozen-In" Hydrocarbon Accumulations or Diagenetic Traps—Exploration Targets[1]

H. HUGH WILSON[2]

Abstract Porosity and permeability of clastic and carbonate reservoir rocks are reduced progressively during burial by plugging of pores with secondary, pressure-dependent, diagenetically derived cements. The depth at which all effective porosity and permeability is lost in water-bearing reservoir rocks varies according to their mineral content. The presence of hydrocarbons in a reservoir inhibits the process of diagenetic plugging with the result that porosities and permeabilities differ greatly above and below an oil/water contact.

This difference in diagenetic evolution within and outside the oil column indicates early emplacement of oil in a trap. Furthermore, diagenetic plugging inhibits the entry of any later generated oil and makes the lateral flushing of oil from a trap progressively more difficult with burial.

When diagenetic plugging below a hydrocarbon paleotrap is complete, the accumulation is sealed in, and deeper burial with attendant pressure and temperature increases results in natural cracking of trapped oil to gas with phase-change expansion causing geopressuring of the depletion-type reservoir.

When a diagenetically sealed trap later is tilted regionally or locally the accumulation will be held in place despite its unfavorable structural position. Such diachronous traps are designated "diagenetic" as opposed to "structural" or "stratigraphic."

Effective search for diagenetic traps requires careful paleostructural analyses coupled with documentation of diagenetic porosity-destruction sequences for each objective reservoir rock during burial.

Because of the lack of present-day structural or primary stratigraphic closure and the unconventional nature of the trapping concept, there probably are many diagenetically trapped hydrocarbon accumulations yet to be discovered, particularly in deeper basin positions.

INTRODUCTION

Seventeen years ago Scholten, in his outstanding paper on oil in synchronous highs, described briefly how hydrocarbon accumulations can be retained in paleotraps despite subsequent structural opening, by cementation of reservoir rock around the edges of a pool, thus "freezing" the accumulation in place prior to redeformation

(Scholten, 1959, p. 1803). This trapping mechanism could be explained logically by Lowry's earlier observation (1956, p. 495-496) that the presence of hydrocarbons in sandstone reservoirs inhibits the process of porosity destruction by cementation during burial which, in water-bearing parts of the reservoir, continues as a normal sequence of diagenetic events.

It seems that the significance of these and subsequent observations on the effect of oil on diagenesis and porosity destruction has not been appreciated as geologists continue to classify hydrocarbon traps into the broad categories of "structural" and "stratigraphic." It is proposed that a third category, here designated "diagenetic" (not in the sense described by Rittenhouse, 1972, p. 22), should be given an equal status with the others, particularly because of its importance to the search for more hydrocarbons in the ultramature exploration areas of North America.

INHIBITION OF DIAGENESIS BY HYDROCARBONS

In an earlier paper (Wilson, 1975a, p. 82) it was recommended that detailed comparative studies be made between diagenetic changes that occur within oil- and water-bearing parts of the same reservoir rock in paleotraps at different burial depths down a basin flank, for this should enable

[1] Manuscript received, June 3, 1976; accepted, August 23, 1976.

[2] Petroleum exploration consultant with Louisiana Land and Exploration Co., New Orleans, Louisiana 70160.

The writer gratefully acknowledges the help of Wendy Kliebert and Warren Boudousquie of The Louisiana Land and Exploration Company who respectively typed the text and drafted the text figure of this paper.

finer pinpointing of the time of primary entrapment of hydrocarbons. Paleotraps with different times of tilt-out were proposed as prime candidates for study (Wilson 1975a, p. 74, Fig. 4; 1975b, p. 2055).

Although such detailed studies have yet to appear in the literature, several published works have emphasized the inhibiting effect of hydrocarbons on diagenesis of both carbonate and clastic reservoirs. The following are significant.

Carbonate Reservoirs

Upper Cretaceous chalk in the North Sea loses porosity and permeability as a direct function of burial depth unless oil entered the rock, reducing or terminating carbonate reactions (Scholle, 1976, p. 719). Comparison of the stage of diagenetic change in the Ekofisk oil reservoir with the equivalent stage in water-bearing chalk, shown by Scholle, indicates early introduction of hydrocarbons with inhibition of diagenetic porosity destruction.

In Iraq, several oil accumulations in anticlinal crests have preserved, by their presence, the original porosity of the reservoir rock, whereas in the synclines and plunges, cementation and like processes (diagenesis) entirely have sealed the aquifer. Adjacent structures are not in pressure communication and pressures in the oil-filled reservoirs are anomalously high, approaching rock pressure (Dunnington, 1960, p. 173).

In Abu Dhabi, porosities in the Lower Cretaceous (Thamama) water-wet carbonate rocks decrease from 30 to 35% in the 0 to 4,000-ft (1,219 m) range to zero below 12,000 ft (3,658 m; Clarke, 1975, p. 16).

The porosity of some Thamama reservoirs is related to cementation and pore filling by calcite diffusing from stylolites. This diagenetic process is strongly inhibited by the presence of oil (Clarke, 1975, p. 1).

In the Fahud field, Oman, southeast Arabia, the reservoir characteristics of the Cretaceous Wasia limestone deteriorate rapidly below the oil-water contact (Tschopp, 1967, p. 249).

The formation of depletion-type reservoirs in deep Smackover accumulations such as Thomasville in Rankin County, Mississippi, has been attributed to the same process of diagenetic sealing below oil/water contacts (Wilson, 1975a, p. 79). Such deep accumulations have become geopressured by natural thermal cracking of oil to gas and phase-change expansion in a diagenetically sealed porosity pod.

Clastic Reservoirs

In Cretaceous sandstone reservoirs in the Powder River basin, Wyoming and Montana, Webb (1974) noted that sandstone saturated with hydrocarbons contained little, if any, authigenic clay, whereas the same sandstone body without hydrocarbons contains much clay. Webb (1974, p. 2248) concluded that hydrocarbons entered the reservoir before or shortly after the precipitation of authigenic clay by reservoir waters began. Webb (1974, p. 2249) then gave a more refined description of the diagenetic trapping mechanism (previously inferred by Scholten, 1959), as follows:

Failure to recognize precipitation of authigenic clay (and other types of authigenic cement) as an event following sand deposition may cause potentially good reservoirs to be disregarded. A clay-filled sandstone may be an excellent reservoir laterally if hydrocarbons migrated into, and were trapped in, its original updip margin prior to clay precipitation. Once the hydrocarbons have accumulated and the surrounding sandstone is filled with clay cement, the trap is sealed. Later structuring may place the trapped oil in a seemingly unfavorable structural position. This is especially true in tectonically active areas such as the Rocky Mountains where the present dip direction may be opposite to the original depositional dip.

In northwest Oklahoma, porosity and permeability in Pennsylvanian sandstones have been destroyed completely by diagenetic plugging although, where gas is present, porosity and permeability have been retained (Adams, 1964, p. 1575).

The same process of destruction of porosities in Tertiary sandstone reservoirs by diagenetic plugging was described by Van de Kamp (1976) as "burial metamorphism," a process which he noted is inhibited by the presence of hydrocarbons.

Russian geologists have been aware of the inhibiting effect of hydrocarbons on diagenesis as was stated so clearly by Yurkova (1970, p. 66) in his study of fields in north Sakhalin Island, eastern USSR.

The investigations have thus established that the sandstones which are least affected by epigenesis are confined, the other geological and hydrogeological conditions being equal, to the areas within the outlines of the oil pools. This fact leads us to connect the delay in the secondary transformation of rocks with the presence of oil in these rocks, since according to the "braking" hypothesis oil is the most unfavorable medium for the processes of epigenetic solution and replacement of the unstable components and for authigenic mineral formation.

In the Denver basin, a detailed study of cementing materials in the Permian Lyons sandstone by Levandowski et al (1973) illustrated how postdepositional diagenetic changes in the reservoir rocks can affect oil migration and retention. It was concluded that early cementation may prevent the accumulation of oil in a trap whereas late cementation may hold oil in a trap during structural movement, and differential cementation may provide the trap itself (Levandowski et al, 1973, p. 2217). The New Windsor accumulation was shown to be restricted to a structural nose by a diagenetic permeability barrier (Levandowski et al, 1973, p. 2230).

In the Andean foothills of northern Argentina and southeastern Bolivia oil often is pooled in Permian-Carboniferous sandstones thousands of feet down the axial plunges of Pliocene-Pleistocene folds when the same sandstones are filled with connate water along the structural culminations and down the opposite plunges (Reed, 1946, p. 595; Weeks, 1958, p. 54). This phenomenon was interpreted by Weeks (1958, p. 56) as being due to the nonadjustment of oil-water relations to the Andean folding which occurred well after the oil had accumulated in the Permian-Carboniferous sands. This interpretation, which implies porosity barriers outside the oil accumulations, is in harmony with Weeks' (1961, p. 5-44) much broader observation that once oil is emplaced in sandstones the pores seem to be protected from cementation to a considerable degree, whereas cementation in the water-bearing parts of the sandstone is quite common.

Of significance also is Weeks' (1958, p. 56) earlier observation that oil/water interfaces that have been tilted as a result of tectonic adjustment maintain their attitude with notable tenacity.

In the San Joaquin basin, California, very late, Pleistocene, movements have opened many earlier Tertiary folds. Although the retention of many paleostructural oil accumulations is explained by updip pinchout or facies barriers, the Pleasant Valley field (Harding, 1976, p. 372, Fig. 9) appears to be diagenetically trapped. In this field the updip barrier is caused by plugging of the Gatchell sandstone by kaolinite (Weddle, 1951, p. 623) on the paleostructural flanks in a manner which seems directly comparable to the porosity destruction by authigenic kaolinite in Rocky Mountain Cretaceous sandstones described by Webb (1974, p. 2249).

Although diagenetic sealing of reservoir rock below oil accumulations has been documented from many basins there are many more fields in which the absence of a water drive (depletion-type reservoirs) allows the assumption that the reservoir is sealed below the oil column.

Thus it may be concluded that diagenetic destruction of porosity takes place during burial of both carbonate and clastic reservoirs. In the former, the plugging effect is wrought mainly by redistribution of calcium carbonate through stylolitization (Dunnington, 1967) whereas, in the latter,

plugging is brought about by silica released by pressure solution at grain contacts as well as by introduction of authigenic clay minerals.

Observations from many hydrocarbon traps establish conclusively that hydrocarbons inhibit the diagenetic process and, as a corollary, that the hydrocarbons are emplaced *early* into those traps.

MISINTERPRETATION OF DIAGENETIC TRAPS AS STRATIGRAPHIC TRAPS

In many descriptions of stratigraphic traps which are available for study in the literature, the updip trapping agent is described as a porosity barrier.

In primary stratigraphic traps, the porosity barrier is formed by a facies change from high-energy carbonate deposits to low-energy mudstones or from sandstones to shale.

However, where the porosity barrier is caused by a change from uncemented high-energy carbonate deposits to cemented high-energy carbonate deposits or by a change from "clean" sand to "dirty" sand, the primary nature of the trap becomes suspect, for the real cause may be diagenetic plugging prior to tilt-out of a paleostructural trap.

Because, in oil search, half the battle in prospect evaluation is to understand the real sequence of events in hydrocarbon entrapment, it is worthwhile to study "apparent" stratigraphic traps in an effort to understand the real mode of entrapment.

In carbonate traps, the common cemented high-energy oolites outside oil pools, for example, the Smackover on the Gulf Coast (Vestal, 1950, p. 7; Bishop, 1971, p. 125), show that primary depositional facies need not be the only porosity barrier in an objective formation. As a follow-up interpretation, the tilted oil/water contact at the Nancy field, Clarke County, Mississippi (Hughes, 1968, p. 324, 328), can be explained as a failure of oil in the Jurassic paleostructure to adjust to a Tertiary tilt southward because of diagenetic plugging of the water leg prior to tilting.

In the Middle East, tilted oil/water contacts in carbonate reservoirs are not uncommon. The tilt of the Jurassic No. 3 limestone oil/water contact in the Dukhan oil field (Qatar Petroleum Co. Staff, 1956, p. 166) is almost certainly due to diagenetic plugging of the limestone below the oil/water contact, resulting in a "frozen," depletion-type reservoir in which the oil was unable to adjust to Tertiary tilt eastward.

Accumulations in sandstone reservoirs which have been recognized as being diagenetically trapped have been discussed previously. However, it seems possible that some traps that have been interpreted as stratigraphic may be, at least partly, diagenetic. Two examples which require thoughtful reappraisal follow.

Pembina Field, Alberta

The stacking of hydrocarbon accumulations at successive levels from Lower Cretaceous Blairmore to Upper Cretaceous Belly River sandstones (Patterson and Arneson, 1957, p. 940) is suggestive of paleostructural control, a suggestion which is enhanced by the isopach thins mapped in the upper shale member of the Cardium Formation (p. 944, Fig. 6). Retention of oil after Tertiary tilting in the basal sandstones of the Belly River Formation, which are described as thick and widespread (p. 947), would seem to be due to secondary-porosity destruction in these sandstones rather than an updip primary facies change such as that described for the Cardium sandstones.

In a paleostructure with multiple growth pulses it is always possible that contemporaneous bathymetry caused winnowing of sands over shoals and ponding of muddy sediments in deeper surrounding water. A careful reconstruction of paleostructural history will help to identify times of active structural growth which so governed local facies developments as to provide stratigraphic traps, from passive periods of more even facies distribution without stratigraphic trapping potential.

Bell Canyon Sandstones, Delaware Basin, Texas

Many accumulations in Permian Bell Canyon sandstones of the Delaware basin are on structural noses (Grauten, 1965, p. 301, 302, Fig. 6). It is clear that the present structural noses result from a late easterly tilting of low-relief paleostructures on the western homocline of the Delaware basin (Grauten, 1965, p. 298, Fig. 2, p. 302). In these circumstances it seems possible that the retention of oil in some Bell Canyon fields may have been affected by pretilt diagenetic cementation below former oil/water contacts such as appears to be the case between the Sullivan and Screwbean Northeast fields (Grauten, 1965, p. 303).

HYDRODYNAMICS AS A FACTOR CONTRIBUTING TO TILTED OIL/WATER CONTACTS

The Pennsylvanian Tensleep Sandstone of the Big Horn basin, Wyoming, has been studied in detail by Todd (1963) and Lawson and Smith (1966). Lawson and Smith (1966, p. 2216-2217) recognized the effect of deep burial on Tensleep porosity reduction. They suggested that, if oil is trapped before secondary cementation, primary porosity may be preserved and, therefore, if deep exploration were directed toward areas where early primary entrapment can be inferred (i.e., pa-

leostructures), the porosity-versus-depth curve (Lawson and Smith, 1966, p. 2219, Fig. 27) may not be valid.

Tilted oil/water contacts in Tensleep fields such as Frannie and Grass Creek were explained by Todd (1963, p. 612-613) as paleohydrodynamic phenomena. In Todd's interpretation, Tensleep porosity has been plugged by secondary dolomite and solid hydrocarbon (oxidized petroleum) at the paleohydrodynamically tilted oil/water interface. This interpretation prompts questions regarding original depth of burial of pre-Laramide paleostructural traps (Todd, 1963, p. 604) and the timing of diagenetic sealing at the oil-water contact relative to time of maximum burial and time of introduction of paleohydrodynamic flow.

In addition, any geologist must address himself to the general questions so admirably set forth by Weeks (1961, p. 5-43–5-45) regarding the ability of present or past hydrodynamic drives to move or tilt emplaced hydrocarbons. It is this writer's view that the theory of hydrodynamic control of oil, like the theory of late oil expulsion from source rocks, is better supported in man's than in nature's laboratory.

The presence of tar seals at the base of some Tensleep fields was interpreted by Todd (1963, p. 612) as due to oxidation of hydrocarbons at the base of the oil column by oxygenated paleohydrodynamic water flow. Again, the question of timing and depth of bitumen precipitation arises, for this presumably must have preceded sealing of the reservoir by secondary dolomite.

Tar mats at the base of oil accumulations are not uncommon and do not appear always to require the oxygenated water flow below an accumulation. Good tar mat examples have been described for two giant oil fields, Burgen in Kuwait (Fox, 1959, p. 102-103) and Prudhoe Bay (Jones and Speers, 1976). In the latter case it was suggested that the tar mat may be partly responsible for preserving the tilted oil/water interface. In the Hawkins field, Wood County, Texas, there is a thick tar layer below the oil column in the Dexter sandstone of the Woodbine Formation (McNabb, 1975, p. 66). This tar layer is impermeable to oil but permeable to water.

Whatever the correct interpretation for porosity destruction by secondary dolomite and tar below Tensleep fields it seems to represent another factor which inhibits the movement of oil from its original place of entrapment.

DEFINITION OF A DIAGENETIC TRAP

The basic assumptions which are founded in geologic observation and are required for the diagenetic trapping of hydrocarbons follow:

1. The porosity of water-wet reservoir rocks is plugged progressively by secondary diagenetic minerals during burial.

2. Emplacement of hydrocarbons inhibits diagenetic processes in reservoir rocks.

3. The inhibited state of diagenesis in hydrocarbon-saturated reservoir rock, when related to the normal diagenetic evolution in the water-bearing parts of that reservoir, is indicative of early hydrocarbon emplacement.

With these basic assumptions a diagenetic trap can be defined as follows:

A diagenetic hydrocarbon trap is a hydrocarbon charged paleotrap in which the paleostructural or paleostratigraphic closure has been rotated from its original attitude. The time of opening of the paleotrap postdates the time of diagenetic plugging of the water leg so that the hydrocarbons are retained by a permeability barrier in what is now a new updip direction.

Thus the creation of a diagenetic trap is diachronous, being the resultant of *late* tilting of an *early* oil-bearing structure. This differs from a stratigraphic trap, *sensu stricto*, in which the updip trapping agent is a contemporaneous facies change of reservoir rock to nonreservoir rock.

EXPLORATION FOR DIAGENETIC TRAPS— GEOLOGIC CRITERIA

The exploration for stratigraphic traps is complicated by the requirement of precise facies control which commonly is absent in sparsely drilled areas. Exploration for diagenetic traps, on the other hand, requires only isopach control of paleostructure and knowledge of maximum depth of burial of objective zones prior to tilt-out.

Because complete diagenetic sealing of water-bearing reservoir rocks may require considerable time and overburden, the time of tilt-out is a critical factor because, if this is too early, partial diagenetic plugging may allow flushing of the accumulation.

For these reasons Mesozoic and Paleozoic paleostructures that have been opened in the Tertiary may be more favorable targets because the greater the time and depth of burial imposed on a paleostructural trap the more likely is it to be sealed diagenetically.

It is a well established fact of petroleum geology that low-relief paleostructures are favorite habitats for hydrocarbon entrapment. Because low-relief paleostructures would be the first to be opened by a later regional or local tilt there may be many apparently weak and unattractive structural noses which will be straddled by diagenetically trapped accumulations.

Exploration for diagenetic traps in a basin should be geared to a logical analytical sequence. Identification of paleostructure and precise dating of paleostructural-trap formation is all important.

A hypothetical sedimentary sequence on a basin flank is shown on Figure 1 and serves as a reference for the following procedural steps.

1. Establish petroliferous sequences from known oil fields and wells in the basin. For the purposes of this example (Fig. 1) it is established that petroliferous sequences with source and reservoir rocks are present in the Lower Triassic, Upper Jurassic, top of the Lower Cretaceous, and Upper Cretaceous. The shale at the base of the Jurassic salt is known to be highly bituminous and conventional subcrop traps have been found.

2. Establish periods of structural activity by identification of unconformities and disconformities. In our hypothetical basin there are three major unconformities—base of Triassic, base of Jurassic, and base of Tertiary, respectively. The last two unconformities followed peneplanation of a block-faulted terrain. The last significant structural event was a major basinward tilt in late Tertiary time.

Less important periods of movement can be identified from isopach relations within the Triassic, as for example, above conventional traps C1 through C6 and within the Upper Jurassic and Cretaceous as a result of diapiric pulses of the Jurassic salt. The dominant salt pillow had its principal period of growth at the end of the Late Cretaceous in harmony with strong Laramide block faulting in the basement. This late diapiric flow masks earlier Jurassic centers of salt uplift which now occupy a flank position on the main dome (Fig. 1, diagenetic traps D4, D5).

3. Prepare isopach maps to define successive generations of paleostructural traps within and immediately above objective sequences.

4. Note relation of all known producing zones to paleostructural closures to enable grading of paleostructural prospects from empirical data.

5. Document all depletion-type reservoirs in basin for each prospective zone and relate these to diagenetic plugging of each reservoir rock.

6. For all tilted-out paleostructures, including subunconformity truncation traps, estimate pre-tilt burial depth in relation to normal sequence of diagenetic porosity destruction in mixed clastic, monogenetic clastic or carbonate reservoir rocks. For example, paleostructural traps D1, D3, and D5 (Fig. 1) retained structural closure until late Tertiary time, but D1 and D3 suffered an offloading of about 7,500 ft (2,286 m) of overburden following Laramide block faulting whereas D5 con-

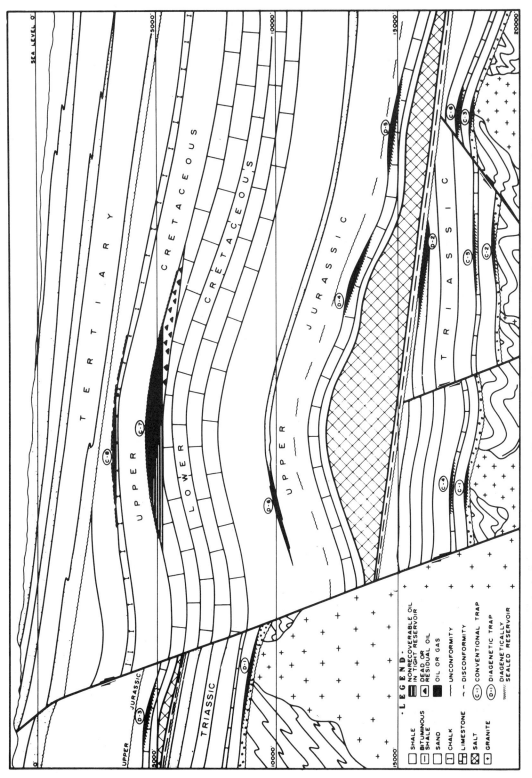

FIG. 1.—Hypothetical cross section over sedimentary basin flank, illustrating habitat of diagenetic hydrocarbon traps. Depths below sea level are in feet.

tinued to be buried to 15,000 ft (4,572 m) before tilt-out.

7. After identifying all tilted-out paleotraps, the most important attributes for high grading prospectivity must be applied. These are paleostructural relief, tilt-out time, and intensity of postaccumulation restructuring. Paleostructural relief must be low and developed during or shortly after the deposition of petroliferous sequence. Tilt-out time must postdate the destruction of effective porosity and permeability in the water leg. Critical tilt-out time will vary between different reservoir rocks depending on their mineral constitution. For example, an immature sand with many feldspar grains probably will seal up with authigenic clay more quickly than a clean-quartz reservoir sand which will depend on grain to grain pressure solution for secondary silica cementation. Too intense postaccumulation restructuring could produce secondary fracture permeability allowing remigration or loss of hydrocarbons.

Varied diagenetic traps that might be present in any sedimentary basin are illustrated on Figure 1. Clearly, this simple presentation does not cover the great variety of trapping possibilities that nature, with her capricious complexities, may have hidden from our view.

Tilted-out truncation trap (Fig. 1, D2)—The origin of the D2 accumulation is downward charging of a truncated Triassic sandstone by oil from the transgressing Jurassic bituminous shale. The original trap configuration would have been similar to the present Prudhoe Bay field (Morgridge and Smith, 1972, p. 500, Fig. 16).

The structural element of the trap remained intact during burial to below 15,000 ft (4,572 m) by which time the sand reservoir was sealed below the oil/water contact. The conventional accumulations C1 through C6 are in geopressured depletion-type reservoirs. Late Tertiary tilt opened the D2 trap structurally so that its expression on a present-day seismic structure map would not draw attention to its real prospective potential.

Tilted-out salt-generated Jurassic paleostructure (Fig. 1, traps D4, D5)—The lateral migration of salt diapiric crests through time is a well-established phenomenon (Hornabrook, 1967, Fig. 5; Wilson, 1976, Fig. 49). This is illustrated schematically on Figure 1.

Isopach thinning in the Jurassic sediments over the carbonate reservoir establishes the position of two paleotraps at that time. Remigration of salt into the main pillow commenced in Late Cretaceous time and reached its climax with a Laramide pulse at the end of the Cretaceous. D4 paleotrap was opened structurally at the end of

Cretaceous time after it had been buried below about 12,000 ft (3,658 m) of Jurassic and Cretaceous sediments, whereas D5 paleotrap remained structurally closed until imposition of the late Tertiary tilt when the reservoir had been buried to a depth of about 15,000 ft (4,572 m). In both cases diagenetic sealing of the carbonate reservoir would have held the accumulations in place although D5 prospectivity rating would be higher than D4.

Both traps would appear as minor structural terraces on the general plunge of the large Late Cretaceous salt pillow. No accumulation is present in the Jurassic reservoir on the main culmination because the trap formation was too late for primary migration and diagenetic sealing prevented remigration from flanking paleotraps.

Tilted-out stratigraphic trap (Fig. 1, trap D6)—During deposition of the Upper Jurassic a primary stratigraphic pinchout formed at D6 in a turbidite sand rich in feldspar grains. After deposition of about 10,000 ft (3,048 m) of Jurassic and Cretaceous sediments the stratigraphic trap was rotated by Laramide salt-pillow growth. Diagenetic sealing by authigenic clay held the hydrocarbons in original trap position on the flank of the salt-generated swell.

Oil distribution in partly sealed diagenetic traps (Fig. 1, trap C7)—The large accumulation in the salt pillow paleotrap formed at the end of the Early Cretaceous was tilted in late Tertiary time after burial below 5,000 ft (1,524 m) of sediments. This burial was not sufficient for complete diagenetic plugging of the carbonate reservoir although a significant reduction in reservoir quality had occurred. Consequently, after tilt-out, the oil/water contact gradually adjusted but recoverability of oil from the partly plugged reservoir that lay below the original oil/water contact will be poor. Traces of dead oil, indicating the original accumulation, will be present in the flushed part of the reservoir.

CONCLUSIONS

In spite of more than 100 years of petroleum search, the decisions of explorationists still are affected by "the seductive influence of the closed contour" (DeGolyer, 1928) or by the deleterious effect of geochemically inspired theories of late oil expulsion and accumulation which defy overwhelming geologic arguments to the contrary. In the desperate search for more hydrocarbons in North America there have been great improvements in data-gathering techniques, but these advances have not been accompanied by imaginative interpretation of the vast amount of data obtained. Still the search goes on for deeper and environmentally more hazardous conventional traps.

The support for diagenetic trapping of hydrocarbons is available in published literature and requires recourse to no magic "black box." The sealing capacity of diagenetic cementation is every bit as effective as primary facies changes and many accumulations on structural noses which today are relatively shallow may have an element of diagenetic trapping.

In deeper basin positions (15,000 to 20,000 ft or 4,572 to 6,096 m) there are likely to be undiscovered diagenetic traps with large oil or gas accumulations because such prospects have been financially unattainable to adventurous independents and conceptually indigestible to the more conservative major operators.

The solution of most geologic problems depends on our ability to recognize in the geologic time scale the correct sequence of events which have combined to form today's end product. Thus the expulsion and entrapment of hydrocarbons and the recognition that diagenetic sequences in reservoir rocks are inhibited by hydrocarbons provide another powerful tool with which to identify the times and depths which constrain the emplacement of hydrocarbons. Evidence from many hydrocarbon habitats so strongly supports the concept of diagenetic trapping that a strenuous exploration effort in search of such traps is warranted—particularly in the ultramature basins of North America.

A fitting postscript is Dickey's (1958, p. 84) erudite observation:

We usually find oil in new places with old ideas. Sometimes, also, we find oil in an old place with a new idea, but we seldom find much oil in an old place with an old idea. Several times in the past we have thought that we were running out of oil, whereas actually we were only running out of ideas.

REFERENCES CITED

Adams, W. L., 1964, Diagenetic aspects of lower Morrowan, Pennsylvanian, sandstones, northwestern Oklahoma: AAPG Bull., v. 48, p. 1568-1580.

Bishop, W. F., 1971, Stratigraphic control of production from Jurassic calcarenites, Red Rock field, Webster Parish, Louisiana: Gulf Coast Assoc. Geol. Socs. Trans., v. 21, p. 125-138.

Clarke, R. H., 1975, Petroleum formation and accumulation in Abu Dhabi: 9th Arab Petroleum Cong., Dubai, v. 120 (B-3), p. 1-20.

DeGolyer, E. L., 1928, The seductive influence of the closed contour: Econ. Geology, v. 23, p. 681-682.

Dickey, P. A., 1958, Oil is found with ideas: Tulsa Geol. Soc. Digest, v. 26, p. 84-101.

Dunnington, H. V., 1960, Some problems of stratigraphy, structure and oil in Iraq: 2d Arab Petroleum Cong., Beirut, v. 2, p. 166-199.

———— 1967, Aspects of diagenesis and shape change in stylolitic limestone reservoirs: 7th World Petroleum Cong., Mexico, Proc., v. 2, p. 339-352.

Fox, A. F., 1959, Some problems of petroleum geology in Kuwait: London Inst. Petroleum, v. 45, no. 424, p. 95-110.

Grauten, W. F., 1965, Fluid relationships in Delaware Mountain Sandstone, in Fluids in subsurface environments: AAPG Mem. 4, p. 294-307.

Harding, T. P., 1976, Tectonic significance and hydrocarbon trapping consequences of sequential folding synchronous with San Andreas faulting, San Joaquin Valley, California: AAPG Bull., v. 60, p. 356-378.

Hornabrook, J. T., 1967, Seismic interpretation problems in the North Sea with special reference to the discovery well 48/6-1: 7th World Petroleum Cong., Mexico, Proc., v. 2, p. 837-862.

Hughes, D. J., 1968, Salt tectonics as related to several Smackover fields along the northeast rim of the Gulf of Mexico basin: Gulf Coast Assoc. Geol. Socs. Trans., v. 10, p. 320-330.

Jones, H. P., and R. G. Speers, 1976, Permo-Triassic reservoirs of Prudhoe Bay field, North Slope, Alaska, in North American oil and gas fields: AAPG Mem. 24, p. 23-50.

Lawson, D. E., and J. R. Smith, 1966, Pennsylvanian and Permian influence on Tensleep oil accumulation, Big Horn basin, Wyoming: AAPG Bull., v. 50, p. 2197-2220.

Levandowski, D. W., et al, 1973, Cementation in Lyons Sandstone and its role in oil accumulation, Denver basin, Colorado: AAPG Bull., v. 57, p. 2217-2244.

Lowry, W. D., 1956, Factors in loss of porosity by quartzose sandstones of Virginia: AAPG Bull., v. 40, p. 489-500.

McNabb, D., 1975, Pressure maintenance to boost Hawkins recovery: Oil and Gas Jour., March 31, p. 65-68.

Morgridge, D. L., and W. B. Smith, 1972, Geology and discovery of Prudhoe Bay field, eastern Arctic Slope, Alaska, in Stratigraphic oil and gas fields: AAPG Mem. 16, p. 489-501.

Patterson, A. M., and A. A. Arneson, 1957, Geology of Pembina field, Alberta: AAPG Bull., v. 41, p. 937-949.

Qatar Petroleum Company Staff, 1956, Symposium on the geological occurrence of oil and gas: 20th Internat. Geol. Cong., Mexico, Proc., v. 2, p. 161-169.

Reed, L. C., 1946, San Pedro oil field, province of Salta, northern Argentina: AAPG Bull., v. 30, p. 591-605.

Rittenhouse, G., 1972, Stratigraphic-trap classification, in Stratigraphic oil and gas fields: AAPG Mem. 16, p. 14-28.

Scholle, P. A., 1976, Diagenetic patterns in chalks (abs.): AAPG Bull., v. 60, p. 719-720.

Scholten, R., 1959, Synchronous highs—preferential habitat of oil?: AAPG Bull., v. 43, p. 1793-1834.

Todd, T. W., 1963, Post-depositional history of Tensleep Sandstone (Pennsylvanian), Big Horn basin,

Wyoming: AAPG Bull., v. 47, p. 599-616.

Tschopp, R. H., 1967, Development of the Fahud field: 7th World Petroleum Cong., Mexico, Proc., v. 2, p. 243-250.

Van de Kamp, P. C., 1976, Inorganic and organic metamorphism in siliciclastic rocks (abs.): AAPG Bull., v. 60, p. 729.

Vestal, J. H., 1950, Petroleum geology of the Smackover Formation of southern Arkansas: Arkansas Resources and Development Comm., Div. Geol. Inf. Circ. 14, p. 19.

Webb, J. E., 1974, Relation of oil migration to secondary clay cementation, Cretaceous sandstones of Wyoming: AAPG Bull., v. 58, p. 2245-2249.

Weddle, H. W., 1951, Pleasant Valley oil field, Fresno County, California: AAPG Bull., v. 35, p. 619-623.

Weeks, L. G., 1958, Habitat of oil and some factors that control it, in L. G. Weeks, ed., Habitat of oil: AAPG, p. 1-61.

———— 1961, Origin, migration, and occurrence of petroleum, in G. B. Moody, ed., Petroleum exploration handbook: New York, McGraw-Hill, p. 5-1—5-50.

Wilson, H. H., 1975a, Time of hydrocarbon expulsion, paradox for geologists and geochemists: AAPG Bull., v. 59, p. 69-84.

———— 1975b, Time of hydrocarbon expulsion, paradox for geologists and geochemists: reply to E. W. Biederman: AAPG Bull., v. 59, p. 2054-2055.

———— 1976, Notes to accompany lectures on evaporite depositional environments and salt tectonics and their relationship to petroleum accumulations: Univ. Tulsa, Dept. Earth Sciences, 15th Short Course in Advanced Petroleum Geology.

Yurkova, R. M., 1970, Comparison of post-sedimentary alteration of oil-, gas- and water-bearing rocks: Sedimentology, v. 15, p. 53-68.

Reprinted by permission of the Society of Petroleum Engineers of AIME. Published in *Formation Evaluation*, V. 1, No. 3, copyright 1986, pp. 295-299.

The Preservation of Primary Porosity Through Hydrocarbon Entrapment During Burial

J.J. O'Brien, SPE, Sohio Petroleum Co.
I. Lerche, U. of South Carolina

Summary. We examine the manner in which trapped pore fluids can inhibit sediment compaction during burial and, hence, can enhance primary porosity at depth. In our model studies, we find that primary porosity can be enhanced through fluid entrapment by a factor of two to three at a depth of 30,000 ft [9144 m]. Preservation of porosity is accompanied by the development of an overpressured condition. The magnitude of these effects depends on the pore fluid, its compressibility, the depth at which the reservoir is sealed, the mechanical properties of the rock matrix, and—in the case of a gas reservoir—the thermal gradient. These results indicate that reservoir formations of significant porosity may be preserved to depths greater than 30,000 ft [9144 m] and that, under favorable conditions, it may be possible to identify such formations seismically.

Introduction

While reservoir porosity is an important consideration in hydrocarbon exploration, it is more critical in deep plays because of the compaction of the rock matrix under the weight of the overburden. Sclater and Cristie[1] modeled porosity as decreasing approximately exponentially with depth for various lithologies in the North Sea. Their North Sea normal-pressured sandstone data yield a best-fit surface porosity of 49%, the porosity decreasing exponentially on a depth scale of ~12,000 ft [~3658 m]. Thus, this relationship would predict a porosity of only 6.6% at a depth of 24,000 ft [7315 m]. At this depth, a reservoir of this porosity may be uneconomical.

The quantity modeled by Sclater and Cristie[1] is the primary porosity of a rock that contains pore fluid under hydrostatic pressure. Other processes also contribute to reservoir porosity, such as secondary porosity and fracture porosity, both of which have yielded excellent reservoirs under favorable conditions. In this paper, we model a third process: the preservation of primary porosity through the entrapment of pore fluid in a sealed reservoir. Because the pore fluid cannot escape, compaction of the rock is limited by the compressibility of the pore fluid; therefore, the decrease of porosity with depth is inhibited. The pore fluid supports an increasing portion of the overburden weight as burial progresses and an overpressure situation develops. We model this situation to determine (1) the extent to which compaction is inhibited and (2) the excess pore pressures that develop when various fluids are trapped.

Oil Trapped in a Sealed Reservoir

In this section, we consider what happens when oil is trapped in the rock pores underneath an impermeable pressure seal. For simplicity, we consider a model in which all the brine in the reservoir is replaced by oil at hydrostatic pressure when the reservoir is buried at a depth D_1

below the sediment surface. At essentially the same geologic time, a pressure seal develops over the reservoir that prevents the oil from escaping. We also assume that no other fluids enter the reservoir at a later time so that we have a closed hydraulic system.

In a normal compaction situation (i.e., if no pressure seal has developed), we assume that the porosity development can be described by a known function $\phi(D)$. At any depth, D, the buoyant weight of the overlying rock matrix is given by the following:

$$W_{ma} = \int_0^D (\rho_R - \rho_{pf})g[1 - \phi(z)]dz, \quad \ldots\ldots\ldots\ldots (1)$$

where ρ_R and ρ_{pf} are the densities of the overlying rock matrix and of the overlying pore fluid, respectively, and are assumed to be independent of depth and g is the acceleration of gravity. In a normal pressure situation, this matrix weight must be supported by the frame pressure, $p_f(\phi)$:

$$p_f(\phi) = \int_0^D (\rho_R - \rho_{pf})g[1 - \phi(z)]dz. \quad \ldots\ldots\ldots\ldots (2)$$

This equation defines the relationship between the frame pressure of the rock and its porosity. The total pressure, p_t, then consists of two components, the frame pressure and the pore-fluid pressure, which in this case is the hydrostatic pressure:

$$p_t = p_f(\phi) + \rho_{pf}gD. \quad \ldots\ldots\ldots\ldots\ldots\ldots (3)$$

Eqs. 2 and 3 are applicable only in a normal pressure situation. We will use them later to define the pressure conditions immediately above our pressure seal.

323

Now let us consider the situation immediately underneath the pressure seal. At the time at which the seal is formed at D_1, we assume that the porosity in the reservoir rock under the seal is $\phi(D_1)$ $(\equiv\phi_1)$; i.e., the reservoir porosity is not altered as the seal is formed, although its later development may be affected. As deposition continues, the pressure seal is buried more deeply. When the seal attains a subsurface depth D, we assume that the porosity underneath the seal is given by $\phi_1+\Delta\phi$, where $\Delta\phi$ is the amount that the rock has compacted and also the amount that the pore fluid has been compressed as the seal moves from D_1 to D. At D, the frame pressure underneath the seal is $p_f(\phi_1+\Delta\phi)$. The pore fluid pressure is p_{pf}. The total pressure is given by

$$p_t=p_f(\phi_1+\Delta\phi)+p_{pf}. \dots\dots\dots\dots\dots\dots(4)$$

The change in p_{pf} since the seal was formed $(p_{pf}-\rho_{pf}gD_1)$ is related to the compression of the pore fluid:

$$p_{pf}-g\rho_{pf}D_1=-c\Delta\phi/\phi_1, \dots\dots\dots\dots\dots(5)$$

where c is the isothermal compressibility of the pore fluid:

$$c=V^{-1}(\partial V/\partial p)_T. \dots\dots\dots\dots\dots\dots(6)$$

Eqs. 2 through 5 define the pressure conditions immediately above and below the pressure seal. Continuity of total pressure across the seal yields

$$p_f(\phi)+\rho_{pf}gD=p_f(\phi_1+\Delta\phi)+\rho_{pf}gD_1-c\Delta\phi/\phi_1. \quad(7)$$

Because the isothermal compressibilities of oils are quite small—typically $\sim15\times10^{-6}$ psi^{-1} [$\sim2.15\times10^{-9}$ Pa^{-1}]—we expect $|\Delta\phi|/\phi_1\ll1$, so we expand $p_f(\phi_1+\Delta\phi)$ in terms of $\Delta\phi$. Assuming that $\phi(D)$ is of the form $\phi_oe^{-D/a}$, Eq. 7 then yields the following expression for $\Delta\phi$:

$$\Delta\phi=[-(\phi_1c)^{-1}-(\rho_R-\rho_{pf})ga_1^{-1}(1-\phi_1)]^{-1}$$

$$\times\int_{D_1}^{D}\{\rho_Rg[1-\phi(z)]+\rho_{pf}g\phi(z)\}dz. \dots\dots\dots(8)$$

Having determined $\Delta\phi$, we can find the excess pore pressure, p_{ex}.

$$p_{ex}=p_{pf}-\rho_{pf}g$$

$$=-c^{-1}\Delta\phi/\phi_1-\rho_{pf}g(D-D_1). \dots\dots\dots\dots(9)$$

To gain some insight into these results, we will examine three sample cases with pressure seals formed at depths of 5,000, 10,000, and 15,000 ft [1524, 3048, and 4572 m], respectively. In these examples, we assume that the porosity above the seals has the same form as that given by Sclater and Cristie[1] for sandstones:

$$\phi(D)=\phi_oe^{-D/a}, \dots\dots\dots\dots\dots\dots(10)$$

where $\phi_o=0.49$ and $a=12,000$ ft [3658 m].

We assume a hydrostatic pressure gradient of 0.436 psi/ft [0.985 Pa/m] (corresponding to a specific density of brine of 1.0), a total pressure gradient resulting from fully compacted sediment of 1.09 psi/ft [2.46 Pa/m] (corresponding to a matrix specific density of 2.5), and oil compressibility of 15×10^{-6} psi^{-1} [2.18×10^{-9} Pa^{-1}].

The results of these calculations are presented in Figs. 1 through 3, which show the dependence of porosity on depth in a normal pressure situation and how trapped oil can inhibit compaction. The porosity just below the seal decreases slowly with increasing depth of burial of the seal; this decrease is dictated by the compressibility of the oil. Accordingly, the frame pressure of the matrix increases only slowly under these circumstances, and an increasing portion of the overburden weight must be supported by the pore-fluid pressure. As shown by the inserts in Figs. 1 through 3, the excess pore pressure rises rapidly. When the pore-pressure differential across the seal exceeds the strength of the seal, the seal fractures and the excess pressure is released. At this point, we no longer have a seal and the porosity/depth relationship will return to the normal pressure curve. Thus pore-fluid entrapment can hold the porosity open, but only over a finite depth range. To estimate this depth range, we note that, by analogy with hydrofracturing of reservoir rocks, fracturing will occur when the excess pore pressure is 0.8 to 1.0 times the frame pressure,[2] assuming that the reservoir rock and the seal have similar strength characteristics. Thus, if a seal is formed at 5,000 ft [1524 m], it will have fractured by the time it is buried to 7,500 ft [2286 m]; a seal formed at 10,000 ft [3048 m] will be fractured by 15,000 ft [4572 m], while a seal formed at 15,000 ft [4572 m] will survive until it is buried to $\sim30,000$ ft [~9144 m]. From these examples, we see that the range of depths over which overpressuring persists increases with increasing depth of seal formation. Two factors contribute to this effect: (1) the mechanical strength of the seal that must be overcome by the excess pore pressure increases as the depth at which the seal is formed increases; and (2) the change in frame pressure for a given change in porosity increases with depth (i.e., the normal pressure porosity/depth curve changes shape), so the corresponding buildup of excess pore pressure is slowed down.

Figs. 1 through 3 show that entrapment of oil can preserve primary porosity to a significant extent over a finite range of depths, particularly if the seal is formed at an intermediate depth of burial. If the seal forms at a shallow depth, it will fracture before being buried much deeper. If the seal forms deep in the subsurface, the effects of oil entrapment are less significant; the preserved porosity is small and, even without the development of overpressure, primary porosity varies only slowly at these depths. Thus oil entrapment is most effective in preserving primary porosity if trapping occurs at an intermediate depth of burial. In the models, we consider this optimum depth to be $\sim15,000$ ft [~4572 m].

In addition to compression under the weight of the overburden, thermal expansion of the pore fluid will also contribute to the excess pore pressure and will inhibit compaction. Therefore, heating of the pore fluid during burial increases the excess pressure, and the seal is fractured sooner than would otherwise be expected. However, for a typical thermal expansion coefficient for oil ($\beta\sim10^{-4}$ K^{-1}) and a temperature gradient of 1°C/100 ft [3.3×10^{-2} K/m], this effect is smaller by at least an

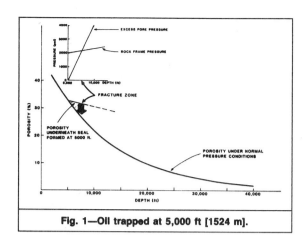

Fig. 1—Oil trapped at 5,000 ft [1524 m].

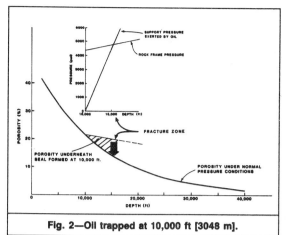

Fig. 2—Oil trapped at 10,000 ft [3048 m].

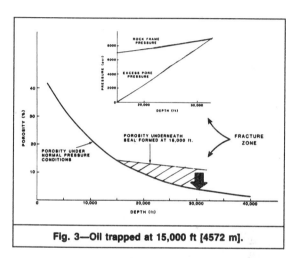

Fig. 3—Oil trapped at 15,000 ft [4572 m].

order of magnitude than that resulting from compression of the pore fluid. Thus, thermal-expansion effects do not significantly alter our previous results.

In this discussion, we have been concerned with the development of overpressure as a result of trapping of oil. It is clear that the same effects will be expected if any other low-compressibility fluid is trapped, including brine.

Gas Trapped in a Sealed Reservoir

If the pore fluid trapped underneath a pressure is a highly compressible one—for example, gas—a different formalism must be used. Once again, we denote the porosity in a normal pressure situation by $\phi(D)$, and the pressure conditions above the seal are described by Eqs. 2 and 3.

In this section, we denote the change in porosity across the pressure seal by $\Delta\phi(D)$; i.e., when the pressure seal is located at D, the porosity above the seal is $\phi(D)$, while the porosity immediately below the seal is $\phi(D)+\Delta\phi(D)$. This should be contrasted with the definition of $\Delta\phi$ in the trapped-oil case. In that model when the seal is buried at D, the porosity immediately below the seal is $\phi(D_1)+\Delta\phi(D)$, where D_1 is the depth at which the seal is formed and the oil is trapped. In the cases of trapped oil and gas, $\Delta\phi$ is defined such that $|\Delta\phi|/\phi(D)<<1$. Hence, different definitions of $\Delta\phi$ are appropriate in the two cases.

In our trapped-gas model, we assume that all brine in the reservoir rock is replaced by gas at hydrostatic pressure when the reservoir is buried at D_1 and that a pressure seal develops over the reservoir at essentially the same time. To illustrate the essential points of the model, we assume that gas emplacement takes place at a single subsurface depth (which we can vary at will) and that no subsequent fluid flow occurs through the reservoir formation. We also assume that the gas can be described by the perfect gas law. This is an excellent approximation for pure methane, for which the critical pressure and temperature are 673 psi [4.64 MPa] and $-116.5°F$ [190.7 K], respectively. When the seal is buried at D_1, gas pressure, p_g, is given by

$$p_g(D)=$$

$$[T(D)/T(D_1)]\{\phi(D_1)/[\phi(D)+\Delta\phi(D)]\}p_g(D_1),$$
$$\dots\dots\dots\dots\dots\dots\dots\dots (11)$$

where D_1 is the depth at which the seal is created, T is the temperature and $p_g(D_1)$ is the hydrostatic pressure at D_1. Thus, in the trapped-gas case, the total pressure underneath the seal is given by

$$p_t=p_f(\phi+\Delta\phi)+p_g(D), \dots\dots\dots\dots (12)$$

where $p_f(\phi+\Delta\phi)$ is again the frame pressure. Continuity of total pressure across the impermeable seal then permits us to write

$$p_f(\phi)+\rho_{pf}gD=p_f(\phi+\Delta\phi)+p_g(D). \dots\dots (13)$$

Because we expect $|\Delta\phi|<<\phi$, we can expand $p_f(\phi+\Delta\phi)$ about $p_f(\phi)$. Retaining only first-order terms, we get the following expression for $\Delta\phi$:

$$\Delta\phi=\{a(\rho_R-\rho_{pf})g[1-\phi(D)]+\rho_{pf}gD\}^{-1}$$

$$\times\rho_{pf}g[D_1\phi_1T/T_1-D\phi(D)]. \dots\dots\dots (14)$$

In Eq. 14, ϕ_1 and T_1 are the values of the porosity and temperature at D_1 at which the seal is formed. Following Sclater and Cristie,[1] we have again assumed that $\phi(D)$ may be modeled by an exponential function $\phi_o e^{-D/a}$.

325

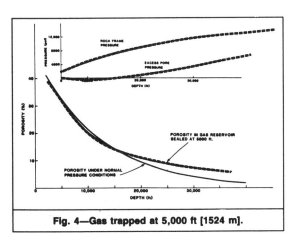

Fig. 4—Gas trapped at 5,000 ft [1524 m].

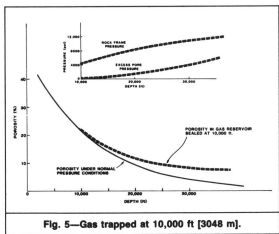

Fig. 5—Gas trapped at 10,000 ft [3048 m].

From Eq. 14, we see that two factors contribute to the numerator on the right side: (1) the increase in temperature (T/T_1) increases the gas pressure—this gives a positive contribution to $\Delta\phi$; and (2) the gas pressure exerted by the overburden decreases both the gas volume and $\Delta\phi$. The resultant effect will depend on the balance between thermal expansion of the gas (and hence on the thermal gradient) and compression of the gas.

To illustrate these effects, let us consider an example in which we have a constant temperature gradient of 10^{-2} °C/ft. [3.3×10^{-2} K/m]. This value is representative of many sedimentary basins. Fig. 4 shows a case where the reservoir is sealed when it is buried at a depth of 5,000 ft [1524 m]. With further burial, porosity essentially follows that of our reference normal-pressure model down to a depth of ~18,000 ft [~5486 m]. Below this depth, the excess pressure developed by the gas inhibits compaction so that by the time the pressure seal is buried at 30,000 ft [9144 m], primary porosity immediately underneath the seal is enhanced by ~50% above that of our reference model. From Fig. 5, we see that a similar result follows if the pressure seal is formed at a depth of 10,000 ft [3048 m]. In both instances, the excess pressures subsequently developed underneath the seal are moderate relative to the frame pressure of the rock matrix over the range of depths considered; hence, the seal is not expected to be fractured by the excess pressure.

To summarize these results, because of its greater compressibility, trapped gas is less effective than trapped oil in preserving primary porosity. Even if gas is trapped at a shallow depth, it can inhibit compaction effectively only during those periods when the reservoir is buried at considerable depth. The preservation of primary porosity is accompanied by overpressure development, but the degree of excess pressure is modest.

Discussion

From the previous sections, we have seen how the entrapment of oil and gas under an impermeable seal can help to preserve primary porosity during burial. Because it is much less compressible, oil is more effective than gas in inhibiting compaction. This also implies that because of this mechanism alone, trapped oil can develop greater excess pressures than trapped gas, even to the point of fracturing the seal. We also found that the greatest en-

hancement of primary porosity occurs at considerable depth for both oil and gas entrapment; in Figs. 3 through 5, primary porosity is enhanced by a factor of 2 to 3 at a depth of 30,000 ft [9144 m]. The implications of these results are significant.

Conclusions

1. The presence of pressure seals can cause considerable deviations from the porosity/depth models of Sclater and Cristie.[1] Thus, even at depths below 30,000 ft [9144 m], sealed reservoirs may be found with porosities considerably greater than predicted by simple exponential-type models. Such reservoirs are also expected to be overpressured and may provide attractive exploration targets at depth.

2. The enhanced porosity of the reservoir will alter the acoustic impedance of that formation and modify the seismic response of the reservoir. This change in response should be greatest at the depth where the enhancement of primary porosity is greatest. The seismic response can be further enhanced by the presence of gas within the reservoir. Thus the preservation of primary porosity within a sealed reservoir provides a distinctive seismic signature by which this condition can be sensed remotely.

3. We note that we have treated only one subsurface process in isolation in this paper. To illustrate the essential points of this effect, we have made many simplifying assumptions. We assumed that hydrocarbon accumulation takes place at a single depth and that all brine is replaced by either oil or gas. We have ignored further maturation after the hydrocarbons have been trapped. We believe, however, that the porosity-preservation mechanism has been accurately described by itself and that the examples presented provide realistic estimates of the magnitude of the porosity-preservation factor and of the subsurface depth range over which it is expected to be important.

Nomenclature

a = scale depth, ft [m]
c = isothermal compressibility, psi^{-1} [kPa^{-1}]
D = depth, ft [m]
g = acceleration of gravity
p = pressure, psi [kPa]
T = temperature, °C [K]

298

V = specific volume, ft^3 [m^3]

W_{ma} = weight of matrix, lbm [kg]

z = depth variable of integration, ft [m]

ρ = density, lbm/ft^3 [kg/m^3]

ϕ = porosity, dimensionless

ϕ_o = surface porosity, dimensionless

Subscripts

ex = excess

f = frame

g = gas

pf = pore fluid

R = rock matrix

t = total

References

1. Sclater, J.G. and Cristie, P.A.F.: "Continental Stretching: An Explanation of the Post-Mid-Cretaceous Subsidence of the Central North Sea Basin," *J. Geophys. Res.* (1980) **85**, No. B7, 3711–39.
2. Fertl, W.H.: "Abnormal Formation Pressures: Implication to Exploration, Drilling, and Production of Oil and Gas Resources," *Development in Petroleum Science,* Elsevier Scientific Publishing Co., Amsterdam (1976) 2.

SI Metric Conversion Factors

ft $\times$ 3.048* E$-$01 = m

psi $\times$ 6.894 757 E$+$00 = kPa

*Conversion factor is exact. **SPEFE**

Original manuscript received in the Society of Petroleum Engineers office Jan. 14, 1985. Paper accepted for publication Dec. 17, 1985. Revised manuscript received Jan. 28, 1986.

American Association of Petroleum Geologists Memoir 16,
Stratigraphic Oil and Gas Fields, copyright 1972, pp. 42-46.

Unconformity Traps[1]

PHILIP A. CHENOWETH

Consulting Geologist, Tulsa, Oklahoma 74119

Abstract Accumulations of oil and gas are closely related to unconformities in almost every oil region of the world. Except for classifications and early attempts to utilize them in regional correlations, the location, origin, and geometry of successive unconformities largely have been ignored.

Unconformities occur in every tectonic and depositional environment. They are most common on the continental platform, where disconformities occur in close rhythmic succession. In the coastal areas between basin and platform, frequent warping has resulted in intersecting low-angle unconformities. Basins have the fewest unconformities, but angular unconformities may occur associated with midbasin uplift and diapirism.

In the typical platform area, traps above disconformities are mainly in quartz sandstones and are long and narrow; carbonate rocks are the principal reservoirs beneath disconformities. Traps in the hinge area between the platform and the basin are associated with low-angle unconformities; they are commonly very large and generally have arenaceous reservoir rocks. In basins not later deformed by tectonism, most unconformity traps are on the upthrown side of growth faults and on and around midbasin ridges, where submarine erosion has produced local unconformities. Reservoir rocks are commonly thin and discontinuous, fine-grained turbidites and residual sandstones.

INTRODUCTION

Surfaces of nonconformity between sequences of sedimentary rocks or between sedimentary rocks and crystalline rocks were recognized by James Hutton as features of considerable significance in geologic history (Hutton, 1788, quoted *by* Adams, 1954). Hutton, however, placed too much emphasis on their importance, for he regarded unconformities as records of "former worlds" which had been destroyed during episodes of violent upheaval. To Hutton and his followers, these revolutions represented the only reliable basis for correlation between widely separated areas. In fact, they were believed to be so important that all the major subdivisions of geologic time were established on the basis of unconformities which were, in turn, regarded as being worldwide. By the beginning of the 20th century, diastrophism was thought to be the "ultimate basis of correlation" (Chamberlin, 1909).

The belief that all unconformities are significant punctuation marks in the geologic time scale was severely discredited when strati-

[1] Manuscript received, February 25, 1971.

graphic studies expanded into regions where no great breaks are present between systems and where major unconformities are present within previously established systems. As the Huttonian concept of great worldwide episodes of diastrophism collapsed under the weight of evidence accumulated in the first decades of this century, major unconformities were discovered to be common. Schuchert (1910), for example, had recognized 11 within the Paleozoic of North America which he regarded as sufficiently important to justify the establishment of new periods. Stratigraphers who sought important structural breaks to use as systemic boundaries became increasingly aware of the local abundance of unconformities. For example, in the southern Mid-Continent the Pennsylvanian-Permian boundary is nowhere marked by an angular discordance, but many separate erosional breaks have at various times been selected as the contact (Branson, 1962).

Geologists are now aware that unconformities are exceedingly common in some geologic settings, less common in some, and virtually absent in others. Moreover, the relation of petroleum accumulations to unconformities has been demonstrated and documented. Levorsen (1934, p. 783) pointed out that:

The majority of the oil fields and by far most of the oil production found in the Mid-Continent region show an intimate association between the producing formation and unconformities. This relationship seems to hold for production found in all ages from Cambro-Ordovician to Tertiary; for production from a sandstone, limestone, dolomite, arkose, or conglomeratic material; where the unconformity has the rank of nonconformity, disconformity, or diastem; whether there has been post-unconformity folding or not; whether the production is oil, gas, or both; whether the time value of the unconformity is relatively long or short; whether the stratigraphic hiatus is great or small; and whether the pre-unconformity deformation was intense and local, gentle and regional, or absent entirely. We may safely conclude that unconformities have an important bearing on petroleum geology in this region.

Likewise, Miller *et al.* (1958, p. 601), in discussing the oil pools of the Maracaibo basin, commented: "Nearly three-fourths of the estimated ultimate reserve is found in . . . rocks associated with an unconformity and an Eocene hinge belt." Similar statements have been made regarding the oil accumulations in many parts of the world.

In view of this close association of petroleum with unconformities, one would assume that concentrated studies of unconformities would occupy much of the time and talents of explorationists. Paradoxically, this is not so, and the unconformity—an extremely common and admittedly important feature—remains imperfectly understood.

Definition and Classification

Generally, classification of unconformities is merely an academic exercise, for it tends to belabor the obvious and clog the literature with cumbersome and unneeded terms. On the other hand, the act of classifying tends to stimulate thought about the various factors responsible for producing the conditions which allow such classification. Unconformities change in character from place to place; the same one may be a high-angle unconformity at one locality and a surface of near-conformity at another. All unconformities mark a hiatus which is represented elsewhere by a stratigraphic unit. Any classification thus becomes arbitrary and any particular name is applicable only locally. There have been numerous attempts to develop an elaborate classification to fit all the possible conditions, but most of these appear to bear little or no relation to the central problems of exploration. Dunbar and Rodgers' simple scheme (1957, p. 118–119) is the most useful:

Nonconformity is that erosion surface between plutonic igneous or metamorphic rocks below and stratified sedimentary rocks above.
Unconformity (usually modified with the adjective "angular") is that variety which separates strata of markedly different structure.
Disconformity has strata above and below essentially parallel yet shows some evidence of erosion between the two sets of strata.
Paraconformity refers to the situation in which no apparent erosional break exists but a time hiatus is nevertheless present.

An unconformity is an erosion surface or a surface of nondeposition overlain by layered rocks or unconsolidated sediments. According to this definition, the base of a lava flow which was extruded upon a land surface is an unconformity. So also is the base of a Pleistocene till lying on Precambrian gneiss, and so is a paraconformity—a bedding plane between two units of disparate ages. Moreover, subaqueous as well as subaerial erosion may produce an unconformity. The magnitude of an unconformity may be expressed by the difference in dip of the strata on opposite sides or by the length of time represented by the missing interval. Thus, magnitude is also an arbitrary and meaningless measurement except for local descriptive purposes.

Distribution of Unconformities

Unconformities of every type occur in nearly every geologic environment. In some environments, however, they are more common than in others. On the continental platform, for example, where beds are thin and alternately of shallow-water marine and nonmarine origin, disconformities occur in close rhythmic succession. Many of these may be traced laterally for long distances. In the coastal area between the basin and the platform, low-angle unconformities abound, commonly in a confusing pattern (Chenoweth, 1967). In geosynclines and depositional basins, sedimentation and biologic evolution have proceeded with fewer interruptions. In the classic geosyncline, furrows and tectonic welts develop which are intermittently emergent (Kay, 1951, p. 15). The close of the geosynclinal cycle produces many relatively local and commonly highly angular unconformities. In a sedimentary basin—whether the basin be on a continental margin, a low spot in a geosyncline, or on the craton—deposition is likely to be interrupted by structural growth and submarine mass movements. Salt or shale intrusions or growth faulting may produce local high areas on the sea floor. Marine currents plane the upper layers, frequently redistributing the sediments and producing local angular unconformities.

Oil and Gas Traps at Unconformities

It has been noted repeatedly that oil and gas traps occur both above and below unconformity surfaces. Places are known where oil pools are present in both situations in the same field. In fact, it is not unreasonable to conceive of many superimposed unconformity traps in a small area. Levorsen (1954) has pointed out that repeated folding is a common phenomenon in many oil regions. Each structural uplift results in a local unconformity; several folded or tilted unconformities may be present, one above another. Residual tectonic forms such as the Central Kansas "uplift"—where movement was actually downward, but subsidence was uneven and slower than that of surrounding elements—were subject to the same frequent development of erosion surfaces.

Martin (1966) outlined many of the trapping conditions which appear both above and below an erosion surface. A thorough understanding of the geomorphology is necessary for efficient exploration, especially in shelf areas such as Western Canada, from which Martin drew many examples.

Along the margins of the shelf, at the hinge between basin and platform, angular unconformities occur in bewildering array—particularly where warping of the surface has occurred (Chenoweth, 1967). In this area a knowledge of subsurface geology and the geometry of the sedimentary bodies is perhaps more important in exploration than is geomorphology.

Unconformities in the basin are less common than on either the shelf or the basin margin, but they may have a profound influence on the location of suitable reservoir rocks. As a consequence of basin downwarping on the surface of the sphere, a midbasin ridge commonly develops (Dallmus, 1958, p. 893) which may become the locus of submarine erosion. Also, the ridge is commonly the site of diapirism which, in a sense, increases the magnitude of uplift and hence enhances erosion. Certain basins subside along series of bordering growth faults. The upthrown side of such faults is subject to erosion by submarine currents. In some places, tilting of strata away from the fault occurs and truncated traps may develop.

UNCONFORMITY TRAPS ON SHELF

The shelf or platform, used in the stratigraphic sense, is that of Horberg et al. (1949): ". . . the area of thinner sediments adjoining a geosynclinal wedge of thicker equivalent beds." On the shelf the sedimentary section is not only thin but contains numerous unconformities, most of which are disconformities—i.e., the strata above and below are nearly parallel. Some warping has occurred in places (on a typical platform), producing a slightly angular relation between the beds on either side of the erosion plane.

Most of central North America, the "hedreocraton" of Kay (1951), was a shelf throughout Paleozoic and Mesozoic time. Locally and at different times on this platform, moderate downwarping resulted in development of basins; the Williston, Michigan, Forest City, and Illinois basins are prime examples. Elsewhere, arching took place, forming the Transcontinental arch—the Chadron, LaSalle, and Cincinnati arches. Angular unconformities developed above and on the flanks of the uplifts and on other residual interbasin highs.

Levorsen (1960) portrayed the major regional unconformities of the Mid-Continent with a series of paleogeologic and subcrop maps. As he pointed out earlier (1934), virtually all the petroleum in this vast region is found to be associated with one or more unconformities, whether they be strictly disconformities, regional low-angular unconformities, or local high-angle unconformities. Even those large oil pools located on structures (Oklahoma City is a good example) are found, when studied in detail, to be closely related to surfaces of unconformity.

Traps in this region of low-angle unconformities and disconformities occur in almost every conceivable rock type below the erosion surfaces, where porosity has been enhanced by leaching, fracturing, and alteration. Above the unconformities, quartz sandstones form the principal reservoir rocks. Regressing seas exposed strata on a broad, flat land to erosion, lithification, and alteration. Stream-channel and deltaic traps are common in places beneath younger transgressive sediments. Generally, lithification due to exposure left little unconsolidated sediment to be reworked by transgressing seas; therefore, wide, blanket-type basal sandstones are relatively scarce. Local sandstone lenses—either reworked stream and residual deposits or beaches and bars—are the rule above the disconformities. Individual traps—such as those found in buttress sandstone bodies, sandstone lenses, and truncated units beneath unconformities—may be fairly large and entirely filled with oil and gas.

Locating traps in such an environment is a task requiring a thorough knowledge of stratigraphy and geomorphology, a maximum amount of subsurface data, a deep understanding of environments of sediment deposition (Busch, 1959), and a superior ability to visualize in three dimensions. That the application of all these related disciplines is effective is demonstrated by the high degree of success in exploration of the Morrow sandstone in western Oklahoma (Forgotson, 1969) and the nearly equal success in the Mesaverde in southwestern Wyoming (Weimer, 1966).

UNCONFORMITY TRAPS ON BASIN MARGIN

The basin margin—the coastal plain and shallow-water region on the side of the platform facing a more or less permanent seaway—is an area of frequent flexing and tilting. In a sense this is a hinge—but a hinge of greatly varied materials and thus subject to rather erratic movements each time it is flexed.

Each deepening of the sea results in the deposition of an overlapping sequence of sediments with a slightly different structural attitude from that of the strata previously laid down and eroded. Ultimately, a coastal plain is produced—characterized by the presence of shallow-marine, gently dipping strata in which there are numerous, repeated, low-angle unconformities. These unconformities commonly are difficult to detect, except through regional studies. They form a highly complex and confusing pattern.

Careful study of basin-flank unconformities can be rewarding. Each unconformity may conceal one or more stratigraphic traps. In order to interpret these unconformities correctly, one must analyze the sequence of events which led to the present configuration and position. For effective exploitation, each must be considered in relation to the others. The subsurface strike, width, rate of thinning, and angle of dip of each potential reservoir are important factors to consider in a successful exploration program. The reconstruction of the history of the unconformity and the determination of the different patterns of outcrop and subcrop related to the unconformity should be studied. This procedure has been termed "unconformity analysis" (Chenoweth, 1967).

Unconformity analysis can be an effective prospecting method in areas which remain relatively undisturbed, such as the inner margin of the Gulf and Atlantic coastal plains. Three conditions of complexity are recognized: where successive transgressions are at only slight angles to previous ones, where anticlinal warping has intervened, and where synclinal warping has occurred between successive transgressions.

Unconformity traps in regions of this type are commonly large. The East Texas and Pembina oil fields and the Clinton gas field are typical examples. A large quantity of the ultimate reserves of the world's major fields is located in this depositional setting (Halbouty *et al.*, 1970); much of that oil is enclosed in unconformity traps.

Unconformity Traps in Basin

As used in this report, the term "basin" refers to a somewhat restricted area with a thickened sedimentary section, either on the shelf or in low spots and sags in the geosyncline. All basins have one characteristic in common: the sediment was deposited more or less continuously with only minor localized and brief interruptions. In those basins which were not subject to later strong tectonic deformation but which have had a history of slow and regular subsidence, probably the only place where uplift and subsequent erosion occurred, and consequently the development of unconformity traps, was above and on the flank of the midbasin ridge. Dallmus (1958) showed that as a consequence of subsidence the basin floor is subject to compression; with continued compression, a ridge rises near the basin center. This compression may be partly responsible for the development of diapirism if sufficient plastic rocks (shale and salt) are present. As the ridge (or salt or shale domes) rises, it pushes bottom sedimentary beds into an area of stronger marine current and wave action where some erosion may occur. The unconformities which develop on the flanks of such domes are commonly very angular; any porous stratum beneath the erosion surface, if truncated and overlapped by sufficiently impervious beds, can form an efficient trap. Such traps are partly or wholly circular in plan. Intermittent and irregular uplift may take place; consequently, several local angular unconformities may be superimposed.

A further consequence of uplift and doming within the basin is the redistribution of porous material. In some basins, particularly in that part where water depth is greatest and which is farthest removed from bordering lands, sediment is brought in and spread on the bottom by turbidity currents. Commonly, the supply of coarse material is meager in the deep part of the basin. Currents sweep around topographic highs on the bottom and, therefore, the crests of such highs may be devoid of porous material. It has been noted frequently that the tops of structural highs formed deep in the basins have little or no porous material; whole provinces have been condemned as having little or no reservoir rock simply because test wells drilled at the crest of anticlines have encountered mainly shale.

Faulting contemporaneous with subsidence and sedimentation is an important, if not the primary, mechanism of basinal subsidence. Shelton (1968) showed that many basins, such as the Los Angeles basin of California, are bounded by positive features which have been affected by significant vertical uplift along faults. Faulting and sedimentation also occurred contemporaneously within the basin. Subsidence, to a large extent, was caused by movement along such growth faults. During fault movement and basin subsidence, erosion of the upthrown blocks and deposition on the down-thrown blocks were common. Growth faulting commonly occurs intermittently, and an upthrown block may be buried for a brief period prior to renewed upward movement. In this manner, unconformity traps are produced on the upthrown blocks if sufficient porous material was deposited over them prior to upward movement.

Searching for unconformity traps in the type of basin described becomes a problem in reconstruction of basin history. It is particularly important to recognize the larger growth faults and the structural uplifts within the basin depths. Moreover, the area from which coarse material may have come and the route by which the currents may have distributed it in the basin should be known. In those basins where sand or other coarse material forms only a small part of the total sediment, it becomes particularly important to reconstruct the manner in which turbidity currents may have distributed this material. Since virtually all traps in such areas are, in some manner, unconformity traps, the least likely place to look for commercial accumulations of hydrocarbons is on structure.

REFERENCES CITED

Adams, F. D., 1954, The birth and development of the geological sciences: New York, Dover Publications, Inc., 506 p.

Branson, C. C., 1962, Pennsylvanian System of the Mid-Continent, in Pennsylvanian System in the United States: Am. Assoc. Petroleum Geologists, 508 p.

Busch, D. A., 1959, Prospecting for stratigraphic traps: Am. Assoc. Petroleum Geologists Bull., v. 43, no. 12, p. 2829–2843.

Chamberlin, T. C., 1909, Diastrophism as the ultimate basis of correlation: Jour. Geology, v. 17, p. 685–693.

Chenoweth, P. A., 1967, Unconformity analysis: Am. Assoc. Petroleum Geologists Bull., v. 51, no. 1, p. 4–27.

Dallmus, K. F., 1958, Mechanics of basin evolution and its relation to the habitat of oil in the basin, in L. G. Weeks, ed., Habitat of oil: Am. Assoc. Petroleum Geologists, 1384 p.

Dunbar, C. O., and Rodgers, John, 1957, Principles of stratigraphy: New York, John Wiley & Sons, 356 p.

Forgotson, J. M., 1969, Factors controlling occurrence of Morrow sandstones and their relation to production in the Anadarko basin: Shale Shaker, v. 20, no. 2, p. 24–38.

Halbouty, M. T., et al., 1970, Factors affecting formation of giant oil and gas fields, and basin classification, in M. T. Halbouty, ed., Geology of giant petroleum fields: Am. Assoc. Petroleum Geologists Mem. 14, p. 528–540.

Horberg, L., Nelson, Vincent, and Church, Victor, 1949, Structural trends in central western Wyoming: Geol. Soc. America Bull., v. 60, p. 183–216.

Kay, Marshall, 1951, North American geosynclines: Geol. Soc. America Mem. 48, 143 p.

Knebel, G. M., and Rodríguez-Eraso, Guillermo, 1956, Habitat of some oil: Am. Assoc. Petroleum Geologists Bull., v. 40, no. 4, p. 547–561.

Levorsen, A. I., 1934, Relation of oil and gas pools to unconformities in the Mid-Continent region, in Problems of petroleum geology: Am. Assoc. Petroleum Geologists, 1073 p.

——— 1954, Geology of petroleum: San Francisco, W. H. Freeman and Co., 703 p.

——— 1960, Paleogeologic maps: San Francisco, W. H. Freeman and Co., 174 p.

Martin, R., 1966, Paleogeomorphology and its application to exploration for oil and gas (with examples from Western Canada): Am. Assoc. Petroleum Geologists Bull., v. 50, no. 10, p. 2277–2311.

Miller, J. B., et al., 1958, Habitat of oil in the Maracaibo basin, Venezuela, in L. G. Weeks, ed., Habitat of oil: Am. Assoc. Petroleum Geologists, 1384 p.

Schuchert, C., 1910, Paleogeography of North America: Geol. Soc. America Bull., v. 20, p. 127–606.

Shelton, J. W., 1968, Role of contemporaneous faulting during basinal subsidence: Am. Assoc. Petroleum Geologists Bull., v. 52, no. 3, p. 399–413.

Weimer, R. J., 1966, Time-stratigraphic analysis and petroleum accumulations, Patrick Draw field, Sweetwater County, Wyoming: Am. Assoc. Petroleum Geologists Bull., v. 50, no. 10, p. 2150–2175.

BULLETIN OF THE AMERICAN ASSOCIATION OF PETROLEUM GEOLOGISTS
VOL. 50, NO. 10 (OCTOBER, 1966), P. 2277-2311, 25 FIGS.

PALEOGEOMORPHOLOGY AND ITS APPLICATION TO EXPLORATION FOR OIL AND GAS (WITH EXAMPLES FROM WESTERN CANADA)[1]

RUDOLF MARTIN[2]
Calgary, Alberta

ABSTRACT

Under the term *paleogeomorphology* are grouped all geomorphological phenomena which are recognizable in the subsurface. Buried relief features with a marked three-dimensional geometry are of importance to the petroleum geologist whenever they lead to the trapping of hydrocarbons. *Paleogeomorphic traps* form a third and distinct group which ranks in importance with stratigraphic and structural traps as a major mechanism for localizing hydrocarbon occurrences. They are not simply another type of stratigraphic trap, but form by themselves a category of considerable economic significance. Paleogeomorphic traps can not be analyzed, nor can their occurrence be predicted, by stratigraphic or structural methods of study, but must be treated as a geomorphological problem. Of the many types of buried relief features, some are of greater interest to the petroleum geologist than others. In this paper, the morphology of buried erosional landscapes is discussed in greater detail. Other types of buried relief features are fossil reefs, barrier beaches, and submarine canyons.

Hydrocarbons may be trapped either directly or indirectly as a result of paleogeomorphological processes. In either case, the traps may occur below or above (against) the morphological surface. Numerous examples of the effects of erosion are known from the Paleozoic and Mesozoic buried landscapes of western Canada and the Mid-Continent area of the United States. Hydrocarbon traps occur below the highs on such erosional surfaces as well as in sandstone bodies deposited in the lows on these surfaces. The rules which govern the formation of ancient landscape forms are worked out in detail, with particular emphasis on the application of quantitative geomorphology to the pattern of the ancient drainage system and on such features as summit levels and the influence of geological factors (erosion-resistant levels, influence of faults and fractures, *etc.*). Weathering and underground solution also play a role in providing both reservoirs and buried traps for oil and gas.

The method of analysis must take into account (1) the structural attitude of the strata below and above the erosional surface, (2) the lithology of the formation overlying the unconformity (if used for isopachous mapping), (3, 4) synsedimentary structural movements and the compaction factor in the formation overlying the unconformity, (5) problems of correlation, (6) reservoir development, (7) presence of a "seat seal," and (8) other factors.

Paleogeomorphology provides an interesting evaluation of "modern" geomorphological thinking. The fossil relief forms "frozen" by the transgression of younger rocks neither disprove the validity of the classical peneplain concept, nor do they support fully the newer ideas regarding slope retreat. The actual conditions found are described by a re-definition of the old term *paleoplain*. The summit level is an intrinsic part of this feature, and not a dissected earlier peneplain. A new geomorphological concept introduced here concerns the alternation of obsequent and resequent interfluve spurs.

Buried landscapes should provide a high percentage of the future oil and gas fields yet to be discovered in North America. The most important stratigraphic levels at which buried landscapes occur are those that formed after major periods of orogenesis. Their geographical locus corresponds to the broad belts of subsequent transgressions.

PALEOGEOMORPHOLOGY AND GEOMORPHOLOGY

The term *paleogeomorphology*, as far as the writer has been able to ascertain, first was used in its proper sense, as a sub-science of geomorphology, by Thornbury (1954, p. 30–33). In an

[1] Read before the Rocky Mountain Section of the Association at Billings, Montana, September 27, 1965. An earlier paper given at the 49th Annual Meeting of the A.A.P.G., Toronto, Ontario, on May 21, 1964, under the title "Techniques of exploration for buried landscapes," is included in the present publication. Manuscript received, January 10, 1966; accepted, May 2, 1966.

[2] Consulting geologist, Rudolf Martin & Associates Ltd. The writer is indebted to J. T. Hack, A. D. Baillie, D. P. McGookey, C. W. Spencer, and M. Berisoff for reading the manuscript and making helpful suggestions.

earlier reference to *paleogeomorphic maps*, Kay (1945) used the word only as a special type of *paleogeographic maps*. Thornbury (1954, p. 582–587) devoted a special section to the "Application of Geomorphology to Oil Exploration," and a considerable part of this subchapter dealt with what the writer subsequently has termed *paleogeomorphic traps*. About the same time that the writer first used the term *paleogeomorphology* in connection with subsurface studies in oil fields (Martin, 1960), Harris (1960) used it in connection with field observations. Since then, the term has been used in two additional publications by the writer (R. Martin and Jamin, 1963; Martin, 1964b) and in another by McKee (1963).

Enlarging on Thornbury's concept, the writer

groups under the term *paleogeomorphology* the study of all geomorphic phenomena which are recognizable in the subsurface and in outcrops of previously buried formations. Geomorphology is the science of the earth's relief features and of the processes which created them. Hence, paleogeomorphology is the science of buried relief features. This includes submarine features, such as submarine canyons and those parts of reefs or volcanic islands that are below sea-level and can not very well be considered separate from their subaerial parts.

The petroleum geologist's emphasis must be on the three-dimensional shape or *form* that has created the hydrocarbon trap. Yet, a proper interpretation of this form, especially from scattered subsurface data, is not possible without an understanding of the *process* which has created the form. Considering how difficult it is in the study of modern landscapes to indicate with absolute certainty the process or processes which created a certain relief feature, one may well doubt the practical value of paleogeomorphological studies, especially as a means of finding accumulations of oil and (or) gas. Nevertheless, the studies made by the writer during the last 7 years, principally in western Canada, have shown that certain deductions can be made from the study of buried landscape forms which lend themselves to geomorphological interpretation; these deductions, then, can serve as a basis for future exploration.

Because of the tremendous number of factors that shape each feature, many of which are of purely local character, only average parameters can be established, and no mathematically exact rules can be formulated. However, once these averages are known, they become determining factors in the reconstruction of a buried landscape from a limited number of control points. Such a reconstruction may then serve as a basis for a lease-acquisition program, geophysical surveying, and eventually for drilling. By defining the objective and limiting the area within which a feature should occur, a paleogeomorphological study can result in considerable savings of effort, time, and money spent searching for these types of hydrocarbon traps.

The geomorphological processes that are involved in paleogeomorphology were defined previously by the writer (Martin, 1960). In that paper, the writer made a broad subdivision into *constructive* and *destructive* geomorphic processes. Apart from the effect of endogene processes, the earth's relief is formed by material constantly being removed in some places and being added elsewhere. On this basis, Lobeck (1939, p. 7, 9), building on an earlier definition by Davis (1884), distinguishes *constructional* and *destructional* forms, and Thornbury (1954, p. 34–35), *aggradation* and *degradation*. The writer's subdivision is similar to, but not exactly the same as, Lobeck's; his depositional forms are included with the first instead of the second group. Thornbury does not include under aggradation such constructive processes as the work of organisms, vulcanism, and the impact of meteorites. The recent trend in geomorphology textbooks is toward emphasis on landforms (mostly destructive processes) alone. To the petroleum geologist, however, the constructive processes (especially organic-reef growth) are of equally great importance.

The most important topics for the petroleum explorationist are the morphology of organic reefs, buried erosional landforms, submarine canyons, and weathering and underground solution. Other features, such as (river) point bars and (marine) barrier beaches, usually are discussed as sedimentary processes, although their truly three-dimensional aspects render them particularly subject to paleogeomorphological analysis. Thornbury (1954, p. 31–32, p. 584–586) devotes several pages to "shoestring sands" and observes that probably no phase of petroleum exploration can be used to better advantage than a knowledge of the detailed characteristics of specific topographic features.

PALEOGEOMORPHIC TRAPPING OF HYDROCARBONS

Thornbury (1954, p. 553) stated[3]: "No claim is made that even the most thorough knowledge of geomorphology equips one to become a . . . petroleum geologist . . ., but it is the author's belief that too often these 'practical geologists' fail to make maximum possible use of basic geomorphic concepts"; and elsewhere (1954, p. 33): "When geomorphologists . . . fully realize . . . this use which can be made of geomorphic principles and knowledge, the subject will become a true working tool in the practical application of geology."

[3] Quotations from Thornbury (1954) are reproduced with permission of John Wiley and Sons, Inc.

LeGrand (1960) emphasized another side of the problem when he stated that ". . . the intricate work of natural agencies operating to form landscapes is poorly understood. This not only has a retarding influence on geomorphic studies, but is damaging to important economic considerations in several fields of geology." Kay (1945, p. 427), discussing "paleogeomorphic maps," did not really do justice to Thornbury's later definition of paleogeomorphology, because he omitted all mention of the use of subsurface data in constructing such maps; his main concern was the separation of seas from lands. Levorsen (1960, p. 22) mentions "paleotopography" in passing, but does not deal with the subject in detail. Thus, although paleogeomorphology plays an important role in the accumulation of hydrocarbons, it is difficult to find specific details on this subject in petroleum geology textbooks, even though many descriptions of oil and gas fields have been published in which paleogeomorphological phenomena are the major factors in entrapment. The literature on reefs, buried hills, and other paleogeomorphic traps is voluminous; what has been lacking thus far is the treatment of the subject as a problem in geomorphology.

The problems posed by paleogeomorphology, especially in areas where buried relief features abound, are as important to the petroleum geologist as those of sedimentation and structure; however, they do not appear thus far to have received equal attention. Most authors, including Levorsen (1954), treat paleogeomorphologic hydrocarbon traps as another type of stratigraphic trap—which they are not. An important difference between stratigraphic and paleogeomorphologic traps is the pronounced three-dimensional aspect of the latter. The writer prefers to limit the use of the term *stratigraphic trap* to those traps which are caused by a lateral change in reservoir properties within a given stratum. Erosion surfaces, reef buildups, and other geomorphic phenomena are bounded by air or water at the time of their formation. Subsequent deposition of younger strata adjacent to but different from those that constitute such a morphological surface does not create a stratigraphic trap but a paleogeomorphic trap. The following new classification of hydrocarbon traps lists the trapping mechanisms in approximately the order in which they originated.

1. *Stratigraphic*—Trap formed by lateral change in reservoir-rock properties.

2. *Paleogeomorphic*—Trap formed by shape of land (or underwater) surface.

3. *Structural*—Trap formed by structural deformation of reservoir rock.

4. *Mixed*—Combination of any two or three of the above. This group includes traps formed by any combination of the preceding together with hydrodynamic effects.

Paleogeomorphic trapping of hydrocarbons caused by the presence of buried relief features may be classified as either direct or indirect in character.

A. *Direct* trapping of hydrocarbons may occur either below the morphological surface or above (against) it. Several examples are given in Figure 1.

I. Accumulations *below* the morphological surface may be caused by the following.

(a) Erosional (destructive) processes, *i.e.,* the formation of traps by alternating buried hills and valleys or by the erosion of a submarine canyon.

(b) Constructive processes, such as organic reef building, grading into bioclastic carbonate bars, and the formation of sand barriers (shoestring sands), dunes, and river terraces.

II. Accumulations *above* or *against* a morphological surface may be caused by the following.

(a) Erosion of a river bed or marine channel, followed by local deposition of sands (point bars).

(b) Deposition of reservoir rocks in an existing valley or against the slope of a hill, a reef, or a volcano (buttress sands).

A combination of I and II (paleogeomorphic surfaces potentially bounding a hydrocarbon accumulation both above and below) may be exemplified by river terraces deposited against the sides of a valley. In this situation, the lower erosional surface is composed of the bottom and sides of the valley in which the reservoir sands were deposited; the upper surface is the valley subsequently cut into these deposits and later filled with clay.

B. *Indirect* trapping of hydrocarbons because of paleogeomorphological process occurs in several forms. As in direct trapping, a distinction is made between accumulations below the morphological surface and those above it (Fig. 2).

I. *Below* the morphological surface, altera-

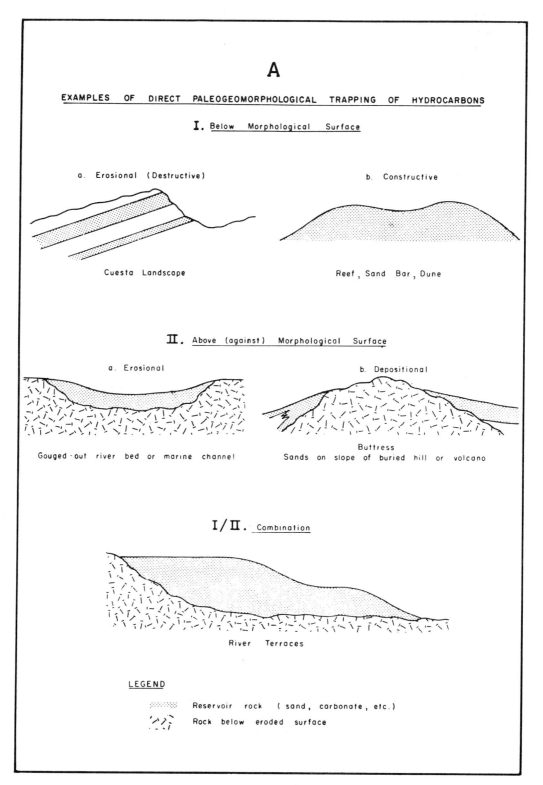

FIG. 1.—Examples of direct paleogeomorphic trapping of hydrocarbons.

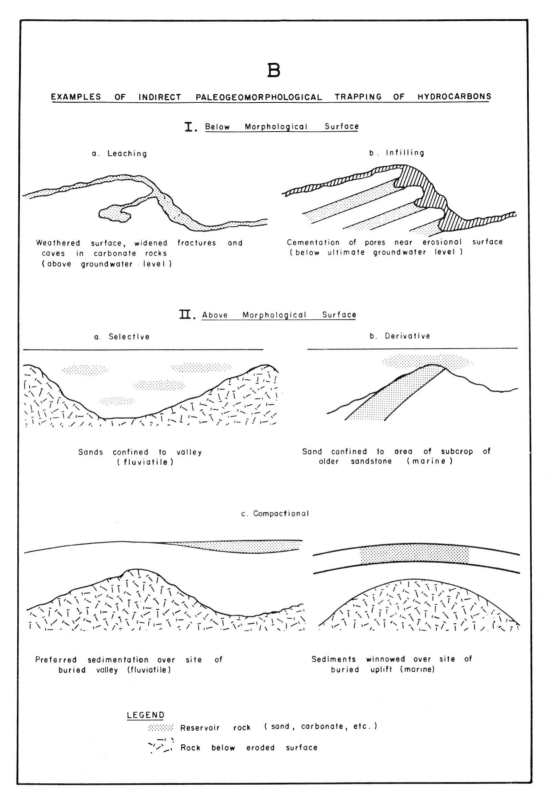

B

EXAMPLES OF INDIRECT PALEOGEOMORPHOLOGICAL TRAPPING OF HYDROCARBONS

I. Below Morphological Surface

a. Leaching

b. Infilling

Weathered surface, widened fractures and caves in carbonate rocks (above groundwater level)

Cementation of pores near erosional surface (below ultimate groundwater level)

II. Above Morphological Surface

a. Selective

b. Derivative

Sands confined to valley (fluviatile)

Sand confined to area of subcrop of older sandstone (marine)

c. Compactional

Preferred sedimentation over site of buried valley (fluviatile)

Sediments winnowed over site of buried uplift (marine)

LEGEND

Reservoir rock (sand, carbonate, etc.)

Rock below eroded surface

FIG. 2.—Examples of indirect paleogeomorphic trapping of hydrocarbons.

tions in the reservoir rock may take place as a result of the action of ground water on carbonate rocks and anhydrite.

(a) While the morphological feature concerned is above ground-water level, there will be leaching, weathering at the surface, solution widening of fractures and formation of cavities below the surface, solution of anhydrite, and formation of collapse breccias.

(b) Below ground-water level (usually at a later stage than (a), because it would result from renewed subsidence), circulating waters carrying dissolved carbonates, sulfates, *etc.* from higher levels will deposit them, causing cementation, infilling of porous rock with secondary calcite, anhydrite, or silica, and formation of a secondary caprock.

II. *Above* the morphological surface, the accumulation of hydrocarbons may be influenced by several factors related to this surface.

(a) Selective deposition of reservoir sands above the valleys of a subsiding old land surface.

(b) Derivation of reservoir sands from subcropping older sandstone bodies during a marine transgression.

(c) Differential compaction of sediments which overlie paleogeomorphological elements. Such compaction will cause the sea floor or a later land surface to be slightly higher above old elevations, and slightly lower above old depressions. Below sea-level, this may lead to winnowing of sediments above the old highs; on land, it will cause rivers to flow over the old lows.

As in studies of sedimentation and structural trends, the problem for the petroleum geologist in making a paleogeomorphological study is to recognize trends and shapes and to find ways of extrapolating these from known points of control. The rules governing trends and forms of earth-relief features are in most respects completely different from those governing either sedimentation or structure; however, in some respects they are interrelated. As Horberg (1952, p. 188) has said, an historical approach to geomorphology ". . . imposes the responsibility of developing a body of principles by which geomorphic history can be interpreted. These principles as such are seldom considered and are yet to be defined and organized."

The following pages deal with several paleogeomorphological phenomena observed in the subsurface, primarily in the oil and gas fields of western Canada. The examples given illustrate subsurface features of purely morphological character which have received too little attention because their true nature was not properly recognized. Comparison with present-day geomorphic phenomena not only leads to better understanding of such features, but also facilitates their analysis and thereby opens the way to extrapolation and prediction. This in turn may result in commercial application of the principles involved. In an earlier paper (Martin, 1960), the writer considered some of the problems of reef morphology. Because this subject will be discussed in a future paper, the present paper is limited to a discussion of buried landscapes.

Buried Erosional Landforms

Buried hills and valleys are important to the petroleum geologist not only because of the hydrocarbons accumulated in the hills, *i.e.*, below the unconformity (type A-I-a), but also because they help determine the occurrence and shape of sand bodies that have accumulated in the valleys (types A-II-a and B-II). Buried erosional landforms are widely known from the Paleozoic and Mesozoic of the Mid-Continent area of the United States and of the provinces of Alberta, Saskatchewan, and Manitoba in western Canada. Surface outcrops also have drawn attention to the existence of unconformities with up to 1,000 feet of paleotopographic relief. Examples include relief observed on the Precambrian surfaces of Wisconsin and northern Michigan, in the Adirondacks, in the Ozarks, and in the Black Hills (Meyerhoff, 1940), and on the Lower Ordovician surface of southwest Virginia and northern Tennessee (Harris, 1960).

Time of Formation of Buried Landscapes

Buried landscapes, as might be expected, are most prevalent at those times in stratigraphic history that followed major periods of mountain building. As a result of each of these orogenic movements, vast areas that previously were part of sedimentary basins were uplifted to positions well above the prevailing base level of erosion (Fig. 3). A further worldwide lowering of this base level must have occurred whenever a period of extensive glaciation followed a major orogeny. Whereas many lesser orogenies also have led to the formation of buried landscapes, it is believed

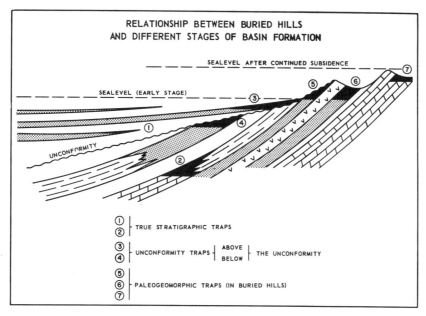

Fig. 3.—Relationship between buried hills and different stages of basin formation.

that the formations in which a search for paleogeomorphic hydrocarbon traps of this type may be most fruitful correspond to the major epochs of landscape formation which coincided in part with widespread glaciation. The major epochs of orogeny—glaciation—landscape formation are tabulated.

Major Orogeny	Glaciation	Landscape Formation
Alpine	Pleistocene	Tertiary-Recent
Hercynian	Permo-Penn.	Post-Miss.
Caledonian	Ord.-Silurian	Ord.-Silurian
Huronian	Precambrian	Post-Precambrian

Sloss (1963) added to these two additional unconformities of continental scope (post-Early Ordovician and post-Early Jurassic). Wheeler (1963) subsequently added a third (post-Devonian). Some of these lesser episodes of landscape formation are more important to the exploration for oil and gas than others. However, a complete list would include many additional episodes not mentioned by these two authors. Until such a complete survey is made, the short list given above will help to define the episodes of landscape formation of major interest.

LOCUS OF BURIED LANDSCAPES

Whereas the preceding considerations help to limit the occurrence of buried landscapes in time, another factor limits their occurrence in space. In the experience of the writer, the areas in which such landscapes occur generally are bounded on the basinward side by a structural hinge line, beyond which the inclination of the strata involved tends to increase markedly into the basin. The hinge line also may be associated with one or more boundary faults, and commonly (but not in every place) coincides with the shoreline of at least one post-unconformity depositional sequence. The locus of buried landscapes formed during any one interval of time therefore tends to be limited on the basinward side by the hinge line corresponding to that time, and on the landward side by the shoreline of the maximum subsequent transgression (Fig. 3). Because the slope of the old pre-transgression landscape, before the time of renewed basin subsidence and transgression, must have been very slight in most cases (i.e., a few feet per mile), the width of the denuded area buried by younger sediments may amount to many hundreds of miles.

Any analysis of the components of a buried landscape must proceed along purely geomorphological lines of reasoning. Serious efforts to prospect for oil and gas traps in buried hills or valleys must be based on such a paleogeomorphological analysis and on the extrapolation of the results according to geomorphological principles.

For a complete analysis, it is necessary to determine the climate that prevailed at the time the

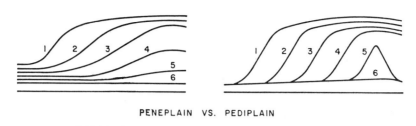

PENEPLAIN VS. PEDIPLAIN

W.M.DAVIS   W. PENCK

YOUTH TO OLD AGE (PENEPLANATION)   SLOPE RECESSION (PEDIPLANATION)

(MODIFIED AFTER W.M.DAVIS)

Fig. 4.—Different stages in slope development according to W. M. Davis and W. Penck.

landscape was formed, the nature of the rocks that formed the landscape, the structure of these rocks, *etc*. In addition, one must deal with the geomorphological parameters that determined the final shape of the landscape, such as drainage pattern, drainage density, slope angles, valley gradients, summit levels, and other factors.

NATURE OF PROCESS OF PLANATION

While making such a paleogeomorphological analysis of certain buried landscapes in western Canada, the writer has found that "modern" geomorphology can gain considerably from the experience gathered from such an analysis of "fossil" or "frozen" landscapes. One of the most important principles involved is that of the process of planation. Here, the classical American geomorphological concept of the *cycle of erosion* or *geographical cycle,* as formulated by Davis (1889), first lost considerable ground to the supporters of the European concept of *scarp retreat* formulated by Penck (1924) (see Fig. 4). Later both theories were tacitly abandoned by most American geomorphologists in favor of quantitative geomorphological concepts and the study of processes. However, neither of these has contributed to a better knowledge of the end product of planation. It is this end product which is represented by the buried landscapes of the past, and it turns out to look neither quite like that which Davis visualized, nor quite like Penck's idea.

According to Davis, a landscape passing from youth through maturity to old age finally would be denuded to a *peneplain* approaching base level. At most, some scattered monadnocks, or disconnected hills, would be left on this surface. Davis did not claim that a peneplain had to be flat; in his description (1899a), he referred to the sub-Cambrian contact (miscorrelated in the quotation) seen in the Grand Canyon of the Colorado River with the following words from Dutton (p. 369 in 1954 Dover reprint): "In the Kaibab division . . . we may observe . . . a few bosses of Silurian strata rising higher than the . . . sandstone which forms the base of the Carboniferous. These are Paleozoic hills, which were buried. . . . But they are of insignificant mass, rarely exceeding *two or three hundred feet* in height" (writer's italics). Hill (1901, p. 363), who first introduced the term *paleoplain* (or buried peneplain), had the following to say about the buried pre-Cretaceous landscape of the Wichita Falls, Texas, area: "That irregularities of configuration once existed in this pre-Cretaceous land is also shown by some of their degraded remnants that still persist, like the Ouachita Mountains, which were not completely buried beneath the Cretaceous sediments, and the Burnet uplift, which was finally buried before the close of the Lower Cretaceous."

The writer's studies have shown that such a paleoplain still contains all the elements of the preceding mature landscape; there is a recognizable drainage pattern, with valleys at a fair gradient, separating a coherent system of ridges that are by no means isolated "monadnocks." Paleogeomorphological data thus might appear to lend support to the theory of Penck (1924), who believed that a slope, once formed, retains its original angle and therewith its identity, while responding to the forces of erosion by retreating gradually parallel with its previous position and losing some of its height. At the foot of such a slope, a footslope or *pediment* develops, which maintains the minimum slope consistent with the local erosional facies. Eventually, when the steep scarps on two sides of an uplifted area have retreated to the point where the ridge between them becomes obliterated and the pediments from adjoining valleys meet, a *pediplain* results which is not really very different as an ultimate landform from Davis' peneplain, but which has reached this end form by very different means. However, the buried landscapes of western Canada definitely have not been reduced to a "pediplain." There is a pronounced difference in slope between the dip slopes and the scarps or "face slopes." Yet, the steep scarps of Penck are gone; the inclination of the buried Mississippian surface in southeastern Saskatchewan ranged from only 15 to 233 ft./mi.

Thus, paleogeomorphology appears to prove that both Davis and Penck were partly right; the slopes, instead of retaining the same angle at all times, do become reduced (to 3° or less), as predicted by Davis; but the framework of the landscape as a whole also retains its general shape, instead of becoming a featureless peneplain. Hill's term *paleoplain* appears well suited to the definition of such a buried landscape. The present writer's re-definition of Hill's term would further reflect the fact that this paleoplain has retained sufficient relief to be an important trapping factor in the accumulation of hydrocarbons, ores, water, *etc.* The closest among modern landscapes to this new definition of paleoplains are the *saucer-shaped valleys* of Tanganyika described by Louis (1964, p. 47), which exhibit slopes of 2–3 per cent (106–158 ft./mi.), locally increasing to 5–10 per cent (264–528 ft./mi.). These are peneplains in the Davisian sense; on the other hand, King (1951, p. 58) cites gradients of only 6 in.–1 ft./mi. for some of the pediplains of South Africa.

CUESTA LANDSCAPES

The paleoplains studied by the writer have retained all or most of the aspects of a cuesta landscape. Here again, the nomenclature of "modern" geomorphology can bear some adjustment. Davis (1899b) and subsequent authors regarded a cuesta landscape as typical of a coastal plain. In reality, a cuesta landscape is bound to develop on the homoclinal flank of any subsiding basin as the tilted strata on its periphery are eroded into parallel, outward-facing, "belted" escarpments. A good example of a cuesta landscape that is not

directly connected with any coastal plain is the Paris basin (Fig. 5); another is the Jurassic "Alb" of Schwaben in southern Germany, which even Davis (1899b) describes as a typical cuesta landscape.

The formation of cuestas is a logical corollary to new basin subsidence after a period of epeirogenic uplift (Fig. 3). However, in areas where structural deformation has taken place on any noticeable scale, a more complex landscape will result. An example of such a complex feature from the Soviet Union, taken from Khutorov (1958), is shown in Figure 6. Though the anticline itself is not productive, a paleotopographical hydrocarbon trap of considerable extent occurs on the topographically highest flank. In addition, an accumulation of "seepage" oil occurs in the younger sandstone beds which cover this buried ridge. The differential compaction above the ridge which has affected these sandstone bodies also should be noted.

A typical cuesta landscape was developed on the northeast flank of the Williston basin in southeastern Saskatchewan, southwestern Manitoba, and northern North Dakota on rocks of Mississippian age (lower Frobisher beds, Alida beds, and Tilston beds), which are overlain by Triassic (?) "Red Beds" of the Lower Watrous Formation in the area studied by the writer (Martin, 1964a, b). The oil has accumulated beneath the unconformity and the accompanying "caprock" wherever the subcrop trend of a reservoir bed crosses a paleotopographical ridge. Accumulations in the upper Frobisher beds and particularly in the overlying Midale beds appear to be less dependent on paleotopography, and are therefore not included in this study; neither are those in the Souris Valley beds which underlie the Tilston beds in southwestern Manitoba.

A map of the southeastern Saskatchewan part of this area, on which the shape of the pre-Triassic (?) landscape has been reconstructed by the use of an isopachous map of the "Red Beds," is shown in Figure 7. The cuestas in this landscape correspond to the subcrops of the beds most resistant to erosion. The regional pre-"Red Beds" dip of the Mississippian strata in southeastern Saskatchewan was about 20–30 ft./mi., which is about 10 times the gradient of the major consequent valley in the area. A schematic cross section at right angles to the general strike of the

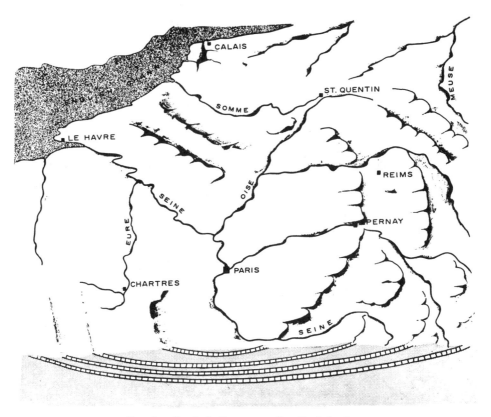

Fig. 5.—Cuesta belts surrounding Paris basin.

Mississippian strata, using as datum the top of the "Red Beds," shows that the cuestas which developed on resistant beds of the Mississippian were sloping at a slightly smaller angle than the beds themselves (Fig. 8, top).

Borrowing a classical diagram from Lobeck (1939), a few words may be said here regarding the nomenclature of valleys and ridges on a homoclinal basin flank (Fig. 9). Streams flowing down the original topographic slope toward the basin (*i.e.*, in this case, down the dip slope) are known as *consequent streams*. As these cut into the stratigraphic sequence, tributaries developed at right angles to them along the exposures of weaker rock, creating *subsequent streams*. Shorter tributaries, parallel with the consequent system but flowing into the subsequent streams, are known as *obsequent* if flowing in a direction opposite to the consequent streams (*i.e.*, down the face slope, or the escarpment of the resistant rock), and as *resequent* if flowing in the same direction as the consequent streams.

On both the map (Fig. 7) and the topographical profile shown in cross section (Fig. 8, top),

the following main features may be noted from north to south.

1. The wedge-out of the Lower Watrous "Red Beds" at their zero isopachous contour line. This does not appear to have any special significance in terms of Mississippian topography.

2. A longitudinal, west-northwest-trending subsequent valley, which is referred to as the Tilston lowland.

3. A parallel ridge, the Tilston cuesta, corresponding to the subcrop of the Tilston beds.

4. A second subsequent valley, here called the Alida lowland.

5. A second ridge, the Alida cuesta, corresponding to the subcrop of the Alida beds.

6. A gradually diminishing southward slope over the Frobisher beds subcrop.

7. A "nickpoint" at the northern end of the Midale beds subcrop, corresponding to a flexure in the "Red Beds" and resulting in an apparent steeper gradient south of this point.

VALLEY GRADIENT

The cross section (Fig. 8, top) shows that the slope between (1) and (2), and the valley bottoms of (2) and (4), are connected with (6) by a smooth gradient, indicating the existence of a graded stream system preceding "Red Beds" time. This would eliminate the possibility of including closed valleys, sink holes, and similar features on "Red Beds" isopachous maps, because the data allow for the construction of a continuous drainage system. The gradient of the main consequent stream was 2 ft. 4 in./mi., which compares with the present Missouri River gradient.

Both the maximum slope observed on the northeast-facing scarps and the maximum slope on the flanks of the northeast-trending obsequent streams are 160 ft./mi. on the "Red Beds" iso-

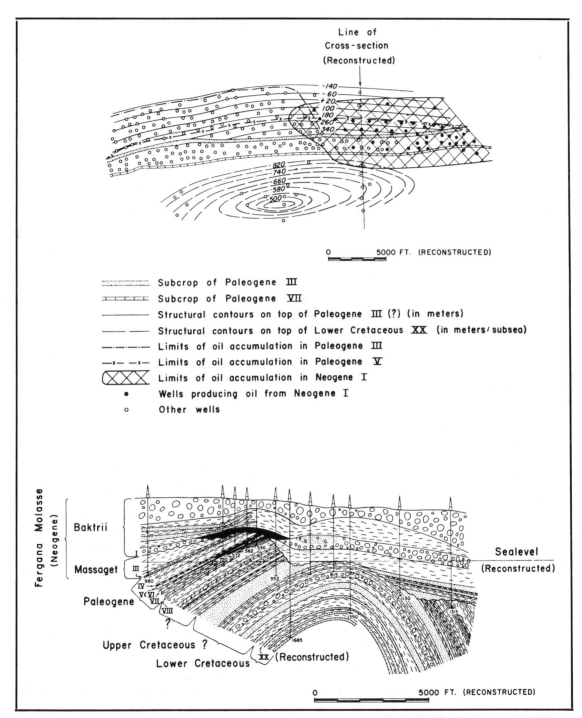

Fɪɢ. 6.—Structure map and cross section of South Alamyshik field, Uzbek S.S.R., Turkestan, U.S.S.R. (after Khutorov, 1958), showing escarpment on the flank of an anticline.

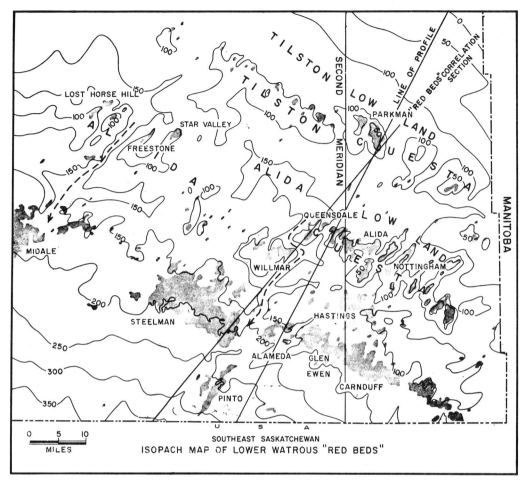

Fig. 7.—Southeast Saskatchewan. Isopachous map of Triassic (?) lower Watrous Formation "Red Beds." Thin isopachous contour values correspond to high elevations of underlying buried Mississippian landscape. Two oblique lines correspond to lines of section of Figure 8. Contours in feet.

pachous map (233 ft./mi. before compaction). This is more than seven times the dip of the Mississippian beds toward the southwest. Therefore, it appears most unlikely that any of the northeast-trending valleys would have been formed by resequent (*i.e.*, in this case, downdip-flowing) streams. The writer thus is confident that a large proportion of the valleys that separate the individual oil fields of the Tilston and Alida cuestas was formed by obsequent streams, which flowed northeast into the subsequent streams of the northwest-trending Tilston and Alida lowlands, and which cut back by headward erosion (scarp recession) into the corresponding cuestas.

COMPACTION FACTOR AND SYNSEDIMENTARY MOVEMENTS

In order to establish the amount of compaction, for which all data based on measured "Red Beds" thicknesses must be corrected, two areas in southeastern Saskatchewan were selected for study, one of 12 townships in the Freestone-Star Valley area, and one of 9 townships around the Alida and Nottingham fields. The method used was to construct a structural contour map across a large area, using the top of the "Red Beds" as datum, and establishing the average dip and strike in the area. Next, the departure of each well-control point from this average (residual structure) was measured and plotted against the local "Red Beds" thickness (Fig. 10). The average compaction value thus established is 32 per cent for both areas. Therefore, pre-compaction thickness of the "Red Beds" was, on the average, 147 per cent of the present thickness. Because of the relatively uniform lithologic character of this formation in most of the area studied, the effects of compaction may be disregarded wherever only

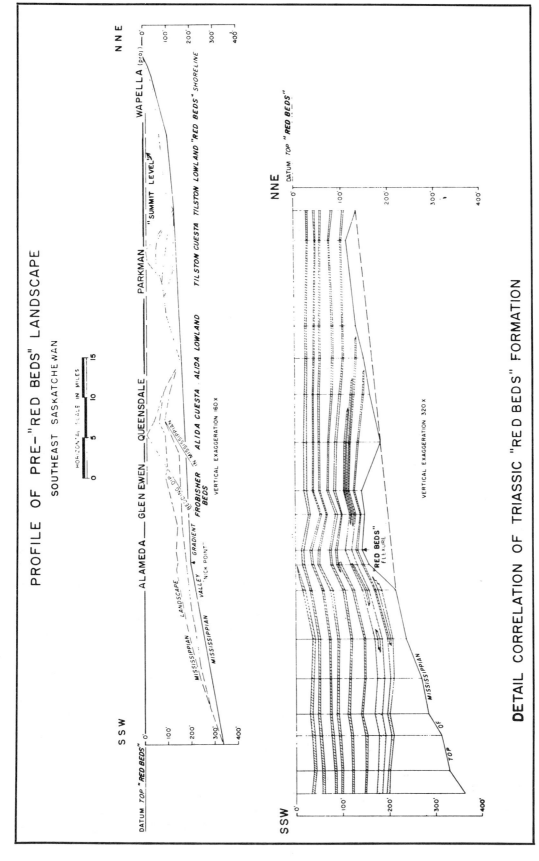

FIG. 8.—Southeast Saskatchewan. Profile of pre-"Red Beds" landscape (top) and detail correlation of Triassic (?) "Red Beds" (bottom). Location of profile and correlation section on Figure 7.

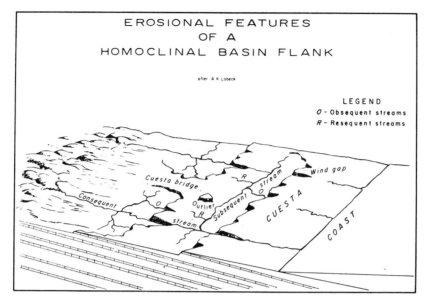

EROSIONAL FEATURES
OF A
HOMOCLINAL BASIN FLANK

after A. K. Lobeck

LEGEND
O - Obsequent streams
R - Resequent streams

FIG. 9.—Erosional features of a homoclinal basin flank, after Lobeck (1939). (Published with permission of McGraw-Hill Book Company.)

relative thicknesses need to be considered (*e.g.*, on maps such as Fig. 7).

A further check, made to establish whether the "Red Beds" had been affected by tilting or other structural processes during their deposition, involved a detailed correlation along the approximate position of the deepest consequent valley shown on Figure 7. Thin siltstone markers within the "Red Beds" can be correlated through surprisingly long distances and exhibit very little change, indicating that these beds were deposited in extremely quiet, or stable, conditions. The cross section (Fig. 8, bottom) shows the bedding planes within the "Red Beds" to be practically parallel with the top of the formation (which is conformable with the overlying Jurassic), except

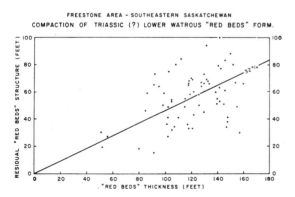

FREESTONE AREA - SOUTHEASTERN SASKATCHEWAN
COMPACTION OF TRIASSIC (?) LOWER WATROUS "RED BEDS" FORM.

FIG. 10.—Freestone area, southeastern Saskatchewan. Compaction of Triassic (?) Lower Watrous "Red Beds."

for a flexure toward the south in the northern part of T. 3, R. 3 W. 2 M. The detailed correlation of all logs in this general area (only a few of which are shown on the cross section) shows that a downwarping occurred during late "Red Beds" time along a hinge line that coincides approximately with the present northern limit of the Midale beds. In order to visualize correctly the pre-"Red Beds" Mississippian topography, the area of the Midale beds subcrop (and even farther south) must be restored to a correspondingly higher original elevation. The apparent steepening of the slope south of the "nickpoint" (7) (Fig. 8, top), resulting from the intra-"Red Beds" downwarping, is unimportant in the discussion of the paleotopographical features north of the area affected by this phenomenon. A real nickpoint, or sudden steepening of the gradient, would indicate the encroachment of a younger, steeper drainage system on an older one; the position of such a nickpoint might have been fixed temporarily in space by the outcrop of a resistant layer in the corresponding valley. In the present case, however, the "nickpoint" is not a Mississippian geomorphological feature, but a post-Mississippian structural feature.

SUMMIT LEVEL

The highest points along the Tilston, Alida, and Frobisher cuestas seem to be at a similar elevation, *e.g.*, 6–10 ft. below the top of the "Red

346

Beds" at Parkman, 0–30 ft. in the Hastings-Alida-Nottingham area, and 40–41 ft. in the Lost Horse Hill-Freestone-Star Valley area (Fig. 7). Figure 11 is an attempted contour map of this "summit level." The classical Davisian geomorphological concept of such a summit level, or "base surface," would be to interpret it as an old peneplain that has been uplifted anew and dissected by a second cycle of erosion (Fig. 12, left). This also was the interpretation expressed by Siever (1951), who described the Mississippian erosion surface which is covered by Pennsylvanian rocks in Illinois. However, it should be noted, first, that the southeastern Saskatchewan summit level, south of the hinge line that coincides approximately with the subcrop trend of the Midale beds, is no longer a near-horizontal surface. Second, the summits of the buried Alida and Frobisher beds hills are lower in the vicinity of the main consequent valley that runs along the west side of the Queensdale field. Nevertheless, although the hills become lower in the direction of this valley in terms of absolute elevation, the amount of relief above the valley floor (80–120 ft.) is approximately the same. Even in the Steelman field, south of the hinge line, where the Mississippian is at a much lower level than farther north, the relief is of the same order. The writer concludes, therefore, that the buried Mississippian landscape of southeastern Saskatchewan supports the idea of Rich (1938, p. 1701) and others that ". . . such structural ridges must develop an even crest line entirely independent of any peneplanation. Briefly, this is because the height of the crest is determined by the meeting of the slopes that rise from the weak-rock areas on either side. Since . . . these slopes tend to have a constant angle, they must meet at a relatively uniform elevation above the grade profiles of the subsequent streams in the adjacent weak-rock area." More recently, Hack (1960, p. 91) observed that ". . . regularity of the landscape and the rather uniform height of the hills owe their origin to the regularity of the drainage pattern that has developed over long periods, by the erosion of rocks of uniform texture and structure." The summit level, according to these ideas, would not be caused during a previous erosion cycle but would be the result of contemporaneous leveling processes (Fig. 12, right). The paleogeomorphological data from southeastern Saskatchewan support this view. It should be stressed in particular that the paleoplain discussed here is essentially equivalent to Davis' peneplain. If a "summit level" were a dissected peneplain, then a paleoplain could not have a summit level. Because it does, the summit level must be an essential and contemporaneous part of the paleoplain.

The summit level is an important concept in the exploration for prospective buried hills. A lowering of this level by 80 ft., as observed at Queensdale, corresponds to a basinward displacement of the escarpment by at least 4 mi. (based on a formation dip of 20 ft./mi.). Therefore, whether the summit level is a remnant of an old erosion level, or whether it is related to the current drainage system of the area, is a problem that has many practical implications.

RESISTANCE OF DIFFERENT LAYERS TO EROSION

In addition to the main escarpments of the Alida and Tilston cuestas, there are numerous scarps that correspond to resistant levels of lesser importance. Figure 13 shows diagrammatically the three subsidiary escarpments which are typical of most of the buried hills along the Alida cuesta and which correspond to resistant layers within the Alida beds. The same features have been described by Vogt (1956, Fig. 4) from the Alida and East Alida (North Nottingham) fields, and also appear on a cross section of the Nottingham field published by Edie (1958, Fig. 7). However, escarpments of this subsidary type do not form the long cuestas which constitute the divides between the principal subsequent valleys. They occur instead as short strike ridges on otherwise southwest-northeast-trending spurs lying between pairs of obsequent or resequent valleys. Where particularly well developed, such strike ridges may give these spurs the shape of "hammerhead hills" (Fig. 18).

In general, the Alida (MC$_2$) shale, as well as most porous rocks (e.g., the Kisbey and other (dolomitic) sandstone beds and the dolomitic sections of the Middle Alida and Lower Frobisher beds), constitute the weaker rocks in which valleys were developed. The dense limestone and evaporite beds appear to have been the more resistant, ridge-forming beds. There is some evidence that lateral facies changes caused variations in resistance of these rock units, with the

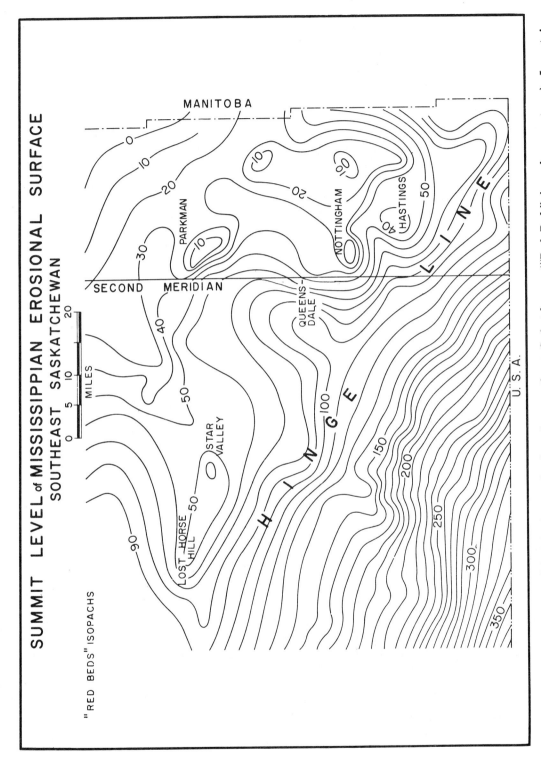

SUMMIT LEVEL of MISSISSIPPIAN EROSIONAL SURFACE
SOUTHEAST SASKATCHEWAN

"RED BEDS" ISOPACHS

FIG. 11.—Summit level of Mississippian erosional surface, southeast Saskatchewan ("Red Beds" isopachous contours). Lowest isopachous contour values correspond to highest summits of buried Mississippian topography. Summit level is tangential to erosional surface illustrated by Figure 7. Contours in feet.

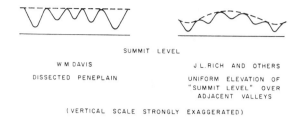

Fɪɢ. 12.—Nature of summit level, according to W. M. Davis and newer theories. Paleogeomorphic data support more recent view.

result that some cuestas and lowlands disappear or reappear along strike. For example, the Alida lowland, which is developed in the Alida shale east of the Queensdale field, is in the basal Middle Alida beds farther west, north of Queensdale (Fig. 18). The most important relief-forming section is in the lower part of the Frobisher beds, just above the Kisbey sandstone or its equivalents (Upper Alida beds). There is some evidence in the South Hastings pool that the Carlevale evaporite is more resistant than the underlying carbonate pay zone, but this zone in the Frobisher beds definitely forms a much lower ridge than the Alida cuesta north of it.

One method of determining which sections are erosion resistant is to construct a frequency graph of the interval thickness between a particular marker (e.g., top of Alida beds) and the top of the Mississippian found in all wells within a

relatively large area, as, for example, between the Freestone and Queensdale fields (Fig. 14). This area was chosen because, at the time of the survey, it contained no major oil fields but only wildcat wells that were fairly uniformly scattered. Where small fields or clusters of wells occurred, only one representative well was used for constructing the graph. The correlation of the resistant intervals with the lithologic character is evident from this graph, but only the graph indicates the relative frequency of occurrence (c.q., relative area of distribution) of each resistant bed. One of the main problems facing the geologist in making such a graph is the exact correlation of logs (commonly of different types) run in wells of relatively shallow penetration into the Mississippian. In addition to the graph, a series of cross sections was constructed from all wells in the same area. Although these cross sections illustrated the concept as well as the graph, it is obviously impossible to incorporate all data from a large area in one cross section; this method, therefore, becomes unwieldy and is less accurate than that of the frequency graph.

QUANTITATIVE GEOMORPHOLOGY

The fact that only two of the many erosion-resistant intervals in the Mississippian of southeastern Saskatchewan formed major cuestas, whereas the remainder form only short strike ridges,

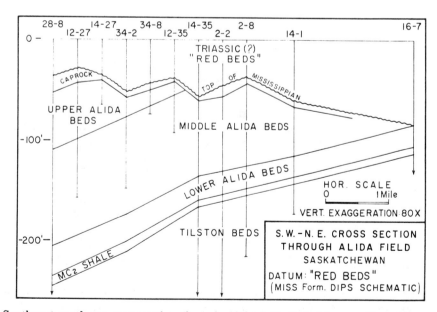

Fɪɢ. 13.—Southwest-northeast cross section through Alida field, Saskatchewan, illustrating occurrence of subsidiary escarpments. Main cuesta at left. Location of cross section on Figure 18.

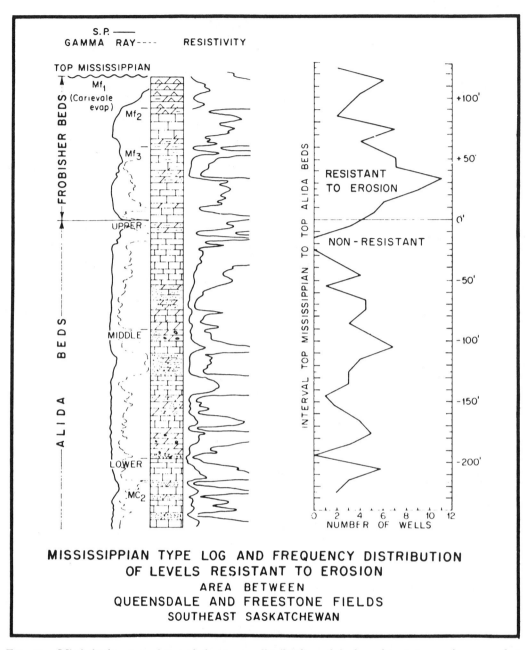

FIG. 14.—Mississippian type log and frequency distribution of beds resistant to erosion, area between Queensdale and Freestone fields, southeast Saskatchewan. Main Alida cuesta corresponds to the +35 ft. level; the —45, —70, and —105 ft. levels correspond to three subsidiary cuestas of Figure 13.

points up the importance of considering this buried landscape from a true geomorphological rather than a limited geological viewpoint. A landscape is formed by the interaction of climatic factors (such as rainfall) and the geological framework. Flowing water may adapt itself to geological factors, but first of all is controlled by hydrophysical laws. These laws and the nature of the underlying geology determine the ultimate shape of the drainage pattern (Fig. 15).

Horton (1945) has formulated several quantitative geomorphological laws which are extremely useful in landscape analyses directed toward the search for hydrocarbon traps. He defines *drainage density* as the sum of the lengths of all streams in a basin, divided by the area of that basin. This factor may be assumed to be about the same for similar rock types under the same climatic conditions. Thus, having determined the drainage density for the Queensdale-Alida-Nottingham area, where obsequent streams appear to have an average spacing of 3.3 mi., it may be assumed that other parts of the Alida cuesta (and probably also of the Tilton cuesta) will have about the same drainage density. Thus, the drainage-density factor may be used to interpret the paleotopography of these areas.

According to the *law of drainage composition,* the number and average length of tributary streams vary in geometrical progression with the stream order. Stream order, in Horton's system, is No. 1 for the smallest unbranched tributaries and increases in number toward the largest (main) stream. (This system of stream-order designations was modified slightly by Strahler, 1952a and 1960.) Horton (1945, p. 298) states that first-order streams usually are more than ⅓ mi. and less than 2–3 mi. long. In the buried Mississippian landscape of southeastern Saskatchewan,

first-order streams correspond to the short tributaries of the obsequent drainage system. These are the tributaries that originate at the foot of the short strike ridges observed in the fields along the Alida cuesta. The average length of these is 1½ mi. These are not difficult to identify where well spacing is ¼–½ mi. The obsequent and resequent streams flowing northeast and southwest between these fields would be second order; the subsequent streams into which they flow would be third order; and the two large consequent streams flowing into the Williston basin sea, fourth order. The total number of streams of one order in a drainage basin tends to approximate closely an inverse geometric series in which the first term is 1 and the ratio is the *bifurcation ratio.* This is the ratio of the total number of branchings of streams of a particular order to that of streams of the next lower order. It usually is constant for all orders of streams in a basin (Fig. 16, right).

Another Horton equation states that the average length of streams of any order tends to ap-

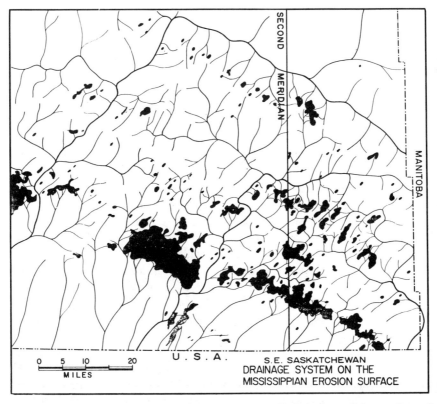

Fig. 15.—Southeast Saskatchewan. Drainage system on Mississippian erosion surface, derived from "Red Beds" isopachous maps.

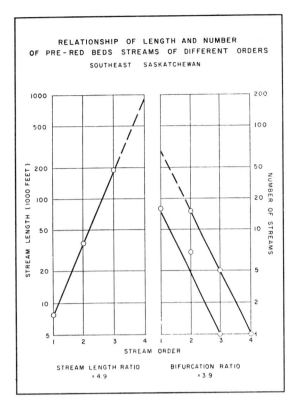

RELATIONSHIP OF LENGTH AND NUMBER
OF PRE-RED BEDS STREAMS OF DIFFERENT ORDERS
SOUTHEAST SASKATCHEWAN

FIG. 16.—Relationship of length and number of pre-"Red Beds" streams of different orders, southeast Saskatchewan. Stream orders 1 to 4 refer to streams of increasing length. Ratio of average length of streams of a given order to that for next lower order is known as *stream length ratio*. *Bifurcation ratio* is ratio of total number of branchings of all streams of a particular order to that for next higher order.

proximate closely a direct geometric series in which the first term is the average length of first-order streams and the ratio is the *stream length ratio*, or the ratio of the average length of streams of a particular order to that of streams of the next lower order (Fig. 16, left). If stream lengths and stream numbers are plotted against the stream orders on a semi-log scale, straight lines result (Fig. 16). These can be used to analyze less densely drilled parts of the same drainage area.

The *law of stream slopes* indicates an inverse geometric relation between average channel slope and stream order. (Strahler, 1952a, also introduced a relation between *basin areas* and stream orders.)

These quantitative concepts are very useful in interpreting an area paleotopographically; *i.e.*, they keep the number of tributaries, lengths of

streams, and channel slopes within specific limits according to geomorphological principles. However, deviations from these "laws" do occur, mostly because of geological factors. During the last 15 years, a tremendous volume of work has been done in the field of quantitative geomorphology, much of it under the direction of A. N. Strahler at Columbia University, and of L. B. Leopold at the U. S. Geological Survey. Their studies have confirmed fully Horton's concepts and have refined further the techniques of this new sub-science. However, many of their findings go far beyond the particular sphere of interest of the paleogeomorphologist. Because this branch of science is so young and its findings are being augmented continuously, only one textbook (Strahler, 1960, p. 376 *ff.*) devotes considerable space to quantitative geomorphology.

The logic of the quantitative approach to geomorphology is so overwhelming that it seems almost unbelievable that its laws were formulated only as recently as 1945. Given a region of simple geological texture, and located within the boundaries of one climatic province, it can be said that about an equal volume of rain will fall per unit of time on each unit of surface area in that region. Working its way down the slope of the original virgin landscape (*e.g.*, immediately after emergence from the sea), this rain water, starting out as sheet flow, gradually combines into several definite courses and carves valleys into the landscape. Where, locally, there are too few valleys in the beginning to handle the rainfall volume per unit area, it is reasonable to assume that additional valleys will form between the first valleys. It is equally reasonable to assume that, where too many valleys form in the beginning, two or more eventually will combine into one. Thus, a certain equilibrium, which can be analyzed statistically, tends to be created. If this equilibrium is upset by external forces, such as a change in climate, uplift or subsidence of the land, or other relative changes in base level, the forces involved will reshape the landscape until it is again in equilibrium. This concept of "dynamic equilibrium" is presented eloquently by Strahler (1950; 1952a, b) and by Hack (1960).

In the buried landscape of southeastern Saskatchewan, other geomorphological observations may be made which reflect this law of dynamic equilibrium. Although these have received practically no attention from "modern" geomorphologists, they are of considerable importance to the paleogeomorphologist in the exploration for oil and gas. One phenomenon, that of the "hammerhead hills," was mentioned earlier in this paper. Another, which may be termed the *law of alternating obsequent and resequent interfluve spurs*, was hinted at briefly by Horton (1945, p. 355–357), and is illustrated in Figure 17. In essence, the principle involved is that, where resequent or obsequent drainage is established on one side of a cuesta, the drainage on the opposite side tends to avoid draining the same part of this ridge, with the result that valleys on one side of the cuesta are opposite spurs on the other side and *vice versa*. Figure 18 illustrates this principle with an example from the area of the Alida and Nottingham fields.

A third principle involves the *oblique angle of entry* of subsequent tributaries into the conse-

quent main valley. This, too, is mentioned briefly in some of the geomorphological literature (*e.g.,* Horton, 1945, p. 349), but assumes considerably greater importance in paleogeomorphology. A practical case in point is illustrated by the Queensdale field (Fig. 18). The subsequent valley along the north side of the Queensdale, Alida, and Nottingham fields is developed in the MC$_2$ shale and lower beds along the eastern part of its course, thus providing an excellent seat-seal for the oil accumulations in the Alida and Nottingham fields. However, because it strikes at an angle to the strike of the beds, this valley cuts back into the lower part of the Middle Alida beds north of Queensdale. As a result, the Queensdale reservoir has an oil-water contact corresponding to its spill-point, whereas farther east the water level occurs much lower because of the presence of the seat-seal.

The main types of non-dynamic interference

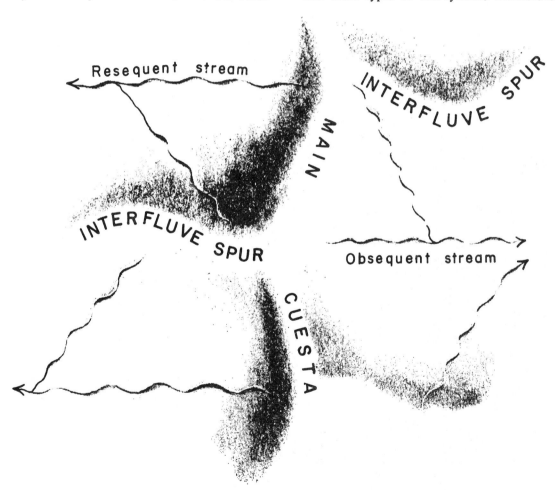

Fig. 17.—Principle of alternating obsequent and resequent interfluve spurs. Modified from Horton (1945).

353

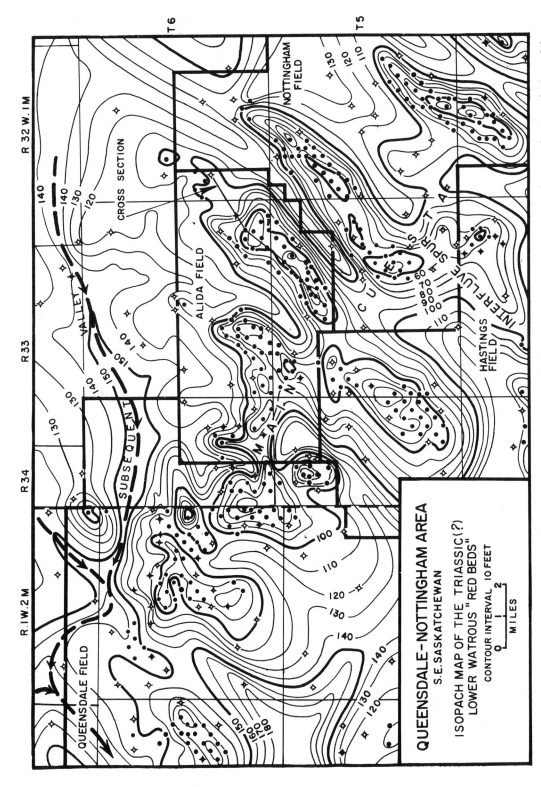

Fig. 18.—Detailed isopachous map of "Red Beds" formation, Alida-Nottingham area, southeast Saskatchewan, showing shape and location of Mississippian buried hills. Note "hammerhead" hills and alternation of interfluve spurs. Line of cross section in Alida field indicates location of Figure 13. Contours in feet.

with the forces of denudation with which geomorphology is concerned are related to the geological nature of the landscape on which these forces act. Interference takes the forms of: (1) differences in resistance to erosion of the underlying rocks (discussed earlier); (2) faults and fractures; a compound form results when a fault separates rocks of different resistance (*cf.* Hack, 1960, Fig. 4); and (3) karst-type drainage in carbonate rocks. The last two types of landscape are discussed below. The geomorphic parameter most strongly influenced by geological factors is the drainage density (Strahler, 1952a, Fig. 4). In an area of relatively simple geology, however, such as the Mississippian of southeastern Saskatchewan or southwestern Alberta, the drainage density should be fairly constant.

FAULTS AND FRACTURES

The Mississippian buried landscape of southwestern Alberta includes good examples both of the influence of faults and fractures and of underground drainage. In this area, a comparable situation exists in the Mississippian Turner Valley and Pekisko Formations to that in southeastern Saskatchewan (Bokman, 1963). The fields in the Sundre-Harmattan-Crossfield-Calgary area are separated by valleys that cut through the Turner Valley into the underlying Shunda Formation; the fields that produce from the Pekisko on the east and north are separated by valleys cut into the Lower Mississippian Banff Group. As a result, within reservoir rocks of the same age, there are large variations in fluid contacts and in oil columns between fields (Bokman, 1963, Fig. 8, top). Here, too, the paleotopography is not the only factor that has determined the trapping of hydrocarbons, but changes in reservoir characteristics of the Turner Valley Formation have contributed to a certain extent to this entrapment.

The Mississippian landscape of southwestern Alberta underwent one more period of erosion than that of southeastern Saskatchewan. In the western part of this area, the Mississippian paleoplain is covered by Jurassic rocks, but these are absent farther east. Pre-Cretaceous erosion has exhumed some of the pre-Jurassic landscape, has left some other valleys untouched and filled with Jurassic sediments, and has created additional valleys. Detailed correlation of the thin Jurassic cover, where present, is required to determine which valleys are pre-Jurassic and which are post-Jurassic in age, and which of the latter may be exhumed pre-Jurassic valleys.

Whereas faults may be suspected to occur and to have played a role in the development of some of the buried valleys of southeastern Saskatchewan, their existence there is more suspected than proved. In southwestern Alberta, on the other hand, the much lower dip of the Mississippian at the time of erosion has resulted in the creation of resistant Turner Valley and Pekisko Formations subcrop belts that are two or more townships wide. This led to the development of a sort of tableland (mesa) landscape in which valley formation could not take place solely by streams concentrating in areas of less resistant rocks, because such areas were too far apart for normal drainage density to develop. Consequently, other geological flaws, such as zones of structural weakness, exerted an important influence on the drainage system which formed (Fig. 19). The straight alignment of some of the major valleys strongly suggests fault or fracture controls. For example, available structural evidence supports the possible presence of a fault with about 100 ft. of vertical throw between the Harmattan-Elkton and Harmattan East fields.

Another subsurface example of a fault-controlled valley is known from the West Edmond field, Oklahoma (McGee and Jenkins, 1946, Fig. 10, *etc.*). A pre-Pennsylvanian stream cut through the Siluro-Devonian Hunton Limestone in this field, at right angles to the subcrop trend of this formation, and generally followed a fault zone trending in the same east-west direction. Whereas the upper and lower reaches of the valley coincide with the fault trace, the middle part of the stream swings ½ mi. north. This swing illustrates the fact that, though a fault or other structural line of weakness may give initial direction to valley development, the stream course may deviate from this direction for unknown subsidiary reasons.

UNDERGROUND DRAINAGE IN CARBONATE ROCKS

Intraformational breccias (caused by roof collapse) and sand-filled cavities, observed in cores or samples from several wells in the area between the Sundre field and Calgary (Fig. 19), indicate

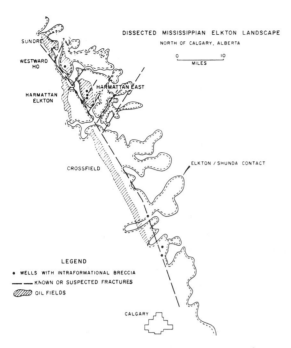

FIG. 19.—Dissected Mississippian Elkton landscape north of Calgary, Alberta. Main subsequent valley trends NNW. along east side of Elkton-Shunda contact. Regional dip is toward west-southwest. Oil and gas are trapped in Elkton salients and outliers (gas accumulations not shown). Some WNW.-trending valleys and their tributaries appear to coincide with fault or fracture trends. Intraformational breccia is indicative of underground drainage, which also may follow fracture trends.

the presence of underground drainage during at least one of the erosion periods that affected the southwestern Alberta Mississippian landscape. Similar phenomena are lacking completely in southeastern Saskatchewan, a fact which indicates a change in climate and therefore of erosion pattern after the Triassic and before the Cretaceous.

Another good example of karst topography accompanied by the formation of sinkholes and a significant amount of underground drainage is in the Cambro-Ordovician Arbuckle Limestone hills of the Central Kansas uplift (Walters, 1946; Walters and Price, 1948). The sinkholes and caves in the Arbuckle are filled with green shale, sandstone, and breccia resembling the "Sooy" or basal conglomerate of the Pennsylvanian Des Moines, a formation which also covers most of these hills. Locally, however, a thin layer of Upper Ordovician Simpson shale, sandstone, and dolomite is between the Arbuckle and the "Sooy," indicating that the landscape was first formed in post-Early Ordovician time, and was exhumed again during

the time interval between the Mississippian and the Pennsylvanian transgression. The fact that there was an earlier period of erosion is confirmed by the discovery of Simpson-filled sinkholes in the Arbuckle of eastern Kansas (Merriam and Atkinson, 1956). The Arbuckle landscape of the Central Kansas uplift consists of a paleoplain of low rolling hills, in which a few canyons several hundred feet deep have been cut, whose walls have slopes as steep as 2,100 ft./mi. The bottoms of the caves filled with "Sooy" clastics, which have been found in wells drilled through sections that otherwise consist only of Arbuckle beds, are higher than the bottoms of the valleys, but lower than the bottoms of the sinkholes on the paleoplain. It thus appears that underground rivers flowed from the sinkholes through the caves into the steep-walled canyons. The latter are aligned in straight patterns which suggest a fault or fracture origin.

FOSSIL CLIMATES AND WEATHERING

The difference between the karst and sinkhole type of buried landscape and that described from southeastern Saskatchewan is striking. Therefore, a discussion of the climatic conditions prevalent at the time and locality of landscape development is in order. An important consideration is whether this climate was humid or arid. In humid areas, a vegetation cover develops that restricts erosion and reduces the rate of runoff, resulting in gentle, more-or-less-rounded landscape forms. In an arid climate, on the other hand, steep escarpments can develop freely. The Mississippian and older formations of the Mid-Continent area and of western Canada, even those covered by Pennsylvanian rocks, may well have been exposed to several erosion periods, beginning in Late Mississippian or Early Pennsylvanian time, and separated by times of relative standstill or even submergence. Some of this erosion may have been during humid and some during arid conditions. The overlying transgressive sediments may give some clue to the climate just before they were deposited. However, unless the last erosion cycle has destroyed completely or altered the effects of all preceding ones, knowledge of the climatic conditions at the time of the final stage of erosion will not be sufficient for a full understanding of the buried landscape. The great importance of this question in the case of carbonate rocks is reflected in the following quotation from Thornbury (1954, p. 56): "In humid regions limestone is usually considered a 'weak' rock. Areas underlain by limestone are generally lower than surrounding areas. This is the result not so much of the physical weakness of limestone as of its susceptibility to solution. In arid regions, however, where moisture is deficient and solution insignificant, we frequently find that limestone is a 'strong' rock and commonly is a cliff or ridge former."

King (1953), however, has pointed out that landscapes developed before the middle Tertiary, when the carpet grasses started spreading over the world, generally should have been of the semi-arid (scarp and pediment) type. Russell (1956, p. 454) also considered that the low forms of plants that covered the earth until mid-Mesozoic time could have created only incipient soils in the modern sense, with the result that topographic profiles did not resemble closely those known today in humid regions. Nevertheless, it must be assumed that the distinction between chemical and mechanical weathering existed even with little soil cover present. As pointed out by Peltier (*in* Thornbury, 1954, p. 58–65), chemical weathering is most intense under conditions of high temperature and abundance of water. Mechanical weathering, on the other hand, is most effective in a temperate, relatively dry climate. As already stated, different climatic conditions may have alternated during long or even during relatively short periods of erosion. The best method to determine the type or types of climate which prevailed during a certain period of landscape formation would appear to be to study the factual information available and to see which concept or concepts best fit the data obtained.

The length of time during which many eroded landscapes of the past have been exposed to atmospheric influences would lead one to believe that weathered surfaces must be a very common adjunct of buried erosional landforms. Weathered surfaces nevertheless appear to be absent in many places where they could be expected to occur. Their apparent absence may be the result of removal of the weathering products either by flowing water while the land was still exposed, or by wave and current action during subsequent marine transgression. In either situation, the coarser weathering products would be expected to collect in low places on the old surface and to become mixed with sediments of the transgressive environment. If the in-mixed sediments were clay or carbonate mud, too little porosity would be left in the original weathering products for them to be of potential interest as reservoirs for oil and gas. If, on the other hand, the in-mixed sediment were a sand or gravel, the resultant product most probably would become part of the overlying transgressive reservoir in the form of a "basal conglomerate." Walters (1946, p. 695–699) has given a good description of "non-marine" and "marine" conglomerate bodies derived from the Arbuckle Limestone of the Central Kansas uplift.

Possible conditions under which weathering products would not be removed from their original place of deposition are the following.

1. Absence of (sufficient) flowing water at the surface
   (a) Arid climate
   (b) Karst landscape (underground drainage)
2. Retention of weathering products in the presence of flowing water
   (a) By gravity (eluvium)
   (b) By plant cover.

In an arid climate, winds may remove the finer weathering products, leaving behind a coarse residuum or eluvium as in case 2-a. A similar situation probably also characterizes karst landscapes. The water, as it enters the subsurface through sinkholes, carries with it the finer particles and soluble material and leaves a residuum of the larger chunks of weathered material (see description by Walters, 1946, p. 697). Gravity has the same effect even when water is free to remove the weathering products, and the residual eluvium consists of the heaviest pieces originally present in the exposed sediment, such as rock fragments, silica, concretions, conglomerate pebbles, *etc*. Unless such an eluvium becomes completely infilled or cemented by sediments during the subsequent transgression, such material can make a very good reservoir rock. It appears to be important that the transgressing sea should not alter such a deposit by transporting it, breaking it down into smaller particles, mixing it with other sediments, *etc*. Optimum conditions for the preservation of the reservoir properties of this deposit thus would be provided by a relatively rapid transgression, quickly moving the new shoreline over the original eroded landscape, exposing it for not more than a short period of time to the effects of waves and longshore currents, and covering it with quiet-water sediments.

Among the fields that produce oil from weathered sediments which may have originated in this manner are the Arbuckle Limestone reservoirs of the Central Kansas uplift. Solution erosion of the karst landscape, developed on the Arbuckle surface during both Late Ordovician and Early Pennsylvanian times, formed a mantle of leached residuum 2–30 ft. thick. This is the present Arbuckle reservoir rock (Walters, 1946, p. 700–701). Another example of this type of reservoir occurs on parts of the TXL structure of western Texas, from which the upper part of the Devonian was removed by pre-Permian erosion. Low areas on the Devonian erosion surface are filled with a highly porous, weathered chert known as the "Tripolite Zone," which forms an oil reservoir separate from those encountered in the Devonian (David, 1946). The tripolite is a weathering product of the Devonian, which here consists of an upper and a lower chert (novaculite) section separated by a middle limestone member. A similar residual mantle occurs on the Mississippian buried hills of southeastern Saskatchewan (limestone rubble) and those of southwestern Alberta (chert), but in these areas a reservoir generally is not developed; this is the "Detrital" of Bokman (1963, p. 258).

The probability that plant cover played a less important role in pre-middle Tertiary times has been mentioned (King, 1953; Russell, 1956). Ancient soils certainly are not conspicuous in the stratigraphic column. Even if they were, they most probably would not form reservoir rocks for oil and gas. In contrast to the coarser eluvium and residuum, most ancient soils, being comparatively fine and generally clayey, can be expected to have been severely altered or even removed by subsequent transgressions, to the point of becoming simply a constituent of the transgressive sediments. Only those parts of the soil which occupied rock crevices or were carried down by burrowing animals and plant roots would be preserved.

Weathering also alters the underlying rocks that remain in place. Mechanical weathering may induce fractures. Chemical weathering, especially of carbonate rocks, may enlarge such fractures by solution and form karst landscapes, sinkholes, caverns, and other phenomena related to underground drainage. In the buried Mississippian hills of southeastern Saskatchewan, the porosity of the reservoir rocks, where exposed along an escarpment, probably at first was increased by such chemical processes. However, as the landscape subsided and the "Red Beds" sea encroached on it, anhydrite filled these enlarged pore spaces and led to the formation of a dense "caprock." This caprock appears to be present in all buried hills of this area, and usually extends about ¼ mi. inward from the subcrop of the reservoir bed.

SAND DEPOSITION IN BURIED VALLEYS

The study of buried landscapes is of importance not only to the exploration for oil and gas in buried hills, but also to the search for hydrocarbons trapped in sandstone beds or other clastic reservoirs that accumulated in the valleys between the hills. In the southeastern Saskatchewan area, where the transgressive "Red Beds" contain practically no sandstone of reservoir quality, there is no economic incentive to search for such sandstone bodies. Farther west along the flank of the Canadian shield, however, a vast paleoplain underlain by Devonian, Mississippian, and Jurassic rocks was covered by Lower Cretaceous (locally Upper Jurassic) sediments which include a high percentage of sandstone which could serve as reservoir, particularly in the basal members, such as the "Basal Quartz," Ellerslie, McMurray, Cantuar, etc. There are indications, also, that the influence of buried topography, through the medium of compaction, is expressed as high in the section as the Upper Cretaceous Cardium Formation. Some of the producing sandstone bodies (e.g., in the Cantuar field, southwestern Saskatchewan) overlie old highs and were derived from subcropping sandstone beds deposited in an environment otherwise characterized by shale; however, most of the sandstone was deposited in valleys of the old paleoplain.

The first person to make a paleotopographical study of this pre-Cretaceous erosion surface in eastern Alberta and western Saskatchewan was Beltz (1953), who presented an isopachous map of the Mannville Group. On the same map, Beltz also showed the pre-Mannville paleogeology (Devonian, Mississippian, and Jurassic subcrops) and the occurrence of oil and gas in the basal Cretaceous. This map was the first to show the obvious relations between these various factors; the more erosion-resistant subcropping formations formed the topographical ridges, whereas the basal Cretaceous hydrocarbon accumulations were concentrated in the sandstone-filled valleys.

Two parts of this paleoplain have been mapped by the writer, using two different methods. Figure 20 shows the northern part of the Province of Alberta; contour datum is sea-level, and the present elevation of the eroded surface of the Paleozoic above or below sea-level is shown. This map extends toward the west an earlier map published by Martin and Jamin (1963). This method of illustrating the buried topography was chosen because the overlying basal Cretaceous would not lend itself to the construction of an isopachous map of the "cast" of this landscape, as did the

"Red Beds" for the Mississippian surface of southeastern Saskatchewan. This is because the sandstone and shale of the basal Cretaceous McMurray Formation of northeastern Alberta, which covers unconformably the old land surface of the Devonian, are in turn cut deeply by Pleistocene to Recent erosion in much of the "Bituminous Sands" area along the Athabasca River and its tributaries. In places, this erosion cuts even into the Devonian. Furthermore, some of the highest areas of the paleoplain west of the present Athabasca River were not covered by McMurray sediments (see cross section, Fig. 21). This contour method of illustrating the old land surface is useful where the regional post-Cretaceous westward tilt is relatively small or even reversed toward the east. However,

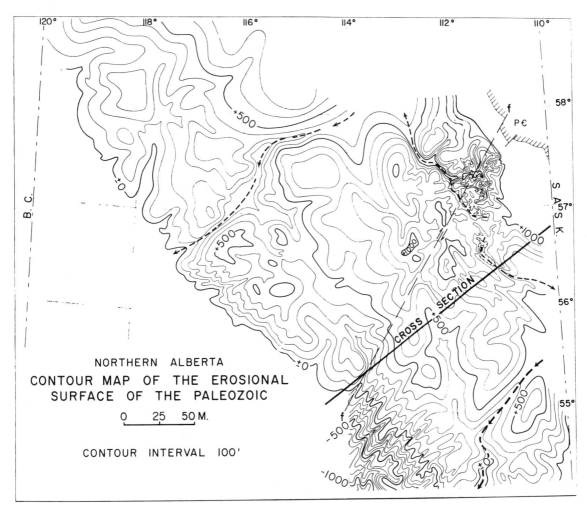

Fig. 20.—Northern Alberta. Contour map of erosional surface on top of Paleozoic. Steepening of dip below zero contour (approximate position of hinge line) caused by subsequent downwarping. Dashed lines with arrows indicate direction of drainage. Athabasca "Bituminous Sands" area occupies northeast quarter of map. Line of cross section indicates location of Figure 21. Contours in feet.

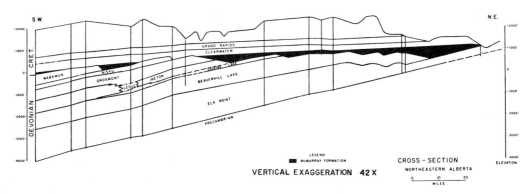

FIG. 21.—Southwest-northeast cross section of Athabasca "Bituminous Sands" area, northeastern Alberta. Note collapse area at right, caused by removal of Elk Point salts by solution. Line of section is indicated on Figure 20.

the drawback is obvious after studying the contours southwest of the basin hinge line, which coincides approximately with the zero (sea-level) elevation of the top of the Paleozoic. Beyond this hinge line, the superimposed steepening of the Cretaceous toward the southwest obliterates most of the original shape of the Paleozoic landscape and thus gives a distorted picture of the paleotopography.

Figure 22 shows the adjoining area of west-central Saskatchewan, by means of an isopachous map, showing the thickness of the "cast," as in southeastern Saskatchewan. The formation used

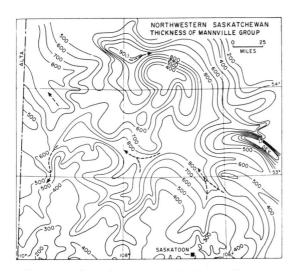

FIG. 22.—Isopachous map of Lower Cretaceous Mannville Group, west-central Saskatchewan, showing post-Paleozoic paleotopography. This figure also illustrates excess thickness of Mannville in center of area (300–400 ft. more than thickness in main consequent valley, at left), which was caused by Devonian salt removal during deposition of Mannville. Contours in feet.

for isopaching is the Lower Cretaceous Mannville Group, which also was used by Beltz (1953). The problem in this instance is that the Mannville was subjected to structural movements during deposition. As a result, the deepest part of the two large subsequent (southeast-northwest-trending) valleys subsided because of salt removal from the underlying Devonian. Selection of a thinner stratigraphic interval (e.g., the basal Lower Cretaceous) might perhaps have given better results. Figure 21 illustrates the thinning that has taken place in the salt-bearing Elk Point Formation near its outcrop; where solution has taken place, the post-Devonian strata have subsided. Nevertheless, the basal Cretaceous sandstone units were deposited by flowing rivers and not in stagnant lakes, as shown by their coarse texture, cross-bedding, and other depositional features. The steeper slopes noted locally on the Devonian paleoplain in the Athabasca "Bituminous Sands" area (360 ft./mi., compared with 233 ft./mi. in southeast Saskatchewan) also appear to be related to later subsidence resulting from salt removal (northeast corner of Fig. 20).

In an earlier paper (Martin and Jamin, 1963), the writer determined the stream-length ratio for the Athabasca "Bituminous Sands" area to be 5.5, which compares well with the ratio of 4.9 established for the Mississippian buried landscape of southeastern Saskatchewan. Applying the principles of quantitative geomorphology, the possible locations of the major consequent valleys that drained this area were calculated. Subsequent work has proved that these valleys are more or less where predicted (Fig. 20). The occurrence in these valleys of bituminous sandstone beds corresponding in age to the McMurray Formation is considered further proof that a properly connected drainage pattern existed during Early Cretaceous time.

Both maps (Figs. 20, 22) show the existence of several northwest-southeast-trending cuestas, which correspond to erosion-resistant layers in the Devonian-Mississippian sequence. Details of the identity of the Devonian cuestas were given by Martin and Jamin (1963), and the cuestas are recognized easily on a cross section (Fig. 21). Extensive gas fields are being developed in some of the buried hills which form these cuestas.

The subsidence which has affected the area east of the Beaverhill Lake subcrop during the Early Cretaceous, especially in west-central Saskatchewan (Fig. 22), makes it difficult to determine from the map alone which was the major consequent valley that drained this area. A massive sandstone section in the basal Cretaceous of the Bellshill Lake oil field and other similar fields

in east-central Alberta can be traced upstream into a broad valley filled with these sediments. This valley crosses the Alberta border into Saskatchewan east of Bellshill Lake and is shown in the southwest corner of Figure 22. This is evidently a third major consequent valley draining this northern area; the three valleys are shown together on Figure 23. On the same map, a major valley is shown in the southwestern corner of Manitoba, which lies outside of the southeastern Saskatchewan area already described. The valley spacing derived from the quantitative geomorphological considerations stated earlier suggests the possible existence of a fifth large consequent stream, the approximate position of which has been indicated by a broken line. Other broken lines illustrate the writer's belief that the origin of some of the largest lakes on the Canadian shield perhaps may be traced to this pre-Cretaceous drainage system. Of special interest is the

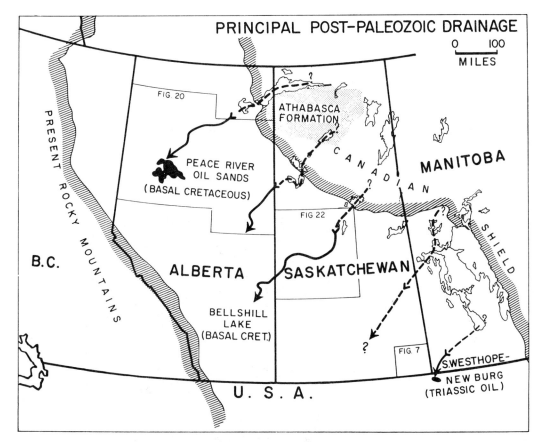

FIG. 23.—Principal post-Paleozoic drainage from Canadian shield. Three western valleys are filled with Cretaceous clastics, the easternmost one with Triassic and younger sediments. Second stream from the east (broken line) is hypothetical. Note outlines of areas covered by Figures 7, 20, and 22. Bend in river above words "Bellshill Lake" is shown in detail on Figure 24. Area of Figure 19 is below word "Alberta." Broken lines on exposed shield area indicate positions of possible pre-Cretaceous streams, now occupied by major lakes.

apparent relation between these major rivers and the occurrence of large oil-bearing sandy deltaic and shoreline deposits of Mesozoic age (Fig. 23).

Busch (1959) published an excellent paper in which he deals in part with sands deposited in buried valleys; he calls them *strike valley sands*. Sandstone bodies of this type are abundant outside of the map areas discussed here, and occur at different levels within the Lower Cretaceous of western Canada. Numerous examples of such sandstone bodies also have been described from United States oil and gas fields.

Less well known is the trapping of hydrocarbons in consequent valleys. Figure 24 shows the details of a small part of the third valley mentioned above, located east of the Bellshill Lake field, Alberta. The outline of a 4–6-mi.-wide valley, indicated by a broken line, is based on the thickness of the Mannville Group (Kb), which is less than 500 ft. on the paleoplain into which the valley has been cut. The cross section (Fig. 25)

shows this valley to be filled with a thick section of "Basal Quartz" (Lower Mannville) sandstone beds, which are shaped into several terraces. The terraces (which are not sand bars (Rudolph, 1959) and certainly not "sand reefs," a term used by Bokman (1963)) were formed by the river cutting down into its own sediments, presumably as a result of the subsidence that affected its source area during Mannville time. When the last meandering river course was filled with clay, the possibility was created for hydrocarbons to be trapped against the updip (northeast) side of these meanders. The final position of the river course was interpreted to be that zone where the top of the "Basal Quartz" is more than 400 ft. below the top of the Mannville Group. Using these criteria, it has been possible to trace this valley upstream, as well as downstream, through a large distance. The meanders of the final river course can be reconstructed by combining the known subsurface data with the mathematical

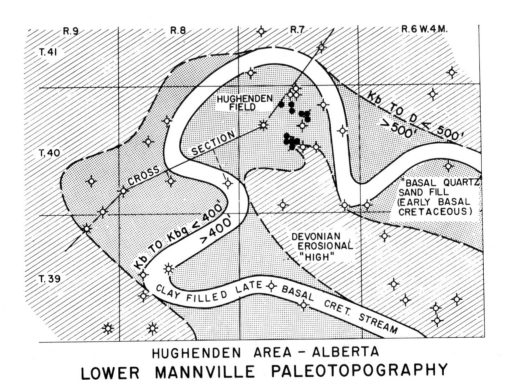

HUGHENDEN AREA - ALBERTA
LOWER MANNVILLE PALEOTOPOGRAPHY

0   5 M.

FIG. 24.—Lower Cretaceous Mannville Group paleotopography, Hughenden area, Alberta. Meanders of a late basal Cretaceous stream have cut into "Basal Quartz" sandstone beds deposited in broad post-Paleozoic valley. Regional dip is toward southwest. Oil is trapped against a clay-filled meander where it intersects regional strike; gas wells shown correspond to sandstone bodies above "Basal Quartz." Line of cross section indicates location of **Figure 25.**

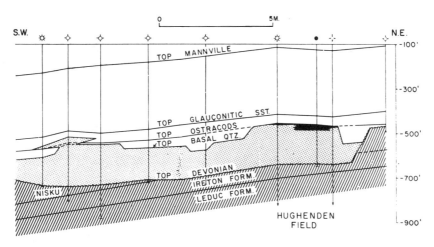

Fig. 25.—Structural cross section, Hughenden area, Alberta, showing incision of late basal Cretaceous stream (now filled with clay) into older "Basal Quartz" sandstone reservoir, creating terraces at two levels. Oil is trapped along updip margin of higher terrace. Line of section is indicated on Figure 24.

formula, derived from quantitative geomorphology, to relate the average dimensions of river meanders.

SUBMARINE CANYONS

Submarine, as opposed to subaerial, erosional topography should be mentioned briefly. Buried submarine canyons occur exclusively in marine sediments originally deposited close to the edge of the continental shelf. Such canyons commonly are several thousand feet deep and many miles wide, and usually are filled with shale. Where a regional tilt is present (*e.g.*, in the area along the Gulf of Mexico), hydrocarbon accumulations may form against the side of the canyon that is regionally downdip (*e.g.*, Yoakum field, Texas; Hoyt, 1959). Other submarine canyons have been described from Louisiana (Bornhauser, 1948) and from California (Frick *et al.*, 1959; B. D. Martin, 1963); the submarine canyons described in the latter two references are filled partly with sandstone which locally is hydrocarbon-bearing. All examples mentioned are Tertiary in age; none is known from the Mid-Continent area or western Canada. The mechanism causing the formation of these channels is very different from that controlling subaerial denudation, and the geomorphological rules governing their occurrence have yet to be formulated in detail.

MAPPING TECHNIQUES

The first step which the geologist usually takes in studying these old land surfaces is to construct a contour map of the present-day attitude of that surface (*e.g.*, the top of the Mississippian erosion surface in southeastern Saskatchewan). In so doing, the geologist recognizes that this unconformity surface has been tilted and possibly affected by other structural influences since its formation. This first map is needed, however, to study factors, such as fluid contacts, that may be related to present-day elevations. Such a map may be all that is required for the study of landscapes that have been disturbed slightly, or not at all, since their formation (*i.e.*, those buried or partly covered by younger sediments whose present structural attitude is still horizontal or nearly so; Fig. 20). Andresen (1962) has called this the *unconformity contour method*.

The second step, or method, involves isopachous mapping. An impression of the morphology of the landscape at the time of burial can be obtained from an isopachous map of the transgressive formation overlying the unconformity. Such a map provides a "cast" of the topography (Figs. 7, 18, 22). Depending on the thickness and lithologic character of the interval selected, such isopachous maps will be influenced to some degree by compaction effects. A thick interval may be preferred, for this reason, as long as it is reasonably certain that the interval represents a time of continuous sedimentation, unaffected by later tilting, folding, faulting, or subsidence during deposition, or by a significant amount of post-depositional erosion. On the other hand, if the stratigraphic interval chosen for isopachous mapping is of relatively uniform lithologic character throughout the area studied, the effects of compaction may be disregarded as long as one is primarily interested in relationships (thick or thin isopachous interval, representing low or high topography) rather than in absolute figures. Andresen (1962) has termed this the *datum plane—valley floor isopach method*. The method used to determine the compaction factor was discussed earlier in this paper (Fig. 10). The effects of synsedimentary movements on the validity of such a "cast" isopachous map also have been discussed. Such effects can be minimized by making an isopachous map only of the formation or member just above the unconformity, provided the upper limit of the isopachous interval is known with reasonable certainty to be a time line.

A third step, or method, which the writer has employed with some degree of success to analyze the Mississippian paleotopography of parts of southwestern Alberta (Fig. 19), is to map the thickness of the formation or member just below the erosion surface. This is meaningful only if the strata concerned were subjected to little or no tilting, or other disturbance, during the period of uplift that preceded the formation of the landscape to be studied. The method was used by Siever (1951, Fig. 5), and constitutes a refinement of the *paleogeologic map method* used by the same author (1951, Fig. 4) and described further by Andresen (1962).

Where tilted formations are present, or the strata were otherwise affected by structural movements, the nature of such deformation is important to determine insofar as the deformation preceded or accompanied the formation of the buried landscape to be studied. A map of the present structural attitude of a marker below the unconformity includes the effects of both pre- and post-unconformity movements. The latter may be eliminated by introducing a fourth step. This consists of mapping the thickness of the interval from the pre-unconformity marker to the unconformity and adding to this thickness that of the post-unconformity "cast," restored to its value before compaction. Siever (1951, Fig. 7) used a similar method, but made no allowance for compaction of the Pennsylvanian. For predominantly sandy beds, like the basal Cretaceous of western Canada, the compaction effect is bound to be much less than for clayey formations such as the "Red Beds" of southeastern Saskatchewan.

A restored pre-unconformity structure map of this type allows one to determine the original spatial position of important reservoir beds as well as of the erosion-resistant, ridge-forming layers, and leads logically to a fifth step. The stratigraphic levels at which the ridge-forming layers occur may be determined from a statistical analysis as outlined in a preceding section (Fig. 14). By means of a summit-level map (Fig. 11), the most likely positions of cuestas and other strike ridges may then be sketched on the restored pre-unconformity structure map. Combining this further with a drainage-pattern map, based on known subsurface data and on the quantitative geomorphological parameters that have been determined for the area of study, a reasonably reliable paleogeomorphic map may be constructed, showing the possible location of unexplored buried hills and valleys in addition to those already known. These prospects may then be defined further by photogeologic and (or) seismic methods of investigation in areas where such methods have proved to be useful in locating buried erosion topography.

Conclusions

Paleogeomorphology is a useful tool for solving problems connected with the exploration for and development of many hydrocarbon accumulations previously classified as "stratigraphic," in particular those with a marked three-dimensional geometry. Included, in particular, are fossil organic reefs and buried landscapes. Barrier beaches, point bars, submarine canyons, and other predominantly three-dimensional phenomena also are considered to form paleogeomorphological rather than stratigraphic traps, because they lack obvious continuity with the laterally adjoining sediments. Deposits resulting from such factors are not shown easily on simple facies maps and do not lend themselves readily to analysis by means of "layer maps" and similar two-dimensional approaches. A combination of isopachous maps, cross sections, panel diagrams, *etc.* is needed for proper understanding, and the analysis itself must be basically of a geomorphological nature. There is thus a wide scope for paleogeomorphological studies with a correspondingly great variety of applications to petroleum geology.

With regard to buried erosional landforms, it is

important to note that buried landscapes may be analyzed according to quantitative geomorphological principles, making it possible to extrapolate a drainage system and its interfluves (buried hills) from a known, thoroughly drilled area. The geometric pattern of a drainage system is determined further by geological factors such as structure, faulting (and fracturing), and the alternation of erosion-resistant and non-resistant beds in homoclinally dipping areas. Hydrophysical factors to be taken into account in outlining paleotopographical prospects are: the presence and position of a summit level (derived from the elevations of buried hills previously discovered in the area; these elevations are approximately those at which the erosion-resistant levels will tend to form escarpments); the angle of entry of tributary valleys; the alternation of obsequent and resequent interfluve spurs; and others.

Sandstone deposited in the valleys of a buried landscape also contains commercial deposits of oil and gas in many areas. The analysis of the corresponding landscapes must proceed along lines of reasoning similar to that used for buried hills. However, additional attention must be given to such factors as the geometry of river meanders, the formation of point bars and terraces, the influence of valley width on sedimentation, *etc.*

Weathered carbonate surfaces may form good reservoir rocks for hydrocarbons. On the other hand, the same surfaces, where filled with anhydrite or calcite, form a tight caprock over the underlying reservoir. Leaching and consequent widening of fractures increase reservoir volume, particularly in otherwise dense carbonate rocks. Fracture distribution, however, is a problem of structural geology. Underground cavities usually result in a loss of reservoir rock where filled with impermeable sediments, but it is most difficult to detect a systematic distribution pattern of such cavity fills.

Submarine canyons locally are sites of accumulation of small pools of oil or gas, but submarine canyon occurrence is sporadic and the laws governing the distribution of such canyons still need to be formulated.

It is not easy to estimate the hydrocarbon reserves ultimately to be recovered from traps formed by buried landforms, for the simple reason that they commonly are complex traps in which, for example, pre-erosion porosity distribution may play as important a role as trapping

below the unconformity. Similarly, the total influence of paleogeomorphological factors in forming hydrocarbon-bearing reservoir rocks above the unconformity may, in some places, be debatable. As a rough approximation, the writer has estimated that about 9 per cent of the ultimately recoverable oil in western Canada occurs in traps related to buried hills and valleys (Martin, 1960). Bokman (1963) gives figures of 9 per cent of oil and 17 per cent of gas reserves for Alberta alone. In southeastern Saskatchewan, this figure is calculated to be as large as 32.5 per cent of oil reserves (Martin, 1964 a, b). For the "free" world as a whole, on the other hand, Knebel and Rodriguez (1956) estimate that "unconformity traps" contain only 6 per cent of all major oil-field accumulations (6.5 per cent, if the Middle East is excluded). These figures indicate to this writer that a very substantial amount of "unconformity" hydrocarbons still are awaiting discovery in areas where exploration for this type of trap hitherto has not received sufficient attention.

For example, the percentage of paleotopographic traps discovered in southeastern Saskatchewan has risen steadily since the discovery of the Alida field in 1955. It is believed that exploration for such traps may benefit substantially in the future from the application of geomorphological principles to the analysis of buried landscapes, as outlined in this paper for some specific areas. New areas that could be analyzed to advantage in this manner include the Trempealeau trend of Ohio and the Mississippian of Kingfisher and neighboring counties in Oklahoma. Other areas should be sought in the broad belts of Paleozoic rocks that have been subjected to one or more periods of erosion on the North American continent. With the increasing use of computers to solve geological problems, the introduction of trend-surface analysis could help to cut down the time necessary for such regional studies.

REFERENCES CITED

Andresen, M. J., 1962, Paleodrainage patterns: their mapping from subsurface data, and their paleogeographic value: Am. Assoc. Petroleum Geologists Bull., v. 46, no. 3 (Mar.), p. 398–405.

Beltz, E. W., 1953, Topography and geology of eastern Alberta and western Saskatchewan during Early Cretaceous time: Alberta Soc. Petroleum Geologists News Bull., v. 1, nos. 1–4 (May), p. 1–3.

Bokman, J., 1963, Post-Mississippian unconformity in Western Canada basin, *in* Backbone of the Americas: Am. Assoc. Petroleum Geologists Mem. 2, p. 252–263.

Bornhauser, M., 1948, Possible ancient submarine canyon in southwestern Louisiana: Am. Assoc. Petroleum Geologists Bull., v. 32, no. 12 (Dec.), p. 2287–2290.

Busch, D. A., 1959, Prospecting for stratigraphic traps: Am. Assoc. Petroleum Geologists Bull., v. 43, no. 12 (Dec.), p. 2829–2843.

David, M., 1946, Devonian (?) producing zone, TXL pool, Ector County, Texas: Am. Assoc. Petroleum Geologists Bull., v. 30, no. 1 (Jan.), p. 118–119.

Davis, W. M., 1884, Geographic classification, illustrated by a study of plains, plateaus and their derivatives: Proc. Am. Assoc. Adv. Science, v. 33. p. 1–5.

———— 1889, Topographic development of the Triassic formation of the Connecticut Valley: Am. Jour. Science, 3d Ser., v. 37, p. 423–434.

———— 1899a, The peneplain: Am. Geologist, v. 23, p. 207–239; also: 1909, Geographical essays, p. 350–380, New York, Dover Publications (repr. 1954), 777 p.

———— 1899b, The drainage of cuestas: Proc. Geol. Assoc., v. 16, pt. 2 (May), p. 75–93.

Edie, R. W., 1958, Mississippian sedimentation and oil fields in southeastern Saskatchewan: Am. Assoc. Petroleum Geologists Bull., v. 42, no. 1 (Jan.), p. 94–126.

Frick, J. D., T. P. Harding, and A. W. Marianos, 1959, Eocene gorge in northern Sacramento Valley (abs.): Am. Assoc. Petroleum Geologists Bull., v. 43, no. 1 (Jan.), p. 255.

Hack, J. T., 1960, Interpretation of erosional topography in humid temperate regions: Am. Jour. Science, v. 258-A, p. 80–97.

Harris, L. D., 1960, Drowned valley topography at beginning of Middle Ordovician deposition in southwest Virginia and northern Tennessee: U. S. Geol. Survey Prof. Paper 400-B, p. 186–189.

Hill, R. T., 1901, Geography and geology of the Black and Grand Prairies, Texas, with detailed description of the Cretaceous formation and special reference to artesian waters: U. S. Geol. Survey, 21st Ann. Rept., pt. 7, p. 362–380.

Horberg, L., 1952, Interrelations of geomorphology, glacial geology, and Pleistocene geology: Jour. Geology, v. 60, no. 2 (Mar.), p. 187–190.

Horton, R. E., 1945, Erosional development of streams and their drainage basins; hydrophysical approach to quantitative morphology: Geol. Soc. America Bull. v. 56, no. 3 (Mar.), p. 275–370.

Howard, A. D., and L. E. Spock, 1940, Classification of landforms: Jour. Geomorphology, v. 3, no. 4 (Dec.), p. 332–345.

Hoyt, W. V., 1959, Erosional channel in the middle Wilcox near Yoakum, Lavaca County, Texas: Trans. Gulf Coast Assoc. Geol. Soc., v. 9, p. 41–50.

Kay, M., 1945, Paleogeographic and palinspastic maps: Am. Assoc. Petroleum Geologists Bull., v. 29, no. 4 (Apr.), p. 426–460.

Khutorov, A. M., 1958, The formation of secondary oil deposits in the Fergana depression: Petr. Geol. (Geologiya Nefti), v. 2, no. 7-B, p. 643–651.

King, L. C., 1951, South African scenery. A textbook of geomorphology: 2d rev. ed., Edinburgh, Oliver and Boyd, 379 p.

———— 1953, Canons of landscape evolution: Geol. Soc. America Bull., v. 64, no. 7 (July), p. 721–752.

Knebel, G. M., and G. Rodriguez E., 1956, Habitat of some oil: Am. Assoc. Petroleum Geologists Bull., v. 40, no. 4 (Apr.), p. 547–561.

LeGrand, H. E., 1960, Metaphor in geomorphic expression: Jour. Geology, v. 60, no. 5 (Sept.), p. 576–579.

Levorsen, A. I., 1954, Geology of petroleum: San Francisco, W. H. Freeman, 703 p.

———— 1960, Paleogeologic maps: San Francisco, W. H. Freeman, 174 p.

Lobeck, A. K., 1939, Geomorphology: New York, McGraw-Hill, 731 p.

Louis, H., 1964, Über Rumpfflächen- und Talbildung in den wechselfeuchten Tropen besonders nach Studien in Tanganyika: Zeitschr. für Geom., N.F., Bd. 8, Sonderheft, p. 43–70.

Martin, B. D., 1963, Rosedale channel—evidence for late Miocene submarine erosion in Great Valley of California: Am. Assoc. Petroleum Geologists Bull., v. 47, no. 3 (Mar.), p. 441–456.

Martin, R., 1960, Principles of paleogeomorphology: Can. Mining and Metall. Bull., v. 53, no. 579 (July), p. 529–538; also: 1960, Oilweek, v. 11, no. 36 (Oct. 22), p. 84–94; also: 1961, Can. Oil and Gas Industry, v. 14, no. 10 (Oct.), p. 28–40.

———— 1964a, The exploration for Mississippian buried hills in the northeastern Williston Basin (abs): Third Int. Williston Basin Symp., p. 285–286.

———— 1964b, Buried hills hold key to new Mississippian pay in Canada: Oil and Gas Jour., v. 62, no. 42 (Oct. 19), p. 158–162 (complete text).

———— and F. G. S. Jamin, 1963, Paleogeomorphology of the buried Devonian landscape in northeastern Alberta, in K. A. Clark Volume, a collection of papers on the Athabaska oil sands: Res. Council Alta., Information Ser. 45, p. 31–42.

McGee, D. A., and H. D. Jenkins, 1946, West Edmond oil field, central Oklahoma: Am. Assoc. Petroleum Geologists Bull., v. 30, no. 11 (Nov.), p. 1797–1829.

McKee, E. M., 1963, Paleogeomorphology, a practical exploration technique: Oil and Gas Jour., v. 61, no. 42 (Oct. 21), p. 140–143.

Merriam, D. F., and W. R. Atkinson, 1956, Simpson filled sinkholes in eastern Kansas: State Geol. Survey Kans. Bull. 119, pt. 2 (Apr.), p. 61–80.

Meyerhoff, H. A., 1940, Migration of erosional surfaces: Ann. Assoc. Am. Geographers, v. 30, no. 4 (Dec.), p. 247–254.

Penck, W., 1924, Die morphologische Analyse. Ein Kapitel der physikalischen Geologie: Geogr. Abh., 2; Reihe, H. 2; Morphological analysis of land forms, a contribution to physical geology (transl. by H. Czeck and K. C. Boswell), 1953: London, Macmillan, 429 p.

Rich, J. L., 1938, Recognition and significance of multiple erosion surfaces: Geol. Soc. America Bull., v. 49, no. 11 (Nov.), p. 1695–1722.

Rudolph, J. C., 1959, Bellshill Lake field, Alberta: Am. Assoc. Petroleum Geologists Bull., v. 43, no. 4 (Apr.), p. 880–889.

Russell, R. J., 1956, Environmental changes through forces independent of man, in Thomas, W. L., Jr. (ed.), Man's role in changing the face of the earth: Chicago, Univ. Chicago Press, p. 453–470.

Siever, R., 1951, The Mississippian-Pennsylvanian unconformity in southern Illinois: Am. Assoc. Petroleum Geologists Bull., v. 35, no. 3 (Mar.), p. 542–581.

Sloss, L. L., 1963, Sequences in the cratonic interior of North America: Geol. Soc. America Bull., v. 74, no. 2 (Feb.), p. 93–113.

Sparks, B. W., 1960, Geomorphology: London, Longmans, 371 p.

Strahler, A. N., 1950, Equilibrium theory of erosional slopes approached by frequency distribution analysis: Am. Jour. Science, v. 248, no. 10, p. 673–696; no. 11, p. 800–814.

———— 1952a, Quantitative geomorphology of erosional landscapes: C. R. XIX Internatl. Geol. Congr., Algiers, fasc. 15, p. 341–354.

———— 1952b, Dynamic basis of geomorphology: Geol. Soc. America Bull., v. 63, no. 9 (Sept.), p. 923–938.

———— 1960, Physical geography: 2d ed., New York,

Wiley & Sons, 534 p.

Thornbury, W. D., 1954, Principles of geomorphology: New York, Wiley & Sons, 618 p.

Vogt, P. R., 1956, Alida field, southeast Saskatchewan: First Williston Basin Symposium, p. 94–100; *also:* 1957, Can. Oil and Gas Industries, v. 10, no. 7 (July), p. 97–101.

Walters, R. F., 1946, Buried pre-Cambrian hills in northeastern Barton County, central Kansas: Am. Assoc. Petroleum Geologists Bull., v. 30, no. 5 (May), p. 660–710.

———— and A. S. Price, 1948, Kraft-Prusa oil field, Barton County, Kansas, *in* Structure of typical American oil fields, v. 3: Am. Assoc. Petroleum Geologists, p. 249–280.

Wheeler, H. E., 1963, Post-Sauk and pre-Absaroka Paleozoic stratigraphic patterns in North America: Am. Assoc. Petroleum Geologists Bull., v. 47, no. 8 (Aug.), p. 1497–1526.

Volume 43   Number 8

BULLETIN

of the

AMERICAN ASSOCIATION OF

PETROLEUM GEOLOGISTS

AUGUST, 1959

SYNCHRONOUS HIGHS: PREFERENTIAL HABITAT OF OIL?[1]

ROBERT SCHOLTEN[2]

University Park, Pennsylvania

ABSTRACT

Synchronous highs are "hills" on the sea floor present during sedimentation. Their origin may be diastrophic, depositional, erosional, or inherited through compaction or draping, or a combination of these. They tend to create conditions favorable to all three stages in the natural history of petroleum —origin, migration, accumulation—and are therefore thought to represent preferential habitats of oil and gas. This thesis is examined in the light of various modern observations and experimental data and exemplified through published descriptions and illustrations of numerous pools. Traps of this type may possess a variety of characteristic structural, topographic, sedimentary, and stratigraphic features which may be found in some cases through field, air-photo, or geophysical methods, but more commonly through careful studies of subsurface samples and logs.

INTRODUCTION: CONDITIONS OF OIL OCCURRENCE

It is generally recognized that the distribution of petroleum and natural gas in the earth's crust is the resultant of a large number of interacting processes operative during the natural history of the hydrocarbons—processes of oil genesis, migration, and accumulation. In the last analysis these processes are, in turn, controlled by the physical nature of the environments in which oil forms, migrates, and accumulates.

This recognition leads to the following logical "ground rules" for oil occurrence.

1. Oil pools accumulate only where a suitably permeable rock (the reservoir) possesses a condition which presents *an obstacle to migrating oil*. Such a condition ("trap") may be structural (anticline, fault, etc.), or textural (the shaling-out of a sandstone, the pinching-out of a reservoir bed), or a combination of these. Superposed may be conditions related to migrating reservoir waters (hydrodynamic conditions).

---

[1] Contribution Number 58-19, Mineral Industries' Experiment Station, The Pennsylvania State University. Manuscript received, September 4, 1958.

[2] Department of Geology, The Pennsylvania State University. The writer is indebted to Professor K. K. Landes of the University of Michigan for helpful critique and suggestions.

2. Such traps are of no use from the point of view of oil accumulation *unless they occur within the path of migrating oil and are present at the time of migration.*

3. The path of migration is determined by structural, textural, and hydrodynamic conditions in the reservoir and by *the location of the oil source environment.*

4. The trap *nearest to the source environment in the path of migration* will be the first to be filled with oil. A trap far away from any source area may be filled if intervening structural or textural obstacles do not block access to it or can spill into it after being completely filled with oil. Gussow (1954, p. 822, Fig. 3) visualizes oil migration as an inverted equivalent of surface water drainage. A logical implication of this view would be that some traps are inaccessible to oil derived from a distant source, as ubiquitous tributary "rivulets" of oil near the source area are in part blocked by local obstacles (traps) and in part combine into a few "mainstreams" which by-pass potential traps in "interstream" areas. Inevitably, therefore, chances for a trap to receive migrating oil would decrease with increasing distance to the source area.

Gussow (1954, pp. 821, 826) nevertheless holds that locally originated oil can never be economic in volume and that only distant traps can become commercial pools. Such an assumption is difficult to prove and moreover rests on the premise that the carrier or reservoir bed in *all* cases possesses continuous permeability along the path of migration—a premise which appears of dubious validity. Furthermore, extrapolation of data from Recent Gulf sediments by Smith (1954, p. 402) indicates these sediments contain volumes of potential crude oil ranging from a conservative 7,000 to a more realistic 16,300 bbls./acre/mile depth. On the basis of the latter figure a pool one mile in radius could contain about 10 million bbls. if it drew all of this oil from overlying, underlying, and surrounding sediments totalling 400 feet before complete compaction and extending only one mile beyond the limits of the pool. Even though part of this oil probably remains in the source sediment, these orders of magnitude nevertheless suggest that local sources may give rise to commercial pools, if the sediments are unusually rich in bituminous matter.

To the writer the only safe and useful conclusion is that the distances between the edges of commercial pools and their source areas depend on structural, textural, and hydrodynamic conditions of the reservoir and on the amount of oil available, and that minimum and maximum distances are therefore likely to vary from one oil district to the next.

PURPOSE OF STUDY

The present discussion is intended to focus attention on one specific environment which appears to create conditions especially favorable to all three stages in the natural history of petroleum—origin, migration, and accumulation. This environment is the "synchronous high," a term here applied to any *local area which was topographically expressed, however gently and for whatever cause, as a high on the sea or lake bottom during the general span of time when sediments were being deposited in the region.* Specifically, we are interested in those highs that were synchronous with the petroliferous sediments.

The basic concept of this oil habitat and the data presented to support it are not original with the writer but are scattered through the literature. Powers (1922, 1926), Athy (1934), and Ferguson and Vernon (1938) were among earlier workers who recognized a close association between petroleum-bearing sediments and sea-floor highs of about the same age in various oil districts without, however, going very far into the genetic relationships between "hills" and oil. Clear statements regarding possible roles such highs may have played in promoting oil pools were offered by several contributors to *Problems of Petroleum Geology* (Adams, 1934; see also Adams, 1930; Clark, 1934; McCoy and Keyte, 1934) and in recent years by Krynine (1948, 1949, 1951) and Weeks (1952). The central idea presented here probably is held today in one form or another by many petroleum geologists. This paper is simply an effort to develop and up-date the theory in its ramifications, to bring the scattered data together into a common framework, and thus to draw the full measure of recognition which the writer believes the concept deserves. In conclusion, ways are explored in which the theory may be made to work in oil finding.

TYPES OF SYNCHRONOUS HIGHS

Synchronous highs as defined in the preceding section may be of various modes of origin. These are listed and illustrated in Figures 1 and 2. In the following paragraphs they are merely described briefly. Subsequent sections are concerned with their theoretical role in encouraging the formation of oil pools, their influence in the genesis of some existing oil pools, and their recognition by surface or subsurface criteria during oil exploration.

For the sake of convenience we may distinguish four principal types of synchronous highs on the basis of their modes of origin, which may be diastrophic, depositional, erosional, or inherited.

SYNCHRONOUS HIGHS OF DIASTROPHIC ORIGIN

Synchronous highs of diastrophic origin are sometimes referred to as "growing structures"—a term indicating that a structural elevation was growing and expressed topographically on the floor of the basin during deposition of the sediments, rather than having formed after all the rocks participating in the structure had been deposited. Examples of growing structures are *early compressional anticlines* raised intermittently in a geosyncline or adjacent cratonic shelf during the long period of sedimentation prior to the final orogenic climax (Fig. 1A).[3]

---

[3] That many folds have such a long history of growth has been recognized by many workers. The importance of progressive folding during sedimentation was stressed as early as 1926 by van der Gracht (in Powers, 1926, p. 427). Recently evidence has been presented by Spieker (1946), Lowry (1957), and Byron Cooper (A.A.P.G. Distinguished Lecture, 1957).

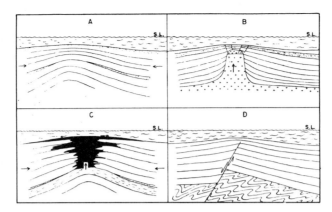

Fig. 1.—Diagram showing four types of synchronous bottom highs of diastrophic origin ("growing structures"). A: Early compressional anticline. B: Salt core structure. C: Mud core structure with intermittent surface extrusion. D: Tilted fault block. S.L. indicates sea-level.

Other examples may be many structures due to or accompanied by flowage of incompetent rock such as *salt anticlines and domes* and *diapiric anticlines with salt or mud cores* (Figs. 1B, 1C), a large number of which show evidence of having risen syngenetic with sedimentation and above the surrounding sea floor. They tend to develop principally in tectonic basin areas of thick sedimentation.

Finally, the growing structure may be due to a vertical uplift such as a *tilted fault block* or *horst* which causes anticlinal arching of overlying strata and of the sea bottom, or which may itself be topographically expressed (Fig. 1D). Growing structures of the latter type are to be expected preferentially in the cratonic environment and adjacent shelf.

SYNCHRONOUS HIGHS OF SEDIMENTARY ORIGIN

Another group of synchronous highs is of depositional origin and may be referred to as sedimentary highs. These, too, are expressed as elevations on the sub-

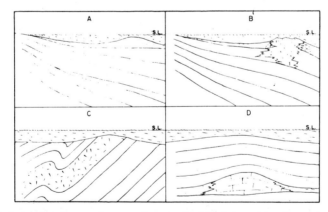

Fig. 2.—Diagram showing synchronous bottom highs of depositional, erosional, and inherited origin. A: Sand bar. B: Bioherm. C: Drowned erosional remnant. D: High due to differential compaction over buried reef. S.L. indicates sea-level.

marine topography. Chief among them are the *sand bar* and the *bioherm* (Figs. 2A, 2B). Both are to be looked for in near-shore environments of the craton or the adjacent basinal shelf.

SYNCHRONOUS HIGHS OF EROSIONAL ORIGIN

Hills on a sea bottom which are simply drowned erosional remnants on a former land surface may be called erosional highs (Fig. 2C). Elevations of this type are to be found behind transgressive shorelines in the cratonic or shelf environments and occur on marked, widespread unconformities.

INHERITED SYNCHRONOUS HIGHS

Inherited bottom highs owe their condition to the arched configuration of a buried stratigraphic surface such as the top of a reef, a sand bar, or an erosional remnant. This arched configuration may be inherited by the bedding surfaces and the top of the burying sediment by differential compaction (Fig. 2D). In some cases, under conditions of rapid submergence inherited highs may form through depositional draping with initial dips of the burying sediment conforming to the submerged topography (Bridge and Dake, 1929; Dake and Bridge, 1932; Landes and Ockerman, 1933; Landes, 1951, p. 226).

ROLE OF SYNCHRONOUS HIGHS IN FORMATION OF OIL POOLS

SYNCHRONOUS HIGHS AND ORIGIN OF OIL

The role of sea-bottom highs in the genesis of hydrocarbon deposits has been emphasized by Adams (1930, 1934), Clark (1934), and other investigators. According to the prevailing concept the chief source organism of oil is marine plankton and in the view of Brongersma-Sanders (1948b, pp. 401, 409) exceptionally abundant production of plankton (hypertrophy) is the main cause of the origin of petroleum. Plankton is dependent for growth on the presence of sunlight, oxygen, and mineral nutrients. The requirement of sunlight restricts abundant planktonic growth to near-surface waters. Oxygen is most abundant in waters constantly disturbed by wave or current action. Nutrients are supplied either from land or by currents welling up from deeper layers where the salts of dead planktonic matter are dissolved (Adams, 1934, p. 361; Brongersma-Sanders, 1948a, 1948b; Yonge, 1951). Prokopovich (1952, p. 882) lists as areas of most abundant planktonic growth the near-shore waters, especially near the mouths of large rivers, and off-shore shoal or bank areas. Both environments are characterized by wave action and both may give rise to upwelling currents. Hence, the localities of most abundant growth of oil-source organisms would be *bottom highs of any of the types described*, which lie at least partly above wave base, not too far from the shore and, in especially favorable conditions, from the mouths of large rivers.

On the other hand it has been repeatedly stated that the same conditions that give rise to abundant planktonic life may also prevent the local accumulation of dead planktonic matter and the preservation of hydrocarbons. Wave and current action inhibit settling of the micro-organisms and promote oxidation of organic matter, and the most abundant source material may, therefore, accumulate not on, but rather in the vicinity of, bottom highs, in localities where the waters are somewhat deeper and more quiet. Here, plankton is allowed to settle to the bottom where the scarcity of oxygen encourages the concentration of hydrocarbons.

Ideal, but probably by no means required, conditions presumably exist if the bottom high is located either in a large "silled" sub-aqueous basin or in the vicinity of a small "silled pit" in a large open basin. The importance of such "silled" conditions on a local or regional scale has been emphasized by Weeks (1952, pp. 2110–12), and Adams (1930, p. 714; 1934, p. 362) who refers to them as "soup-bowls." Silled depressions can be created by the same conditions that caused the bottom high. Thus, both may be due to submarine structural movements, and Clark (1934, p. 313) actually suggests that "structure plays its most important part in producing favorable localities for the accumulation of essential organic remains." Alternatively, the depressions may be lagoons or marshes behind sedimentary highs like reefs and sand bars. Modern examples of such silled basins with sediments high in hydrocarbons include the tectonic basins in the Channel Island area off the California coast (Emery and Rittenberg, 1952), Gulf Coast lagoons such as Laguna Madre, Texas (Smith, 1954, pp. 392–94), and the marshes behind sand barriers along the New Jersey coast (Biederman, 1958). Trask (1927, pp. 1229–30) noted an abnormally high organic content in samples from basins in the Gulf of Maine and the Pacific bottom west of Los Angeles. The sediments in these depressions are plausible sources of future oil pools.

However, it is doubtful that the stagnant bottom of the silled basin provides the only environment in which hydrocarbons can be preserved and allowed to concentrate in quantity. For one thing, as pointed out by Brongersma-Sanders (1948b) anaerobic conditions may be caused simply by hypertrophy, as in areas with upwelling currents, without persistent stagnation. If, in such an environment, the water over the bottom high is not too shallow and disturbed, permitting dead plankton to settle out, potential source beds could accumulate not only in adjacent low areas, but across the crest as well. Furthermore, sedimentation tends to bring organic matter into a reducing environment even if the bottom water were oxidizing, and thus may play an important role in the genesis of oil source deposits (Weeks, 1952, p. 2112; Adams, 1930, p. 714; 1934, p. 362). In shallow cores of recent basin sediments off the California coast taken by Emery and Rittenberg (1952, pp. 773–75, Figs. 5–17) the oxidation-reduction potentials (E_h), though positive at the sediment surface, in most cases decreased with depth and commonly became negative a few feet below the top. In a core from the continental slope the E_h decreased almost to zero. This suggests that, if burial is not too slow, hydrocarbons could probably be protected from oxidation soon enough to permit oil generation not only in well silled, truly euxinic basins, but in less restricted depressions or even the unsilled sea bottom as well.

This brings to mind another way in which a synchronous high may perhaps promote nearby oil generation. Fairbridge (1946, pp. 90–91) has pointed out that quick burial of organic matter can be effected by submarine slumping, thus helping to prevent destruction of hydrocarbons. Such slumping could occur on the submarine slopes of large sea bottom highs. Kuenen (1950, p. 246) refers to studies by Archanguelsky showing that slumping on the Black Sea bottom occurs on slopes as low as one degree. Potter (1957, pp. 2702–08) relates large-scale slide and slump structures in the Eastern Interior Coal Basin, in Pope County, Illinois, to contemporaneous activity along an anticlinal trend. Conceivably, turbidity currents could play a similar role (a large-scale example would be the postulated current off the Grand Banks submarine high in 1929), or even mud flows from submarine mud volcanoes associated with structural-topographic highs (Fig. 1C).

In view of these considerations and of the "ground rules" outlined in the Introduction it appears, then, that *synchronous highs are greatly significant from the point of view of oil occurrence in that they are the most likely to be located in the vicinity of oil-source environments and hence to be the first to receive migrating oil.*

SYNCHRONOUS HIGHS AND ACCUMULATION OF OIL

From the point of view of oil accumulation the several types of synchronous highs described offer a variety of conditions which tend to render them superior as potential traps by comparison with highs formed long after deposition of the source sediments. These conditions include the time of the trap and the nature of the reservoir.

Synchronous highs and time of accumulation.—McCoy and Keyte (1934, pp. 298–301) emphasized the importance of proper *timing* of a trap with regard to the time of migration and accumulation of the hydrocarbons. A trap which forms after oil has ceased to migrate through a given area may remain barren, having "missed the boat." Exactly how soon oil migrates after deposition of the source materials is uncertain. Gussow (1955) has summarized criteria for determining the time of migration. He believes "flush" migration due to compaction does not begin until the source sediment is buried at least 2,000 to 3,000 feet, but it is difficult to prove this. Even if it is true, an overburden of such magnitude may, in areas of rapid and continuous sedimentation (geosynclines) represent no more than several million to less than one million years, though in tectonically more stable regions a much longer period would be necessary. Levorsen (1954, p. 524) presents evidence that oil may migrate less than a million years after deposition and believes the minimum time may be measurable in thousands, or even hundreds of years.

Some observations by Smith (1954) are relevant to this matter. He has demonstrated the presence of liquid hydrocarbons in various modern marine sediments in quantities potentially large enough to form commercial oil pools after migration. This suggests that potentially commercial oil may begin to form very shortly after deposition. Moreover, he observed that sandy layers in a shallow core of Recent Gulf bottom sediment near Grande Isle, Louisiana, contain eight times as much hydrocarbon matter as interbedded clayey layers. In a 2,300-foot core near Pelican Island the average content of non-asphaltic hydrocarbons in sands is about 30 times that in clays according to his figures, in spite of the fact that the clays contain an average of 45 per cent more total organic matter (Smith, 1954, pp. 397–98). This could conceivably be ascribed to either original deposition or subsequent transformation of the organic matter, but it is much more plausible that primary migration of hydrocarbons from clay to sand has taken place at a very early date.

Secondary migration would follow as soon as sufficient dip or hydraulic gradient (due to compaction of clayey beds or to hydrodynamic conditions) is attained to cause oil slugs to move through the inter-pore capillaries of the reservoir sediment. Upon deeper burial the migratory forces and, presumably, the mobility of the oil would tend to increase but, as pores decrease in size, the opposing entry pressures would do likewise and opportunities for migration would not necessarily become better. Thus, one may expect that the time of secondary migration could vary greatly between areas, being late in some (Gussow, 1955), but early in others. Opportunities for early migration and accumulation may well be especially favorable around structural highs with a history of recurrent bottom elevation. For one thing, the sediments in such localities obtain a local dip soon after deposition. Furthermore, differential loading between crest and adjacent low areas causes differential compaction, which, in turn, sets up a hydraulic gradient toward the crest. Goubkin (1934, p. 652) even calls upon such differential load pressures near bottom highs to account for rock flowage and resulting diapiric structures in the Caucasus area.

Obviously, the sooner primary and secondary migration take place in an area, the greater the role of synchronous highs in controlling oil accumulation. They become available as potential traps as soon as the reservoir is sealed by an impervious cap rock, whereas non-synchronous structures may not develop until a long time after migration has ceased. An additional advantage of early migration into a synchronous high is that the presence of the hydrocarbons may prevent subsequent water circulation and thus inhibit later precipitation of cementing material such as may elsewhere make the rocks impermeable to migrating oil and unsuitable as commercial reservoirs (Lowry, 1956, p. 495).

Synchronous highs and reservoir porosity.—One of the most important aspects of the synchronous high is the superior porosity and permeability of the rocks associated with it. Here we may differentiate cogenetic porosity, produced by the same process that created the high, from epigenetic porosity which is caused by virtue of the prior existence of a high.

Obvious examples of the first category include the openings in a porous reef and the intergranular pores in sand bars formed by currents or surf action. Also included are tension fractures in the crests of synchronous highs produced by diastrophic uplift or differential compaction. Although such fractures may occur equally well on late diastrophic uplifts, the early ones have the advantage of being available for early migrating hydrocarbons or leaching waters which may then cause epigenetic porosity improvement.

The second category of conditions making for superior reservoir porosity involves depositional and post-depositional processes over pre-existing bottom highs—epigenetic porosity in the sense defined. In an offshore area of clastic sedimentation the turbulence of water over a sea-bottom elevation, the crest of which lies above effective wave base, inhibits the settling of clay and silt-size particles, while allowing the deposition of coarser grains. This "winnowing" action results in local deposition of a relatively coarsened sediment with improved sorting as compared with deposits in nearby, deeper waters (Fig. 1A). This, in turn, makes for greater porosity of the clastic sediment and thus for a better potential petroleum reservoir on the crest as compared with "dirtier" sands and more silty or shaly deposits laid down in the vicinity or, more generally, in areas where uplifts will not rise until later.

Sandy deposits of this type are called "screened" sands by Weeks (1952, p. 2107). Krynine (1948, 1949, 1951) presents a more extensive description of the phenomenon, the petrographic character of the sands, and the implication to oil occurrence. He applies the term "specific reservoir" to these selectively concentrated sand bodies and compares them to placer deposits.[4] A study of reservoir sand textures in the United States and Trinidad by Griffiths confirms that reservoir sands are "among the best sorted sediments" and can be recognized through quantitative analysis of grain-size distributions (Griffiths, 1952, pp. 214, 226).

In environments of clastic carbonate deposition, the result of such conditions may be a porous, medium to coarsely clastic or oölitic limestone over the bottom high, or an accumulation of highly porous reef debris on a bioherm. Moreover, pre-existing submarine hills of tectonic origin may offer ideal conditions for organic sedimentation by reef growth and thus be indirectly responsible for a reservoir with cogenetic porosity.

If the synchronous high is a growing structure, the process of winnowing may repeat itself with each new diastrophic pulsation, and screened sands may be found at several stratigraphic horizons. Erosional highs influence deposition only in superjacent beds, an excellent example being afforded by the "granite wash" type of sandstone. When an erosional or depositional high is exhumed by differential compaction of overlying sediments and becomes an inherited high, screened sands can again be repeated at higher levels. Because these sands are less compressible than surrounding deposits, the sea-floor high may maintain itself for a considerable length of time (Athy, 1934, p. 822).

[4] A word of caution applies to submarine channel sands, deposited and sorted by current action. Clean sands of this type would occur in bottom depressions instead of over synchronous highs. For example, Ley (1924, pp. 446, 452) states that Lower Pennsylvanian Bartlesville sands in southeastern Kansas filled old synclinal valleys in the submerged topography on Mississippian rocks, so that although pre-Cherokee oil prospects are best in the buried anticlinal "hills," the Cherokee prospects are best in-between.

A common corollary of the greater "sandiness" and improved sorting of clastic sediments over bottom highs is the thickening of time-rock units into low areas. This is the result of the preponderance of silt and clay influx over sand influx in most depositional basins, or of temporary interruptions of sand deposition over the high areas while they are above effective wave base for sand. In the latter case the sediments on the crest may be characterized by diastems, and under certain conditions screened sand lenses may be confined to the flanks, wedging out toward the crest (Fig. 1A), thus providing excellent texture (stratigraphic) traps. Further contributing causes of thinning over the crest of a synchronous high may be penecontemporaneous slumping of sediments off the crest, and postdepositional differential compaction.

The crest of a synchronous high on the sea bottom also offers opportunities for further increase in reservoir porosity by post-depositional diagenetic processes. The formation of pinpoint or vuggy pores in carbonate sediments may accompany penecontemporaneous dolomitization or recrystallization, thought to be favored by relatively shallow-water conditions and slow dissipation of CO_2 from decomposing organic matter such as might be expected over less extensive bottom highs (Weeks, 1952, p. 2123; Twenhofel, 1950, p. 393). At times when the feature is temporarily emergent due to a diastrophic pulsation or a eustatic or intertidal drop in sea-level, the sediment may be subjected to porosity improvement by leaching (Adams, 1930, p. 711; 1934, p. 362). Weathering may also help to increase porosity of the rocks of an erosional high on a land surface. Still another way in which rocks on a pre-existing sea-bottom high may attain superior porosity is by virtue of tension fractures in the semi-consolidated sediments, formed as a result of subaerial drying or submarine radial slippage away from the crest. Finally, these highs not only promote porosity increase, but, by encouraging early migration of oil, may escape later loss of porosity by post-diagenetic cementation.

After a sedimentary bed has been laid down across the basin with superior porosity over elevated parts due to one or more of the foregoing processes, a period is likely to occur sooner or later when the entire sea bottom becomes temporarily submerged below wave base. Silt and clay particles are allowed to settle out over the bottom highs, and a cap rock may be formed. Accumulation may follow not only because of a trap in an updip structural position from the source sediments, but perhaps also because of a favorable capillary gradient from fine pores in the low area to coarser pores in the high, which may help to propel oil toward the trap.

The overlapping sediments themselves may, under the proper conditions, become source beds of oil, and all that is needed in such a case is downward expulsion, perhaps with slight lateral migration through the reservoir. Landes (pers. comm.) also pointed out to the writer that for every sea-floor high of adequate size to become a commercial oil pool there are likely to be many smaller highs. If reservoir permeabilities are sufficiently continuous these could become transient traps which upon later spilling could contribute oil into nearby larger highs. Such "secondary enrichment" may help to form truly large pools in synchronous highs. Furthermore, it may also occur that several potentially commercial highs are distributed across a synchronous uplift of regional extent. In such a case the best prospects may be those traps close to the margin of the uplift or on its flank, for if there is insufficient oil those near the center remain inaccessible to oil migrating from depressions adjacent to the uplift. In the writer's experience, conditions of this type prevail in the Cretaceous traps of the Denver-Julesburg basin.

It may be that later deformation causes a shift in the crest of the high, in which case the superior reservoir conditions are no longer located at the crest of the structure but somewhere along its flank. In such cases the oil pool may remain in its original place in the rock, thus becoming a flank pool held in place by the capillary entry pressures in the finer pores higher in the structure. The same effect occurs if cementation around the edges of a pool "freezes" the oil in place prior to re-deformation.[5] If, on the other hand, the forces of migration (buoyancy or hydraulic gradient) are great enough to overcome the entry pressures, remigration may occur to create a new pool in the crest. A third possibility is that only part of the oil leaks through the permeability barrier to the crest, until the original pool is reduced in size to a point where its pressure by buoyancy or hydraulic gradient becomes too small to overcome the entry pressures in the finer pores.[6] In such a case there would be both a flank and a crest pool, one texturally and the other structurally controlled. In any event, it is clear that, to quote Weeks (1952, p. 2103), "one must go back to the original deposition basin to understand oil occurrence."

In summary, synchronous highs rate highly in terms of the "ground rules" here outlined insofar as they apply to migration and accumulation of oil. *Of all highs they are the most likely to be filled with oil because they are available for early migrating oil (and themselves tend to encourage oil to migrate early), and because they promote local development of textures which offer superior reservoir conditions as well as a capillary gradient tending to propel oil in their direction.*

EXAMPLES OF OIL FIELDS ASSOCIATED WITH SYNCHRONOUS HIGHS

Published descriptions of many of the world's oil fields reveal conditions indicating the existence of synchronous highs during the general time-span when oil-bearing sediments were laid down. Some of the best and most obvious examples are cited in the following paragraphs. A systematic search of the petroleum literature would undoubtedly reveal many more possessing such conditions. For still others the evidence may be buried in company files, or detailed studies may be lacking to reveal what could be rather subtle evidence.

---

[5] Landes (pers. comm.) states that such "freezing in place" is more common than realized and adds that K. C. Heald once cited many examples of this phenomenon, which apparently has not been adequately described in the literature.

[6] This would be similar to the concept by Hill (1956) that cap rocks may act as sieves or seals, depending in part on the size of the pool.

379

POOLS ASSOCIATED WITH GROWING STRUCTURES

Examples of commercial oil pools with a history of diastrophic uplift during deposition of the oil-bearing sediments abound in the literature. Common lines of evidence include interval thinning over domal crests, improved sandiness and other porosity factors on or near the crests, increased throw of faults with depth, and slumped deposits on the flanks of domes. Van Tuyl and Parker (1941, pp. 161–62) state: "In several petroliferous areas where barren domes and anticlines occur, a greater thickness of certain formations has been observed over these sterile folds than over adjacent productive folds of similar magnitude, thus implying that the latter experienced certain early positive impulses which the others did not. It is generally agreed among students of Mid-Continent geology that the structures of more complex history are not only more often productive but when oil-bearing contain more oil than those of relatively simple origin."

U. S. Mid-Continent region.—The U. S. Mid-Continent region offers a large number of excellent examples, as has been frequently pointed out in the literature. For example, Athy (1934, p. 822) observes that "the oil-producing structures of the Mid-Continent area are known to be associated with lines of weakness in basement rocks," with evidence of recurrent uplift in Devonian (?), Late Mississippian, Pennsylvanian, and Permian time.

The oil-rich Oklahoma City uplift is a case in point. Jacobsen (1949, pp. 715–19) states that the thinning of the Viola limestone, the Sylvan shale, the Hunton group, and various Carboniferous units toward the uplift shows doming started as early as Middle Ordovician and proceeded with continuous readjustments into Pennsylvanian time, when sharp uplift and faulting occurred. Earlier, McGee and Jenkins (1946, p. 1803) had reached a similar conclusion; their cross section (Fig. 3) illustrates the relations. The main reservoirs in the fields on this uplift are Ordovician and Siluro-Devonian in age.

Also in Oklahoma, the net thickness and total number of sand bodies in the Confederate and Devils Kitchen zones (Pennsylvanian) of Carter County are related to an anticlinal trend (Sloss, 1955, pp. 16–18, Figs. 7–9), as shown in Figure 4. As early as 1922 Monnett (1922, p. 199, Fig. 21) had noted a general increase in the percentage of sandstone toward higher parts of the structure in the Garber field, Oklahoma (Fig. 5) and attributed the structure in part to lesser compaction of the sand.

In Kansas, McCoy and Keyte (1934, pp. 298–301, Figs. 2–4) show that the Voshell and Valley Center oil-field structures were active in Late or post-Mississippian time (Fig. 6). Similar conditions are stated to be prominent throughout Kansas and Oklahoma. The Beaumont dome in Kansas, on the other hand, does not show prominent evidence of early growth and is dry. The authors believe this dome rose too late to catch the oil migrating through the "Wilcox" sandstone, which is the productive unit in the early structures.

Other cases have been described in northern Texas. Klinger (1941, pp. 561–62) states that the reservoir sands are thickest on the high parts of producing struc-

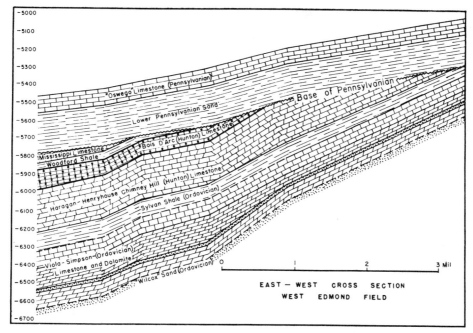

Fig. 3.—Cross section through West Edmond field, Oklahoma, showing thinning of various Paleozoic units and increased stratigraphic hiatus of unconformities toward Oklahoma City uplift on right. After McGee and Jenkins, 1946, p. 1803.

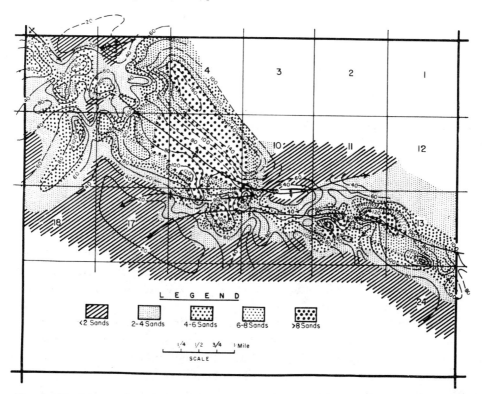

Fig. 4.—Facies map showing net sand isopachs and number of sands in Devil's Kitchen zone, Carter County, Oklahoma. Production is from large lenses in Secs. 5 and 6, Sec. 14 (W.½), and Secs. 15 (NW.¼) and 16 (NE.¼), and from small lens in Secs. 13 and 24. Structural closure exists only over part of first lens. Redrawn after Sloss, 1955, Fig. 9, with main anticlinal axes added by present writer as inferred from Sloss' structure-contour map, Fig. 7.

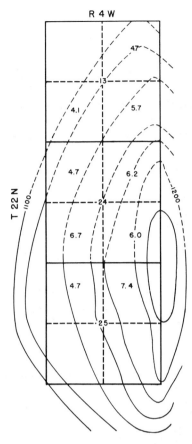

Fig. 5.—Structure-contour map of part of Garber field, Oklahoma. Numerical values in quarter sections show average percentages of sand, increasing toward crest. After Monnett, 1922, Fig. 21.

tures in the Cross Cut-Blake district, Texas, and that in most synclinal areas the sands are thin or absent, or shaly. Similar relations are described by Thompson and Hubbard (1929, p. 432) for the oil fields of Archer County, Texas (Fig. 7). In the Cross Cut-Blake district some isolated oil and gas pools occur in localities which convergence and superior sand development show to have been individual highs during part of Pennsylvanian time, but which today lie on the flanks of structural domes, lacking closure. These relations indicate that the domal crests were in some places shifted by later movements, but the oil was retained in the earlier synchronous crest.

U. S. Eastern Interior region.—Growing structures appear to have had a distinct influence on oil or gas accumulation in some pools in the Michigan and Illinois basins. As early as 1934 Max Ball had suggested that the lenses of the Michigan stray sand "might be offshore bars caught on anticlinal shoals in a subsea topography so young that it still reflected structural deformation"; Ball *et al.* add that "these topographic highs were partly structural and partly erosional,

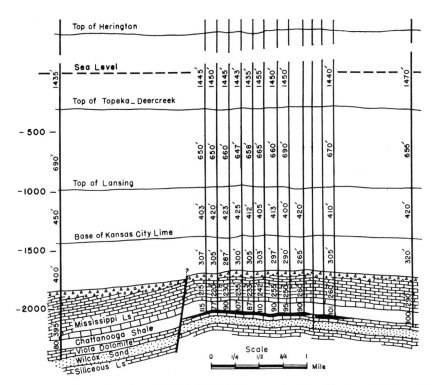

FIG. 6.—Cross section through Voshell field, Kansas, showing thinning of units and unconformable overlap across uplift. After McCoy and Keyte, 1934, Fig. 2.

the relative influence of structure and erosion doubtless varying from place to place" (Ball *et al.*, 1941, pp. 238–42).

In the western Kentucky part of the Illinois basin, Jacobsen (1953, p. 1532) finds that oil in the Chester sandstone is closely associated with local areas of anomalously large amounts of sandstone which coincide with structural uplifts. In the Indiana part of the same basin, Wigginton (1958) describes four pools in the Odon East oil field, Daviess County, which occur on structural noses or with slight offset on structural crests (Fig. 8). The two largest pools are on a "paleo-anticlinal ridge," revealed by pronounced thinning of the Meramecian and

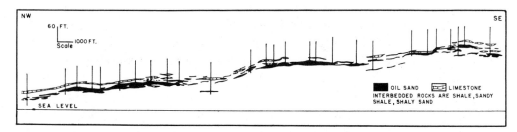

FIG. 7.—Cross section of the producing zone in Litchfield, Ragle, and Carey pools, Archer and Young counties, Texas, showing relation of oil sand development to structure. After Thompson and Hubbard, 1929, Fig. 6.

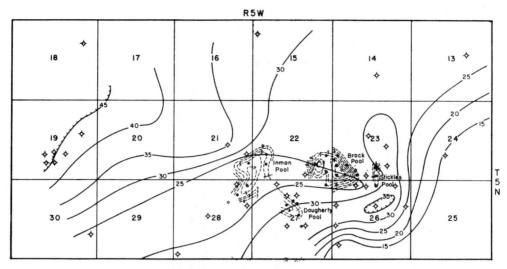

Fig. 8.—Isopach map of several Mississippian units in Odon East area, Indiana. Drawn-out contours represent combined thickness of basal St. Louis and upper Salem limestone members, revealing "paleo-anticlinal ridge." Dashed contours represent thickness of upper reservoir zone in underlying Salem member. Two wells in Daugherty pool produce from a lower Salem member. Brock pool lies at crest, others on flanks and noses of present structural domes (not shown). After Wigginton, 1958, Pl. 5 and Fig. 11.

Chesterian rocks overlying the reservoir. The reservoir in these pools consists of lenses of well sorted, medium- to coarse-grained bioclastic calcarenite in the upper Salem limestone (Mississippian), interfingering with "dirtier" and finer-grained calcareous sands north and south of the shoal ridge. Oil accumulation in this area seems clearly related to "old" structural features.

Appalachian region.—According to Finn (1949, p. 314) there are some areas in New York and Pennsylvania, notably in the Woodhull, Tuscarora, and Tioga County pools, where a relation exists between structure and porosity in the Oriskany sandstone (Devonian), the gas reservoir in this region. "The sand is generally porous only on the high parts of the anticline, and much of the pore space is filled with gas, with only a narrow rim of salt water surrounding it downdip. Farther down the flanks the sand appears to be non-porous and devoid of both gas and water." Finn ascribes this to early movement and resulting improved sorting along the anticlinal trends. The Bradford pool in western Pennsylvania and New York (Fig. 9) may be another example. The distribution of the Bradford Third sand and of the oil in it roughly follows two merging anticlinal crests but extends well down their southwesterly plunge and into intervening and adjacent synclines (Fettke, 1938, Pl. E).

A case of porosity caused by carbonate solution on an early high is mentioned by Krynine (1948, p. 33), who states that the Cambrian Gatesburg formation was leached by Middle Ordovician weathering on an uplift in central Pennsylvania, leaving a porous sandy zone "analogous in all respects to the Wilcox sand

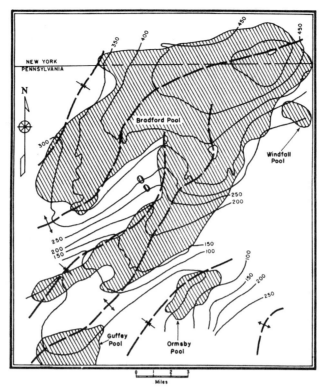

Structure Contours on Top of Bradford Sand

FIG. 9.—Map of Bradford field, Pennsylvania and New York, after Fettke, 1938, Pl. E. Distribution of oil sands (indicated by cross-hatching) is roughly related to two merging anticlinal axes. Redrawn from Landes, 1951, Fig. 89.

of the Oklahoma City field." However, no commercial quantities of oil or gas have yet been found in the Gatesburg.

Rocky Mountain region.—An example in the Rocky Mountain region is the Bisti oil field, New Mexico, where Tomkins (1957, p. 915) describes a thinning of the interval from the base of the Greenhorn limestone to the top of the Gallup sandstone (Fig. 10), suggesting the existence of a topographic bottom high at the location of the structural crest prior to Gallup deposition. Accompanying this is a localized increase in sand content in the otherwise silty facies of the Gallup formation across the structural high. The Gallup is the petroleum reservoir. According to Love *et al.* (1945), local thinning of the Morrison-Cloverly and lower Thermopolis intervals in central Wyoming indicates that gentle structures began to grow during, or at the close of, the Late Jurassic or Early Cretaceous in the area of the Lost Soldier and Big Sand Draw fields and in the vicinity of the Alcova and Lost Cabin pools. They suggest that the distribution of oil and gas in the Morrison and Cloverly may be related to these early trends.

A very instructive example is afforded by the South Glenrock oil field in Wyoming (Curry and Curry, 1954, pp. 2153–54, Figs. 4, 6, 12, 30, 32). This field

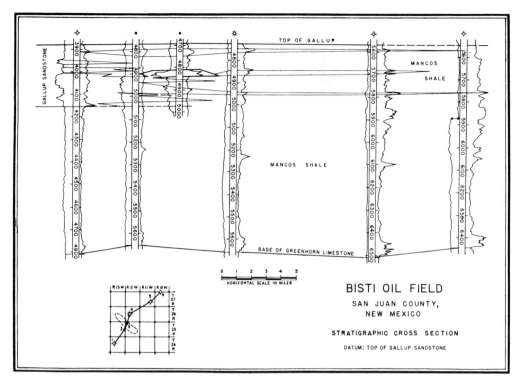

Fig. 10.—Electric-log cross section through Bisti oil field, New Mexico, and surrounding area showing slight thinning of interval from base of Greenhorn to top of Gallup, as well as increased sand development in silty Gallup formation across anticlinal pool. After Tomkins, 1957, p. 215.

lies in an elongate trend of clean Dakota sandstone representing a shoal ridge on the Dakota sea floor and including also the Big Muddy, South Cole Creek, Cole Creek, and Sage Spring Creek fields. Optimum Dakota sand deposition in the South Glenrock area occurred close to the present structural axis (Fig. 11) which evidence shows to have been active at the time. Later Laramide deformation caused the South Glenrock area to become a terrace and structural nose without closure on the flank of the larger Big Muddy anticline at the west, but low-permeability rocks between those two areas prevented the South Glenrock oil from moving into the Big Muddy trap. Hence, Dakota oil distribution at South Glenrock is related to a former synchronous high rather than to any modern structural closure. That lensing of clean sands is not invariably related to bottom highs is illustrated by the clean sands in the Muddy formation in this same area, which also produces oil. These sands occur in a trend winding around the South Glenrock structural nose. They are not the product of deposition on a bottom high but rather in a bottom channel, and they would have been present if there had been no high, though their detailed distribution was influenced by the growing structure. Still another instructive feature is offered by this same field. One of the lines of evidence of its synchronous history is the presence of slumped zones in the Dakota sands on the flanks of the South Glenrock nose (Curry and Curry,

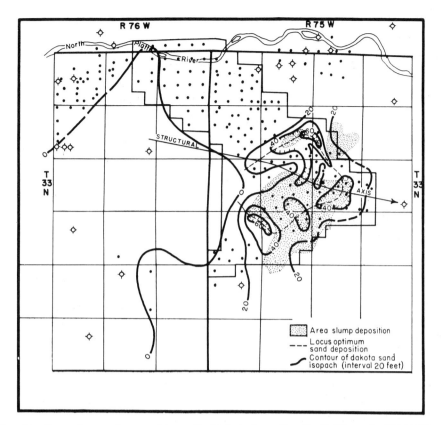

FIG. 11.—Isopach map of permeable sand in Dakota formation, South Glenrock oil field, Wyoming, showing relation of sand development and slump deposits to structure. After Curry and Curry, 1954, Fig. 12.

1954, pp. 2131–32, Figs. 15, 16). This may well have contributed to the quick burial and preservation of hydrocarbons in the intercalated black shales adjacent to the high, in the manner suggested by Fairbridge (1946, pp. 90–91).

In the Canadian Rocky Mountain region the Turner Valley oil field, Alberta, has significant features related to early structural movements. According to Goodman (1945, p. 1166), leaching and accompanying increase in dolomite percentage in the productive Rundle limestone (Mississippian) during pre-Jurassic emergence is responsible for reservoir conditions. Adams (1934, p. 348) states that porosity dies out on the flanks at the point where the dolomite of the pool area grades into limestone; in his opinion this pool, as well as several other carbonate pools in the Rocky Mountain and Mid-Continent areas, was topographically high at the time of burial.

U. S. West Coast region.—The West Coast province offers other striking examples. Hoots *et al.* (1954, Fig. 2) published a cross section through the area of the Midway-Sunset field in the San Joaquin Valley, California, which shows repeated uplift with resulting unconformities over the domal structure during Miocene and Pliocene time (Fig. 12). Production is from Miocene and Pliocene sands. They also

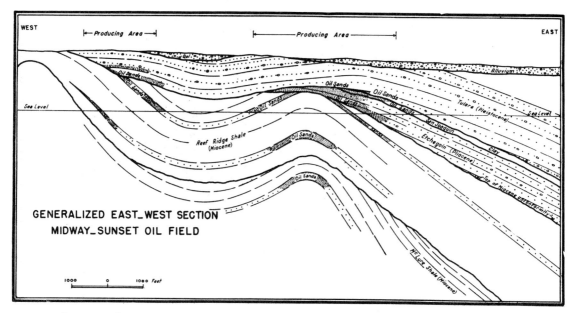

FIG. 12.—Generalized structure section across the North Midway-Sunset oil field, California, showing recurrence of domal uplift by increased stratigraphic hiatus of unconformities over crest. After Hoots *et al.*, 1954, Fig. 2.

state (p. 29): "The oil-producing Etchegoin sands of Buena Vista Front and the eastern part of Elk Hills occur in a southwestward thinning section of beds that were deposited along the northeast flanks of anticlines as these folds were growing from the floor of the Etchegoin sea."

In the Santa Maria basin, California, structures of several oil fields likewise show evidence of growth during Miocene or Pliocene time by thinning of the sections of that age, which also contain the oil. The Orcutt (Fig. 13) and West Cat

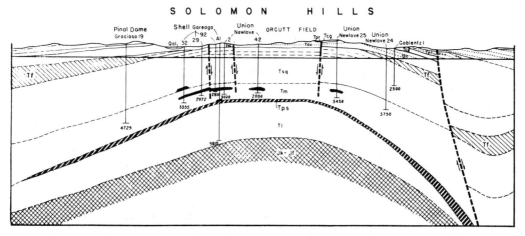

FIG. 13.—Structure section of Orcutt field, Santa Maria basin, California, showing thinning of Miocene shale reservoir (T_m) and overlying units across anticline. After Woodring and Bramlette, 1950, Pl. 2.

Canyon fields illustrate this (Woodring and Bramlette, 1950, pp. 117–27, Pl. 2).

West Texas Permian basin.—An excellent example of a "growing structure" and its bearing on the origin and accumulation of oil is the Yates pool, Pecos County, Texas, discussed by Adams (1930, pp. 706–07, 711–15). Interesting features are universal convergence of formations above the Permian Big Lime (the main reservoir) across the pool, cavernous reservoir porosity restricted to the higher parts of the dome with several "pay zones" separated by fairly impervious limestones, and highly organic sediments in the adjacent Crockett County "soup-bowl" basin, said to have been 1,200–1,800 feet below the top of the Yates high at the time of deposition. The evidence indicates recurrent folding, repeated leaching of the Big Lime during periods of local emergence, and accumulation of source materials in a nearby depression.

Gulf Coast region.—A history of synchronous uplift is especially common for diapiric structures of various kinds, including the salt domes in the Gulf Coast region. Thinning or disappearance of sedimentary units is a characteristic phenomenon over many of the domes (Levorsen, 1954, pp. 270–71, Figs. 7–19 to 7–22; Currie, 1956, p. 8). Figure 14 taken from Reedy (1949, Fig. 8) illustrates this point. Furthermore, the common increase in the throw of associated faults

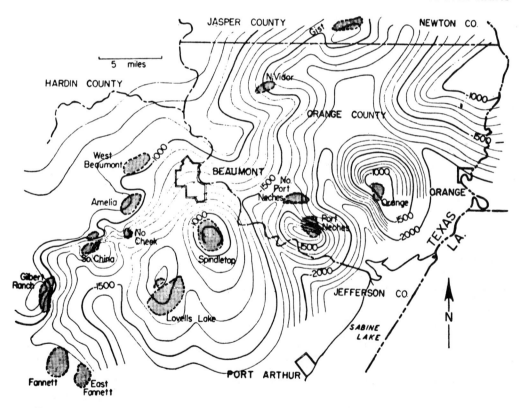

FIG. 14.—Isopach map of upper and middle Frio formation, Oligocene, in Beaumont area, Texas, showing prominent thinning over or near crests of productive salt domes (shaded). After Reedy, 1949, Fig. 8, as reproduced in *Geology of Petroleum* by A. I. Levorsen, W. H. Freeman and Company, San Francisco, 1954.

with depth (Wallace, 1944, p. 1302) is additional evidence of structural movement during sedimentation. The common association between salt domes and oil, not only in the Gulf Coast but in many other parts of the world, may well exist for reasons more fundamentally genetic than the simple fact that the domes provide excellent traps.

An outstanding illustration from outside the Gulf area but showing all of the features characteristic of Gulf Coast domes is shown in Figure 15, a cross section through the Reitbrook field, Germany. Good examples in the Gulf area itself in-

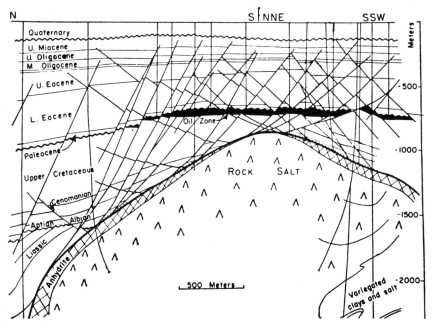

Fig. 15.—Cross section through Reitbrook salt-dome oil field, Germany, showing thinning of units over dome and increase in fault throw with depth. After Behrmann, *British Intelligence Objectives Sub-Committee, 1939–1945*, Pt. III, Fig. 95. Reproduced from *Geology of Petroleum* by A. I. Levorsen, W. H. Freeman and Company, San Francisco, 1954.

clude the Hawkins field, Texas, and the Erath field, Louisiana (Wendlandt *et al.*, 1946; Steig *et al.*, 1951). A contrast between synchronous and non-synchronous domes is provided by the Van and Kelsey structures in eastern Texas (McCoy and Keyte, 1934, p. 302). The Van dome shows marked thinning in the Woodbine, Eagle Ford, and Austin formations and produces from the Woodbine sandstone. The Kelsey dome shows no thinning in these units and is dry.

According to Halbouty and Hardin (1951, pp. 1970, 1976–77), Gulf Coast piercement-type salt domes with strong relief, such as the South Liberty dome, Texas, commonly are surrounded by porous, permeable, and highly productive multiple sands, whereas minor, deep-seated, low-relief domes are more likely to have hard and thin sands with low productivity. They speak in terms of a few to several tens of feet of sea-bottom relief for the South Liberty dome. In this field the sands in the Cook Mountain and Yegua formations are cleaner and more productive around the northern part of the dome than farther south, indicating a shift in the structural crest.

An example of a growing structure not associated with a salt core in the Gulf Coast region is the great Schuler field in Arkansas (Fig. 16). Weeks and Alexander (1942, p. 1512) find the Jurassic Cotton Valley formation thins across the dome, although the Jones sand at its base, which forms the main producing reservoir, is best developed over the crest. Further evidence of structural growth and topographic expression is the presence of a local erosional unconformity with gravel patches on the underlying Smackover limestone across the structure. The crest has shifted slightly since Jurassic times and now lies a little southeast of the thinnest Cotton Valley development. For the Frio sand reservoir of the Reynosa field, Mexico, Alvarez (1951, p. 76) accepts an origin similar to that suggested by Ball *et al.* (1941, p. 238) for the Michigan stray sand in the Michigan basin.

Middle East.—In the Middle East region early growth of anticlines, revealed by crestal convergence of strata, is believed to be an important reason for the abundance of oil in the Persian Gulf area, in particular for the pre-Tertiary pools southwest of the Gulf and the River Tigris (Law, 1957, pp. 63–64, Fig. 5b). Henson (1950, pp. 224–27, 236–37, Fig. 11) states that during Late Cretaceous and Tertiary time tectonic movements in the Persian Gulf foreland area raised bottom highs on which detrital, shelly, and foraminiferal limestones accumu-

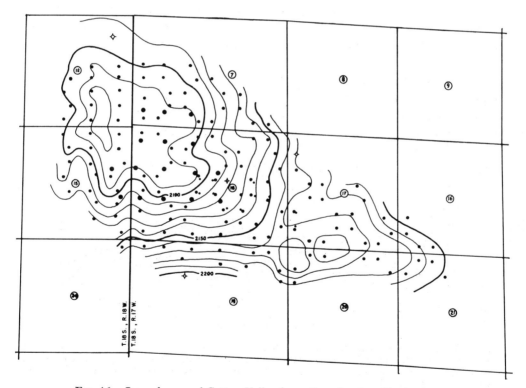

FIG. 16.—Isopach map of Cotton Valley formation, showing thinning across Schuler oil field, Arkansas. After Weeks and Alexander, 1942, p. 1512.

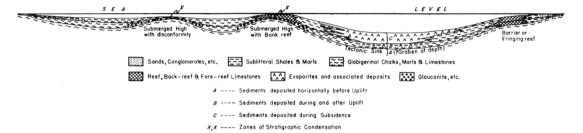

Fig. 17.—Diagram illustrating influence of Late Cretaceous and Tertiary tectonic bottom highs in Middle East foreland shelf on sedimentation, including development of bank reefs, detrital limestones, unconformities, and zones of "stratigraphic condensation" (indicated by concentrations of relatively coarse bioclastic and authigenic constituents such as glauconite, phosphatic grains and nodules, fish teeth, and microfossils). After Henson, 1950, Fig. 11.

lated or "bank-reefs" grew, which interfinger with a globigerinal, organically rich chalk-marl-shale facies in the intervening tectonic sinks with deeper water (Fig. 17). The high-bottom sediments are marked by disconformities, and associated faults increase in throw with depth, attesting to early growth. Examples described include Pir-i-Mugrun and Aqra in northeastern Iraq, both of which crop out but are still highly bituminous, the bitumen content decreasing away from the reef. Baker and Henson (1952, pp. 1895, 1898–1900) consider that possibly all of the oil-producing anticlines in the foreland region of the Persian Gulf may have risen intermittently during, and perhaps before and after, Cretaceous and Eocene time, and mention the possibility that oölitic and detrital limestones are related to these rising highs, though this requires further proof. Henson (1951, p. 122) attributes the folds of the Middle East foreland shelf primarily to faulting of basement blocks as well as to salt plugs. Weeks (1952, p. 2119, Fig. 25) stresses the importance of growing fault structures directly expressed in the Miocene sea-bottom topography of the "stable" Gulf of Suez area (Fig. 18). Oil occurrence is thought to be due to the creation of Miocene source conditions in intervening subsidiary sinks.

Specific instances of early growth of oil structures in the Middle East fore-

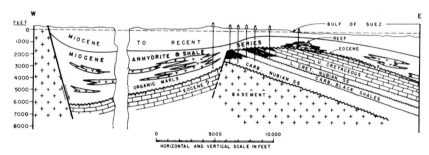

Fig. 18.—Diagram illustrating "growing" fault block in semi-stable Gulf of Suez area and its influence on sedimentation and oil occurrence. Miocene shale and evaporite sequence may include source beds, overlapping across reservoirs in topographically expressed fault block and related reef. After Weeks, 1952, Fig. 25.

land region include the Dukhan, Kirkuk, and Ain Zalah fields, discussed by Daniel (1954, pp. 806, 782, 797). In the Dukhan field, Qatar, there is evidence of persistent downflank thickening of several formations (Fig. 19), showing that the structure has been developing from at least latest Jurassic time. Production is from the Upper Jurassic. At Kirkuk, Iraq, there are vague indications in the lithologic character and thickness of some of the rocks associated with the Oligocene and Miocene oil that gentle uplift occurred at that time. In the anticlinal Ain Zalah field, Iraq, the best producing wells in the fractured Upper Cretaceous limestone reservoirs are grouped in a pear-shaped area which coincides almost exactly with the area where isopachs of the lowest Eocene and Paleocene show

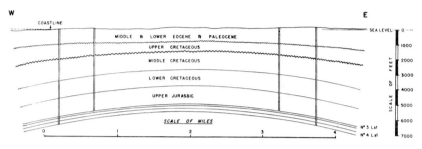

FIG. 19.—Structure section through Dukhan field, Qatar, showing unconformities with increased stratigraphic hiatus across anticline and accompanying downflank thickening of sedimentary units. After Daniel, 1954, Fig. 11.

maximum thinning and where erosion of the Upper Cretaceous is likewise believed to have been greatest. This area lies toward the western end of the present structural crest, indicating a shift in the position of the crest since Early Tertiary time. The superior production in this area is attributed to locally superior reservoir fracture porosity, caused either by bending over an old, buried high, or to penecontemporaneous radial slippage off a bottom uplift. Fractures increase with age of the beds, also indicating contemporaneous growth of the structure. According to Lees (1950, p. 30) gentle pre-Maestrichtian fold movement has occurred as far west as the great oil field in Kuwait, which produces from Middle to Upper Cretaceous rocks.

Some believe that not only the oil structures in the shelf region, but also those of the orogenic zone in southwestern Iran and northeastern Iraq have suffered intermittent uplift. Powers (1926, p. 430, Fig. 3) describes uplift, partial erosion, and dolomitization of the Asmari limestone reservoir at Maidan-i-Naftun, Iran, before deposition of the overlapping Fars series. If it is true that most, or all, of the Middle East oil fields occur in "growing structures," it may be concluded that the bulk of known world oil reserves are in traps of this kind.

Caucasus region.—Diapiric structures with cores of incompetent sediment other than salt are common in parts of the Caucasus region and, like salt domes, many are associated with oil accumulations. Goubkin (1934, pp. 652, 667) shows that most of the major oil fields in the Apsheron Peninsula are structures of this type, commonly characterized by mud volcanoes and dikes. Mud flows intercalated between the sedimentary rocks, and thickening of the sedimentary rocks away from the crest indicate the synchronous history of sedimentation, diapirism, uplift, and the existence of bottom highs. The greater load adjacent to the structure provides the mechanism for flowage, diapirism, and mud extrusion at the crest at various intervals up to and including the present. Examples cited are the Balakhany-Sabunchy-Ramany, Bibi-Eibat, Lok Batan, Puta, Surakhany, Kara Chukhur, and Kela oil fields (Figs. 20, 21).

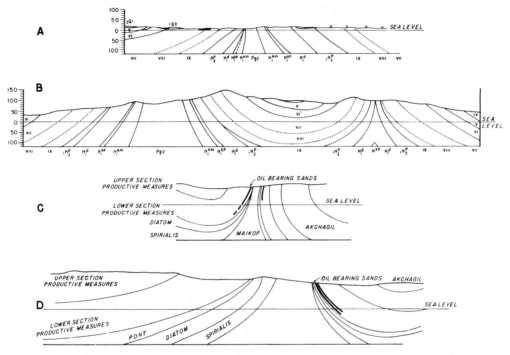

FIG. 20.—Cross section through several oil-field structures of Apsheron Peninsula, Caucasus area, USSR, showing mud-core diapirs and thinning of sedimentary units toward anticlinal crests. After Goubkin, 1934, p. 652.

Trinidad.—A cross section through the West Barrackpore-Wilson oil fields of Trinidad by Higgins (1955, Fig. 7) reveals conditions similar to those in the Caucasus area. Sandy units in the upper Miocene are shown to be best developed on the flanks around the crestal mud core of the Barrackpore anticline, shaling out at greater distances from the crest. Lower Miocene shales thin over the crest (Fig. 22).

POOLS ASSOCIATED WITH SEDIMENTARY HIGHS

Sedimentary highs include sand bars and bioherms. Occurrences of oil-bearing features of this type are too well known to require detailed discussion at this place. Good examples of the former are the oil pools in the sand lenses embedded

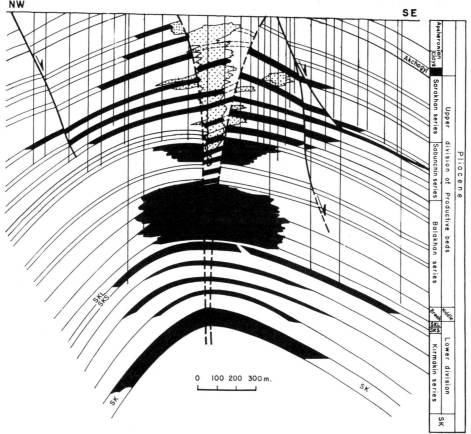

FIG. 21.—Cross section through Bibi-Eibat oil field, Caucasus area, USSR, showing "fossil" mud volcano (cross-hatched) in core of dome. Interbedded mud layers prove volcano was active during sedimentation. Oil pools shown in black. After Trust "Stalinneft," *XVIIth Int. Geol. Cong.*, Vol. 4, 1940, Fig. 12. Redrawn from *Geology of Petroleum* by A. I. Levorsen, W. H. Freeman and Company, San Francisco, 1954.

in the Cherokee shale of Greenwood and adjacent counties, southeastern Kansas, interpreted by Bass (1936, p. 105, Pls. 19–21) to represent offshore bars in the Cherokee sea (Fig. 23). The Music Mountain pool of northwestern Pennsylvania is said to be another example (Fettke, 1941, p. 506). The great importance of bioherms as exceptionally prolific oil and gas traps has become particularly evident after World War II, with developments in Alberta, West Texas, New Mexico, and the Middle East. An exposed example of a typical reef is shown in Figure 24.

Among the reasons why buried sand bars favor oil accumulation are, undoubtedly, the ideal reservoir and trap conditions and early trap timing of these synchronous highs, which were ready to receive and retain oil as soon as an impervious cover was deposited on top. Moreover, the rocks in the area of the Greenwood County pools also show evidence that the proximity of source areas of oil played an additional role. Apparently, the shale surrounding the sand

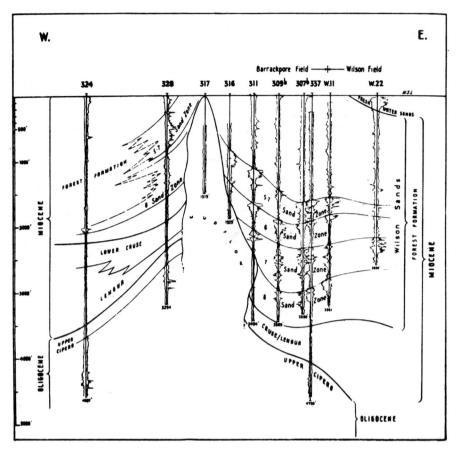

Fig. 22.—Well section through West Barrackpore-Wilson fields, Trinidad, showing thinning of lower Miocene and increased sand development in upper Miocene around mud core in anticlinal crest. Reproduced from Higgins, 1955, Fig. 7.

lenses, especially that between the former bars and the former coastal plain on the northwest, is abnormally rich in organic matter. This indicates that the synchronous highs helped to create their own oil source conditions in nearby waters, particularly in the lagoons which they isolated (Bass, 1936, p. 123; Landes, 1951, Fig. 69).

The same favorable factors (excellent reservoir, excellent trap, early trap origin, and nearby oil sources) are thought by many to have played essential roles in the case of reef pools. On the other hand, nearby sources are discounted by Gussow as insignificant contributors to commercial accumulations. He presents evidence that the pools in the Rimbey Homeglen-South Westerose-Westerose-Bonnie Glen-Wizard Lake-Glen Park-Leduc Woodbend-Acheson-St. Albert-Morinville-Redwater reef trend of the Leduc D_3 zone are solely the result of distant origin and migration over as much as 100 miles by the mechanism of successive trap spilling and differential entrapment (Gussow, 1954, pp. 825–27, Figs. 4, 5). Certain aspects of the oil and gas distribution in this trend might, how-

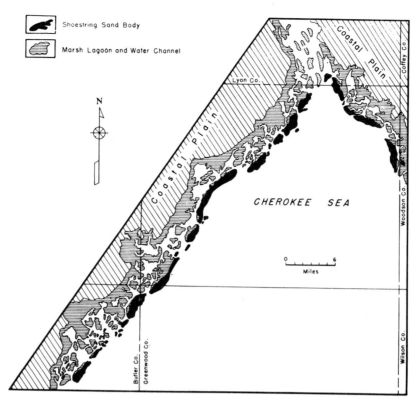

Fig. 23.—Map of reconstructed Cherokee shore in southeastern Kansas, showing offshore sand bars (synchronous highs of depositional origin) which later trapped oil. After Bass, 1936, Pl. 20.

ever, be interpreted as violating the exclusive operation of this mechanism; these are defended by Gussow on the basis of various unprovable assumptions, or remain unexplained. Thus, "downward draining" of oil out of the gas cap or renewed migration of oil after deeper burial is assumed to explain oil columns at Rimbey Homeglen and South Westerose, where only gas would be expected.

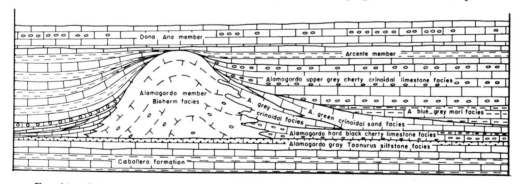

Fig. 24.—Cross section through typical bioherm (synchronous high of depositional origin) and contemporaneous Mississippian sediments in Lake Valley formation, Sacramento Mountain region, New Mexico. Lagoonal side is on the right. After Laudon and Bowsher, *Bull. Amer. Assoc. Petrol. Geol.*, Vol. 25, 1941, Fig. 10.

"Splitting" of the migration path beyond South Westerose and again beyond Leduc Woodbend is called upon to explain the large gas cap at Bonnie Glen, where little or no gas would be expected, and the oil at both Acheson and Redwater, one of which should otherwise be dry. Unreported gas accumulations are assumed in the higher Nisku D_2 zone south of Wizard Lake to account for oil and gas pools at Wizard Lake and beyond. The one factor which is never assumed as important is local migration. Unexplained remains how oil could have spilled out of Leduc Woodbend (unless it has a spillpoint several hundred feet higher than shown in Gussow's cross section) and how oil could have spilled into St. Albert from Acheson, where the spillpoint lies more than 500 feet below the present oil-water level (unless a three-way split of the oil path beyond Leduc Woodbend allowed oil to spill directly into St. Albert). Moreover, according to Irwin (1955, pp. 262–63) the inter-reef areas in this trend seem to lack carrier beds between and below the reefs, and porosity conditions in the Nisku are "definitely unfavorable to long distance migration." Such conditions also may be indicated by the fact that initial reservoir pressures in the Leduc at Bonnie Glen and in the Nisku at Leduc Woodbend are not adjusted to present ground-water level (Gussow, 1955, p. 566), suggesting there may be a lack of hydrostatic connection between these traps and other parts of the reservoir because of low permeability.

While acknowledging that spilling between reefs may well have played an important role in the Alberta reef trends, it seems a mistake to consider this principle as virtually the sole factor responsible for the oil and gas distribution. The detailed complexities probably can be explained only by taking other factors into account as important. These other factors include: (1) depth of burial, with more volatile hydrocarbons produced at greater depth during evolution of the oil, and with greater opportunity for gas to escape through fractures in the shallower reefs (similar to modern oil leakage from the shallow Norman Wells reef, North West Territories); and (2) local migration from the surrounding fine-grained sediments into the reef. In this the writer is in agreement with Irwin (1955, p. 263) and with Boos (1950, p. 313) (and others before him), who believe that much of the oil now found in reefs originated from organisms which lived and died in abundance near these synchronous bottom highs, and accumulated and were buried on the surrounding bottom. Irwin mentions the isolated Golden Spike bioherm of Alberta as an example of local, or even *in situ* accumulation of oil. The pontic or basin facies of reefs, commonly containing dark, bituminous shales and dense limestones, is most often mentioned as a likely source of hydrocarbons, but there seems to be no good reason why argillaceous and calcareous sediments and marls, in places interbedded with evaporites in the near-reef facies of the silled lagoonal area could not likewise have generated petroleum.

The location of some reefs has evidently been predetermined by the presence of some earlier sea-bottom high of a structural nature. For example, Ellisor (1926, pp. 976–85) describes Oligocene reefs around two Gulf Coast salt domes, Damon Mound and West Columbia, Texas. These domes grew near the close of

Jackson time, bringing the Jackson sediments above the surface of the sea as islands around which the reefs then formed. Similar reefs are said to occur around the Nash, Boling, and Barber's Hill salt domes. Weeks (1952, Fig. 25) shows a Miocene reef on a paleotopographically expressed fault block in the Gulf of Suez area (Fig. 18). Rojas (1949, p. 1341) states that, in the opinion of most geologists in the area, the Golden Lane El Abra (Early Cretaceous) reef trend formed along the crest of a pre-existing fold or the faulted boundary of a continental block.

POOLS ASSOCIATED WITH EROSIONAL HIGHS

Sproule (1957, pp. 852–53, 858) has recently emphasized the importance of erosional paleotopographic highs on the surface of the Precambrian in the North American interior in controlling the occurrence of oil source conditions and of porous reservoir sand lenses of the "granite-wash" type at the bottoms of transgressing seas (Fig. 25). Cited examples of such "granite-wash" occurrences are

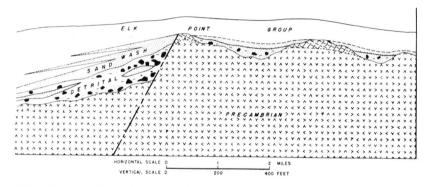

FIG. 25.—Diagram illustrating theoretical type occurrence of "granite wash" associated with paleotopographically expressed erosional and structural highs on Precambrian surface of Peace River area, Alberta. After Sproule, 1957, Fig. 7.

the oil fields of the Peace River area in Alberta, including the Clear Hills, Red Earth Creek, Amerada Crown, Iroquois, Gilwood, Gulf Iosegun, and Giroux Lake fields. Somewhat similar relations appear in oil fields of the Algerian Sahara. The accumulations in the "granite wash" derived in Early Pennsylvanian time from the exposed core of the Amarillo "mountains," Texas, as well as in associated Pennsylvanian limestones are another illustration of the influence of erosional highs (Fig. 26). Rogatz (1939, p. 986) speaks of exposed "monadnocks" scattered plentifully over the top of the ridge.

The Central Kansas uplift provides other good examples of this type of oil field. At Kraft-Prusa, Kansas, for instance, production comes in part from porous, sandy sediments located on the flanks of three buried hills on a pre-Pennsylvanian erosion surface (Walters and Price, 1948, pp. 257, 273–75). These sands were derived through weathering of a sandy dolomite member at the base of the Arbuckle limestone, which cropped out around the Precambrian core of the erosional high. Some of the sands may actually be Cambro-Ordovician shoreline

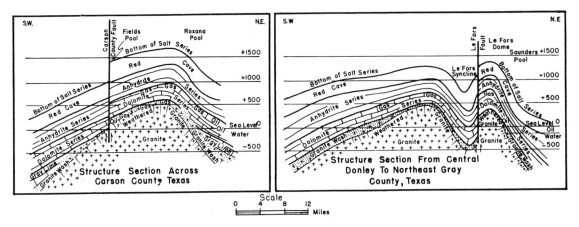

Fig. 26.—Cross sections through buried Amarillo "hills," Texas, showing occurrence of hydrocarbons (1) in "granite wash" derived from synchronous erosional high, and (2) in anticline of Permian rocks, probably first expressed as an inherited high. After Cotner and Crum, 1935, Figs. 2B and 2C.

facies of the sandy dolomite, as there is evidence that this member becomes increasingly sandy and shaly toward the largest Precambrian hill, which was then also an erosional high (Fig. 27). Additional oil occurs in younger Arbuckle carbonate units leached in pre-Pennsylvanian time around the paleotopographic hill, and some also occurs in the Precambrian rocks exposed at that time in the central part of the hill. Thus, it seems clear that the synchronous erosional highs

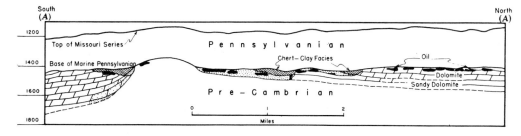

Cross Sections through KRAFT—PRUSA FIELD

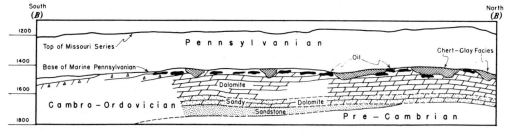

Fig. 27.—Cross sections through Kraft-Prusa field, Kansas, showing oil accumulation in (1) weathered synchronous erosional high on Precambrian surface, (2) sandy sediments related to such highs, and (3) Arbuckle limestone and dolomite leached in pre-Pennsylvanian time around related paleotopographic hills. After Walters and Price, 1948, Fig. 4.

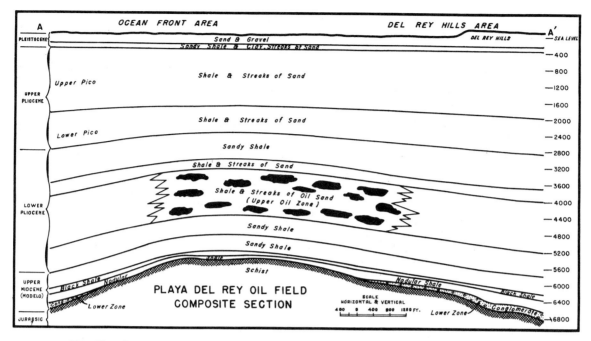

Fig. 28.—Composite section through Playa del Rey oil field, California, showing oil zones around buried erosional ridge of schist and in younger sediments domed partly by differential compaction. After Metzner, 1943, Fig. 123.

in this area have acted in several ways and at different times to bring about superior reservoir conditions. The oil could have come from the surrounding and overlapping Pennsylvanian sediments (which also yield commercial oil), with source conditions influenced by the nearby sea-bottom highs.[7]

In Oklahoma, Powers (1922, p. 243) described a local unconformity in the great Cushing oil structure between Lower Pennsylvanian oil sands and Cambro-Ordovician limestone and postulated the existence of a local erosional "ridge" which was buried by the Pennsylvanian sediments.

The Playa Del Rey field, California (Fig. 28), shows similar relations. Here, part of the production comes from basal sandstone and conglomerate skirting an erosional ridge composed of schist, and derived from it in late Miocene time

---

[7] In a recent article Walters (1958, pp. 2162–72) presents evidence that the oil and gas of the Arbuckle pools on the Central Kansas uplift (including Kraft-Prusa) came from sources, presumably Cambrian or Ordovician, south of the uplift and were distributed in traps according to the principle of differential entrapment. This provides a good explanation for the fact that several traps on the northeast flank of the uplift are dry, although otherwise quite similar to the producing traps—a fact difficult to account for on the basis of local sources adjacent to local highs. One could accept Walters' explanation without necessarily following him in deriving the hydrocarbons from the deep part of the Anadarko basin in south-central Oklahoma, as much as 200 miles away. It seems at least as plausible that a rich source existed near the uplift down its south or southwest flank. In that case the role of the individual highs on the post-Precambrian and pre-Pennsylvanian erosion surfaces over the uplift was restricted to promoting the development of local traps with superior reservoir porosity. The source of oil could be tied not to these highs separately, but to the presence of the entire uplift as a regional paleotopographic feature.

(Metzner, 1943, p. 292). This reservoir is overlain by nodular phosphatic and black shales which probably represent source sediments onlapping out of adjacent lows.

Erosional highs are also known which influenced development of secondary depositional highs, specifically bioherms. Hume (1921, pp. 408–09) stated that coral reefs at the base of the Miocene at the southern end of the productive zone along the Red Sea in Egypt rest on igneous rock without faulting, and that the reefs fringed island ridges on the inundated basement erosion surface, finally overtopping them. Depressions between were filled with clays, marls, and evaporites. It seems probable that these depressions became the source localities for the oil

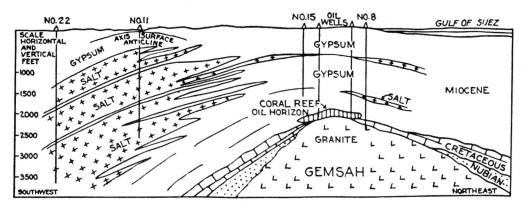

Fig. 29.—Cross section through the Gemsah oil field, Egypt, showing buried granite ridge surmounted by oil-bearing Miocene coral reef. After Powers, 1926, Fig. 5.

that later moved into the nearby highs. The old Ras Gemsah field appears to be of this type (Fig. 29). Powers (1926, pp. 436–37, Fig. 5) stated that "the coral reef appears to have grown on all sides of and over the granites, when the granite formed a hill in the Miocene sea."

POOLS ASSOCIATED WITH INHERITED HIGHS

The U. S. Mid-Continent region includes several oil fields in structural highs caused by differential compaction over pre-existing bottom highs, and for some of these it can be demonstrated that the inherited high was also expressed topographically on the sea bottom. Athy (1934, p. 822) stated that pre-Pennsylvanian structures in this area persisted as topographic highs or islands on the earliest Pennsylvanian sea floor and were maintained by differential compaction of the Pennsylvanian sediments, thus causing coarse sediments to be deposited above the crests. In the Oklahoma City field minor relief existed on the pre-Pennsylvanian erosion surface according to McGee and Clawson (1932, pp. 978–83, Figs. 9–12). These authors doubt that the structure in the overlying, oil-bearing Pennsylvanian-Permian sediments was due to differential settling, but Gussow (1955, p. 556) considers this the most important factor.

In Kansas, production from Pennsylvanian rocks at Kraft-Prusa (Fig. 27) comes mostly from domes over pre-Pennsylvanian hills of Precambrian quartzite on the Central Kansas uplift (Walters and Price, 1948, pp. 261–65, Figs. 5–7). In this field, and in the Augusta and Eldorado fields on the Nemaha "ridge," differential compaction probably has played an important role in the formation of the structures. At Augusta structural domes in the productive Ordovician rocks overlie and reflect hills on the crystalline rock surface, and various formations, including the Simpson and Arbuckle, thicken markedly away from the crest (Berry and Harper, 1948, Fig. 5, p. 216). An inherited bottom high seems to have controlled reservoir conditions and thus oil occurrence; this is indicated by the fact that the Simpson formation is absent on the highest structural crests and that the Ordovician sands are tight in the main structural saddle on the uplift. The Eldorado field is similar to the Augusta field (Landes, 1951, p. 383).

In the Amarillo district, Texas, the bulk of the gas occurs in Permian dolomites of the "Big Lime" series which, in most places, overlaps the granite "ridge" of the Amarillos (Fig. 26). Uplift and gentle folding occurred after Big Lime deposition, but differential compaction is believed to have occurred also (Rogatz, 1939, p. 1039; Cotner and Crum, 1935, p. 393, Fig. 2), and is probably mainly responsible for the arching of the sediments over the granite core.

In the Playa Del Rey field, California (Fig. 28), the Pliocene sediments were deposited on a surface which was anticlinal in shape, and their domal structure resulted in part from compaction over a buried ridge of schist (Metzner, 1943, p. 294). Oil occurs in lower Pliocene lenses of sand at and near the crest of the structural high.

An interesting case of compaction over a buried reef and apparent influence on sedimentation by the resulting bottom high is the North Snyder pool in the Scurry-Snyder County reef area of West Texas. A cross section through this pool published by Keplinger and Wanenmacher (1950, p. 182) shows arching of the Canyon sediments over the reef, attributable in part to differential compaction (initial dip may have played a role also). The Canyon sediments contain more sandstone units above the reef than in adjacent surrounding areas, suggesting the presence of an inherited sea-bottom high (Fig. 30).

Reefs may also be built on pre-existing bottom highs caused by differential compaction. Some of the oil-bearing lower Miocene reefs in Egypt rest on differentially compacted sediments over old granite ridges; an example is the Hurghada field (Lalicker, 1949, p. 245; Landes, 1951, p. 595; Hume, 1921, p. 408).

Finally, bottom highs inherited from a submerged surface by depositional draping rather than differential compaction are described from the subsurface of Kansas by Landes and Ockerman (1933), where the Cretaceous sediments reflect a buried Permian surface. The structures, however, are barren.

PROSPECTING FOR SYNCHRONOUS HIGHS

The following is a brief summary of criteria that may lead to recognition of synchronous as distinct from non-synchronous uplifts. Several of the characteristics listed may, under favorable circumstances, be discovered by means of

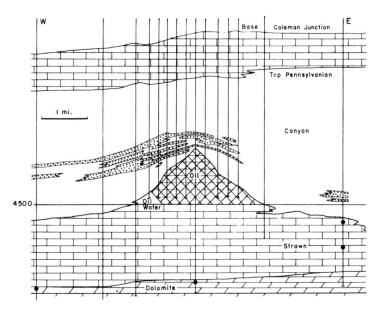

FIG. 30.—Cross section showing increased sand development in differentially compacted sediments over buried reef in North Snyder pool, Texas. After Keplinger and Wanenmacher, 1950, Fig. 2.

surface geologic work, and others may be found through geophysical work. These two methods have the obvious virtue of being applicable prior to exploratory drilling. Ordinarily, however, subsurface geologic methods are necessary to explore for synchronous highs. This does not mean that such conditions can, in most cases, be found only after the oil has already been discovered. A structure which was being developed during the deposition of the petroleum-bearing sediments is likely to continue active long afterward. Consequently, drilling for shallow objectives or by exploratory slim holes would uncover the tell-tale signs. In general, shallow indications of synchroneity encourage deeper drilling.

Lithologic changes associated with structural uplifts.—A fairly reliable criterion is local increase in sorting and mean grain size of a sedimentary unit approximately over a structural crest, as revealed by subsurface samples and expressed on facies maps of several types superposed on structure contour maps. In exceptionally clear cases the change may show up on clastic ratio or sand/shale ratio or percentage maps. Commonly however, the changes are likely to be more subtle and to appear only on facies maps with contours expressing mean grain size or percentile or quartile sorting in a sandstone. Other possible changes might be an increase in oölite or dolomite content in a carbonate rock or a local gradation into reef-type limestone toward the crest of a structural high. Henson (1950, Fig. 11) indicates concentrations of glauconite, phosphatic grains and nodules, fish teeth, microfossils, and other relatively coarse-grained constituents over submerged bottom highs in the Middle East subsurface, attributed to winnowing. Such changes likewise are subject to expression on facies maps.

The change may not always culminate to coincide with the present structural crest, but rather may occur a little farther down the flank. In such cases there may be commercial oil trapped in the flank as well as in the crest or, if the change is a major one, all commercial oil may be trapped in the flank of the synchronous uplift.

Changes of this type are not confined to isolated bottom highs; they may be regional, related to a nearby land area undergoing erosion, or to deposition in a delta or a channel. A submarine channel sand may be elevated later into a high structural position, although it may also occur that the channel winds around the structure (as at Glenrock, Wyoming), in which case it may be interpreted as corroborative evidence of a synchronous high. In any case, it would be possible by subsurface work to recognize lithologic changes related to channels, deltas, regional relations, etc. It may be said generally, that, if some change of the type described occurs in an isolated lens more or less coinciding with a locally high position of the bedding planes, a synchronous high is indicated.

Thickness changes associated with structural uplifts.—The most common criterion is the thinning or disappearance of units over a structural uplift. Such a thinning may indicate a "growing structure" or an inherited high with an erosional or sedimentary high in the core. In either case an early structural feature is indicated. It does not *a priori* follow that a bottom high must have existed, though in many instances this will have been the case.

Thinning may be picked up by seismic or subsurface geologic data or, less commonly, through geologic field work. It may be shown on cross sections and isopach maps, but it is important to choose the right sort of unit for contouring. Ideally, they should be time-rock units. Because of facies changes near synchronous highs, time-rock units may not coincide with rock units. For example, where a sandy shale on the flank fingers into a sandstone on the crest it would be misleading to contour the thickness of the sandstone. If this were done the isopach map would show thickening over the crest rather than thinning. Obviously, correct detailed correlation is of great importance. In the case of sedimentary highs (bars and reefs), isopachs of the time-rock unit to which they belong would show a local increase in thickness, but overlying units would show thinning because of differential compaction.

A corollary of thinning over the crest is that dips on the flanks of the structure increase with stratigraphic depth. However, increased dip by itself is not a sufficient criterion, for it may simply result from non-synchronous concentric folding of strata with constant thickness. In any case, it may be necessary to correct vertical thicknesses for dip in order to pick up slight changes in true thickness over a structural uplift. Finally, it needs to be pointed out again that maximum thinning does not necessarily coincide with maximum height of the present structure.

Changes in throw of faults.—On faulted structures a corollary of thinning over the high is that the throw of the faults increases with depth. This, again, may be revealed by subsurface geological or geophysical methods. On curved faults the stratigraphic throw will vary with depth even though the faults were not active during deposition.

Diastems associated with structural uplifts.—The presence of numerous local diastems in the section on a structural high as opposed to complete conformity in adjacent lows is also indicative of synchronous uplift and high bottom conditions. This may be indicated by exceptional concentration of mud-cracks, intraformational breccias, solution cavities, cross-bedding, ripple marks, etc.

Slump structures associated with structural uplifts.—An unusually large number of slump structures in the sediments down the flank of a structural high may be another suggestion of synchronous growth. Turbidity-current deposits in adjacent lows may have a similar significance.

Increase of fractures with depth.—Subsurface samples may indicate that deeper strata on a structure are more strongly fractured than shallower beds. This may be the result of recurrent or continuous folding during sedimentation, though by itself it is not a reliable criterion. Tension fractures filled with sediment from the overlying bed on a structural crest also suggest the high was synchronous.

Diapirism associated with structural uplifts.—Diapiric phenomena supply an indirect criterion in view of the fact that so many diapiric structures have a history of synchronous uplift. The intruding material may be argillaceous or evaporitic sediment, and associated features may be mud dikes and mud flows interbedded with the sedimentary rocks or extruded at the surface on the flanks of the high.

Miscellaneous indirect indications.—Experience in an oil district may indicate that synchronous highs in that area commonly have certain distinguishing features not characteristic of late structures. Though these features have no direct relation to the fact that a high existed during sedimentation and can therefore not be a criterion of synchroneity, the association may be so strong as to be useful in prospecting. Diapiric phenomena actually belong within this group. Another example might be a case where synchronous highs occur in certain trends (early tectonic lines, shorelines, etc.) which differ from trends of non-synchronous structures. Thus, the salt domes of northern Germany occur in trends representing fracture zones.

Still another case might occur where the early structures have basement cores as opposed, perhaps, to later, strictly orogenic structures characterized by *décollement*. Basement participation could be established by geophysical work. The later structures might also have a different shape from the early ones, e.g., more elongate. The early structures may also continue active longer than the late ones; if they continue into Recent time they may then be expressed in topography and by drainage patterns even in areas of young sediments. In particular, many uplifts over deep-seated basement faults show recurrent rejuvenation. Air-photo studies of anomalous geomorphic features, of fracture patterns and lineaments expressed by soil tone and vegetation, etc., therefore take on added significance as a prospecting method.

Conclusion

The thesis has been advanced that oil traps which were in existence and were topographically expressed at the time the petroliferous sediments were deposited are a preferential habitat for commercial accumulations of oil and gas. Examples from oil districts in different parts of the world have been quoted as illustrations. The writer is aware that much more evidence is needed to demonstrate conclusively that the hypothesis is valid, i.e., that this habitat is really preferential. It is hoped that the present contribution will stimulate detailed examination or re-examination of many other oil fields with this concept in mind.

In any case, the writer does not wish to be interpreted as advocating that otherwise favorable prospects should be rejected if they lack evidence of synchronous growth. Synchronous highs may be a preferential habitat, but they emphatically are not the only habitat. Accordingly, they may be preferential drilling prospects, but they are not the only prospects. In fact, to the extent that non-synchronous traps may be greatly predominant in a region, most of the oil and gas may actually occur in such traps. It is simply suggested that in a choice between two or more potential traps, one of which displays clear evidence of synchronous growth, that trap is likely to be the more favorable exploration prospect. Thus, wildcat drilling could be concentrated on those traps *most likely to contain oil.*

BIBLIOGRAPHY

ADAMS, J. E., 1930, "Origin of Oil and Its Reservoir in Yates Pool, Pecos County, Texas," *Bull. Amer. Assoc. Petrol. Geol.*, Vol. 14, No. 6, pp. 705–17.

——, 1934, "Origin, Migration, and Accumulation of Petroleum in Limestone Reservoirs in the Western United States and Canada," *Problems of Petroleum Geology*, pp. 347–363, A.A.P.G.

ALVAREZ, MANUEL, JR., 1951, "Geological Significance of the Distribution of the Mexican Oil Fields," *Proc. Third World Petrol. Cong.*, Sec. I, pp. 73–85.

ATHY, L. F., 1934, "Compaction and Its Effect on Local Structure," *Problems of Petroleum Geology*, pp. 811–23, A.A.P.G.

BAKER, N. E., AND HENSON, F. R. S., 1952, "Geological Conditions of Oil Occurrence in Middle East Fields," *Bull. Amer. Assoc. Petrol. Geol.*, Vol. 36, No. 10, pp. 1885–1901.

BALL, M. W., WEAVER, T. J., CRIDER, H. D., AND BALL, D. S., 1941, "Shoestring Gas Fields of Michigan," *Stratigraphic Type Oil Fields*, pp. 237–66, A.A.P.G.

BASS, N. W., 1936, "Origin of the Shoestring Sands of Greenwood and Butler Counties, Kansas," *State Geol. Survey Kansas Bull. 23*, pp. 1–135.

BERRY, G. F., JR., AND HARPER, P. A., 1948, "Augusta Field, Butler County, Kansas," *Structure of Typical American Oil Fields*, Vol. III, pp. 213–24, A.A.P.G.

BIEDERMAN, E. W., JR., 1958, *Shoreline Sedimentation in New Jersey.* Ph.D. Thesis, The Pennsylvania State University.

BOOS, C. M., 1950, "Source Beds for Oil near Coral Reefs," *Bull. Amer. Assoc. Petrol. Geol.*, Vol. 34, No. 2, p. 313.

BRIDGE, JOSIAH, AND DAKE, C. L., 1929, "Initial Dips Peripheral to Resurrected Hills," *Missouri Bur. Geol., Mines Bienn. Rept., 1927–1928*, pp. 93–99.

BRONGERSMA-SANDERS, MARGARETHA, 1948a, "The Importance of Upwelling Water to Vertebrate Paleontology and Oil Geology," *Verhandelingen der Koninklijke Nederlandsche Akademie van Wetenschappen*, Tweede Sectie, Deel XLV, No. 4, pp. 1–112.

——, 1948b, "On Conditions Favouring the Preservation of Chlorophyll in Marine Sediments," *Proc. Third World Petrol. Cong.*, Sec. I, pp. 401–13.

CLARK, F. R., 1934, "Origin and Accumulation of Oil," *Problems of Petroleum Geology*, pp. 309–35, A.A.P.G.

COTNER, VICTOR, AND CRUM, H. E., 1935, "Geology and Occurrence of Natural Gas in Amarillo District, Texas," *Geology of Natural Gas*, pp. 385–415, A.A.P.G.

CURRIE, J. B., 1956, "Concurrent Deposition and Deformation in Development of Salt-Dome Graben," *Bull. Amer. Assoc. Petrol. Geol.*, Vol. 40, No. 1, pp. 1–16.

Curry, W. H., Jr., and Curry, W. H., III, 1954, "South Glenrock, a Wyoming Stratigraphic Oil Field," *ibid.*, Vol. 38, No. 10, pp. 2119–56.

Dake, C. L., and Bridge, Josiah, 1932, "Buried and Resurrected Hills in Central Ozarks," *ibid.*, Vol. 16, No. 7, pp. 629–52.

Daniel, E. J., 1954, "Fractured Reservoirs of Middle East," *ibid.*, Vol. 38, No. 5, pp. 774–815.

Ellisor, A. C., 1926, "Coral Reefs in the Oligocene of Texas," *ibid.*, Vol. 10, No. 10, pp. 976–85.

Emery, K. O., and Rittenberg, S. C., 1952, "Early Diagenesis of California Basin Sediments in Relation to Origin of Oil," *ibid.*, Vol. 36, No. 5, pp. 735–806.

Fairbridge, R. W., 1946, "Submarine Slumping and Location of Oil Bodies," *ibid.*, Vol. 30, No. 1, pp. 84–92.

Ferguson, J. L., and Vernon, J., 1938, "The Relationship of Buried Hills to Petroleum Accumulation," *Science of Petroleum*, Vol. I, pp. 240–43.

Fettke, C. R., 1938, "The Bradford Oil Field, Pennsylvania and New York," *Pennsylvania Geol. Survey, 4th Ser., Bull. M21*, pp. 1–154.

———, 1941, "Music Mountain Oil Pool, McKean County, Pennsylvania," *Stratigraphic Type Oil Fields*, pp. 492–506, A.A.P.G.

Finn, F. H., 1949, "Geology and Occurrence of Natural Gas in Oriskany Sandstone in Pennsylvania and New York," *Bull. Amer. Assoc. Petrol. Geol.*, Vol. 33, No. 3, pp. 303–35.

Goodman, A. J., 1945, "Limestone Reservoir Conditions in Turner Valley Oil Field, Alberta, Canada," *ibid.*, Vol. 29, No. 8, pp. 1156–68.

Goubkin, I. M., 1934, "Tectonics of Southeastern Caucasus and Its Relation to the Productive Oil Fields," *ibid.*, Vol. 18, No. 5, pp. 603–71.

Griffiths, J. C., 1952, "Grainsize Distribution and Reservoir Characteristics," *ibid.*, Vol. 36, No. 2, pp. 205–29.

Gussow, W. C., 1954, "Differential Entrapment of Oil and Gas: a Fundamental Principle," *ibid.*, Vol. 38, No. 5, pp. 816–53.

———, 1955, "Time of Migration of Oil and Gas," *ibid.*, Vol. 39, No. 5, pp. 547–74.

Halbouty, M. T., and Hardin, G. C., 1951, "Types of Hydrocarbon Accumulation and Geology of South Liberty Salt Dome, Liberty County, Texas," *ibid.*, Vol. 35, No. 9, pp. 1939–77.

Henson, F. R. S., 1950, "Cretaceous and Tertiary Reef Formations and Associated Sediments in Middle East," *ibid.*, Vol. 34, No. 2, pp. 215–38.

———, 1951, "Observations on the Geology and Petroleum Occurrences of the Middle East," *Proc. Third World Petrol. Cong.*, Sec. I, pp. 118–34.

Higgins, G. E., 1955, "The Barrackpore-Wilson Oil Field of Trinidad," *Jour. Inst. Petrol.*, Vol. 41, No. 376, pp. 125–47.

Hill, G.A., 1956, "Trap Barriers—Sieves or Seals?", *Program of Annual Meeting*, pp. 35–36, A.A.P.G.

Hoots, H. H., Bear, T. L., and Kleinpell, W. D., 1954, "Stratigraphic Traps for Oil and Gas in the San Joaquin Valley," *California Div. Mines Bull. 170*, Chap. IX, pp. 29–32.

Hume, W. F., 1921, "The Geology of the Egyptian Oil Field," *Jour. Inst. Petrol. Tech.*, Vol. 7, pp. 394–421.

Irwin, J. S., 1955, "Differential Entrapment of Oil and Gas," *Bull. Amer. Assoc. Petrol. Geol.*, Vol. 39, No. 2, pp. 260–64.

Jacobsen, Lynn, 1949, "Structural Relations on East Flank of Anadarko Basin, Cleveland and McClain Counties, Oklahoma" *ibid.*, Vol. 33, No. 5, pp. 695–719.

———, 1953, "Lateral Variation in the Amount of Sandstone in the Chester Series of Western Kentucky and Its Relation to the Occurrence of Oil," *Bull. Geol. Soc. America*, Vol. 64, No. 12, p. 1532.

Keplinger, C. H., and Wanenmacher, J. M., 1950, "The New Reef Fields of Texas," *World Oil*, Vol. 131, No. 4, pp. 181–88.

Klinger, E. D., 1941, "Cross Cut-Blake District, Brown County, Texas," *Stratigraphic Type Oil Fields*, pp. 548–63, A.A.P.G.

Krynine, P. D., 1948, "Petrologic Aspects of Prospecting for Deep Oil Horizons in Pennsylvania," *Producers Monthly*, Vol. 12, No. 3, pp. 28–33.

———, 1949, "Current Mineralogical Research in Oil Findings at The Pennsylvania State College," *ibid.*, Vol. 13, No. 6, pp. 36–37.

———, 1951, "Reservoir Petrography of Sandstones," in "Geology of the Arctic Slope of Alaska," *U. S. Geol. Survey Map OM 126*, Sheet 2, Oil and Gas Inv. Ser.

Kuenen, Ph. H., 1950, *Marine Geology*. John Wiley & Sons, Inc., New York.

Lalicker, C. G., 1949, *Principles of Petroleum Geology*. Appleton-Century-Crofts, Inc., New York.

Landes, K. K., 1951, *Petroleum Geology*. John Wiley & Sons, Inc., New York.

———, and Ockerman, J. W., 1933, "Origin of Domes in Lincoln and Mitchell Counties, Kansas," *Bull. Geol. Soc. America*, Vol. 44, No. 6, pp. 529–40.

Law, J., 1957, "Reasons for Persian Gulf Oil Abundance," *Bull. Amer. Assoc. Petrol. Geol.*, Vol. 41, No. 1, pp. 51–69.

Lees, G. M., 1950, "Some Structural and Stratigraphical Aspects of the Oilfields of the Middle East," *Proc. 18th Internat. Geol. Cong.*, Pt. VI, pp. 26–33.

Levorsen, A. I., 1954, *Geology of Petroleum*. W. H. Freeman and Company, San Francisco.

Ley, H. A., 1924, "Subsurface Observations in Southeast Kansas," *Bull. Amer. Assoc. Petrol. Geol.*, Vol. 8, No. 4, pp. 445–53.

Love, J. D., Thompson, R. M., Johnson, C. O., Sharkey, H. H. R., Tourtelot, H. A., and Zapp, A. D., assisted by Nace, H. L., 1945, "Stratigraphic Sections and Thickness Maps of Lower Cretaceous and Non-Marine Jurassic Rocks of Central Wyoming," *U. S. Geol. Survey Prelim. Chart 13*, Oil and Gas Inv. Ser.

Lowry, W. D., 1956, "Factors in Loss of Porosity by Quartzose Sandstones of Virginia," *Bull. Amer. Assoc. Petrol. Geol.*, Vol. 40, No. 3, pp. 489–500.

———, 1957, "Implications of Gentle Ordovician Folding in Western Virginia," *ibid.*, Vol. 41, No. 4, pp. 643-55.

McCoy, A. W., and Keyte, W. R., 1934, "Present Interpretations of the Structural Theory for Oil and Gas Migration and Accumulation," *Problems of Petroleum Geology*, pp. 253–307, A.A.P.G.

McGee, D. A., and Clawson, W. W., Jr., 1932, "Geology and Development of Oklahoma City Field, Oklahoma County, Oklahoma," *Bull. Amer. Assoc. Petrol. Geol.*, Vol. 16, No. 10, pp. 957–1020.

McGee, D. A., and Jenkins, H. D., 1946, "West Edmond Oil Field, Central Oklahoma," *ibid.*, Vol. 30, pp. 1797–1829.

Metzner, L. H., 1943, "Playa Del Rey Oil Field," *California Div. Mines Bull. 118*, Pt. 3, pp. 292–94.

Monnett, V. E., 1922, "Possible Origin of Some of the Structures of the Mid-Continent Oil Field," *Econ. Geol.*, Vol. 17, No. 3, pp. 194–200.

Potter, P. E., 1957, "Breccia and Small-Scale Lower Pennsylvanian Overthrusting in Southern Illinois," *Bull. Amer. Assoc. Petrol. Geol.*, Vol. 41, No. 12, pp. 2695–2709.

Powers, Sidney, 1922, "Reflected Buried Hills and Their Importance in Petroleum Geology," *Econ. Geol.*, Vol. 17, No. 4, pp. 233–59.

———, 1926, "Reflected Buried Hills in the Oil Fields of Persia, Egypt, and Mexico," *Bull. Amer. Assoc. Petrol. Geol.*, Vol. 10, No. 4, pp. 422–42.

Prokopovich, Nickola, 1952, "Primary Sources of Petroleum and Their Accumulation," *ibid.*, Vol. 36, No. 5, pp. 878–90.

Reedy, Frank, Jr., 1949, "Stratigraphy of Frio Formation, Orange and Jefferson Counties, Texas," *ibid.*, Vol. 33, No. 11, pp. 1830–58.

Rogatz, Henry, 1939, "Geology of Texas Panhandle Oil and Gas Field," *ibid.*, Vol. 23, No. 7, pp. 983–1053.

Rojas, A. G., 1949, "Mexican Oil Fields," *ibid.*, Vol. 33, No. 8, pp. 1336–50.

Sloss, L. L., 1955, "Location of Petroleum Accumulation by Facies Studies," *Proc. 4th World Petrol. Cong.*, Sec. I, pp. 315–35.

Smith, P. V., 1954, "Studies on Origin of Petroleum: Occurrence of Hydrocarbons in Recent Sediments," *Bull. Amer. Assoc. Petrol. Geol.*, Vol. 38, No. 3, pp. 377–404.

Spieker, E. M., 1946, "Late Mesozoic and Early Cenozoic History of Central Utah," *U. S. Geol. Survey Prof. Paper 205-D*.

Sproule, J. C., 1957, "Clastic Reservoirs on Precambrian Surface in North America," *Bull. Amer. Assoc. Petrol. Geol.*, Vol. 41, No. 5, pp. 848–60.

Steig, M. H., Nichols, I. K., Shapleigh, G. G., and Denham, R. L., 1951, "Geology of Erath Field, Vermilion Parish, Louisiana," *ibid.*, Vol. 35, No. 5, pp. 943–87.

Thompson, W. C., and Hubbard, W. E., 1929, "Relation of Accumulation to Structure in the Oil Fields of Archer County, Texas," *Structure of Typical American Oil Fields*, Vol. I, pp. 421–39, A.A.P.G.

Tomkins, J. Q., 1957, "Bisti Oil Field, San Juan County, New Mexico," *Bull. Amer. Assoc. Petrol. Geol.*, Vol. 41, No. 5, pp. 906–22.

Trask, P. D., 1927, "Results of Distillation and Other Studies on the Organic Nature of Some Modern Sediments," *ibid.*, Vol. 11, No. 11, pp. 1221–31.

Twenhofel, W. H., 1950, *Principles of Sedimentation*. McGraw-Hill Book Company, Inc., New York.

Van Tuyl, F. M., and Parker, B. H., 1941, "The Time of Origin and Accumulation of Petroleum," *Quar. Colorado School Mines*, Vol. 36, No. 2, pp. 1–180.

Wallace, W. E., Jr., 1944, "Structure of South Louisiana Deep-Seated Domes," *Bull. Amer. Assoc. Petrol. Geol.*, Vol. 28, No. 9, pp. 1249–1312.

Walters, R. F., 1958, "Differential Entrapment of Oil and Gas in Arbuckle Dolomite of Central Kansas," *ibid.*, Vol. 42, No. 9, pp. 2133–73.

Walters, R. F., and Price, A. S., 1948, "Kraft-Prusa Oil Field, Barton County, Kansas," *Structure of Typical American Oil Fields*, Vol. III, pp. 249–80, A.A.P.G.

Weeks, L. G., 1952, "Factors of Sedimentary Basin Development that Control Oil Occurrence," *Bull. Amer. Assoc. Petrol. Geol.*, Vol. 36, No. 11, pp. 2071-2124.

Weeks, W. B., and Alexander, C. W., 1942, "Schuler Field, Union County, Arkansas," *ibid.*, Vol. 26, No. 9, pp. 1467–1516.

Wendlandt, E. A., Shelby, T. H., Jr., and Bell, J. S., 1946, "Hawkins Field, Wood County, Texas," *ibid.*, Vol. 30, No. 11, pp. 1830–56.

Wiggington, W. B., 1958, *Subsurface Stratigraphy and Structure of Mississippian Sediments of the Odon East Oil Fields, Daviess County, Indiana.* M. S. Thesis, The Pennsylvania State University.

Woodring, W. P., and Bramlette, M. N., 1950, "Geology and Paleontology of the Santa Maria District, California," *U. S. Geol. Survey Prof. Paper 222*, pp. 117–27.

Yonge, C. M., 1951, "Oceanography and Petroleum," *Discovery. The Magazine of Scientific Progress*, Vol. 12, No. 4, pp. 113–14. London.